EMBEDDED SYSTEMS CIRCUITS and PROGRAMMING

EMBEDDED SYSTEMS CIRCUITS and PROGRAMMING

Julio Sanchez
Minnesota State University, Mankato
Professor Emeritus

Maria P. Canton
Software Engineer, SKIPA Software Associates

CRC Press
Taylor & Francis Group
Boca Raton London New York

CRC Press is an imprint of the
Taylor & Francis Group, an **informa** business

CRC Press
Taylor & Francis Group
6000 Broken Sound Parkway NW, Suite 300
Boca Raton, FL 33487-2742

First issued in paperback 2017

© 2012 by Taylor & Francis Group, LLC
CRC Press is an imprint of Taylor & Francis Group, an Informa business

No claim to original U.S. Government works

Version Date: 20120224

ISBN 13: 978-1-4398-7904-7 (hbk)
ISBN 13: 978-1-138-07406-4 (pbk)

Library of Congress Cataloging-in-Publication Data

Sanchez, Julio, 1938-
    Embedded systems : circuits and programming / Julio Sanchez, Maria P. Canton.
    p. cm.
    Includes bibliographical references and index.
    ISBN 978-1-4398-7904-7 (hardcover : alk. paper)
    1. Embedded computer systems--Design and construction. 2. Embedded computer
systems--Programming. I. Canton, Maria P. II. Title.

TK7895.E42S263 2012
006.2'2--dc23   2012000494

**Visit the Taylor & Francis Web site at
http://www.taylorandfrancis.com**

**and the CRC Press Web site at
http://www.crcpress.com**

Table of Contents

Preface

Embedded systems are everywhere in our modern world. You can find them in automobiles, toys, kitchen appliances, computers, airplanes, TVs and digital recording devices, cell phones, gaming machines, nuclear power plants, space telescopes; in every electronic device that furnishes an intelligent, independent, or controllable functionality. By definition an embedded system is a computer designed to perform one or a few specialized functions, most often in real-time. The embedded system is usually a part of a hardware or mechanical device that includes other physical components. For example, a household dishwasher contains a cabinet with trays, one or more motors, sprayers and jets, deposits for holding soap and other chemicals, and an embedded computer system that determines cycles, modes, and timing of the device. It is this embedded controller that actually operates the machine, and what our book is about.

The authors conceived *Embedded Systems: Circuits and Programming* as a reference and a resource kit for engineers, scientists, and electronic enthusiasts who need to develop systems and boards that contain digital components and controls sometimes combined with analog devices. Our book is also intended as a tutorial on microcontroller programming and on the basics of embedded design. The focus is on the needs of working professionals in the fields of electrical, electronic, computer, and software engineering.

Many engineered products require the development of one or more embedded systems. At the commercial fabrication stage the embedded system or systems are refined and optimized by specialists in order to facilitate production and minimize cost. However, during the development and prototyping phases, the developer often needs to create an embedded system that demonstrates the operation of the device, proves its viability, or exercises its functionality. This initial product is often called a *prototype*. The availability of off-the-shelf components, the abundance of design and prototyping tools, and the ease with which digital controllers can be programmed make it possible for the non-specialist to develop these boards and controllers. In this book we focus on the following development tools and resources:

- The use of standard or off-the-shelf components such as input/output devices, integrated circuits, motors, and programmable microcontrollers

- The development of circuit prototypes and their implementation using breadboards, followed by the in-house fabrication of test-time PCBs, and finalized by the manufactured board

- The availability and use of development tools that facilitate the design and prototyping of embedded systems, such as electronic design programs and software utilities for creating printed circuit boards

- Sample circuits that can be used as part of the targeted embedded system

- The selection and programming of microcontrollers suitable for use in the circuit at hand

We have aimed at a book that is functional and hands-on. The resources furnished to the reader include sample circuits with their corresponding programs. The circuits are depicted and labeled clearly in a way that is easy to follow and reuse. For some critical circuits we provide tested PCB files. The sample programs are matched to the individual circuits but general programming techniques are also discussed in the text. There are appendices with hands-on information and the book's online software package includes tools, resources, and code listings. These materials are available at

http://www.crcpress.com/product/isbn/9781439879047.

Julio Sanchez

Maria P. Canton

Chapter 1

Real-Time Computing

1.0 Defining the Embedded System

An embedded system can be defined as a computer system that performs one or more dedicated or specialized functions usually with Real-Time constraints. It is the digital controller that is embedded as part of system, which also includes other hardware and mechanical elements. A modern dishwasher contains an embedded control system. On the other hand, a general-purpose computer such as a PC is designed to meet a variety of user needs. The control system in the dishwasher does one task: wash dishes, while the PC on your desktop can be used in word processing, browsing the Web, balancing your checkbook, and many other applications.

Embedded systems are controlled by one or more digital processing devices, typically a microcontroller or *digital signal processor* (DSP). The key feature of the embedded system is that it is dedicated to a specific task. Sometimes a specialized or dedicated computer system, such as the one handling a dam, or the traffic in a major city, or the operation of a nuclear power plant, is considered an embedded system. Although it is a matter of semantics, we prefer to exclude specialized or dedicated computer systems from the embedded category. We also exclude devices based on nanotechnology because in this case it is the molecular size of the component that defines the system. But even excluding specialized computing devices and molecular-scale components, embedded systems can range widely in size and complexity; from handheld or portable devices such as a digital watch to complex controllers such as the one operating a space exploration robot.

A good example of an embedded system, as considered in this context, is a programmable thermostat controlling the operation of household air conditioning and heater systems. In real-time the thermostat receives information from a sensor that reads the house temperature at predetermined intervals. The controller logic then compares this temperature value with the one programmed for the current time-of-day and turns on or off the AC or the heater component as necessary.

1.1 Embedded Systems History

One of the first recognizably modern embedded systems was the Apollo Guidance Computer, developed at the MIT Instrumentation Laboratory in 1961. It was the Apollo Guidance Computer that first explored the viability of monolithic integrated circuits which have because become commonplace. Another early embedded system was the Autonetics D-17 guidance computer for the Minuteman missile, which was based on transistors and contained a hard disk drive as its main memory. In 1996, the D-17 controller was replaced with a new computer that pioneered the development of integrated circuits. One result from these projects was the reduction in cost of the quad nand gate ICs from $1,000 each to less than $3. Commercial applications soon followed.

The invention and development of microcontrollers relates closely to the evolution of embedded systems. The integration of processing functions, memory, and programmable input and output operations in a single circuit has simplified the design and development of embedded systems, to the point that today it is difficult to find an embedded system that does not include at least one microcontroller. Furthermore, the low cost of microcontrollers compared to full-fledged computer ICs has further increased the type of applications for which embedded systems are commercially viable. All embedded systems discussed in this book contain a microcontroller although some systems also include other digital components that provide specialized functionality. For example, an embedded system that deals in Real-Time units (such as years, months, days-of-the-week, hours, minutes, and seconds) will typically include a Real-Time clock IC such as the NJU6355, which will furnish this functionality.

1.2 Hardware Complexity

To the developer or engineer an embedded system consists of hardware and software. The hardware typically includes a circuit board that houses and interfaces the components, one or more integrated circuits that provide the processing capabilities (usually a microcontroller), and one or more input and output devices that communicate with the outside world. But not all embedded systems are stand alone devices. For example, an embedded system that turns on and off a stepper motor and controls its rotational speed and direction may be designed to communicate with another system that provides the logic and command signals. In this case the embedded motor control system is actually a component of a larger control system and does not contain a user interface, processing logic, or other command and control functions.

On the other hand, some embedded systems come close to the functionality of a conventional computer. For example, an embedded system may include a computer screen with a graphical user interface, several buttons and control devices, LEDs, network connections, or even a touch screen. In this case it is the fact that the embedded system is dedicated to a particular task that differentiates it from a computer. But keep in mind that there is no clear demarcation line between a sophisticated embedded system and a specialized computer. One is free to classify such borderline systems in either category. In this sense an ATM machine, with its complex processing and IO capabilities, can be considered either as an embedded system or as a specialized computer.

1.2.1 Processor

It is virtually impossible to conceive of an embedded system that does not provide some processing, although, here again, the demarcation line is not always clear. For example, the circuit that operates a household doorbell often consists of a simple pushbutton control and a buzzer or bell. Pressing the pushbutton closes the circuit that activates the sound-producing element. Because there is no logic or processing such a doorbell circuit could hardly be classified as an embedded system. However, if we add a processor that selects a different sound pattern or chime according to the time of day, then the doorbell driver could be classified as an embedded system. Here, it is the presence of a processing operation that defines the system.

Embedded processors are usually microcontrollers although it is conceivable that an embedded system can be based on a microprocessor. The principal difference between the microprocessor and the microcontroller is that the latter includes peripheral functions that simplify its application and reduce the cost of the IC. Additionally, it is usually assumed that a microprocessor system contains more complex and powerful logic; however, the advent of the Digital Signal Processor (DSP) has also diminished this perception because some DSPs match or even exceed the processing power of microcomputers.

1.2.2 Microcontrollers

In this book we do not cover embedded systems based on microprocessors or on DSPs because the complexity of such systems clearly exceeds our present scope and intention. But even excluding these high-end devices, dealing with the many sources, types, and complexity of currently available microcontrollers can be a daunting proposition. Chapter 5 contains an overview of microcontrollers from Intel, Microchip, Motorola, and Atmel often found in embedded systems. However, for space constraints and practical didactic considerations, in this book we limit our coverage to a few popular and well-supported microcontrollers from Microchip, all based on its mid-range architecture.

1.2.3 Hardware and Software

The hardware components that can be found in an embedded system are difficult to enumerate. They include (but are not limited to) the following elements:

- Circuit boards
- Processors (usually one or more microcontrollers)
- Transistors
- Logic gates
- Power supplies
- Clocks and flip-flops
- Frequency dividers, timers, and counters

- Multiplexers and demultiplexers

- Memories and data storage facilities

- Communications controllers

- Switches and keypads

- LEDs, seven-segement LEDs, and LCDs

- Motors and motor control devices

- Sensors

- Solenoids and actuators

- AC converters and output devices

Books can be and have been written on many of these elements. In fact, titles on some of them (such as microcontrollers) fill several good-sized shelves. Fortunately, others are rather simple devices, and their understanding and use is straightforward and direct. For example, from an embedded systems developer point of view, there is not much more that can be done with an LED than to turn it on or off.

By a stretch of the imagination we can conceive of an embedded system that operates solely on the off-the-shelf logic contained in its digital devices. However, a more common scenario is that the processing operations of an embedded system must be customized by the developer to meet a specific design functionality. This means that a program must be developed for the system's controller. The reference in this book's title to circuits and programming indicates that its contents relate to both the physical electronics and the software component of an embedded system.

1.3 Execution in Real-Time

Real-time or reactive computing refers to systems that are subject to a time constraint. By definition, a real-time computing system is one that interacts with the real world or that has timing requirements related to this interaction. In general, the response of a Real-Time system must take place within a specific time frame. For example, in order to be effective, the air-bag system in the steering wheel of a car must deploy within a certain time period after the event that activates it. If the air bag deploys earlier or later than this ideal moment, its action would be useless or even counterproductive. Note that the concept of execution in real-time does not mean that the system's interaction must take place as fast as possible.

By definition a real-time system includes not only those that must execute within a temporal constraint, but also any system that interacts with the real world. In embedded systems this interaction with the real world usually takes place through sensors and actuators. For example, the doorbell system previously mentioned executes in Real-Time because it is expected that the buzzer or chime will sound within a short time after the pushbutton is depressed.

1.3.1 Hard and Soft Real-Time Systems

Real-Time systems are often classified as hard and soft. Systems that must meet strict timing requirements are classified as *hard* while those in which there is a more flexible tolerance of the timing requirement are *soft* RTSs, (Real-Time Systems). In this context, the air-bag system mentioned previously completely fails if its strict timing requirement is not met; therefore, it is a *hard* RTS. The doorbell system, on the other hand, has a much larger tolerance in its timing requirement and can be classified as a *soft* RTS. The timeliness requirement is one of the main issues in the design of a hard real-time system, but must also be considered in soft RTSs.

Design considerations in Real-Time systems include the following topics:

- Real-Time operating systems
- Scheduling
- Real-Time communications
- Analysis of Real-Time components
- Testing and debugging Real-Time systems

In this context our discussion of Real-Time systems is limited to specific circuits with a specific Real-Time constraint. Most of the topics listed above fall within the field of theoretical computer science.

Chapter 2

Circuit Fundamentals

2.1 Electrical Circuit

An *electrical network* is a collection of electrical elements that can include voltage sources, lines, connectors, resistors, inductors, capacitors, switches, controllers, and a host of other standard and special components. The *circuit* is a closed-loop electrical network that provides a return path for the current and that contains at least one active electronic component.

Electronic circuits are classified as analog, digital, or mixed-signal, the latter being a combination of the analog and digital types. Circuits contain discrete components often purchased off-the-shelf. These include but are not limited to resistors, capacitors, inductors, diodes, processors (such as microcontrollers), as well as input and output devices such as pushbuttons, toggle switches, light-emitting diodes, liquid crystal displays, actuators, and motors. During the development stage, the circuit components are typically assembled on breadboards, perfboards, or stripboards that allow for easily making changes to the circuit. Figure 2-1 is a photograph of a breadboard used in testing a motor driver circuit developed later in the book.

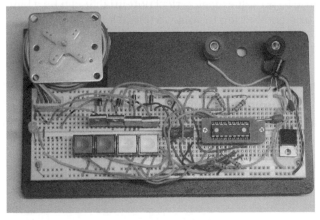

Figure 2-1 Breadboard Circuit for a Motor Controller.

Once the circuit has been tested and debugged, the components are soldered on a printed circuit board that provides a finished product. If the product is to be manufactured in large numbers, it is usually modified into a production version that is cheaper to make, smaller, or both. Figure 2-2 shows a printed circuit board with components.

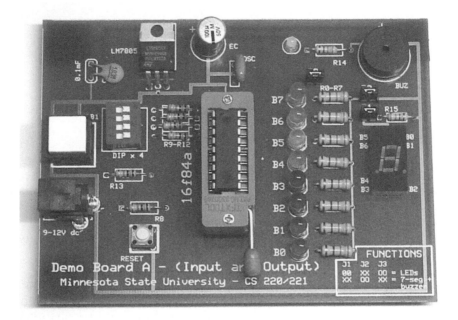

Figure 2-2 Assembled Printed Circuit Board.

2.2 Circuit Concepts and Components

All electrical circuits have at least some of the following elements.

- A power source to produce an electrical potential; it can be a battery, an alternator, a generator, or a connection to some external source.

- Conductors. These include wires and circuit boards that provide a path for the current.

- Loads, in the form of devices that use electrical energy to provide some form of work. These include lamps, motors, actuators, solenoids, and many others.

- Control elements such as switches and potentiometers. These components serve to regulate if any current flows through the circuit or its magnitude.

- Devices to protect the circuit from overloads. These include fuses and circuit breakers.

- A common ground.

Figure 2-3 shows a simple circuit that includes connectors, a light bulb, a potentiometer, a fuse, a battery, and a common ground.

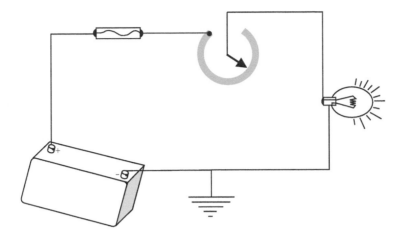

Figure 2-3 Unconventional Schematics of a Simple, Analog Circuit.

2.3 Digital Electronics

The circuit in Figure 2-3 is classified as an *analog circuit*. The circuit is analog because the angular position of the potentiometer arm conveys continuous information regarding the resulting brightness of the light bulb. Digital devices, on the other hand, use individual analog levels to represent the circuit's action. For example, the potentiometer in Figure 2-3 could be replaced with several sets of in-line resistors so that the selector arm could only be placed in one particular level of resistance, as shown in Figure 2-4. Nevertheless, circuits are neither digital nor analog in themselves but in how the circuit values and parameters are interpreted.

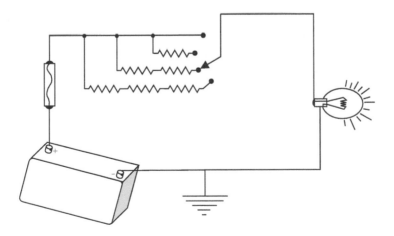

Figure 2-4 A Simple Digital Circuit.

Digital circuits are prevalent in electronics because it is easier to get a device to switch into one of a number of known states than to reproduce a continuous range of values. Digital circuits are made from logic gates which model simple electronic representations of Boolean logic functions. Digital circuits offer the following advantages over their analog counterparts:

- Digital signals can be represented and transmitted without degradation due to wear or noise as long as the 1s and 0s encoding the binary values can be accurately reconstructed.

- The accuracy of a digital system can be improved by increasing the number of binary digits used to encode it. The accuracy or resolution of an analog system requires improvements in the linearity and noise characteristics of the signal chain.

- Digital systems can be controlled by software.

- Information can be stored digitally more conveniently than analog.

Digital circuits are the basic building blocks from which microprocessors, microcontrollers, computer systems, and virtually all digital electronic devices are constructed. These building blocks are essentially simple and perform elementary functions. The complication sometimes arises from the fact that a single device can contain thousands of these primitive components.

Understanding these components requires viewing them at the proper abstraction level. To understand a simple digital device it is essential to know how the simpler components that make up the device operate. For example, to understand how a shift register works, it is useful to visualize it in terms of the logic gates from which it is built. Once one understands how counters and registers work, it is easy to grasp how a complex large-scale integrated circuit, such as a serial port, operates.

Fortunately, at any given level of abstraction it is not necessary to consider every single possible device of that class. For example, once we understand the operation of the few primitive logic gates, the operation of devices that contain more than one of these primary components can be easily grasped. In this chapter we start by explaining the basic facts about diodes and transistors as basic digital electronics devices; then we discuss the primitive logic gates that are the building blocks of all digital circuits, followed by the more complex circuits that are constructed from the elementary logic gates.

2.4 Diode

In basic electronics we learn that the diode acts as a one-way valve for electrical current, and that diodes are one of the most powerful inventions of semiconductor physics. The physical and electrical principles that make a diode work are discussed in basic electronics and are therefore outside the scope of this book.

Reviewing fundamental concepts, we can state that the diode is a semiconductor and all semiconductors are made primarily of silicon. The combination of silicon and phosphorous, with an extra phosphorus electron, is called an *n-type* silicon. In this case the n stands for the extra negative electron. The extra electron donated by the phosphorus atom can easily move through the crystal; therefore *n-type* silicon can carry an electrical current. When a boron atom combines in a cluster of silicon atoms, there is a deficiency of one electron in the resulting crystal. Silicon with a deficient electron is called *p-type* (*p* stands for positive). The vacant electron position is sometimes called a "hole." An electron from another nearby atom can "fall" into this hole, thereby moving the hole to a new location. In this case the hole can carry a current in the *p-type* silicon.

When voltage is applied to the diode it makes the n-type end more positive than its p-type. Consequently, electrons flow in the *n* to *p* direction but not in the *p* to *n* direction. This fact makes the diode behave as a one-way filter that allows electrons to flow in one direction but not in the other one. Figure 2-5 shows the *p-n* junction in a diode and its electrical symbol.

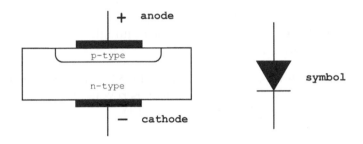

Figure 2-5 Diode Elements and Symbol.

The general convention is that current flows from positive to negative, although in reality electrons flow is from negative to positive. Benjamin Franklin is usually blamed for the erroneous convention. Current in the diode in Figure 2-5 flows from the anode to the cathode, but not vice versa.

The electrical symbol for a diode resembles an arrow pointing in the direction of current flow. When the anode voltage of a diode exceeds the cathode voltage, the diode is said to be forward biased. A forward-biased diode acts like a short circuit. To prevent too much current from flowing, a resistor is usually inserted in series with the diode, as in Figure 2-6.

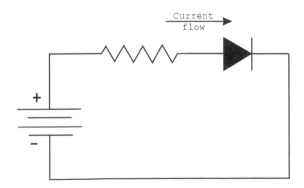

Figure 2-6 Diode and Resistor in a Circuit.

The typical diode, such as the ones used in logic and display circuits, can handle a current of 10 to 20 milliamps. For a 5-volt supply, a 300-ohm series resistor limits the current through the diode to a reasonable value.

2.4.1 Light-Emitting Diode (LED)

One of the most common types of diodes is an LED (*light-emitting diode*). The LED is a semiconductor device that emits incoherent light when forward biased. The color of the light depends on the chemical composition of the semiconducting material. The first practical LEDs were developed in 1962. LEDs are sometimes used in electronic devices to signal the presence of an electric current. Like any diode, the LED consists of a chip of semiconducting material impregnated with impurities to create a *p-n* junction. As in all diodes, current flows easily from the *p*-side, or anode to the *n*-side, or cathode, but not in the reverse. The first LEDs were made of gallium arsenide. Today, LEDs are made of a variety of materials so as to produce light of different colors. The most common LEDs have a distinctive red color, although they can also be obtained in white, amber, green, and blue.

Because LEDs are diodes, they will only light with positive electrical polarity, that is, when forward biased. When the polarity is reversed, very little or no light is emitted by the LED. Figure 2-7 shows a typical LED.

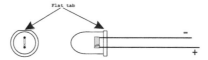

Figure 2-7 Polarity in a Typical LED.

The polarity of a new LED can usually be determined by observing that the longer terminal is the anode. If the terminals have been altered, then it is possible but risky to try to determine polarity by observing the LEDs internals. In most LEDs the larger internal tab is the cathode, but there are some in which it is not. A more dependable clue to the LEDs polarity is the flat tab on the LEDs base, which indicates the cathode, as in Figure 2-7.

Ratings vary among the different sizes and types of LEDs. Most are rated to operate between 1.7 and 3.8 volts and at currents of 10 to 40 mA. The light-emitting capacity of an LED is measured in *megacandela* or mcd. Small commercial LEDs range from 10 to about 5000 mcd. Once the LED's rating is known, and given the circuit's voltage, it is necessary to calculate the value of a series resistor so that the current does not exceed the LED's capacity. For example, the series resistor for wiring a commercial red LED rated at 2.6 VDC and 28 mA on a 5-volt circuit can be calculated as follows:

STEP 1: Calculate the voltage across the resistor by subtracting the LED's forward voltage from the supply voltage, in this case.

STEP 2: Apply Ohm's law to calculate the required resistor.

The electronic symbol for a LED is somewhat similar to that for a diode, as shown in Figure 2-8.

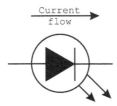

Figure 2-8 Electrical Symbol for LED.

A simple experiment is to connect an LED in series with a 330-ohm resistor to a 5-volt power supply and then test to see that light is emitted for one orientation of the diode and not for the other. This small circuit makes a convenient probe for determining polarity in logic circuits. If the LED's cathode is touched to some point in a circuit, the LED lights up if the voltage at that point is less than about a volt or two. The LED remains dark if the voltage is greater than this value. The circuit acts as a 1-bit binary digital voltmeter. A commercial version of this device, usually with several different-colored LEDs, is called a *logic probe*.

In addition to LEDs, other diodes, such as the 1N4148, can be used to ensure that the current in a circuit flows only in one direction. Heftier diodes are used to make the DC power supplies found in computers and other electronic equipment. In this application the alternating polarity of the 110-volt source is converted by the diodes into a unidirectional DC flow.

2.5 Transistors

The transistor is a solid-state semiconductor device that is used for signal amplification, voltage stabilization, switching, signal modulation, and many other functions. In a typical circuit, a voltage or current is applied to a pair of terminals in the transistor so as to change the current level through another pair of terminals. Because the power of the output terminals can be much greater than that of the controlling terminals, the transistor can be used to amplify a signal. Transistors were developed and first used in the early 1950s. They are manufactured as individual components or as part of integrated circuits. Transistors come in two basic types: bipolar and MOS.

2.5.1 Bipolar Transistor

The *bipolar transistor* was the first type of transistor to be commercially mass-produced. The terminals of a bipolar transistor are called the *emitter*, the *base*, and the *collector*. Physically, the bipolar transistor consists of two n-type regions separated by a thin p-type region or, alternatively, by two p-type regions separated by a thin n-type. A transistor with two n-type regions is called an *NPN* transistor. One of the n-type regions is called the collector, the other the emitter, and the central p-type region the base. The NPN bipolar transistor is shown in Figure 2-9.

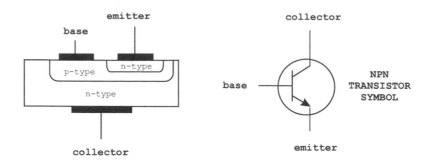

Figure 2-9 Bipolar, NPN Transistor, and Symbol.

A common simplification is to consider a bipolar transistor as two diodes connected back to back so that they share a common end. Because the central base region between the collector and emitter is very thin, the device has the unique property of serving as an amplifier. When the transistor's base-to-emitter p-n junction is forward biased (this could be called the p-n diode), it creates a low resistance in the thin base region that allows a much larger current to flow from the collector to the emitter. If the base-emitter current is turned off, then the collector-emitter current is also completely turned off. In this case the transistor is said to be *cut off*.

Over a given range, the collector-emitter current is directly proportional to the base-emitter current. In this implementation the transistor amplifies small currents into larger ones. Transistors used in radios and other sound-amplifying applications are of this type. For larger base currents, the transistor acts as if there is nearly a short circuit between the collector and the emitter. In this case the transistor is said to be in *saturation*. The effect is that a positive voltage on the base turns on the transistor and pulls the output low (to about 0.5 volts). When this voltage is removed, the transistor is turned off and the output is high (+5 volts). The action of a transistor is that of a current-controlled switch, as shown in Figure 2-10.

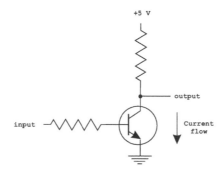

Figure 2-10 NPN Transistor Used as a Current-Controlled Switch.

The circuit in Figure 2-10 operates as follows: If the input voltage is held at zero volts, the *p-n* base-emitter junction has no current flowing through it and thus the output voltage is +5 volts. However, if the input voltage is raised to any value between +2 and +5 volts, a base-emitter current flows. This, in turn, allows a collector-emitter current to flow and the output voltage is pulled down to ground (typically between 0.5 and 1 volts). An alternative architecture for a bipolar transistor is called PNP. In this case, the *n*-type silicon is sandwiched between two *p*-types, as shown in Figure 2-11.

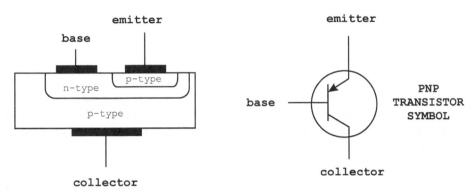

Figure 2-11 PNP Transistor and Symbol.

The PNP transistor in Figure 2-11 works in the same way as the NPN transistor, except that in the PNP design the base must have a negative voltage with respect to the emitter in order to turn on the transistor.

2.5.2 MOS Transistor

The second major type of transistor is the *metal oxide semiconductor* transistor, or MOS. It consists of two separated *n*-type regions embedded in *p*-type silicon. Alternatively, the MOS can consist of two *p*-type regions embedded in *n*-type silicon. In the first case the device is called an *n-channel MOS* (or NMOS) transistor, in the second case it is called a *PMOS*.

One of the two *n*-type regions is called the *source*, and the other is called the *drain*. An area between the source and the drain consists of a metal contact separated from the *p*-type body by a thin layer of nonconductive silicon dioxide. This area is called the *gate*. When a positive voltage is applied to the gate, the electric field attracts a thin layer of electrons into the *p*-type region underneath the gate. This provides a low resistance path between the two *n*-type regions. Figure 2-12 shows the construction of an NMOS transistor and the symbols for the NMOS and PMOS.

During construction of the MOS transistor, the body is connected internally to the source. In the electrical symbol this is indicated by the central wire with an arrow. In the NMOS transistor the direction of the arrow indicates that electrons in the body are attracted to the gate when a positive voltage is applied. This same voltage repels electrons in the PMOS.

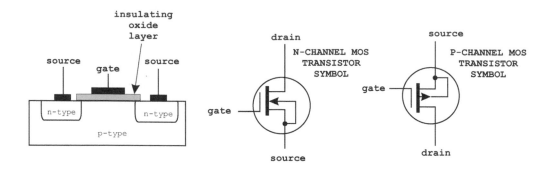

Figure 2-12 MOS Transistor and Symbols.

One of the most valuable features of MOS transistors is that they require very small currents to turn on. This makes MOS transistors behave like voltage-controlled switches in a digital circuit while bipolar transistors operate as current-controlled switches.

Chapter 3

Logic Gates and Circuit Components

3.1 Logic Gates

A logic gate can be a virtual or a physical device. In either case the logic gate takes one or more binary signals as input and produces a binary output as a logical function. The basic logical operations of AND, OR, XOR, and NOT are described in basic electronics and Boolean algebra texts. Although logic gates can be made from electromagnetic relays, mechanical switches, or optical components, nowadays they are normally implemented using diodes and transistors.

Charles Babbage's Analytical Engine, devised around 1837, used mechanical logic gates based on gears. Electromagnetic relays were later used for logic gates, and these were eventually replaced by vacuum tubes, as Lee De Forest's modification of the Fleming valve can be used as an AND logic gate. In 1937, Claude E. Shannon wrote a thesis paper that introduced the use of Boolean algebra in the analysis and design of switching circuits. The first modern electronic gate was invented by Walther Bothe in 1924, for which he received part of the 1954 Nobel prize in physics.

The primitive types of gate are the AND, OR, and NOT. Additionally, the XOR gate offers an alternative version of the OR. All other Boolean operations can be implemented by combining the three primitive types. However, for convenience, other secondary types have been developed. These are called NAND (NOT plus AND), NOR (NOT plus OR), and XNOR (XOR plus NOT). The advantage of these secondary logic gates is that they require fewer circuit elements for a given function. In fact, the NAND gate is the simplest of all gates, except for the NOT gate. Furthermore, a NAND can implement both a NOT and an OR function; therefore it can replace AND, OR, and NOT. This means that the NAND gate is the only type actually needed in a real system. Programmable logic arrays will very often contain nothing but NAND gates. The symbols for logic gates are shown in Figure 3-1.

The notion of a binary signal is based on it being in only one of two states: high or low. Conventionally we represent a high signal with binary digit "1" and a low with a binary digit "0." True and false and high and low are also associated with binary signals, binary 0 representing false or low, and binary 1 representing true or high. In digital electronics voltage is used to encode binary 0 and 1. A voltage of about 0.5 volts (actually 0 to 0.8 volts) is interpreted as logic 0 and a voltage of about 3.5 volts (actually 2.4 to 5.0 volts) is interpreted as logic 1. Voltages from 0.8 to 2.4 volts are not allowed. This voltage convention is referred to as TTL (*transistor-transistor logic*).

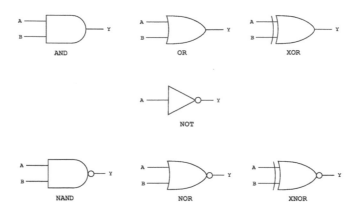

Figure 3-1 Symbols for Logic Gate.

3.2 Power Supplies

Standard logic circuits usually require a power source of +5 volts DC; batteries are one possible source. A D cell battery produces 1.5 volts, so three of them connected in series produce 4.5 volts DC. An alternative power source can be a standard wall outlet. To obtain +5 volts DC from 110 volts AC requires scaling down the voltage and converting alternating current to direct current. In addition, most power supplies include a voltage regulator component that ensures that the circuit voltage is exactly +5 volts. The circuit in Figure 3-2 is a regulated 5-volts DC power supply.

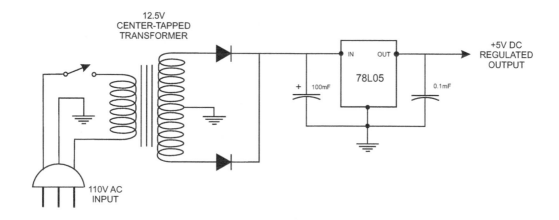

Figure 3-2 Regulated, +5 Volts DC Power Supply.

In the circuit of Figure 3-2 the transformer reduces the household voltage from 110 to about 12 VAC. The diodes rectify the input to an oscillating signal of about +12 VDC. The 100mF electrolytic capacitor smoothes out the oscillation producing a largely DC voltage with little ripple. The 7805 is a voltage regulator that accepts an input voltage from about 8 volts to about 35 volts and produces a constant 5V output. Voltage regulator ICs are Zener diodes with a precise, reverse-biased breakdown voltage. The 7805 is usually mounted on a metal base with a drilled hole so that a heat sink can be attached to it. With a heat sink, the 7805 can produce up to 1 amp output. Figure 3-3 shows a 7805 voltage regulator IC.

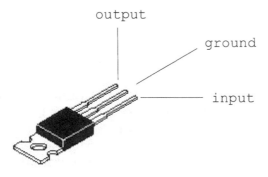

Figure 3-3 7805 Voltage Regulator IC.

3.3 Clocked Logic and Flip-Flops

In the circuit diagrams considered so far, the outputs are entirely determined by the inputs to these circuits. If the inputs change, so do the outputs. However, we often need a digital component whose output remains unchanged even if there is a change in input. One use for such a circuit is to store a binary number. A flip-flop is a circuit that performs as a 1-bit memory that stores either the value 0 or 1.

3.3.1 RS Flip-Flop

A circuit with only two stable states is said to be *bistable*. A toggle switch that can be either OPEN or CLOSED is a bistable device. A *flip-flop* is an electronic circuit with two stable states, as its output is either 0 or +5 VDC. In digital terms, a flip-flop is set if it stores a binary 1 and reset otherwise; the RS designation refers to these terms. The flip-flop can also be said to have memory because its output will remain set or reset until it is intentionally changed. When the flip-flop output is 0 VDC it can be regarded as storing a logic 0 and when its output is +5 VDC as storing a logic 1, flip-flops can be constructed using primary logic gates. One possibility is to use two NAND gates, as in Figure 3-4.

The fact that a NAND gate is equivalent to a negative logic OR gate makes the flip-flop easier to understand. First consider that the Set input is pulled low by flipping the switch counterclockwise and sending the input to ground. In this case the output of the upper gate (1) is forced high because the gate's output goes high if either input 1 or input 2 is low. Because the Reset input to the lower gate is high (4), then neither input of the lower gate (3 or 4) is low and its output is low.

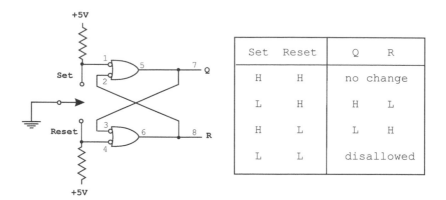

Figure 3-4 NAND Gate-Based RS Flip-Flop.

Note that input 3 is low because the bubble on the lower OR gate inverts the value fed back from the upper OR gate. Now the feedback line from the lower gate (6) sends low to input 2 on the upper gate, which is inverted by the upper gate bubble. So both inputs to the upper gate are high, which determines that the upper gate's output remains high even when the Set input returns to a logic high, as would be the case if the switch were turned back to the neutral position. Thus, the Q output of the flip-flop stays high (and the inverted Q output remains low). When the flip-flop is in this state, it is said to be set. The flip-flop is placed in the cleared state by momentarily pulling the Reset input low. This forces the lower gate's output to be high and the upper gate's to be low.

The action of the flip-flop in Figure 3-4 is consistent with the description of a device with two steady states that is controlled by two corresponding input lines labeled Set and Reset. Also, once in either state the device will remain in that state until the opposite state is enabled, thus "remembering" its set or reset status. The condition of two low input lines is not allowed in the flip-flop, as shown in the truth table.

All mechanical switches used in electronic devices contain a spring of some sort. It is this spring that maintains the switch's contact in either position, but it also makes the switch electrically "bounce" whenever it is activated. Although the bounce only takes a few milliseconds, it is highly undesirable because the logic level can change between high and low several times during this period. If an RS flip-flop is connected to the switch, the first contact switches the flip-flop and subsequent ones have no effect, thus effectively "debouncing" the switch.

3.3.2 Clocked Circuits

So far the circuits discussed are examples of combinatorial logic, also called *asynchronous*. In combinatorial circuits, if we ignore a few nanoseconds of propagation delay, the outputs change as soon as the inputs change. Although in theory you can build complex logic circuits using combinatorial logic, it is more convenient to use clocked logic pulses to ensure high reliability and noise immunity. Circuits that use clocked impulses are said to use *synchronous* logic.

In synchronous circuits, changes in logic outputs are not allowed to propagate unconstrained. The synchronous circuit is designed so that logic-level changes can progress through the circuitry, one stage at a time, under the control of a clock. Between the clock pulses that cause changes to take place, the temporary state of the system is stored in memory elements or flip-flops.

In clocked or synchronous logic, all the gates in the system change outputs at the same time. The output state of each gate depends only on the state of the gate inputs at the time of the clock pulse. In combinatorial circuits, the gates may briefly see the wrong logic level and cause incorrect operation of the circuit. Clocked logic ensures that gate outputs settle down during the time between clock pulses; thus, only valid logic levels are present by the time the next clock pulse arrives.

The RS pushbutton in Figure 3-4 cannot be used in a clocked logic circuit because its output changes immediately whenever the Set or Reset inputs change. However, the circuit can be made into a clocked RS flip-flop by adding two NAND gates, as shown in Figure 3-5.

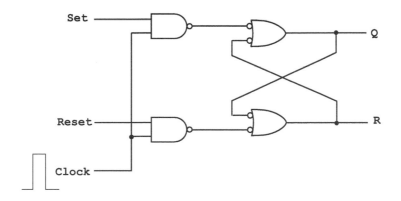

Figure 3-5 A Clocked RS Flip-Flop.

In the clocked flip-flop of Figure 3-5, the Set and Reset inputs can change at any time, but those changes will be ignored by the flip-flop except during the interval when the logic high of a clock pulse is present. It is the NAND gates that ensure that the Set and Reset lines are read by the circuit only when the clock signal is high. This determines that the state of the Set or Reset line is stored by the flip-flop only during the high phase of the clock pulse.

3.3.3 D Flip-Flop

One objection to the flip-flops in Figures 3-4 and 3-5 is that they require two data input lines (which are labeled Set and Reset in the illustrations). In many applications a more convenient scheme is to have a single input line that is read as Set if it is high and Reset otherwise. To implement this circuit so that a single input line is used we can connect an inverter between the Set line and the Reset input. The circuit for a D (from data) flip-flop is shown in Figure 3-6.

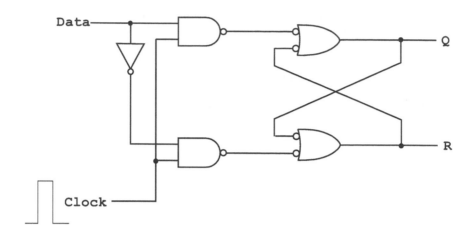

Figure 3-6 The D Flip-Flop.

The D flip-flop is also called a *transparent latch*, or a *D latch*. In the D flip-flop the state of the input line, called the D (or data) input, is stored in the flip-flop when a clock pulse occurs. An additional advantage of the D latch is that the disallowed state (see Figure 3-4), in which both Set and Reset are simultaneously low, cannot be reached accidentally.

We have mentioned that a flip-flop can be used for storing binary data. To visualize how this can be done, imagine four D flip-flops driven by the same clock signal. When the clock goes high, input data is loaded into the flip-flops and appears at the output. When the clock goes low, the output retains the data. For example, suppose four data inputs, as follows:

$$D_0 D_1 D_2 D_3 = 0101$$

When the clock signal goes high, these four bits are loaded into the D latches, resulting in the output

$$Q_0 Q_1 Q_2 Q_3 = 0101$$

This operation is represented in Figure 3-7.

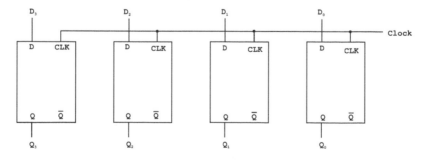

Figure 3-7 Four Data Bits Stored in D Latches.

In the 4-bit D latch of Figure 3-7, the output data is stored as soon as the clock goes low. For as long as the clock is low, the D values can change without affecting the Q values. The 7475 IC contains four D flip-flops and is called a quad *bistable* latch. This circuit is well suited for handling 4-bit data bits simultaneously (one nibble).

3.3.4 Edge-Triggered D Flip-Flop

The D flip-flop or transparent latch is available in several versions. Although the pure D flip-flop is a useful IC, it has the drawback that outputs follow the D input during the entire time that the clock line is high. In some circuits it is preferable for the flip-flop to store data at a single and unique point in time. The *Edge-Triggered D-type flip-flop* approaches this behavior. In this device, the flip-flop only stores the state of the data line at the instant the clock signal makes a transition from low to high. Figure 3-8 shows an Edge-Triggered D flip-flop.

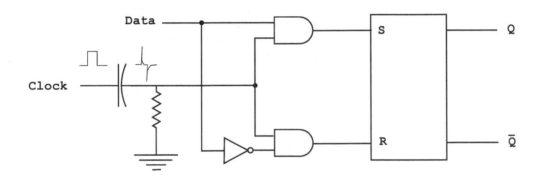

Figure 3-8 Edge-Triggered D Latch.

The circuit in Figure 3-8 is sometimes called an *RC differentiated clock input latch*. In this case, RC stands for the resistor/capacitor pair at the input of the D latch. By design, the RC time constant is made smaller than the clock's pulse width. This determines that the capacitor fully charges when the clock goes high, which produces a narrow positive voltage spike across the resistor. Later, the trailing edge of the pulse results in a narrow negative spike, which in turn enables the AND gates for a brief period. The effect is to activate the AND gates only during the positive spike; the negative spike does nothing in this circuit. The result is equivalent to sampling the value of D for an instant. At this point in time, D and its complement hit the flip-flop inputs, forcing Q to set or reset (unless Q is already equal to D).

3.3.5 Preset and Clear Signals

The use of flip-flops in digital circuits usually requires some way of placing the signals in a known state. In this sense a Preset signal is used to make sure that the Set line is high, and a Clear signal to make sure that the Reset line is high. Alternatively, sometimes these signals are referred to as *Preset R* and *Preset S*. Figure 3-9 shows how the Preset and Clear functions can be implemented in an RS flip-flop.

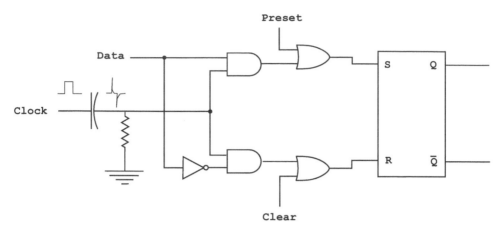

Figure 3-9 Implementing Preset and Clear.

The OR gates in the circuit of Figure 3-9 allow selectively setting the S or the R lines of the Edge-Triggered D flip-flop. The Preset and Clear signals are called asynchronous inputs because they activate the R or S lines of the flip-flop independently of the clock. The D input, on the other hand, is synchronous because it has an effect only when the clock edge signal is high. Figure 3-10 shows the electrical symbol for a positive Edge-Triggered flip-flop with active high Preset and Clear lines.

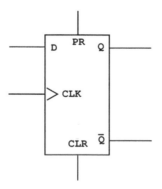

Figure 3-10 D-Type Edge-Triggered Flip-Flop Symbol.

The standard mode of operation for a D-type flip-flop is to have the Set and Clear inputs high (not active), so that a transition of the clock input from low to high (called a positive edge) clocks the value of D into Q and the inverse of D into not-Q. The clock transition is required because nothing happens to Q and not-Q until a positive edge occurs on the clock line.

3.3.6 D Flip-Flop Waveform Action

An easy way to understand the interaction of the various signals in a clocked RS flip-flop is by means of a *waveform diagram*. The reference circuit is the one in Figure 3-9, which includes a clock signal, a data input line, Preset and Clear lines, Set and Reset input lines, and Q and not-Q output lines. The signals are described as follows:

1. The clock signal (CLK) is a square wave that oscillates between a high and a low state. It provides a synchronized beat that coordinates the various digital devices present in the circuit.

2. The data signal is used as a single input line into the flip-flop. Setting the data signal high also sets high the flip-flop's Set line. A low data signal makes the flip-flop Reset line high.

3. The Set signal or line is one of the two inputs into the flip-flop. The other one is the Reset line.

4. The Preset line is used to make the flip-flop Set line active. The Clear signal has the effect of setting high the Reset line into the flip-flop.

5. The Q and not-Q lines provide the flip-flop output. Q is high if the Set line is high, otherwise the not-Q line is high.

Figure 3-11 is a waveform diagram for a clocked RS flip-flop.

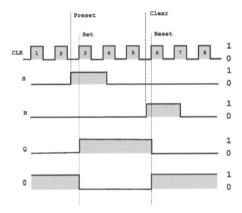

Figure 3-11 Waveform Diagram for Clocked RS Flip-Flop.

In Figure 3-11 note that the clock input (at the top of the illustration) provides the synchronization beat for the flip-flop inputs (R and S) and the outputs (Q and not-Q). However, the Preset and Clear signals are asynchronous; that is, they operate independently of the clock pulse. Therefore, when the Preset line is set high, the S input line into the flip-flop immediately follows. However, the Q output line must wait until the next rising clock pulse, which corresponds to the dot-dash line labeled Set in the illustration. Similarly, the Clear signal immediately sets the R line; however, the not-Q output is not set until the next rising clock pulse. Notice that during clock pulse number 4, both the R and S lines are held low. This corresponds to a hold state during which the output on lines Q and not-Q remains unchanged.

3.3.7 Flip-Flop Applications

The D-type flip-flop finds many uses in digital technology. Perhaps the most obvious one is as a memory. We have seen that the flip-flop stores the value clocked into it from the D line; its value can be read on the output lines Q and not-Q. A type of memory known as *static* RAM is implemented as a large array of flip-flops with address decoding circuitry. This design allows selecting which flip-flop is being accessed by a read or write operation. Processors, microprocessors, and microcontrollers contain many flip-flops, usually in the form of *registers*. These registers are just a group of 8, 16, 32, or 64 flip-flops. Flag registers are also flip-flops that are set or cleared by the results of the CPU's internal operations.

Digital devices interface with the outside world by means of input and output ports. These elements are implemented as flip-flops. For example, the logic in a given circuit may require turning on an LED so as to signal that some event has occurred. To achieve this, a data line from the digital device can be connected to the D input of a flip-flop. Then a pulse is sent on another line to the clock input. When the clock pulse goes from low to high, the state of the data line at that instant is clocked into the flip-flop. This state remains on the Q output until a new value is clocked in. Another example is the 74374 IC, which contains 8 flip-flops in a single 20-pin DIP package. The chip is called an *octal latch* because the data is latched into all eight flip-flops all at once by a single clock line.

D-type flip-flops are also used in implementing more complex digital interfaces, such as an interrupt system. For example, a digital device that reads in data from an external source, such as a switch. Each time a new data state is produced by the switch, a flip-flop is set and the output of this flip-flop is connected to an *interrupt request line* (IRQ) on the device. When the IRQ line goes high, the microcontroller saves its current state and branches off to an input routine that takes some action according to the state of the switch, for example, turns on an LED if the switch is high. To prevent the microcontroller from getting interrupted again by the same input, the same signal is also used to clear the flip-flop until the next data byte comes along.

3.4 Digital Clocks

A *clock signal* consists of a sequence of regularly spaced pulses, typically in the form of a square wave. Digital devices use the rising or the falling edges of the square wave to operate logic circuits. Clocks provide the heartbeat, without which the system would cease to function.

3.4.1 Clock Waveforms

In a digital device, such as a microcontroller, the clock furnishes a periodic waveform that is used as a synchronizing signal. Although the typical clock waveform is depicted as a square wave (as in Figure 3-11), it should be noted that the wave need not be perfectly symmetrical. In fact, a series of positive or negative edges can serve as a timing pulse in a digital circuit. The one requirement of a clock pulse is that it be perfectly periodic.

The basic timing interval for a digital circuit, which is equal to one full waveform period, is called the *clock cycle*. This determines that all logic elements in the circuit, including gates and flip-flops, must complete their transitions in a time period no longer than a complete clock cycle.

We can assume that the ideal clock produces a perfectly square waveform that is absolutely stable, as is the one shown in Figure 3-12.

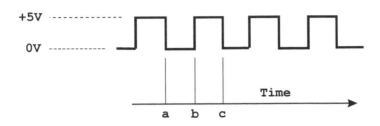

Figure 3-12 Ideal Clock Waveform.

A stable and uniform waveform reaches exactly the same voltage every time it is high, for example, +5 volts. By the same token, every time the clock signal goes low, the voltage level must be the same, typically 0 volts. In addition, the clock signal must remain at the high and low levels for the same time, and the time between each high and low cycle must be exactly the same. This last element is usually called the *frequency stability* of the clock. In Figure 3-12 the frequency stability refers to the time it takes for the signal to transition from point *a* to point *c* during each clock cycle. In practice, the stability and uniformity of the clock signal are more important than the absolute value. For example, it is usually acceptable that the high voltage level of the clock signal be 4.8 volts instead of 5 volts, as long as the 4.8-volt level is exactly reproduced at every clock cycle.

Another characteristic of the clock signal is the time required for clock levels to change from high to low and vice versa. Ideally, this transition could be represented by a vertical line, as in Figure 3-12. This means that the transition is instantaneous, which is not achievable in actual circuits. In practice, some time is required for the waveform to transition from low to high, and vice versa. The actual graph of the waveform, as can be seen on an oscilloscope, shows a slightly sloping side. Customarily, the actual measurement of the transition time is referred to as the 10 and 90 percent points. For example, in a 5-volt waveform, the rise-time is the time it takes for the voltage to go from 0.5 to 4.5 volts, which are the 10 and 90 percent points for that waveform.

3.4.2 TTL Clock

A much-used TTL-compatible clock can be built around a 7404 hex inverter IC such as the one in Figure 3-13. The idea is to use two inverters to build a two-stage amplifier with an overall shift of 360 degrees. The output signal at one of the inverters is fed back, through a crystal, to the first inverter, which determines that the circuit oscillates at a frequency determined by the crystal. Thus, the frequency of this clock signal is determined by the crystal: values between 1 and 20 MHz are common. The TTL clock circuit is shown in Figure 3-13.

The crystal in the circuit of Figure 3-13 makes the frequency of oscillation very stable. The third inverter is used as an output buffer and allows driving the load simulated by the RC circuit.

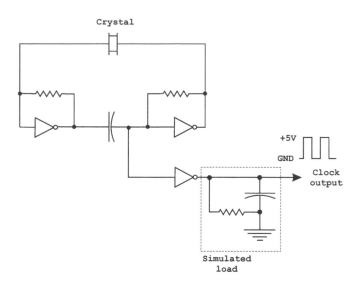

Figure 3-13 TTL Clock Circuit.

The clocks used in digital systems need to be stable and uniform so that the frequency is the same and each pulse is identical to every other one. To achieve this, a very narrowband frequency-selective filter whose center frequency does not change is required. Quartz crystals are a good choice for this purpose because they provide a stable, precision oscillator. A *quartz crystal* is actually a thin piece of polished crystalline quartz with contacts plated on each surface and a lead attached to each contact. Quartz is a piezoelectric material, which means that there is one particular electrical frequency that excites the crystal's resonance. It is this very narrow resonant frequency that is used to build a frequency-selective filter whose center frequency changes very little as the components age or with changes in temperature. Crystal oscillators are available with frequencies that range from 10 KHz up to 600 MHz. They are typically housed in small metal cases with the frequency printed on the outside.

3.4.3 555 Timer

One of the most versatile timer ICs is the TTL-compatible 555 timer. This chip can be used to make many different kinds of oscillators, pulse generators, and timers. As an oscillator, the 555 can be made to produce square, sawtooth, or triangle waves, and its frequency can be modulated by an external input. Although the 555 is not a TTL part, its output is TTL compatible when it is used with a 5-volt power supply.

The 555 has two distinct output levels, which continuously switch back and forth between these two unstable states. Because of this oscillation, the circuit output is a periodic, rectangular waveform. The fact that neither output is stable accounts for the circuit being classified as astable or bistable. The frequency of oscillation as well as the duty cycle are accurately controlled by two external resistors and a single timing capacitor. Figure 3-14 shows the logic symbol for a 555 timer as well as the wiring to implement an asymmetric square-wave generator.

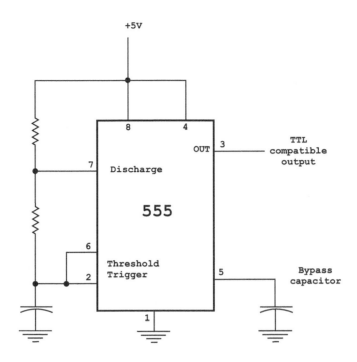

Figure 3-14 555 Timer as a Square Wave Generator.

3.4.4 Microcontroller Clocks

Microcontrollers, like most digital components, require a synchronizing timing pulse provided by some form of clocking device. There are five common ways of implementing a timer in a microcontroller:

1. Internal clock
2. RC network
3. Crystal oscillator
4. Ceramic resonator
5. External oscillator

The selection depends on the specific microcontroller, the circuit requirements, and the cost of each available option. The cheapest option is the resistor/capacitor oscillator circuit (RC network). The disadvantages are its slow speed and inherent inaccuracies. Some of the newer generations of microcontrollers come equipped with an internal RC oscillator that operates as a programmable timer. Typical speeds are 4 MHz with a 1.5 percent error. The actual use and implementation of microcontroller clocks are discussed in relation to each specific device.

3.5 Counters and Frequency Dividers

Counters and *frequency dividers* are actually the same circuitry, used in different ways. Counters are one of the most useful and versatile digital devices. They can be used to keep track of the number of clock cycles and as an instrument for measuring time and therefore period or frequency. There are two different types of counters: *synchronous* and *asynchronous*.

3.5.1 Frequency Dividers

Circuit designers often need to reduce the frequency of a wave clock signal. One easy way of doing it is to divide the frequency by two by feeding back the not-Q output of a D-type flip-flop to its data line. Figure 3-15 shows a divide-by-two circuit and its effect on the resulting wave.

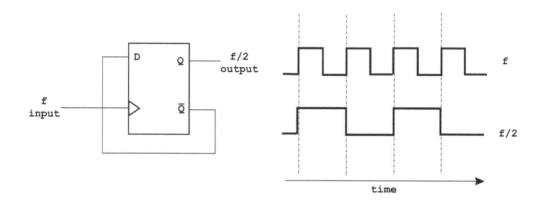

Figure 3-15 A Divide-by-Two Circuit.

In the circuit of Figure 3-15 the frequency division occurs because the rising edge of each clock input toggles the flip-flop's output. This results from the fact that when the Q output goes low, the not-Q line goes high and the high feedback signal is fed back to the data line, thus canceling out the next high wave of the *f* signal.

3.5.2 JK Flip-Flop Counter

One type of specialized flip-flop is the *JK flip-flop*. The JK flip-flop is an ideal component to build a circuit that keeps track of the number of positive or negative clock edges of the input clock. The name of this flip-flop relates to the two variables, J and K, that are used as inputs to the circuit. Figure 3-16 shows one possible circuit implementation for the JK flip-flop.

In Figure 3-16 the RC components convert the rectangular wave clock pulse into a narrow spike. The three-input AND gates make the circuit positive-Edge-Triggered. When J and K are both low, both AND gates are disabled; therefore, clock pulses have no effect. This corresponds with the first entry in the truth table. When J is low and K is high (second entry in the truth table), the upper gate is disabled, so the flip-flop cannot be set, which means that it must be reset. When Q is high, the lower gate passes a RESET trigger as soon as the next positive clock edge arrives. This forces Q to become low (the same second entry in the truth table). Therefore, J low and K high means that the next positive clock edge will reset the flip-flop.

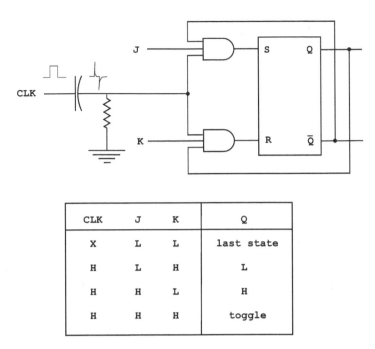

CLK	J	K	Q
X	L	L	last state
H	L	H	L
H	H	L	H
H	H	H	toggle

Figure 3-16 The JK Flip-Flop Circuit.

When J is high and K is low (third entry in the truth table), the lower gate is disabled, so it is impossible to reset the flip-flop. However, the flip-flop can be reset when Q is low because in that case not-Q is high; therefore the upper gate passes a SET trigger on the next positive clock edge. This drives Q into the high state (the third entry in the truth table). As you can see, J = 1 and K = 0 means that the next positive clock edge sets the flip-flop (unless Q is already high). When J and K are both high it is possible to set or reset the flip-flop. If Q is high, the lower gate passes a RESET trigger on the next positive clock edge. On the other hand, when Q is low, the upper gate passes a SET trigger on the next positive clock edge. Either way, Q changes to the complement of the last state (see last entry in the truth table). Therefore, J = 1 and K = 1 means the flip-flop will toggle on the next positive clock edge.

3.5.3 Ripple Counters

The simplest of all counters is called a *ripple counter*. A two-bit ripple counter can be constructed by wiring together two divide-by-two circuits, as in Figure 3-17. Stringing together two divide-by-two circuits produces a divide-by-four circuit. Stringing together three flip-flops produces a divide-by-eight circuit, four flip-flops create a divide-by-sixteen circuit, and so on. The counting action of the connected flip-flops is based on the fact that each flip-flop changes state before it triggers the next one in line. Thus, each stage performs as a bit in a binary counter, the first stage being the LSB and the last state the MSB. Because the preceding flip-flop acts as a clock for the next one in line, the flip-flop to the right will toggle each time its neighbor to the left goes low. In Figure 3-17 the signal labeled Q_0 is the LSB of a two-bit counter, while the signal labeled Q_1 is the most significant bit.

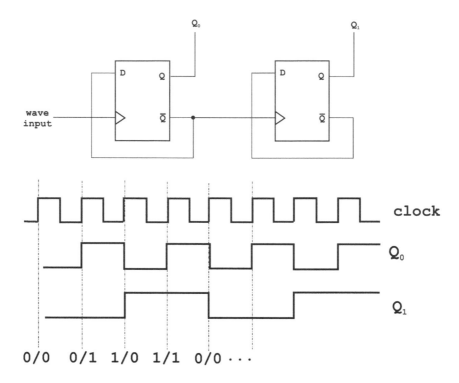

Figure 3-17 Two-Bit Ripple Counter.

In this design, each flip-flop is triggered by the previous one; thus the count is said to "ripple" down the device. One objection to the ripple counter is that the change in each output is determined by the previous output in the flip-flop chain, which produces a few nanoseconds' time lag from output line to output line. This cumulative settling time is why these counters are called serial or asynchronous.

Note that the ripple counter of Figure 3-17 uses the not-Q line to drive the next flip-flop in the chain. If a ripple counter is wired so that the Q line drives each next stage, then the transitions take place not when the previous waveform goes low, but when it goes high. The result is that the counter counts down instead of up. In other words, in the down counter, the count is reduced by one during each clock transition. Commercial counters, such as the 74193, can be made to operate as an up-counter or a down-counter by selecting the corresponding input line.

3.5.4 Decoding Gates

A decoding gate is a way of connecting the output of a counter so that it will signal a given state. For example, if four D-type flip-flops are wired so as to produce a four-bit ripple counter, similar to the one in Figure 3-18, the counter will represent binary digits 0000 to 1111. If we wanted to detect the value 1101 (16 decimal), the resulting circuit could be designed as in Figure 3-18. This circuit uses a NOR gate to invert the value of bit number 1. The AND gate serves to trigger the output when bits 0, 2, and 3 are high and bit 1 is low. This state corresponds to the binary value 1101.

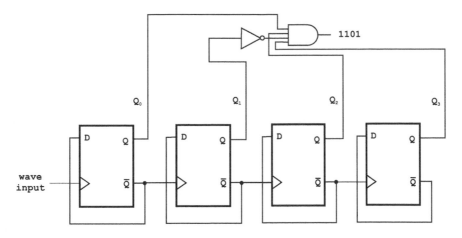

Figure 3-18 A Decoding Gate.

3.5.5 Synchronous Counters

Although the ripple counter is the simplest one, it has the previously mentioned disadvantage that each flip-flop has to wait for its neighbor to switch states. This determines that in a ripple counter, the delay times are additive, and the total "settling" time for the counter is approximately the delay multiplied by the total number of flip-flops. In addition, with ripple counters the resulting delay creates the possibility of glitches occurring at the output of decoding gates. These problems can be overcome by the use of a synchronous or parallel counter.

A counter in which each flip-flop is triggered at every clock beat can be built by observing how counting takes place in binary numbers. Binary counting has the property that when a bit changes from high to low (0 to 1), it sends a toggle command to its neighbor to the left. So assuming that the low-order bit changes consecutively from one state to its complement, and starting from all bits initialized to 0, binary counting can be visualized as in Figure 3-19.

0000

0001

0010

0011

0100

0101

0110

0111

.

.

.

1111

Figure 3-19 Visualization of Binary Counting.

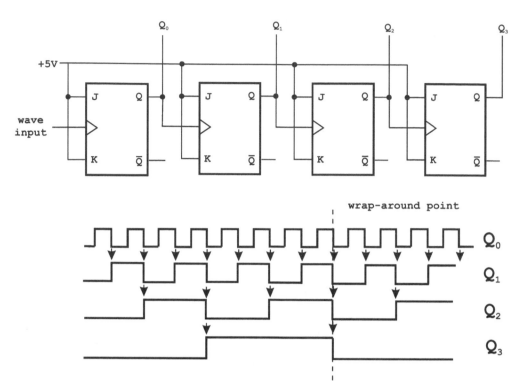

Figure 3-20 A Synchronous 4-Bit Up Counter.

In Figure 3-19 the arrows indicate the transition from high to low, which is the command for the column to the left to change to its complement (toggle). Using this property of binary counting, it is possible to wire four JK flip-flops so that every high-to-low transition of a flip-flop triggers its higher-order neighbor to toggle its state. Figure 3-20 shows such a system.

In Figure 3-20, the first flip-flop (Q_0 output line) has a positive-edge triggered clock input. The lowest-order flip-flop toggles with each rising edge of the clock signal (not shown in the illustration). The second flip-flop to the right toggles with every falling edge of the signal from its neighbor to the left. And so on to the last flip-flop in the chain. The arrows in the waveform portion of Figure 3-20 show that each succeeding output bit is toggled by the transition from high to low of its lower-ordered neighbor. Also note the dashed line that marks the point where all counters are transitioning from high to low. At this point, all four counters wrap-around to zero and a new count begins.

Observe that the not-Q output line transitions opposite to the Q output. That is, when the Q output line goes high, not-Q goes low, and vice versa. Therefore, if the pulse into each successive flip-flop originated in the not Q-line, instead of the Q line, then the resulting circuit would be a synchronous counter that transitions on the positive edge (low-to-high) instead of on the negative edge. If we observe that the not-Q lines also provide a set of negated outputs in reference to the Q lines, then it is possible to come up with a circuit that serves both as an up- and down-counter according to the selected set of outputs. Such a circuit is shown in Figure 3-21.

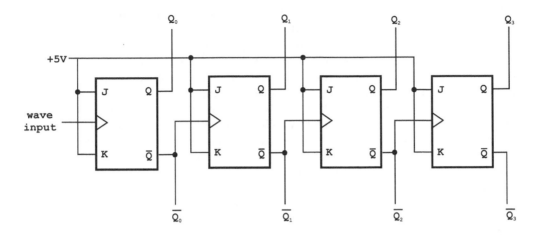

Figure 3-21 Synchronous 4-Bit Up- and Down-Counter.

In the counter of Figure 3-21, the Q outputs generate the up-count series while the not-Q outputs produce the down-count series.

3.5.6 Counter ICs

Counters are available as standard TTL components. The 7493 is an asynchronous 4-bit ripple counter that counts from 0 to 15. The 7490 is another version of the ripple counter, called a decade counter, as the count output is in the range 0 to 9. The 74193 is a 4-bit synchronous up/down counter in the range 0 to 15. Figure 3-22 is a pin diagram of the 74193.

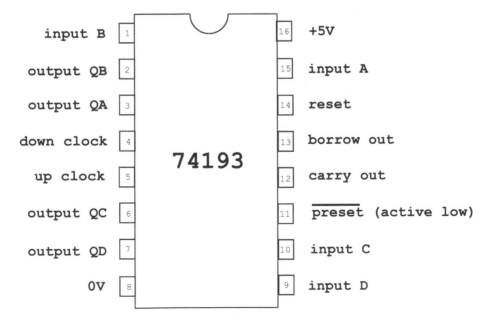

Figure 3-22 Pin Diagram of the 74193 Synchronous Up/Down Counter.

The 74193 is a synchronous counter so its output changes precisely at each clock pulse. This is convenient because it allows for connecting its output to other logic gates and avoids the glitches associated with ripple counters. In Figure 3-22 you can see that the 74193 has separate clock inputs for counting up and counting down. The count increases as the up clock input becomes high (on the rising-edge). The count decreases as the down clock input becomes high (on the rising-edge). In both cases the other clock input should be high. For normal operation the preset input should be high and the reset input low. When the reset input is high, it resets the count to zero, that is, lines QA to QD are low. The counter can be preset by placing any desired binary number on inputs A to D and making the preset input low. These inputs may be left unconnected if not required. Several 74193 counters can be chained by wiring a common reset line, connecting the carry to the up clock line of the next counter and the borrow to the down line.

3.5.7 Shift Registers

The hardware implementation of binary shift and rotate operations are called *shift counters* or *shift registers*. Shift counters are often based on the D-type flip-flop. Actually, several D-type flip-flops can be chained together so that the D output of one goes into the D input of the next one. If all the flip-flops are driven by the same clock signal, then the effect would be to shift the bits from one flip-flop into the next one at each rising clock pulse. A common implementation of a shift counter is called a *parallel-in/serial-out shift register*, as the one in Figure 3-23.

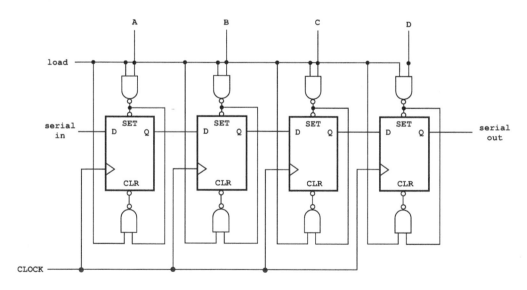

Figure 3-23 4-Bit Parallel-in/Serial-out Shift Counter.

The circuit in Figure 3-23 shows four flip-flops connected so that the output of one feeds into the input of the next one. Also, a set of NAND gates allow parallel data input. When the load signal is set high, the flip-flops in the shift register are loaded simultaneously with the logic values at the inputs A, B, C, and D. The 74165 IC is an 8-bit parallel-in/serial-out shift register with asynchronous parallel load and two OR-gated clock inputs. Figure 3-24 is a pin diagram of the 74165 IC.

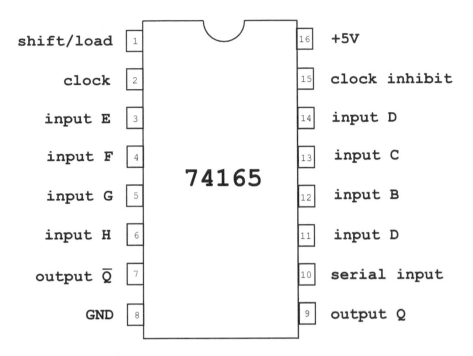

Figure 3-24 Pin Diagram of 74165 IC.

The serial input line in the diagram of Figure 3-23, and in the 74165 IC in Figure 3-24, allow cascading multiple chips.

Parallel-in/serial-out shift registers find common use in the implementation of serial ports. In serial communications, data is sent 1 bit at a time over a single wire. In order to accomplish this, data is first loaded into a parallel-in/serial-out shift register. The individual bits are then shifted out one at a time. The frequency of the driving clock in this case corresponds to the baud rate being used. A second type of shift register is used to receive the data on a serial communications line. In this case the operation is *serial-in/parallel-out*. The circuit that accomplishes this is based on D-type flip-flops in which the Q outputs are connected to the D input lines.

The 74164 IC is one such device. In actual serial ports, the transmitting and receiving shift registers are contained in a single device called a *UART*.

3.6 Multiplexers and Demultiplexers

It is often the case in digital electronics that different signals must be sent out on a single output line, or several signals must be received in a single input line. The digital circuits that perform these operations are called *multiplexers* and *demultiplexers*. Multiplexers and demultiplexers are TTL analogs of the many-to-one and one-to-many mechanical switches.

3.6.1 Multiplexers

Multiplexing (also called *muxing*) is a way of combining the data of two or more input channels into a single output channel. The hardware multiplexer, also called a *mux*, combines several electrical signals into a single one. The multiplexer performs a many-into-one function. Sometimes multiplexers and demultiplexers are combined into a single device, which is still referred to as a "multiplexer." Figure 3-25 shows the schematics diagram of a multiplexer.

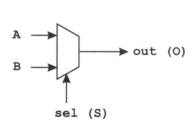

A	B	S	O
0	0	0	B (0)
0	0	1	A (0)
0	1	0	B (1)
0	1	1	A (0)
1	0	0	B (0)
1	0	1	A (1)
1	1	0	B (1)
1	1	1	A (1)

Figure 3-25 Multiplexer Schematics.

The truth table in Figure 3-25 describes the multiplexer operation. The line labeled "sel" in the illustration is the selector line. If the selector is low (S = 0), then input line B is mirrored in the output line O. Otherwise, input line A is vectored to the output. The Boolean expression for the multiplexer in Figure 3-25 is

$$O = (A \wedge S) \vee (B \wedge \neg S)$$

Often, a multiplexer circuit is preceded by a decoder circuit so that input can be compressed into fewer lines. For example, a four-to-one multiplexer receives a binary value in the range 0 to 3 (binary 00 to 11) on two input lines and sets high one of four output lines accordingly. Figure 3-26 shows the circuit diagram for such a device.

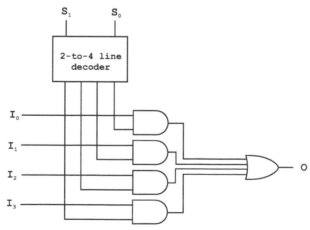

Figure 3-26 Two-Bit One-of-Four Multiplexer.

In the circuit of Figure 3-26 there are four input lines. The binary value in the two S lines determines which one of the four lines is copied to the multiplexer output (labeled O). The 2-to-4 line decoder converts this value into one of four selector lines, which will be in one of these four states:

LLLL LLHL LHLL HLLL

Similarly, an eight-input multiplexer has eight data inputs and three binary selection inputs, which are converted into one-of-eight selection lines by the decoder, while a sixteen-input multiplexer requires four binary digits in the decoder input, which are converted into one-of-sixteen selection lines. Alternatively, the decoder circuit can be simplified using multiple AND gates and negating the input signals, as shown in Figure 3-27.

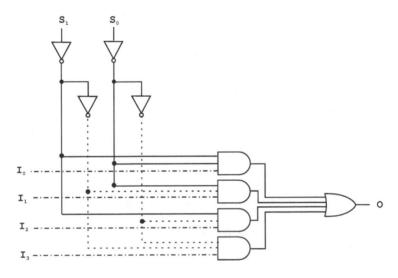

Figure 3-27 Multiplexer with Multiple AND Gates.

In the multiplexer in Figure 3-27, one could assume that the input bits are both low, that is, $S_1 = 0$ and $S_2 = 0$. In this case the first-level NOR gates would change the L signals to H. The two high signals will go into the first multiple AND gate, as shown by the solid lines in the illustration. This will determine that the first input line (I_0) will be copied to the circuit output. In fact, the four inverters at the top of the illustration will perform the function of the two-to-four decoder in the circuit of Figure 3-26.

3.6.2. Demultiplexers

A demultiplexer takes one data input and a number of selection inputs, and returns multiple outputs. While the multiplexer performs a many-into-one operation, the demultiplexer performs a one-into-many. For example, a four-output demultiplexer has one data input line, two selection inputs, and four data output lines. Figure 3-28 shows such a circuit.

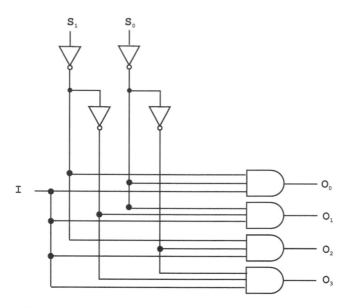

Figure 3-28 Two-Bit into Four-of-One Demultiplexer.

A demultiplexer can be made to act as a decoder by holding the input line high. For example, the circuit in Figure 3-28 performs as a binary one-to-four-line decoder if the I line is held high. In this case, the binary bit patterns on the two input lines are converted into a single output in one of the four output lines. Thus, if there were four devices, each one connected to one of the output lines, the demultiplexer circuit would select which one is enabled according to the binary value of the input.

3.6.3 Multiplexer and Demultiplexer ICs

Several ICs are available that perform multiplexing and demultiplexing operations. For example, the 74138 is a three-line to eight-line decoder and demultiplexer. With this IC any of eight inputs can be selected by placing the corresponding 3-bit number on the device's three address lines. The 74151 is a 1-of-8 data selector/multiplexer. This device routes data from eight sources to a single output line. Here again, a 3-bit selector is used to determine which of the eight inputs is routed to the output.

An important use of the multiplexer is to encode row and column addresses onto the address lines of dynamic RAM, although more often tristate buffers such as the 74541 are used for this. Another important use of multiplexers is in implementing *dual-port memories* for video displays.

Chapter 4

Input and Output Devices

4.1 Obtaining Input

Electronic devices, including computers and microcontrollers, must often receive the data and commands required for their operation. In computer technology a common input device is the keyboard, which allows for entering text data as well as keystroke orders. Alternate computer input devices are the mouse, trackballs, light pens, graphical tablets, scanners of several designs, speech recognition devices, optical character recognition hardware and software, and many others. Although these devices are not excluded from use in microcontroller-based systems, a more typical case is that microcontroller input devices are much simpler and limited. In this chapter we discuss the two most commonly used devices for microcontrollers: the switch and the keypad. However, dedicated systems often use special input devices; for example, a radio receiver could be the input device for a radio-controlled system.

4.2 Switches

The *electrical switch* is a device for changing current flow in a circuit. Although mechanical switches find use in fields such as railroads and fluid flow control, here we refer to switches used in controlling electrical power or electronic telecommunications. In abstract terms, the switch is sometimes referred to as a *gate*, in the same sense as the logic gates discussed in Chapter 3. The simplest electrical switch has two components, called *contacts*, that touch to *make the circuit* and separate to *break the circuit*. The terms "make" and "break" are commonly used in this context. The selection of material for the contacts is important because corrosion can form an insulating layer that prevents the switch from performing its function. One possible solution is plating the contacts with noble metals, such as gold or silver.

In a switch the *actuator* is the part that applies the operating force to the contacts. Common switch types are rocker, toggle, pushbutton, DIP, rotary, tactile, slide, keylock, snap-action, thumbwheel, and several others. Figure 4-1 shows several switches commonly found in microcontroller circuit boards.

Figure 4-1 Common Electrical Switches.

The switch contacts are said to be "closed" when there is no space between them, thus allowing electricity to flow. When the contacts are separated by a space, they are said to be "open." In this case no electricity flows through the switch. Switches are classified according to the various contact arrangements. In the normally-open switch, the contacts are separated until some force causes them to close. In the normally-closed switch, the contacts are held together until some force separates them. Some switches can be selected to operate as either normally open or normally closed. The term "pole" is used in reference to a single set of contacts on a switch. While the term "throw" refers to the one or more positions that a switch can adopt. Figure 4-2 shows some common switch designs and their electrical symbols.

Diagram	Electronic abbreviation	Description
	SPST	Single pole, single throw. On-off switch such as a household light switch.
	SPDT SPCO	Single pole, double throw. Single pole, changeover. Changeover switch. C is connected to either L1 or L2.
	DPST	Double pole, single throw. Equivalent to two SPST switches operated by the same mechanism.
	DPDT	Double pole, double throw. Equivalent to two SPDT switches operated by the same mechanism.

Figure 4-2 Switch Symbols and Types.

A *multi-throw switch* can have two possible transient behaviors as it transits from one position to the other one. One possibility is that the new contact is made before the old one is broken. This *make-before-break* action ensures that the line is never an open circuit. Alternatively, there is a *break-before-make* action, where the old contact is broken before the new one is made. This mode of switch operations ensures that the two fixed contacts are never shorted. Both designs are in common use.

A *biased switch* is one in which the actuator is automatically returned to a certain position, usually by the action of a spring. A *pushbutton* switch is a type of biased switch, of which the most common type is a push-to-make switch. In this case the contact is made when the button is pressed and breaks when it is released. A *push-to-break* switch, on the other hand, breaks contact when the button is pressed. Many other special-function switches are available, for example, tilt switches, such as the mercury switch, in which contact is made by a blob of mercury floating inside a glass bulb as the switch is tilted. Other specialized switches are activated by vibration, pressure, fluid level (as in the float switch), linear or rotary movement, the turning of a key, a radio signal, or a magnetic field.

4.2.1 Switch Contact Bounce

Switch contact bounce is a common problem of electrical switches. Switch contacts are metal surfaces that are forced into contact by an actuator. Due to momentum and elasticity, the striking action of the contacts causes a rapidly pulsating electrical current instead of a clean transition from zero to full current. Parasitic inductance and capacitance in the circuit can further modify the waveform, resulting in a series of sinusoidal oscillations.

Switch contact bounce sometimes causes problems in circuits that are not designed to cope with oscillating voltages, particularly in sequential digital logic circuits. Several methods of *switch debouncing* have been developed. These can be divided into *timing-based* schemes and *hysteresis-based* schemes. Timing-based techniques operate by adding sufficient delay before reading the switch so as to prevent the bounce from being detected. The main advantage of using timing to control bouncing is that it does not require any special switch design. Alternatively, it is possible to use hysteresis to separate the positions where the make and break actions are detected. Hysteresis refers to systems in which the output depends not only on the input, but also on the system's internal state.

The actual hardware circuits used in hysteresis-based switch debouncing belong to three common types: RS flip-flops, CMOS gate debouncers, and integrated RC circuit debouncers. The debouncing action of the RS flip-flop is obvious from its operation: that is, when the key is in a position in which neither contact is touched (key bouncing), the inputs are pulled low by the pull-down resistors. In this case, the key will appear as being pressed.

Alternatively, switch debouncing can be accomplished by means of a CMOS buffer circuit with high input impedance. One such circuit is the 4050 hex buffer IC, which has eight input and eight output gates. When the switch is pressed, the input line of the 4050 chip is grounded, and output is forced low. The output voltage is also kept low when the switch is bouncing by means of the internal resistor. The effect is that the switch action is debounced.

Finally, switch debouncing can also be implemented by means of a simple resistor-capacitor circuit. The circuit action is based on the rate at which the capacitor recharges once the ground connection is broken by the switch. As long as the capacitor voltage is below the threshold level of the logic zero value, the output signal will continue to appear as logic zero.

4.2.2 Keypads

In the world of microcontroller-based circuits, a *keypad* (also called a *numeric keypad*) is a set of pushbutton switches sometimes labeled with digits, mathematical symbols, or letters of the alphabet. In this sense, a calculator keypad contains the decimal (or occasionally hexadecimal) digits, the decimal point, as well as keys for the mathematical features of the calculator. Although in theory the computer keyboard is a keypad, the keypad is usually a smaller arrangement of buttons, or refers to a part or area of a computer keyboard.

By convention, the keys on calculator-style keypads and keypads on computer keyboards are arranged such that the keys 123 are on the bottom row. On the other hand, telephone keypads have the 123 keys on the top row, as shown in Figure 4-3.

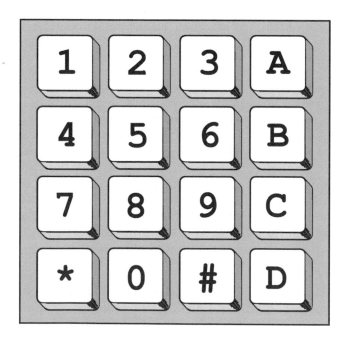

Figure 4-3 A 16-Key Telephone-Style Keypad.

Keypads are usually implemented as pushbutton switches located in a row and column *matrix*. For example, the location of any key on the keypad in Figure 4-3 can be based on two coordinates: the row and column position for that key. This determines that in processing keypad data only eight outputs are required: four to define the row and four to define the column.

Determining which switch on a keypad has been activated can be done either by *polling* or by an *Interrupt-Driven routine*. In the polling approach, the controller checks the status of each switch in a loop. In the Interrupt-Driven routine, the action on a key notifies the processor of a keystroke and a dedicated routine proceeds to execute the required action. Keypads, like the switches they incorporate, require debouncing. The three methods of switch debouncing described previously apply to keypads.

4.3 Output Devices

Electronic devices, including computers and embedded systems, must often provide data output in a humanly readable form. Here again, computer technology uses many different types of output devices, including video displays, printers, plotters, film recorders, projectors, sound systems, and even holographic devices. Although these output devices cannot be excluded from use in some embedded systems, a more typical case is that they often make do with much simpler and limited output means. In this section we discuss two common output devices used in embedded systems: the *Seven-Segment LED* and the *liquid crystal display*. Also note that simple devices, such as LEDs and buzzers, are sometimes used as output devices. Because LEDs were covered in Chapter 2 we de not include them in this discussion. Buzzers are such simple components that their operation does not require a detailed explanation.

4.3.1 Seven-Segment LED

Digital devices often need to output a numeric value. Although individual LEDs can be combined to represent binary, decimal, or hexadecimal digits, a far more convenient device is one consisting of seven built-in LEDs that can be turned on or off to represent all ten decimal digits and even the six letters of the hex character set. Such a circuit is furnished in a single IC called a Seven-Segment LED, common in clocks, watches, calculators, and household appliances.

The classic scheme of a Seven-Segment display consists of placing lighted bars in a figure-eight pattern. By selecting which bars are lighted, all the digits and some letters of the alphabet can be represented. In addition, Seven-Segment LEDs are usually capable of displaying one or two decimal points. Figure 4-4 shows the layout of a Seven-Segment LED and the combinations to generate the decimal and hex digit sets.

Note, in Figure 4-4, that two of the letters (b and d) of the hexadecimal set are displayed in lowercase while the others are in uppercase. This is a limitation of the Seven-Segment LED because an uppercase letter "D" would match the one pattern for the digit "0," and an uppercase letter "B" would match the digit "8."

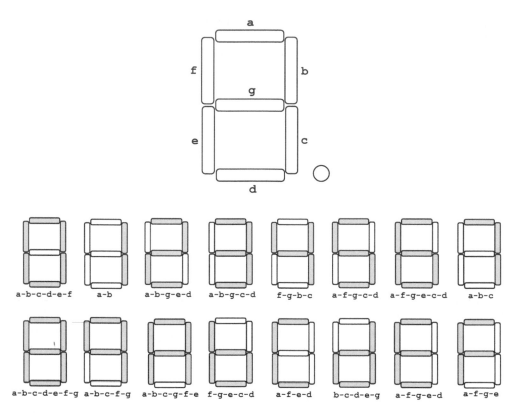

Figure 4-4 Layout and Digit Patterns in a Seven-Segment LED.

Some Seven-Segment LED displays are slanted to make the digits appear in italics. One reason for the scheme is used in clock displays where the two digits are inverted so that the decimal points appear like a colon between the digits. In addition, Seven-Segment displays are packaged in several different ways. Sometimes several digits are combined in a single IC. Another variation is in the form of a 14-pin DIP.

Seven-Segment displays are also furnished using display technologies other than LEDs. Many line-powered devices and home appliances, such as clocks and microwave ovens, use fluorescent Seven-Segment displays. Battery-powered devices, such as watches and miniature digital instruments, use Seven-Segment liquid crystal displays. Liquid crystal technologies are covered in sections that follow.

The LEDs in a Seven-Segment display are interconnected. The two interconnection modes are to wire together the cathodes of all individual LEDs. Another mode is to wire together the anodes. In one case the device is said to have a *common-cathode* and in the other one a *common-anode*. This scheme simplifies the circuit and reduces the number of connections, as only one line is necessary for controlling each LED. There is no intrinsic advantage to either system because each one is suited to different applications. Figure 4-5 shows the pin diagram for a common-cathode Seven-Segment LED in a DIP package.

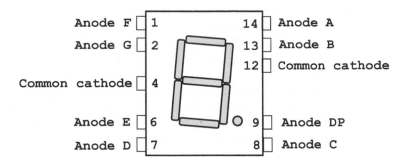

Figure 4-5 Pin Diagram for a Common Cathode Seven-Segment LED

4.3.2 Liquid Crystal Displays

A liquid crystal display (LCD) is a pixilated output device capable of displaying ASCII characters and dot-based graphics. LCDs can be color or monochrome according to their construction. One of the advantages of LCD displays is their very small consumption of electric power, which makes them suitable for battery-powered embedded systems.

In operation the liquid crystal display consists of two pieces of polarized glass with perpendicular axes of polarity. Sandwiched between the polarizers is a layer of nematic crystals, as shown schematically in Figure 4-6.

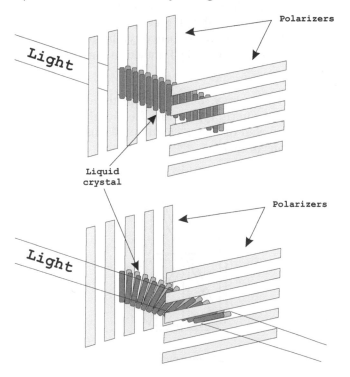

Figure 4-6 Schematic Representation of an LCD Display.

In the top image of Figure 4-6, light cannot pass through the system because the liquid crystal layer preserves the original angle of vibration of the light. In the lower image, the various molecule layers of the liquid crystal are twisted approximately 90 degrees. This twisting of the liquid crystal also changes the light's pane of vibration. So when light reaches the second polarized filter, it vibrates at the same angle as the final layers of molecules of the liquid crystal and can pass through the polarizer. Note that the electrical current applied to the crystals has the effect of straightening the various molecule layers. When the current is released, the various molecule layers resume their twisted form. By varying the amount of twist in the liquid crystals the amount of light that passes through can be controlled.

4.3.3 LCD Technologies

Depending on the positioning of the light source, an LCD can be either *transmissive* or *reflective*. A transmissive LCD is illuminated from the back and viewed from the front. This type is common in applications that require high levels of illumination, as is the case with computer displays and television sets. Reflective LCDs, on the other hand, are illuminated by an external source. This type finds use in digital watches and calculators. Reflective technology produces a darker black color than the transmissive type, as light is forced to pass twice through the liquid crystal layer. Because reflective LCDs do not require a light source they consume less power than the transmissive ones. A third type, called *transflective* LCDs, works either as a transmissive or a reflective LCD, depending on the ambient light.

LCDs can be color or monochrome. In color systems each individual pixel consists of three cells, which are colored red, green, and blue. These cells, sometimes called *subpixels*, are controlled independently to yield thousands (or even millions) of possible colors for screen dots. Most LCDs used in embedded systems are monochrome.

According to display technology, LCDs can be divided into *alphanumeric* or *dot-addressable*. The alphanumeric type, most frequently used in embedded applications, use a matrix composed of linear elements. Figure 4-7 shows several possible electrode configurations of LCDs.

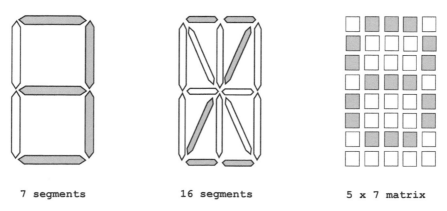

7 segments   16 segments   5 x 7 matrix

Figure 4-7 Electrode Configurations in LCD Displays.

The first two electrode configurations in Figure 4-7 are based on linear segments similar to the ones in Seven-Segment LEDs. Segmented electrodes are suitable for simple alphanumeric displays, as is often required in small digital devices such as watches or calculators. To display entire character sets, or graphics, a dot-addressable matrix arrangement of electrodes is necessary. This setup is shown in the rightmost image in Figure 4-7. However, such power comes at a price, as the more addressable elements in the display, the greater the number of connections and the more complex the driver logic required to operate the system. Note that the 5 x 7 matrix display in Figure 4-7 actually contains eight dot rows. The reason is that the lowest row is used for displaying the cursor. Most popular LCD displays for embedded systems use the 5 x 7 matrix format.

One way of reducing the number of electrical connections in an LCD is by a method called a *passive matrix display*. Here the pixels to be lighted are determined by the crossing points between the row and the column selector electrodes. For example, in the 5 x 7 matrix display in Figure 4-7, the pixel at the center of the character is selected by picking row number 4 and column number 3. The name passive matrix originates in the fact that each pixel must retain its state between refresh cycles. As the number of pixels to be refreshed increases, so does the time required for the refresh cycle. As a consequence of their design, passive matrix displays usually have slow response times and poor contrast.

In high-resolution and color LCDs, an *active matrix* is used. In this design, a grid of thin-film transistors is added to the polarizing and color filters. Each pixel contains its own dedicated transistor, and each row line and column line is addressed individually. During the refresh cycle, each pixel row is activated sequentially. Active matrix displays are brighter and sharper and have quicker response times than a passive matrix. Active matrix displays are also known as *thin-film-transistor* or TFT displays.

Chapter 5

From Circuit Schematics to PCB

5.1 Circuit Diagram

Circuits are often represented in a simplified manner by means of a circuit diagram, also called an *electronic schematic*. One possible approach is a pictorial representation of the circuit components to a given degree of realism. Figure 2-3, in Chapter 2, is an example of a pictorial, or quasi-pictorial, circuit representation. Pictorial circuit representations have some application in tutorials and education but are not very useful to the circuit designer. Block diagrams are another simplification in which components are abstracted into a box that may show some connections but is lacking most electronic details. Figure 5-1 compares the pictorial and schematic drawings of the same circuit.

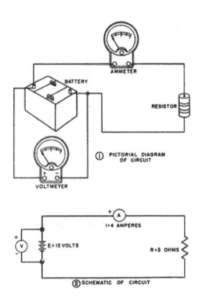

Figure 5-1 Pictorial and Schematic Drawings of an Electrical Circuit.
(Image from Wikimedia Commons)

The schematic diagram depicts the circuit using simplified, standard symbols that show components as well as connections for power and signals. The schematic circuit diagram is the engineer's and circuit designer's basic tool. Unlike the block diagram, the circuit schematics show the actual connections between components, however, the physical arrangement of the individual components need not match the actual board. This gives the circuit designer the freedom of locating components in their logical positions in the diagram, rather than having to abide by the physical constraints of the actual board.

5.1.1 Symbols

The actual symbols used in electronic schematics have changed over the years and some have become obsolete as the components they represented are no longer used; for example, these days the symbols for vacuum tubes have mostly a historical interest. In the United States, IEEE Standard 315-1975 provides graphics symbols and class designation letters for electrical and electronic components used in diagrams and drawings. In Britain, the corresponding standard is designated as IEC 60617. The International Electrotechnical Commission has also approved a standard for graphical symbols that is designated as IEC 60617. This database currently includes 1,750 different symbols. Figure 5-2 shows some common symbols in the US standard. More complete lists can be obtained online.

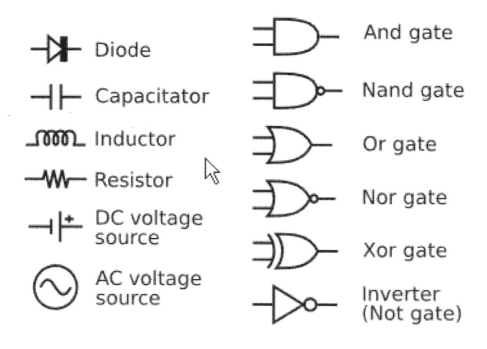

Figure 5-2 Electronic Symbols in the United States Standard.
             (Image from Wikimedia Commons)

Circuit design software and individual companies often introduce extensions and variations to the standard symbol set. In any case, introducing major changes to the standard symbols is almost certainly a bad idea.

5.1.2 Tools for Electronic Circuit Design

Not long ago, integrated circuits and embedded system boards were designed by hand and laid out manually. In the 1970s and 1980s, some circuit designers started using photoplotters and geometric software to facilitate the process but still most of the process was done manually. In the late 1970s, the first placement and routing tools for circuit design were developed. The composition and fabrication of complex *VLSI systems* furnished the incentive for automating and simulating system design. The approach still in use today is to develop textual programming languages that receive as input the desired behavior and generate a detailed physical design of the IC. Some of the best-known tools for electronic design automation were UNIX utilities developed at the University of California, Berkeley.

Along these same lines, a consortium of universities and circuit fabricators developed the *Metal Oxide Semiconductor Implementation Service*, (or MOSIS.) The system was used to train students by producing real, integrated circuits that were low-cost, reliable, and based on relatively simple technologies. In the early 1980s, electronic design tools ceased to be developed internally by the circuit development companies and became a separate business and technology. The term EDA, for *Electronic Design Automation*, was coined at this time and the first design automation conference took place in 1984. Several specialized programming languages were developed during this period for the hardware design and description of electronic circuits. *Verilog*, developed by a private enterprise, and *VHDL*, by the US Department of Defense, are among the best known. Simulators based on these languages also became available.

Today the most sophisticated and advanced circuit design tools are used in the field of *integrated circuits*, which are outside the scope of this book. In this context we are interested in tools used in the design, layout, and fabrication of *circuit boards* with nonminiaturized components.

5.2 Circuit Board Design

The schematics for an electronic circuit can be designed with a pencil and eraser on a paper napkin; many circuits have started in this manner. Computers can help in this effort through specialized software and general-purpose drawing programs. Computer Assisted Design (CAD) packages, such as Autocad, are powerful drawing tools and usually provide dedicated files and attachments for circuit design. Specialized circuit design programs are also available, some as freeware, others are furnished by printed circuit board manufacturers, and still others are commercial products intended for engineers and design specialists.

Each category of circuit design program has its advantages and drawbacks. For example, commercial technical drawing programs, such as Autocad, are powerful and well-tested tools, and many professional designers rely on them. However, these programs are expensive and come with a steep learning curve. If one already has and uses a commercial technical package, then adopting this package for circuit design is a practical option. However, purchasing and learning one of these programs to use it exclusively in circuit design seems less attractive.

There are a number of free circuit design programs available online. A dedicated web site at http://www.opencircuit.com serves as a repository for many of these freeware programs. Some of these programs (such as Magic) are VLSI layout tools of little use to the embedded systems developer. Other programs, such as XCircuit, are suitable for general-purpose circuit design. Some commercial programs have free demo versions. An electronic circuit simulation program that has become a classic in the field is named SPICE. Its initial release in 1973 was written in Fortran but current versions are in C. However SPICE is more a circuit simulator than a design tool and is mostly suited for use in the development of integrated circuits.

Another type of board design program are those furnished by board manufacturing firms. These programs are usually compatible with the specific manufacturing technology of the company that furnishes them. In most cases the schematic design application serves as a front end to a board layout application that generates the actual board drawings. One set of such programs is the *Express PCB* and *Express SCH* package. Express SCH is the circuit design component and Express PCB is the board layout program. These programs are available for free download at http://www.expresspcb.com. Other similar programs can be found on the Internet websites of the board manufacturers.

5.2.1 Board Design Standards

Several industry standards relate to the design of printed circuit boards. The *Association Connecting Electronics Industries* is an organization devoted to standardizing the assembly and production of electronic equipment and assemblies. It was founded in 1957 as the Institute for Printed Circuits (hence IPC) and later named the Institute for Interconnecting and Packaging Electronic Circuits. In 1999, the name was finally changed to its present form (Association Connecting Electronics Industries) but the IPC moniker was kept. The IPC publishes the most widely accepted standards for the electronic industries. There is an IPC standard for every aspect of PCB design, manufacture, and testing. The major document that covers PCB design is IPC-2221, "Generic Standard on Printed Board Design."

5.2.2 Gerber File Format

The *Gerber file format* is a PCB file convention used by the electronics industries. The format includes copper layers, connections, solder masks, and legends. Although milling and drilling data can be present in the Gerber file, it is more common to provide this information in the *Excellon Format*. The Gerber format is the de-facto standard for the specification and transfer of PCB data. The current version of the format is RS-274X, also known as *Extended Gerber* or X-Gerber. An *RS-274X file* is a humanly readable ASCII file that contains commands and coordinates. The following is an example (from Wikimedia Commons) of a file in Gerber RS-274X format.

```
G04 Film Name:    paste_top*
G04 Origin Date:  Thu Sep 20 15:54:22 2007*
G04 Layer:  PIN/PASTEMASK_TOP*
%FSLAX55Y55*MOIN*%
%IR0*IPPOS*OFA0.00000B0.00000*MIA0B0*SFA1.00000B1.00000*%
%ADD28R,.11X.043*%
%ADD39O,.07X.022*%
...
%AMMACRO19*
21,1,.0512,.0512,0.0,0.0,45.*%
%ADD19MACRO19*%
%LPD*%
G75*
G54D10*
X176250Y117500D03*
Y130000D03*
Y163750D03*
...
G54D39*
X496250Y142500D03*
Y137500D03*
Y132500D03*
Y127500D03*
M02*
```

Gerber files are typically generated by circuit development software products of the EDA or CAD types discussed previously. The Gerber file can be fed into a PCB fabricator that is equipped to process the image and component data by means of *Computer-Assisted Manufacturing* (CAM).

5.3 Developing the Circuit Prototype

The implementation and development of most electronic circuits follows several stages in which the product progresses through increasingly refined phases. This progression ensures that the final product meets all the design requirements and performs as expected. The designer or the engineer cannot be too careful in avoiding manufacturing errors that later force the scrapping of multiple components or, at best, force costly or unsightly circuit repairs. A common norm followed by electronic firms is not to proceed to fabrication until a finished and unmodified prototype has been exhaustively tested and evaluated. Even the text and labels on the circuit board should be checked for spelling errors and to make sure that the final placement of hardware components will not hide some important information printed on the board.

The methodology that uses a cyclic process of prototyping, testing, analyzing, and refining a product is known as *iterative design*. The practicality of iterative design results from the fact that changes to a product are easier and less expensive to implement during the early stages of the development cycle. The iterative design model, also called the *spiral model*, can be described by the following steps:

1. The circuit is designed on paper, usually with the support of software as previously described.

2. The paper circuit is checked and evaluated by experts other than the designer or designers.

3. A primitive hardware prototype is developed and tested. *Breadboarding* and *wire-wrapping* are the most common technologies used for this first-level prototype.

4. The breadboarded or wire-wrapped prototype is tested and evaluated. If changes to design are required, the development process restarts at Step 1.

5. A second-level prototype is developed, usually by means of printed circuit boards. This PCB prototype is evaluated and tested. If major modifications are required, the development process is restarted at Step 1.

6. A final, third-level, prototype is developed using the same technology and number of signal layers as will be used in the final product. If modifications and changes are detected during final testing, the development process is restarted at Steps 1, 3, or 5 according to the nature of the defect to be remedied.

7. If the final prototype passes all tests and evaluations, a short production run is ordered. This short run allows finding problems in the manufacturing stage that can sometimes be remedied by making modifications to the original design or by changing the selected component or components.

The preceding steps assume a very conventional circuit, and a simple and limited development process. The mass production of electronic components has to consider many other factors.

5.3.1 Breadboard

One of the most useful tools in developing a circuit prototype is a *breadboard*. The name originated in the early days of radio amateurs who would use a wooden board (sometimes an actual bread board) with inserted nails or thumbtacks to test the wiring and components of an experimental circuit. The modern breadboard is usually called a solderless breadboard because components and wires can be connected to each other without soldering them. The term "plugboard" is also occasionally used.

The main component of a modern solderless breadboard is a plastic block with perforations that contact internal, tin-plated spring clips. These clips provide the contact points. The spacing between holes is usually 0.1", which is the standard spacing for pins in many nonminiaturized electronic components and in ICs in dual inline (DIP) packages. Capacitors, switches, resistors, and inductors can be inserted in the board by cutting or bending their contact wires. Most boards are rated for 1 amp at 5 volts. Figure 5-3 shows a populated breadboard for a motor driver circuit developed later in this book.

Solderless breadboards come in different sizes and designs. The one in Figure 5-3 has the plastic interconnection component mounted on a metal base and includes female banana plug terminals suitable for providing power to the board. Simpler boards consist of nothing more than a perforated plastic block with the corresponding spring clips under the perforations. These can sometimes serve as modules that can be attached to each other in order to accommodate a more complex circuit.

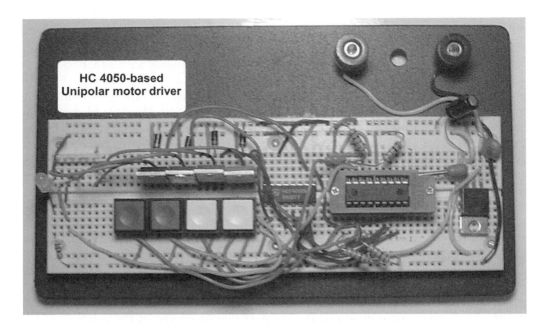

Figure 5-3 A Populated Breadboard Circuit.

On the other hand, more sophisticated devices, sometimes called *digital laboratories*, include with the breadboard several circuits for providing power in different intensities as well as signals with the corresponding adjusters and selectors and switches of several types. Figure 5-4 shows the IDL_800 Digital Lab.

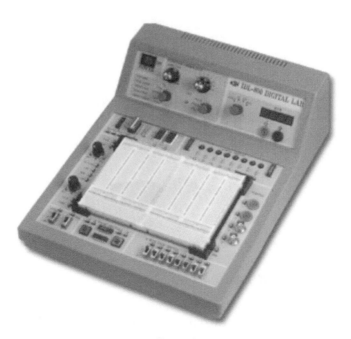

Figure 5-4 IDL-800 Digital Lab Breadboard.

Limitations of Breadboards

Although breadboards are valuable tools in prototyping, the circuit design community is divided regarding their value. One of the common problems with breadboards is faulty contacts. The spring-loaded clips that are designed to provide contact between components often fail. This leads the operator to believe that there is something wrong with the circuit when it is just a faulty contact. Another problem is that wire-based connectors are often longer than necessary, which introduces problems that may not be related to the circuit itself. In some sensitive circuits, the routing required by the board hardware may produce interference or spurious signals. The circuit developer must take all these limitations into account when testing breadboard circuits.

Another major limitation is that a solderless breadboard cannot accommodate surface mount components. Many other standard components are not manufactured to meet the 0.1" spacing of a standard breadboard and are also impossible to connect. Sometimes the circuit developer can build a breakout adapter as a small PCB that includes one or more rows of 0.1" pins. The board-incompatible component can then be soldered to the adapter and the adapter plugged into the board. The need to solder components to the adapter negates some of the advantages of the breadboard, but with components that are likely to be reused this may be a viable option.

For example, a 6P6C male telephone plug connector does not fit a standard breadboard. In this case we can build a breakout adapter such as the one shown in Figure 5-5.

Figure 5-5 Telephone Plug Breadboard Adapter.

Breadboarding Tools and Techniques

Several off-the-shelf tools are available to facilitate breadboarding and others can be made in-house. One of the most useful ones is a set of jumper wires of different lengths and colors. These jumper kits are usually furnished in a plastic organizer. Longer, flexible connectors are also available and come in handy when wiring large circuits.

Wiring mistakes are easy to make when breadboarding integrated circuit components. In this case the operator must frequently look up the circuit diagram or the component schematics to determine the action on each IC pin. One solution is to use a drawing program to produce a labeled drawing of the component's pin-out. The drawing is scaled to the same size as the component and then printed on thick paper or cardboard. A cutout is then glued, preferably with a nonpermanent adhesive, to the top part of the IC so that each pin clearly shows a logo that is reminiscent of its function. Figure 5-6 shows a portion of a breadboard with a labeled IC used in developing a circuit described later in this book.

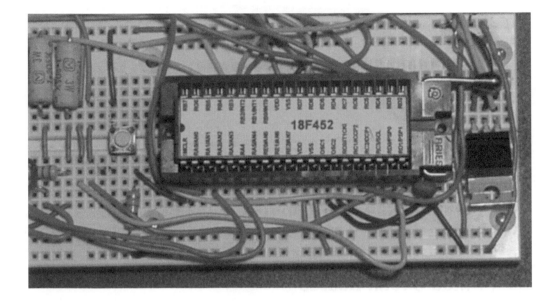

Figure 5-6 Labeled IC in a Breadboard.

Note in Figure 5-6 that the 18F452 microcontroller is inserted in a device called a ZIF (*zero insertion force* socket). When the ZIF socket handle is lifted, the IC can be easily removed. ZIF sockets are often used in prototypes and demo boards so that the component can be easily replaced or re-programmed.

5.3.2 Wire-Wrapping

Another popular technique used in the creation of prototypes and individual boards is wire-wrapping. In wire-wrapped circuits, a square, gold-plated post is inserted in a perforated board. A silver-plated wire is then wrapped seven turns around the post, which results in 28 contact points. The silver- and gold-plated surfaces cold weld, producing connections that are more reliable than the ones on a printed circuit board, especially if the board is subject to vibrations and physical stress. The use of wire-wrapped boards is common in the development of telecommunications components but solderless breadboards have replaced wire wrapping in conventional prototype development.

5.3.3 Perfboards

Thin sheets of isolating material with holes at 0.1" spacing are also used in prototype development and testing, and in creating one-of-a-kind circuits. The holes in the *perfboard* contain round or square copper pads on either one or both sides of the board. In the perfboard each pad is electrically insulated. Boards with interconnected pads are sometimes called *stripboards*. Components, including ICs, sockets, resistors, capacitors, connectors, and the like, are inserted into the perfboard holes and soldered or wire-wrapped on the board's back side. Figure 5-7 shows a circuit on a perfboard.

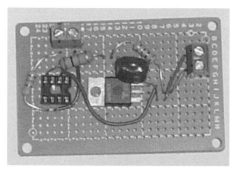

Figure 5-7 Circuit on a Perfboard.

In this section we have omitted circuit prototyping methods and techniques that have become obsolete, such as point-to-point wiring and through-hole construction.

5.4 Printed Circuit Boards

The methods and techniques described so far, including breadboards, wire-wrapping, and perfboards, are used in developing and testing the electronics of the circuit itself. Once the circuit prototype has been tested and evaluated, the next step is usually the production of a circuit board that can house the components in a permanent manner and thus becomes a prototype of the final product. This typically requires a *printed circuit board*, or PCB, where the components can be mechanically housed and electrically connected. The PCB is also called a *printed wiring board* (PWB) and a *printed circuit assembly* (PCA). Very few commercially made electronic devices do not have at least one PCB.

The base of a conventional PCB is a nonconductive laminate made from an epoxy resin, with etched conductive copper traces that provide pathways for signals and power. PCBs can be produced economically in large or small volumes and even individually. Production operations on the PCB include etching, drilling, routing, creating solder-resistant layers, screen printing, and installation of components. All of these operations can be automated in a production setting or done by hand by the hobbyist or when creating a prototype. PCB technology has flourished because the final product is inexpensive and reliable. Standards regarding PCB design, assembly, and quality control are published by IPC. Figure 5-8 shows two images of a PCB.

Figure 5-8 PCB Drawing and Populated Board.
(Image from Wikimedia Commons)

In Figure 5-8 the image on the left-hand side shows the board as it appears in the design software. In this example the board is double sided, so there will be another image for the reverse side. These image files are used in manufacturing the board. The image on the right-hand side of Figure 5-8 shows the finished board populated with surface-mount and through-the-hole components.

5.4.1 PCB Layers

The simplest PCBs have all copper traces and connections on a single layer. Most home-made and in-house boards are of this type, and many circuits can be easily designed for single-layer boards. Single-layer boards are easy to manufacture and therefore less expensive but the majority of commercial boards are multilayered. The following are common layers found in commercial PCBs:

- Top copper layer
- Bottom copper layer
- Inner copper ground plane layer
- Inner copper power plane layer
- Top solder mask layer
- Bottom solder mask layer
- Silkscreen (text) layer

The top and bottom copper layers and the silkscreen layer are present in most commercial boards. In this case the bottom solder layer is where the through-the-hole components are soldered. Surface-mount components are soldered to the top layer.

The solder mask layer is a coating applied to the top or bottom layers so as to prevent soldering except on the pads furnished for this purpose. These masks prevent undesired solder connections and make soldering easier.

The inner copper layers are only present in boards with four or more layers. During board manufacturing, any through-the-hole pad can be connected to or isolated from these inner layers. The inner layers make the boards more compact and improve the *noise immunity* of the circuit.

5.4.2 PCB Connectors

Several elements in PCBs serve as connectors between components; these are

- Traces and planes
- Pads
- Vias
- Jumpers

Traces are etched copper lines that conduct power and signals from one point in the board to another one. Most PCB layout programs allow controlling the width and style of traces. A tracer width of 0.10" is a good default value for traces that conduct digital and analog signals. Traces that conduct power and ground signals are often much wider. In fact, ground signals are usually carried by ground planes that are large, polygonally shapes copper-filled areas. These planes are typically used as conductors for ground and power signals.

Individual pads are used to provide a solder surface for a component element. PCB design programs provide standard pads that define the pad's diameter, its shape, and the hole diameter and tolerance. A typical pad is circular in shape, with an outside diameter of 0.062" and a hole of 0.029". Square pads are often used in pin number 1 of an IC component.

Vias are pads used to pass signals between layers. Via pads are typically round. The connection between layers is made during manufacturing according to the information in the PCB drawing.

Jumpers are connectors that can be opened or closed on the board. Breaking a trace and placing a jumper allows the board user to cut off the signal to a specific portion of the circuit. Figure 5-9 is a screen snapshot of the PCB drawing developed with the ExpressPCB application available online.

5.5 Making Your Own PCB

Several methods are available for making printed circuit boards on a small scale. This approach is usually convenient for the experimenter and prototype developer who needs to make a single board, often as soon as possible. Electronic supply catalogs list convenient PCB-making kits based on different technologies and using varying levels of complexity. The method we describe in the sections that follow is one of the simplest ones because it does not require a photographic process. The process consists of the following steps:

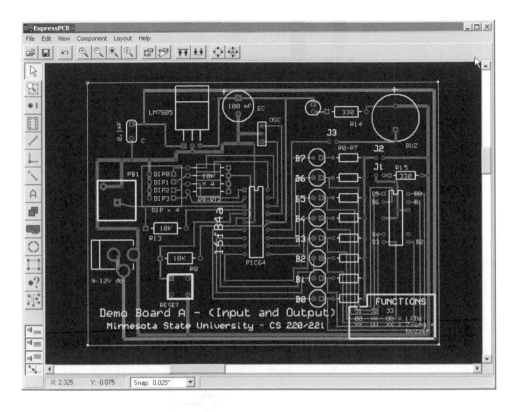

Figure 5-9 Screen Snapshot of a PCB Design Application.

1. The circuit diagram is drawn on the PC using a general-purpose or specialized drawing program.

2. A printout is made of the circuit drawing on photographic paper using a standard laser printer.

3. The printout is transferred to a single-sided, copper-clad circuit board blank by ironing over the backside with a household clothes iron.

4. The resulting board is placed in an etching bath that eats away all the copper, except the circuit image ironed onto the board surface.

5. The board is washed of etchant, cleaned, drilled, and the components soldered to it in the conventional manner.

6. Optionally, another image can be ironed onto the backside of the board to provide component identification, logos, etc.

5.5.1 Drawing the CPB Circuit

Any computer drawing program serves to draw the PCB. In the samples provided in this chapter we used CorelDraw but there are several specialized PCB drawing programs available on the Internet that can also be used. Figure 5-10 is a circuit board drawing used for a PIC flasher circuit described later in this book.

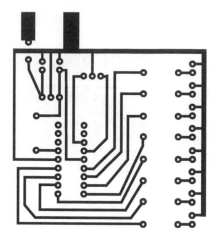

Figure 5-10 PIC Flasher Circuit Board Drawing.

The small, circular objects in the drawing are the solder pads. Figure 5-11 zooms into the lower right-hand corner of Figure 5-10 to show the details of the solder pads.

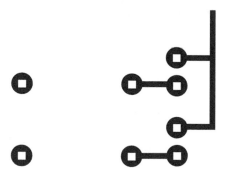

Figure 5-11 Detail of Circuit Board Pads.

Quite often in a PCB drawing it is necessary for a trace to cross between two pads. In this case the two pads can be modified so that the trace will not contact either one. The modified pads are shown in Figure 5-12.

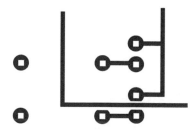

Figure 5-12 Modified Circuit Boards Pads.

5.5.2 Printing the PCB

The circuit diagram must be printed using a laser printer. Inkjet toners do not produce an image that resists the action of the etchant. Although in our experiments we used several models of LaserJet printers, it is well documented that virtually any laser printer will work. Laser copiers have also been used successfully for creating the PCB circuit image.

With the laser printer method the width of the traces can become an issue. The traces in the PCB image of Figures 5-11 and 5-12 are two points, which is 0.027". Traces half that width and less have been used successfully with this method but as the traces become thinner the entire process becomes more critical. For most simple circuits, 0.020" traces can be considered a practical limit. In the process of printing the PCB image, one must be careful not to touch the glossy side of the paper or the printed image. Note that the pattern is drawn as if you were looking from the component side of the board.

5.5.3 Transferring the PCB Image

One of the most critical elements in this method of producing PCBs is the paper used in printing the circuit. Pinholes in some papers can degrade the image to the point that the circuit lines (especially if they are very thin) do not etch correctly. Another problem relates to removing the ironed-on paper from the board without damaging the board surface.

Glossy, coated inkjet printer paper works well. Even better results can be obtained with glossy photo paper. We use a common high-gloss photographic paper available from Staples and sold under the name of "picture paper". The 30 sheets, 8-by-10 size, has Staples number B031420197 1713. The UPC barcode is

7 18103 02238 5.

It has been mentioned that Hewlett-Packard toner cartridges with microfine particles work better than store-brand toner cartridges.

Transferring the image onto the board blank is done by applying heat from a common clothes iron, set on the hottest setting, onto the paper/board sandwich. In most irons the hottest setting is labeled "linen." After going over the back of the paper several times with the hot iron, the paper becomes fused to the copper side of the blank board. The board/paper sandwich is then allowed to soak in water for about 10 minutes, after which the paper can be removed by peeling or light scrubbing with a toothbrush.

5.5.4 Etching the Board

Once the paper has been removed and the board washed with soap and rinsed in clean water, it is time to prepare the board for etching. The preliminary operations consist of rubbing the copper surface of the board with a Scotchbrite plastic abrasive pad and then scrubbing the surface with a paper towel soaked with acetone solvent.

The etching can proceed once the board is rubbed and clean. The etching solution contains ferric chloride and is available from Radio Shack as a solution and from Jameco Electronics as a powder to be mixed by the user. PCB Ferric Chloride etchant should be handled with rubber gloves and rubber apron because it stains the skin and utensils. Also, concentrated acid fumes from ferric chloride solution are toxic and can cause severe burns. These chemicals should be handled according to cautions and warnings posted on the containers.

The ferric chloride solution should be kept and used in a plastic or glass container, never metal. Faster etching is accomplished if the etching solution is first warmed by placing the container in a tub of hot water. Once the board is in the solution, face up, the container is rocked back and forth. It is also possible to aid in copper removal by rubbing the surface with a rubber-gloved finger.

5.5.5 Finishing the Board

The etched board should be washed well, first in water and then in lacquer thinner or acetone; either solvent works. It is better to gently rub the board surface with a paper towel soaked in the solvent. Keep in mind that most solvents are flammable and explosive, and also toxic. Once the board is clean, the mounting holes can be drilled using the solder pads as a guide. A small electric drill at high revolutions, such as a Dremmel tool, works well for this operation. The standard drill size for the mounting holes is 0.035". A #60 drill (0.040") also works well. Once all the holes are drilled, the components can be mounted from the backside and soldered at the pads.

5.5.6 Backside Image

The *component side* (backside) of the PCB can be printed with an image of the components or with logos or other text. A single-sided blank board has no copper coating on the backside so the image is just ironed on without etching. Probably the best time to print the backside image is after the board has been etched and drilled but before mounting the components. Because the image is to be transferred directly to the board, it must be a mirror image of the desired graphics and text. Most drawing programs contain a mirroring transformation so the backside image can be drawn using the component side as a guide, and then mirrored horizontally before ironing it on the backside of the board. Figure 5-13 shows the backside image of the sample circuit board, before and after mirroring.

Notice on the left-side image in Figure 5-13 that a lighter copy of the circuit diagram was used to lay out the image of the backside. Once drawn, the backside drawing was mirrored horizontally, as shown in the right-side image.

5.6 Surface-Mount Components

Surface-mount technology (also known as SMT) is an electronic building technology in which the components are mounted directly onto the surface of a board without having to drill holes. Avoiding the holes and lead wires has led to smaller, more compact, and lighter components that have smaller leads or no leads at all. Electronic manufacturing industries have largely replaced through-the-hole components with surface mount.

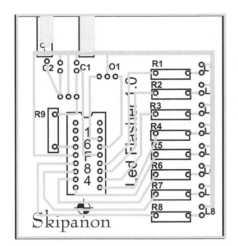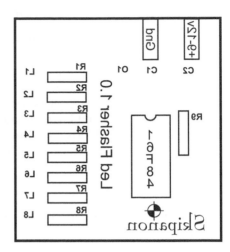

Figure 5-13 Graphics and Text for Board Backside Image.

Because of their size and the dimensions of their solder pads, surface-mount components are considerably more difficult to solder by hand and to position on the PCB than the through-the-hole type. This difficulty is often a factor when dealing with individual production or short run boards that cannot be conveniently adapted to production-scale technologies. The prototype maker is often in this situation, to which we sometimes must add the complication that many components are only available in surface-mount designs.

5.6.1 SMT Adapters

In many cases it is possible to purchase or make an adapter board that allows using a surface mount component on a standard breadboard or a perfboard. Figure 5-5 shows an adapter for a telephone connector. Similarly, it is possible to make or purchase adapters for surface-mount formats.

One popular adapter technology is based on flexible, adhesive pads that can be attached to a standard perfboard. This adapter provides connections between the surface-mount component and the 0.100" holes and the standard board. Figure 5-14 shows a flexible component adapter by the Protoflex company.

Figure 5-14 Protoflex SMT to Standard Adapter.

5.6.2 Soldering SMT Components

Although more difficult than through-the-hole parts, surface-mount components can also be hand-soldered to boards and adapters. For this purpose, the general-purpose soldering iron is not suitable and its use will only lead to damaged components. The soldering must be performed with a temperature-controlled station with a temperature readout. Additionally, it is necessary to use special soldering tips that are designed for SMT pins and have the adequate form and dimensions. These are usually furnished with the soldering station or are available separately from the station manufacturer. The following items would be required in a minimal toolkit for soldering SMT components:

- Temperature-controlled soldering station
- Soldering tips appropriate for the SMT type
- Solder
- Soldering flux and flux remover
- Special heat gun for desoldering components
- Desoldering wick (braided)
- Exacto knife
- Water in a spray bottle
- Tin foil heat shield
- Magnifying glasses and 10X jeweler's loupe
- A steady hand

The soldering tip must be well tinned and must be frequently cleaned on a wet sponge. Tips must be replaced as soon as they show signs of oxidation or at least once per month. The soldering temperature should be held to about 700° F so as to avoid damaging the pads. Probably the most difficult part is to get the SMT component correctly aligned with the pad and tacked down. The actual soldering is easier than it appears because any solder that flows across the pins can be easily removed by wiping it with the de-soldering wick in the direction of the pins.

5.7 Troubleshooting the Circuit Board

Circuit board prototypes may not perform correctly when first tested. Circuit design errors and logical flaws must be eliminated first. Then the board must be tested for mechanical defects. These include broken connections, missing pads, problems with vias and jumpers, damaged or defective components, and a host of other possible problems and defects.

Troubleshooting a PCB requires systematically testing all functions and operations for which it was designed and making sure that all circuit components are performing their expected functions. It is usually a good idea to follow a testing methodology that proceeds from the simplest to the most complicated and eliminating problems and defects in that order.

5.7.1 Circuit Testing Tools

Many tools and techniques have been developed for testing and troubleshooting defective boards and components. The fundamental ones are the logic probe, the multimeter, and the oscilloscope.

The *logic probe* is a pen-like device that is used in determining the Boolean state (1 or 0) of a digital circuit. Most probes use the circuit's power but some are battery operated. A typical probe has three different-colored LEDs. The red LED indicates a high state, the green LED a low state, and an amber or yellow LED indicates a pulse. Some logic probes have audible output tones. A more complicated device is the *logic analyzer*, which allows testing multiple or complicated logical signals. Figure 5-15 shows a logic probe.

Figure 5-15 Logic Probe.
            (Image from Wikimedia Commons)

The *multimeter* or *multitester* is a measuring instrument that combines several functions, including the measurement of voltage, current, and resistance. Multimeters can have analog or digital readouts. In the analog version, a needle or pointer moves over a calibrated scale. In the digital version, the reading is displayed as decimal digits. Multimeters can be handheld devices or precise bench instruments costing thousands of dollars. Figure 5-16 shows a handheld digital multimeter.

Figure 5-16 Digital Multimeter.
            (Image from Wikimedia Commons)

Multimeters are used in testing the voltage, current, and resistance in different parts of a circuit. While troubleshooting, the multimeter's most frequent use is in testing continuity. For this use, the instrument is set to emit an audible beep whenever the probes are placed on a closed circuit. The operator then proceeds to touch various points in the circuit using the instrument's probes. If there is an open circuit, the continuity beep will not be heard.

The most refined and powerful of the three common testing instruments is the *oscilloscope*. The oscilloscope allows the observation of wave shapes by displaying a two-dimensional graph of the signal. It is typically used to observe electrical events that do not change or that change slowly. Figure 5-17 shows the front panel of a basic oscilloscope.

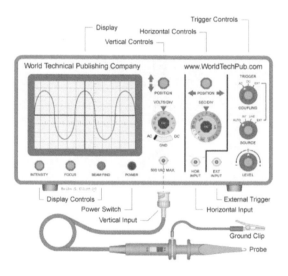

Figure 5-17 Front Panel of Basic Oscilloscope.
(Image from Wikimedia Commons by Brian S. Elliot – World Technical Publishing)

The classical oscilloscope uses a *cathode ray tube* (CRT) to display the two-dimensional graph of the signal. More modern instruments use a *liquid crystal display* (LCD). The vertical control knob controls the amplitude of the signal and the horizontal control knob defines the "sweep" or time base. Oscilloscopes are furnished with a probe that connects to the instrument's input port and to the signal source on the circuit.

In recent years computer-based oscilloscopes have become available. In this case the user's computer performs as a virtual oscilloscope with a probe connected to a machine port. Virtual oscilloscopes are usually less expensive than their hardware counterparts. In any case, the use of an oscilloscope in signal analysis is outside the focus of this book.

Chapter 6

Introducing the Microcontroller

6.1 A Computer on a Chip

A microcontroller is a type of microprocessor furnished in a single integrated circuit and needing a minimum of support components. Its name is sometimes abbreviated μC or MCU. It is in fact a small computer that contains a processing element, memory, and peripherals for input and output operations. The principal feature of microcontrollers are self-sufficiency and low cost. It is not intended to be used as a computing device in the conventional sense; that is, a microcontroller is not designed to be a data processing machine, but rather an intelligent core for a specialized dedicated system.

Microcontrollers are embedded in many control, monitoring, and processing systems. All the basic circuits discussed in this book contain at least one microcontroller. A few microcontrollers are designed as general-purpose devices, especially the type called *digital signal processors* (DSPs). However, most microcontrollers are used in specialized and embedded systems such as washing machines, telephones, microwave ovens, automobiles, and weapons of many kinds. In addition to a processor, memory, and input/output facilities, a microcontroller usually includes an internal clock and one or more peripherals such as timers, counters, analog-to-digital converters, serial communication facilities, and Watchdog circuits. Figure 6-1 shows a PIC 18F8720 microcontroller in an 80-pin TQFP surface-mount package.

Figure 6-1 PIC 18F8720 Microcontroller on a Board.
(Image from Wikimedia Commons)

More than two dozen companies in the United States and abroad manufacture and market microcontrollers. They range from 8- to 32-bit devices. Those at the low end are intended for very simple circuits and provide limited functions and program space, while those at the high end have many of the features normally associated with microprocessors. The most popular ones include several from Intel (such as the 8051), from Zilog (derivatives of their famous Z-80 microprocessor), from Motorola (such as the 68HC05), from Atmel (the AVR), the Parallax (the BASIC Stamp), and many from Microchip. In this book we focus on microcontrollers from Microchip, popularly called PICs.

Due to their low cost and easy availability, some microcontrollers have become very popular among amateur system designers. Large hobby communities have developed around the most popular ICs with many dedicated websites with thousands of examples of circuits and programs currently online. Their popularity, support, and availability of technology samples have been factors that determined our adoption of the PIC microcontrollers for this book.

6.2 PICMicro Microcontroller

PIC is a family of microcontrollers made by Microchip Technology. The original one was the PIC1650 developed by General Instruments. This device was called PIC for "Programmable Intelligent Computer" although it is now associated with "Programmable Interface Controller." Microchip does not use PIC as an acronym; instead, they prefer the brand name PICmicro. Popular wisdom relates that PIC is a registered brand in Germany and Microchip is unable to use it internationally. The original PIC was built to be used with General Instruments's CP1600 processor, which had poor I/O performance. The PIC was designed to take over the I/O tasks from the CPU, thus improving performance. In 1985, the PIC was upgraded with EPROM to produce a programmable controller. Today a huge variety of PICs are available with many different on-board peripherals and program memories ranging from a few hundred words to 32K.

The PIC primitive instruction set varies in length from about 35 instructions for the low-end devices to more than 70 for the high-end devices. The *accumulator*, which is known as the *work register* in Microchip documentation, is part of many instructions because the PIC contains no other internal registers accessible to the programmer. The PICs are programmable in their native Assembly language, which is straightforward and not difficult to learn. C language and BASIC compilers have been developed and are available from several sources. Open-source Pascal, JAL, and Forth compilers are also available. Storing a program in a PIC microcontroller is popularly referred to as "blowing" a PIC.

One of the reasons for the commercial success of the PIC is the support provided by Microchip. This includes a professional-quality development environment called MPLAB which can be downloaded free from the company's website. The MPLAB package includes an assembler, a linker, a debugger, and a simulator. Microchip also sells a low-cost in-circuit debugger called MPLAB ICD 2. Other development products intended for the professional market are also available from Microchip. The

Microchip website furnishes hundreds, if not thousands, of free support documents, including data sheets, application notes, and sample code.

In addition to the documents and products on the Microchip website, the PIC microcontrollers have gained the support of many hobbyists, enthusiasts, and entrepreneurs who develop code and support products and publish their results on the Internet. This community of PIC users is an invaluable source of information and know-how easily accessible to the beginner and useful even to the professional. One such Internet resource is an open-source collection of PIC tools named GPUTILS, which is distributed under the GNU General Public License. GPUTILS includes an assembler and a linker. The software works on Linux, Mac OS, OS/2, and Windows. Another product named GPSIM is an open-source simulator featuring PIC hardware modules.

6.2.1 Programming the PIC

Programming a PIC microcontroller requires the following tools and components:

1. An Assembler or high-level language compiler. The software package usually includes debugger, simulator, and other support programs.

2. A computer (usually a PC) in which to run the development software.

3. A hardware device called a *programmer* that connects to the computer through the serial, parallel, or USB line. The PIC is inserted in the programmer and "blown" by downloading the executable code generated by the development system. The hardware programmer usually includes the support software.

4. A cable or connector for wiring the programmer to the computer.

5. A PIC microcontroller.

6. PIC programmers.

The development system (assembler or compiler) and the programmer driver are the software components. The computer, programmer, and connectors are the hardware elements. Figure 6-2 shows a commercial programmer that connects to the USB port of a PC. The one in the illustration is made by MicroPro.

Figure 6-2 USB PIC Programmer by MicroPro.

Many other programmers are available on the market. Microchip offers several high-end devices with *in-circuit serial programming* (ICSP) and *low-voltage programming* (LVP) capabilities. These devices allow the PIC to be programmed in the target circuit. Some PICs can write to their own program memory. This makes possible the use of so-called *bootloaders*, which are small resident programs that allow loading user software over the RS-232 or USB lines. Programmer/debugger combinations are also offered by Microchip and other vendors.

Development Boards

A *development board* is a demonstration circuit that usually contains an array of connected and connectable components. Their main purpose is as a learning and experiment tool. Like programmers, PIC development boards come in a wide range of prices and levels of complexity. Most boards target a specific PIC microcontroller or a PIC family of related devices. Lacking a development board, the other option is to build the circuits oneself, a time-consuming but valuable experience. Figure 6-3 shows the LAB-X1 development board for the 16F87x PIC family.

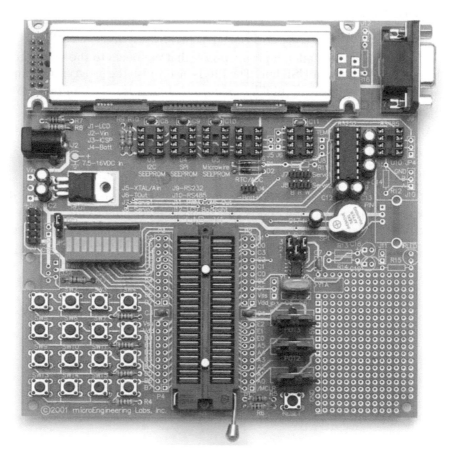

Figure 6-3 LAB-X1 Development Board.

The LAB-X1 boards, as well as several other models, are products of microEngineering Labs, Inc. Some of the sample programs developed for this book were tested on a LAB-X1 board. Development boards from Microchip and other vendors are also available.

6.2.2 Prototyping a PIC Circuit

It is a risky proposition to write a PIC program and assume that it works correctly without testing. Testing a program is not difficult with a development board but only if the board is compatible with the microcontroller, and if it provides the hardware that we need to test. But often one of these elements is missing, which means that we must build the circuit for which the program was designed. Prototyping options and techniques, including breadboards, perfboards, and commercial PCBs, were discussed in Chapter 5.

6.3 PIC Architecture

PIC microcontrollers are classified by Microchip into three groups: *baseline*, *mid-range*, and *high-performance*. Within each of the groups, the PIC is reclassified based on the first two digits of the PIC's family type. However, this sub-classification is not very strict because there is some overlap. For this reason we find PICs with 16X designations that belong to the baseline family and others that belong to the mid-range group. In the following sections we describe the basic characteristics of the various sub-groups of the three major PIC families with 8-bit architectures.

6.3.1 Baseline PIC Family

This group includes members of the PIC10, PIC12, and PIC16 families. The devices in the baseline group have 12-bit program words and are supplied in 6- to 28-pin packages. The microcontrollers in the baseline group are described as being suitable for battery-operated applications because they have low power requirements. The typical member of the baseline group has a low pin count, flash program memory, and low power requirements. The following types are in the baseline group.

PIC10 Devices

The PIC10 devices are low-cost, 8-bit, flash-based CMOS microcontrollers. They use 33 single-word, single-cycle instructions (except for program branches, which take two cycles.) The instructions are 12-bits wide. The PIC10 devices feature power-on and device reset, an internal oscillator mode that saves having to use ports for an external oscillator. They have a power-saving SLEEP mode, A Watchdog timer, and optional code protection.

The recommended applications of the PIC10 family range from personal care appliances and security systems to low-power remote transmitters and receivers. The PICs of this family have a small footprint and are manufactured in formats suitable for both through-hole and surface-mount technologies. Table 6.1 lists the features of some members of the PIC10 family.

Table 6.1

PIC10F Devices

	10F200	10F202	10F204	10F206
Clock:				
Maximum Frequency of Operation (MHz)	4	4	4	4
Memory:				
Flash Program Memory	256	512	256	512
Data Memory (bytes)	16	24	16	24
Peripherals:				
Timer Module(s)	TMR0	TMR0	TMR0	TMR0
Wake-up from Sleep	Yes	Yes	Yes	Yes
Comparators	0	0	1	1
Features:				
I/O Pins	3	3	3	3
Input Only Pins	1	1	1	1
Internal Pull-ups	Yes	Yes	Yes	Yes
In-Circuit Serial Programming	Yes	Yes	Yes	Yes
Instructions	33	33	33	33
Packages:				
6-pin SOT-23				
8-pin PDIP				

Two other PICs of this series are the 10F220 and the 10F222. These versions include four I/O pins and two analog-to-digital converter channels. Program memory is 256 words on the 10F220 and 512 in the 10F222. Data memory is 16 bytes on the F220 and 23 bytes in the F222.

PIC12 Devices

The PIC12C5XX family are 8-bit, fully static, EEPROM/EPROM/ROM-based CMOS microcontrollers. The devices use RISC architecture and have 33 single-word, single-cycle instructions (except for program branches which take two cycles). Like the PIC10 family, the PIC12C5XX chips have power-on reset, device reset, and an internal timer. Four oscillator options can be selected, including a port-saving internal oscillator and a low-power oscillator. These devices can also operate in SLEEP mode and have Watchdog timer and code protection features.

The PIC12C5XX devices are recommended for applications ranging from personal care appliances, security systems, and low-power remote transmitters and receivers. The internal EEPROM memory makes possible the storage of user-defined codes and passwords as well as appliance setting and receiver frequencies. The various packages allow through-hole or surface-mounting technologies. Table 6.2 lists the characteristics of some selected members of this PIC family.

Table 6.2

PIC 12Cxxx and 12CExxx Devices

	12C508(A) 12C509A 12CR509A	12C518	12CE519	12C671 12C672	12CE674
Clock:					
Maximum Frequency of Operation (MHz)	4	4	4	10	10
Memory:					
EPROM Program Memory	512/1024/1024 x12	512x12	1024x12	1024/2048/ 1024x12	2048x14
RAM Data Memory (bytes)	25/41/41	25	41	128	128
Peripherals:					
EEPROM Data Memory (bytes)	—	16	16	0/0/16	16
Timer Module(s)	TMR0	TMR0	TMR0	TMR0	TMR0
A/D Converter (8-bit) Channels	—	—	—	4	4
Features:					
Wake-up from SLEEP on pin change	Yes	Yes	Yes	Yes	Yes
Interrupt Sources	—	—	—	4	4
I/O Pins	5	5	5	5	5
Input Pins	1	1	1	1	1
Internal Pull-ups	Yes/Yes/No	Yes	Yes	Yes	Yes
In-Circuit Serial Programming	Yes/No	Yes	Yes	Yes	Yes
Number of Instructions	33	33	33	35	35
Packages	8-pin DIP SOIC	8-pin DIP JW,SOIC	8-pin DIP JW. SOIC	8-pin DIP SOIC	8-pin DIP JW

Two other members of the PIC12 family are the 12F510 and the 16F506. In most respects these devices are similar to the other members of the PIC12 family previously described. The one difference is that the 12F510 and 16F506 both have flash program memory. Table 6.3 lists the most important features of these two PICs.

Two other members of the PIC12F are the 12F629 and 12F675. The difference between these two devices is that the 12F675 has a 10-bit analog-to-digital converter while the 629 does not. Table 6.4 lists some important features of both PICs.

Several members of the PIC12 family (12F635, 12F636, 12F639, and 12F683) are equipped with special power-management features (called *nano-watt technology*). These devices are specially designed for systems that require extended battery life.

Table 6.3

PIC12F510 and 12F506

	16F506	12F510
Clock:		
Maximum Frequency of Operation (MHz)	20	8
Memory:		
Flash Program Memory	1024	1024
Data Memory (bytes)	67	38
Peripherals:		
Timer Module(s)	TMR0	TMR0
Wake-up from Sleep on Pin Change	Yes	Yes
Features:		
I/O Pins	11	5
Input Only Pin	1	1
Internal Pull-ups	Yes	Yes
In-Circuit Serial Programming	Yes	Yes
Number of Instructions	33	33
Packages	14-pin PDIP, SOIC, TSSOP	8-pin PDIP SOIC, MSOP

Table 6.4

PIC12F629 and 12F675

	12F629	12F675
Clock:		
Maximum Frequency of Operation (MHz)	20	20
Memory:		
Flash Program Memory	1024	1024
Data Memory (SRAM bytes)	64	64
Peripherals:		
Timers 8/16 bits	1/1	1/1
Wake-up from Sleep on Pin Change	Yes	Yes
Features:		
I/O Pins	6	6
Analog comparator module	Yes	Yes
Analog-to-digital converter	No	10-bit
In-Circuit Serial Programming	Yes	Yes
Enhanced Timer1 module	Yes	Yes
Interrupt capability	Yes	Yes
Number of Instructions	35	35
Relative addressing	Yes	Yes
Packages	8-pin PDIP, SOIC, DFN-S	8-pin PDIP SOIC, DFN-S

PIC14 Devices

The single member of this family is the PIC14000. The 14000 is built with CMOS technology, which makes it fully static and gives the PIC an industrial temperature range. The 14000 is recommended for battery chargers, power supply controllers, power management system controllers, HVAC controllers, and for sensing and data acquisition applications. Table 6.5 lists the most important characteristics of this PIC.

Table 6.5

PIC14000

Clock:	
Maximum Frequency of Operation (MHz)	20
Memory:	
Flash Program Memory	4096
Data Memory (SRAM bytes)	192
Peripherals:	
Timers (16 bits with capture)	1
Wake-up from Sleep on Pin Change	Yes
Features:	
I/O Pins	22
Analog-to-digital converter	2 channels
On-chip temperature sensor	1
On-chip comparator modules	2
In-Circuit Serial Programming	Yes
Interrupt capability:	
Internal	6 sources
External	5 sources
I2C-compatible serial port	1
Number of Instructions	35
Relative addressing	Yes
Packages	22-pin PDIP, SOIC, SSOP, Windowed CERDIOP

6.3.2 Mid-Range PIC Family

The mid-range PICs include members of the PIC12 and PIC16 groups. According to Microchip, all mid-range PICs have 14-bit program words with either flash or OTP program memory. Those with flash program memory also have EEPROM data memory and support interrupts. Some members of the mid-range group have USB, I2C, LCD, USART, and A/D converters. Implementations range from 8 to 64 pins. In the following sub-sections we list the basic characteristics of some mid-range PICs.

PIC16 Devices

This is by far the most extensive PIC family. Currently, over 80 versions of the PIC16 are listed in production by Microchip. We selected a few of the most prominent members of the PIC16 family to list their most important features. The Microchip website has detailed information on any of these devices. Table 6.6 lists the features of some members of the PIC16 group.

6.3.3 High-Performance PIC Family

The high-performance PICs belong to the PIC18 group. They have 16-bit program words, flash program memory, a linear memory space of up to 2 Mbytes, as well as protocol-based communications facilities. They all support internal and external interrupts and a much larger instruction set than members of the baseline and mid-range families.

Table 6.6

PIC16 Devices

	16C432	16C58	16C770	16F54	16F84A	16F946
Clock:						
Maximum Frequency MHz	20	40	20	20	20	20
Memory:						
Program memory type	OTP	OTP	OTP	Flash	Flash	Flash
K-bytes	3.5	3	3.5	0.75	1.75	14
K-words	2	2	2	0.5	1	8
Data EEPROM	0	0	0	0	64	256
Peripherals:						
I/O channels	12	12	16	12	13	53
ADC channels	0	0	6	0	0	8
Comparators	0	0	0	0	0	2
Timers	1/8-bit	1/8-bit 1/16-bit	2/8-bit	1/8-bit	1/8-bit	2/8-bit 1/16-bit
Watchdog timer	Yes	Yes	Yes	Yes	Yes	Yes
Features:						
ICSP	Yes	No	Yes	No	Yes	Yes
ICD	No	No	No	No	0	1
Pin count	20	18	20	18	18	64
Communications	-	-	MPC/SPI	-	-	AUSART
Packages	20/CERDIP, 20/SSOP 208mil	18/CERDIP 18/PDIP 18/SOIC 300mil	20/CERDIP 20/PDIP 20/SOIC 300mil	18/PDIP 18/SOIC 300mil	18/PDIp 18/SOIC 300mil	64/TQFP

PIC18 Devices

The PIC18 family has over 70 different variations currently in production. The PIC18 family uses 16-bit program words and are furnished in 18 to 80 pin packages. Microchip describes the PICs in this family as high-performance with integrated A/D converters. They have 32-level stacks and support interrupts. The instruction set is much larger and starts at 79 instructions. The PICs in this family have flash program memory, a linear memory space of up to 2 Mbytes, 8-by-8 bit hardware multiplier, and communications peripherals and protocols. Table 6-7 lists some members of the PIC18 family.

Table 6.7

PIC18 Devices

	18F222	18F2455	18F2580	18F4580	18F8622
Clock:					
Maximum Frequency MHz	40	48	40	40	40
Memory:					
Program memory type	flash	flash	flash	flash	flash
K-bytes	4	24	32	32	64
K-words	2	12	16	16	321
Data EEPROM	256	256	256	256	1024
Peripherals:					
I/O channels	25	23	25	36	70
ADC channels	10	10	8	11	16
Comparators	2	2	0	2	2
Timers	1/8-bit 3/16-bit	1/8-bit 3/16-bit	1/8-bit 3/16-bit	1/8-bit 3/16-bit	2/8-bit 3/16-bit

(conitnues)

Table 6.7

*PIC18 Devices **(continued)***

	18F222	18F2455	18F2580	18F4580	18F8622
Watchdog timer					
	Yes	Yes	Yes	Yes	Yes
Features:					
EUSART	Yes	Yes	Yes	Yes	2
ICSP	Yes	Yes	Yes	Yes	Yes
ICD	1	3	3	3	3
Pin count	28	28	28	44	80
Communications					
	MPC/SPI	MPC/SPI/USB	MPC/SPI	MPC/SP	2-MPC/SPI
Packages					
	28/PDIP,	28/SOIC	28/QFN	40/PDIP	80/TQFP
	28/SOIC	28/PDIP	20/PDIP	44/QFN	
	300mil	300mil	300mil	44/TQFP	

Chapter 7

Architecture and Instruction Set

7.1 Mid-Range PIC Architecture

In Chapter 6 we analyzed the three major PIC families of 8-bit devices. From the descriptions and functionality one can see that as the PIC architecture increases in complexity and power, so does the size, intricacy, and cost of the devices. For the embedded systems covered in this book, the PICs of the mid-range family provides sufficient power and functionality. It can be argued that the baseline PICs do find extensive use and are quite practical for many applications; in fact, some of the embedded systems covered in the text could have been implemented with baseline devices. On the other hand, the baseline PICs are quite similar in architecture and programming to their mid-range relatives. In most cases the difference between a baseline and a mid-range device is that the low-end one lacks some features or has less program space or storage. We have not covered the implementation and programming of the baseline PICs due to the belief that someone familiar with mid-range devices could easily re-design a circuit and port the code in order to use a simpler baseline microcontroller.

For all these reasons we have opted to limit our coverage to the mid-range family of PICs. Within this family we have concentrated our attention on the most used, documented, and popular PICs: these are the 16F84 and 16F84A, the 16F877, and the newer 16F684. Although other PIC microcontrollers are mentioned in the text, the programming examples and applications refer only to these three devices.

We start by mentioning several general characteristics of the PIC: Harvard architecture, RISC processor design, single-word instructions, machine and data memory configuration, and characteristic instruction formats.

7.1.1 Harvard Architecture

The PIC microcontrollers are not based on the *von Neumann architecture* that has been almost exclusively adopted in the computer world. The model of hardware design adopted for the PIC family is called the *Harvard architecture*. Harvard architecture refers to a computer design in which data and instruction used different signal paths and storage areas. Data and instructions are not located in the same memory area but in separate ones. One consequence of the traditional von Neumann architecture is that the processor can either read or write instructions or data but cannot do both at the same time, as both instructions and data use the same signal lines. In a machine based on the Harvard architecture, on the other hand, the processor can read and write instructions and data to and from memory at the same time. This results in a faster, albeit more complex machine. Figure 7-1 shows the program and data memory space in a mid-range PIC.

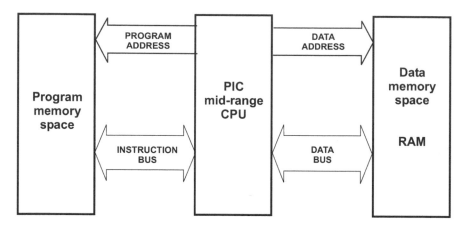

Figure 7-1 Mid-Range PIC Memory (Harvard Architecture).

One of the strongest arguments in favor of the Harvard architecture is based on the speed of access to main memory. Making a CPU faster while memory access remains at the same speed represents little total gain, especially if many memory accesses are required. This situation is often referred to as the *von Neumann bottleneck* and machines that suffer from it are said to be *memory bound*.

Many microcontrollers, including the Microchip PICs, are based on the Harvard architecture. These devices have separate storage for program and data and a reduced instruction set. The mid-range PICs in particular have 8-bit data words but either 12-, 14-, or 16-bit program instructions. Becausee the instruction size is much wider than the data size, an instruction can contain a full-size data constant.

7.1.2 CISC versus RISC

The CISC (*Complete Instruction Set Computer*) design is based on each low-level instruction performing several operations. For example, one Intel 80x86 opcode can decrement a counter register, determine the state of a processor flag, and execute a jump instruction if the flag is set or cleared. Another CISC instruction moves a number of bytes of data contained in a *counter register* from an area pointed at by a source register, into another area pointed at by a *destination register*. The Intel 80x86 CPU contains about 120 primitive operations in its instruction set.

The original design idea of the CISC architecture was to provide high-level instructions, to facilitate the implementation of high-level languages. Supposedly this would be achieved through complex instruction sets, multiple addressing modes, and primitive operations that performed multiple functions. However, in many cases the CISC architecture did not result in better performance because the more complex the instruction set resulted in greater decoding time. At the same time, implementing large instruction sets required more silicon space and considerably more design effort. Some CISC processors developed in the 1960s and 1970s are the IBM System/360, the PDP-11, the Motorola 68000 family, and the Intel 80x86 family of CPUs.

In contrast, a RISC (*Reduced Instruction Set Computer*) machine contains fewer instructions and each instruction performs more elementary operations. One consequence of this is a smaller silicon area, faster execution, and reduced program size with fewer accesses to main memory. The PIC designers have followed the RISC route. Other CPUs with RISC design are the MIPS, the IBM Power PC, and the DEC Alpha.

7.1.3 Single-Word Instructions

One of the consequences of the PIC's Harvard architecture is that the instructions can be wider than the 8-bit data size. Because the device has separate buses for instructions and data, it is possible that instructions are sized differently than data items. Being able to vary the number of bits in each instruction opcode makes possible the optimization of program memory and the use of single-word instructions that can be fetched in one bus cycle.

In the mid-range PICs, each instruction is 14-bits wide and every fetch operation brings into the execution unit one complete operation code. Because each instruction takes up one 14-bit word, the number of words of program memory in a device exactly determines the number of program instructions that can be stored. Because von Neumann instructions can span multiple bytes, there is no assurance that each program memory location contains the first opcode of a multibyte instruction. For these reasons, instruction storage and fetching in a von Neumann machine becomes a much more complicated issue.

As is the case with conventional processors, the PIC architecture has a two-stage instruction pipeline. However, because the fetch portion of the current instruction and the execution of the previous one can overlap in time, one complete instruction is fetched and executed at every machine cycle. This is known as *instruction pipelining*. Because each instruction is 14 bits and the program memory bus is also 14-bits wide, each instruction contains all the necessary information so that it can be executed without any additional fetching. The one exception is when an instruction modifies the contents of the *Program Counter*. In this case a new instruction must be fetched, which requires an additional machine cycle.

The PIC clocking system is designed so that an instruction is fetched, decoded, and executed every four clock cycles. In this manner, a PIC equipped with a 4-MHz oscillator clock will beat at a rate of 0.25 ms. Because each instruction executes at every four clock cycles, each instruction takes 1 ms.

7.1.4 Instruction Format

All members of the mid-range family of PICs have 14-bit instructions in a set of 35 instructions. The format for the instructions follows three different patterns: *byte-oriented, bit-oriented,* and *literal and control instructions.* Figure 7-2 shows the bitmaps for the three types.

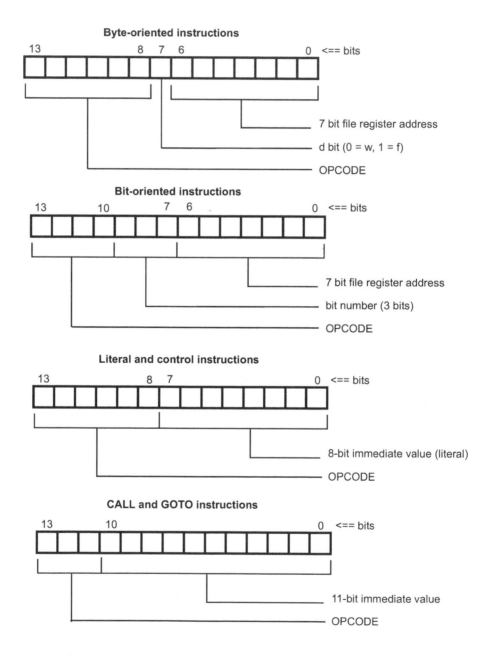

Figure 7-2 Mid-Range Instruction Formats.

In Figure 7-2 note that, in the PIC instruction set, the opcode field can be 6- 4- or 3-bits wide. It is this scheme that allows implementing all 35 different instructions with the 14 available opcode bits. Also note that instructions that reference a file register do so in a 7-bit field. Because the numerical range of seven bit is 128 values, the mid-range PICs that address more than 128 data memory locations must resort to *banking*. In this case a bit or bit field in the STATUS register serves to select the bank currently addressed.

A similar situation arises when addressing program memory with an 11-bit field. Eleven bits allow 2048 addresses; so if a PIC is to have more than 2K program memory, it is necessary to adopt a *paging* scheme in which a special register is used to select which memory page the instruction is located. Paging is only required in devices that exceed the 2K program space limit that can be encoded in 11 bits.

7.1.5 Mid-Range Device Versions

The names used by Microchip use different code letters to indicate the various versions of a device. In this case the first letter following the family affiliation designator represents the memory type of the device, as follows:

1. The letter C, as in PIC16Cxxx, refers to devices with EPROM type memory.

2. The letters CR, as in PIC16CRxxx, refer to devices with ROM type memory.

3. The letter F, as in PIC16Fxxx, refers to devices with flash memory.

The letter L immediately following the affiliation designator refer to devices with extended voltage range. For example, the PIC16LFxxx designation corresponds to devices with extended voltage range.

7.1.6 Arithmetic-Logic Unit

In a digital system the *central processing unit* (CPU) is the component that executes the program instructions and processes data. It provides the fundamental functionality of a digital system and is responsible for its programmability. In the PIC architecture the CPU is a portion of the device that fetches and executes the instructions contained in a program. The *arithmetic-logic unit* (ALU) is the CPU element that performs arithmetic, bitwise, and logical operations. It also controls the bits in the STATUS register as they are changed by the execution of the various program instructions. For example, if the result of executing an instruction is zero, the ALU will set the zero bit in the STATUS register.

7.2 Data Memory Organization

The structure and organization of data memory in the PIC hardware also has some unique and interesting features. The programmer accustomed to the flat, addressable memory space of the von Neumann computer with its multiple machine registers may require some time to gain familiarity with the PIC's data formats.

7.2.1 w Register

Conventional machine registers do not exist in PICs. Instead, there is a single addressable register called the *work register*, or the *w register*. The CISC low-level programmer used for multiple general purpose registers into which data can be moved and later retrieved will need to become accustomed to working with a single machine register that takes part in practically every instruction. Add to this the lack of an addressable stack and you will see that mid-range PIC programming is based on a different paradigm.

7.2.2 Data Registers

The PIC's data memory consists of registers, also called *file registers*. The file registers behave more like conventional variables. They can be addressed directly and indirectly. In mid-range PICs, all data registers are 8 bits and come in two types: *general-purpose registers* (GPRs) and *special-function registers* (SFRs).

Memory Banks

Figure 7-2, previously in this chapter, shows that the PIC instruction format devotes 7 bits to the address field. A 7-bit address allows access to 128 memory locations. Because many PICs of the mid-range family have more than 128 bytes of data memory, an addressing scheme based on *memory banks* must be implemented. The memory *banking* mechanism adopted by the PICs is effective, although not very user-friendly.

The number of banks varies according to the amount of available RAM, always in multiples of 128 bytes. All mid-range PICs have banked memory. Banking is accomplished through the special bank select bits in the STATUS register, described later in this chapter. Not all banking bits are implemented in all devices. For example, the 16F84/16F84A contain two memory banks. In this case, bank shifting requires a single bank select bit (RP0). In devices with more than two memory banks, bank selection is as shown in Table 7.1.

Table 7.1

Mid-Range Bank Selection Options in Direct Addressing

BANK ACCESSED	STATUS REGISTER BITS (RP1:RP0)
0	0 : 0
1	0 : 1
2	1 : 0
3	1 : 1

Figure 7-3 shows how banked memory is accessed in direct addressing. The illustration refers to a mid-range PIC with four banks, as is the case with the 16F87x.

SFRs

The special function registers are defined by the device architecture and have reserved names. For example, the TMR0 register is part of the system timer, the STATUS register holds several processor flags, and the INTCON register is used in controlling interrupts. Some SFRs can be written and read, and others are read-only. Reserved and not-implemented SFR bits always read as zero. Two SFR registers, which are used in indirect addressing, have special characteristics: one of them (the indirect address register) is not a physical register, and the other one (the FSR register) is used to initialize the indirect pointer. The SFRs are allocated starting at the lowest RAM address (address 0).

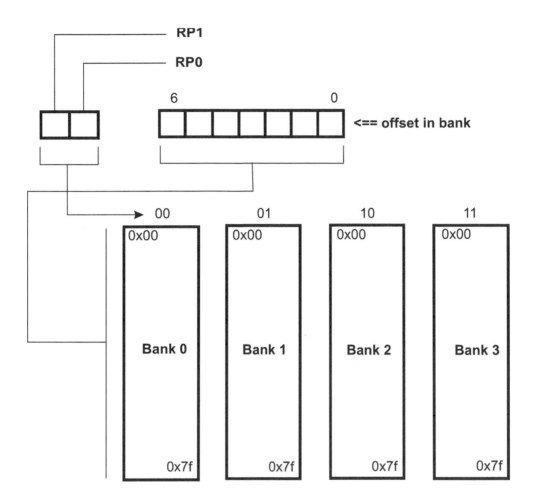

Figure 7-3 Memory Access in Direct Addressing.

Figure 7-4 is a map of the register file in the 16F87x family. Observe, in Figure 7-4, that the general-purpose registers do not start at the same address offset in each bank. However, there is a common area that extends from 0x70 to 0x7f that is accessible no matter which bank is selected. In applications that require frequent bank switching, this 16-byte area is very valuable real estate because user variables created in it are accessible no matter which bank is currently selected. GPRs created outside this common area are only accessible when the corresponding bank is selected.

The registers in boldface in Figure 7-4 are accessible from any bank. These registers, such as STATUS and the indirect addressing registers FSR and INDF, must be bank independent. Also, some registers are mirrored in more than one bank. For example, the PORTB register is accessible in bank 0 and in bank 2, and the TRISB register in bank 1 and bank 3. The mirrored registers were implemented to simplify data access and minimize bank changes in applications.

Bank 0		Bank 1		Bank 2		Bank 3	
INDF	0x00	**INDF**	0x80	**INDF**	0x100	**INDF**	0x180
TMR0	0x01	OPTION*	0x81	TMR0	0x101	OPTION*	0x181
PCL	0x02	**PCL**	0x82	**PCL**	0x102	**PCL**	0x182
STATUS	0x03	**STATUS**	0x83	**STATUS**	0x103	**STATUS**	0x183
FSR	0x04	**FSR**	0x84	**FSR**	0x104	**FSR**	0x184
PORTA	0x05	TRISA	0x85		0x105		0x185
PORTB	0x06	TRISB	0x86	PORTB	0x106	TRISB	0x186
PORTC	0x07	TRISC	0x87		0x107		0x187
PORTD	0x08	TRISD	0x88		0x108		0x188
PORTE	0x09	TRISE	0x89		0x109		0x189
PCLATH	0x0a	**PCLATH**	0x8a	**PCLATH**	0x10a	**PCLATH**	0x18a
INTCON	0x0b	**INTCON**	0x8b	**INTCON**	0X10b	**INTCON**	0x18b
PIR1	0x0c	PIE1	0x8c	EEDATA	0x10c	EECON1	0x18c
PIR2	0x0d	PIE2	0x8d	EEADR	0x10d	EECON2	0x18d
TMR1L	0x0e	PCON	0x8e	EEDATH	0x10e	Reserved	0x18e
TMR1H	0x0f		0x8f	EEADRH	0x10f	Reserved	0x18f
T1CON	0x10		0x90		0x110		0x190
TMR2	0x11	SSPCON2	0x91				
T2CON	0x12	PR2	0x92				
SSPBUF	0x13	SSPADD	0x93				
SSPCON	0x14	SSPTAT	0x94				
CCPR1L	0x15		0x95				
CCPR1H	0x16		0x96				
CCP1CON	0x17		0x97				
RCSTA	0x18	TXSTA	0x98	General Purpose Registers		General Purpose Registers	
TXREG	0x19	SPBRG	0x99				
RCREG	0x1a		0x9a				
CCPR2L	0x1b		0x9b				
CCPR2H	0x1c		0x9c				
CCP2CON	0x1d		0x9d				
ADRESH	0x1e	ADRESL	0x9e				
ADCON0	0x1f	ADCON1	0x9f				
	0x20		0xA0				
General Purpose Registers		General Purpose Registers					
			0xef		0x16f		0x1ef
Common area 0x70-0x7f		Common area 0x70-0x7f	0xf0	Common area 0x70-0x7f	0x170	Common area 0x70-0x7f	0x1f0
	0x7f		0xff		0x17f		0x1ff

* Actual name is OPTION_REG

Figure 7-4 16F87x File Register Map.

Other members of the mid-range PIC group, such as the 16F84 and 16F84A, have a different memory footprint. Figure 7-5 is a bitmap of the 16F84A.

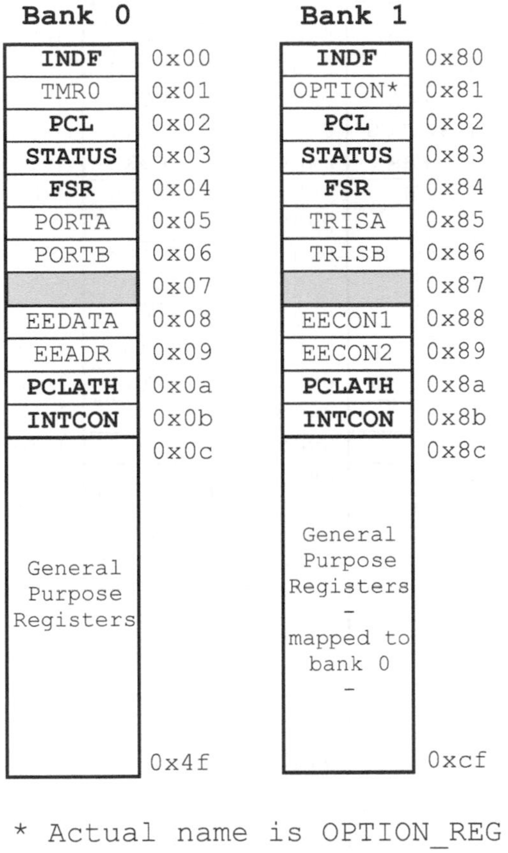

Bank 0   **Bank 1**

INDF	0x00
TMR0	0x01
PCL	0x02
STATUS	0x03
FSR	0x04
PORTA	0x05
PORTB	0x06
	0x07
EEDATA	0x08
EEADR	0x09
PCLATH	0x0a
INTCON	0x0b
	0x0c

General
Purpose
Registers

	0x4f

INDF	0x80
OPTION*	0x81
PCL	0x82
STATUS	0x83
FSR	0x84
TRISA	0x85
TRISB	0x86
	0x87
EECON1	0x88
EECON2	0x89
PCLATH	0x8a
INTCON	0x8b
	0x8c

General
Purpose
Registers
-
mapped to
bank 0
-

	0xcf

* Actual name is OPTION_REG

Figure 7-5 16F84A File Register Map.

Here again, the general-purpose registers do not start at the same address offset in each bank. Also note that all GPRs are mapped to bank 0. This determines that, in the 16F84A, user-defined registers created in bank 0 are accessible no matter which bank is currently selected. Figure 7-6 is a register map of the 16F684 PIC.

GPRs

General-purpose registers are created and named by the programmer and must be allocated in the reserved memory space. In the 16F84A, all GPRs are mapped to the same memory area, no matter in which bank they are defined. The GPR memory space extends from 0x0c to 0x4f (68 bytes). A much different situation exists in the 16F87x PICs, in which only 16 bytes of GPR space are mirrored in all three banks. This is the memory referred to as the common area in Figure 7-4. In the 16F87x, the total available GPR space is as follows:

BANK 0	BANK 1	BANK2	BANK3
96 bytes	80 bytes	96 bytes	96 bytes

Total = 368 bytes

Bank 0

Register	Address
INDF	0x00
TMR0	0x01
PCL	0x02
STATUS	0x03
FSR	0x04
PORTA	0x05
	0x06
PORTC	0x07
	0x08
	0x09
PCLATH	0x0a
INTCON	0x0b
PIR1	0x0c
	0x0d
TMR1L	0x0e
TMR1H	0x0f
T1CON	0x10
TMR2	0x11
T2CON	0x12
CCPR1L	0x13
CCPR1H	0x14
CCP1CON	0x15
PWM1CON	0x16
ECCPAS	0x17
WDTCON	0x18
CMCON0	0x19
CMCON1	0x1a
	0x1b
	0x1c
	0x1d
ADRESH	0x1e
ADCON0	0x1f
General Purpose Registers 96 Bytes	0x20 ... 0x7f

Bank 1

Register	Address
INDF	0x80
OPTION*	0x81
PCL	0x82
STATUS	0x83
FSR	0x84
TRISA	0x85
	0x86
TRISC	0x87
	0x88
	0x89
PCLATH	0x8a
INTCON	0x8b
PIE1	0x8c
	0x8d
PCON	0x8e
OSCCON	0x8f
OSCTUNE	0x90
ANSEL	0x91
PR2	0x92
	0x93
	0x94
WPUA	0x95
IOCA	0x96
	0x97
	0x98
VRCON	0x99
EEDAT	0x9a
EEADR	0x9b
EECON1	0x9c
EECON2	0x9d
ADRESL	0x9e
ADCON1	0x9f
General Purpose Registers 32 Bytes	0xa0 ... 0xbf ... 0xff

* Actual name is OPTION_REG

Figure 7-6 16F684 File Register Map.

7.2.3 Indirect Addressing

The instruction set of most processors, including the PICs, provides a mechanism for accessing memory operands indirectly. Indirect addressing is based on the following capabilities:

- The address of a memory operand is loaded into a register. This register is called the pointer.

- The pointer register is then used to indirectly access the memory location at the address it "points to."

- The value in the pointer register can be modified (usually incremented or decremented) so as to allow access to other memory operands.

In the PIC architecture, indirect addressing is implemented using two registers: INDF and FSR. The INDF register, always located at memory address 0x00 and mirrored in all banks, is not a physical register; therefore it cannot be directly accessed by code. The FSR register is the pointer register that is initialized to the address of a memory operand. Once a memory address is placed in FSR, any action on the INDF register takes place at the memory location pointed at by FSR. For example, if the FSR register is initialized to memory address 0x20, then clearing the INDF register has the effect of clearing the memory location at address 0x20. The action on the INDF register actually takes place at the address contained in the FSR register. Now if FSR (the pointer register) is incremented and INDF is again cleared, the memory location at address 0x21 will be cleared. Indirect addressing is covered in detail in the programming chapters later in the book.

7.3 Mid-Range I/O and Peripherals

Mid-range devices contain special modules to implement peripheral and I/O functions. The more complex the device, the more peripheral modules are likely to be present. For example, a simple mid-range PIC like the 16F84A contains few peripheral modules, specifically EEPROM data memory, I/O ports, and a timer module. The 16F87x PICs, on the other hand, in addition to I/O ports, EEPROM, and three individual timers, also have a parallel slave port, a capture and compare (PWM) module, a master synchronous serial port (MSSP) module, a universal asynchronous/synchronous receiver and transmitter (USART) module, and an analog-to-digital converter (A/D) module.

Other members of the mid-range family have additional or different peripheral and I/O modules. In the following sections we briefly describe the architecture of the most common peripheral modules. The programming details are covered in context elsewhere in the book.

Implementation of many different functions in a device with a small footprint requires multiplexing many of the PICs access connections. Figure 7-7 shows the pinout of the 16F84A and the 16F877 as well as the multiple functions of many pins in either device.

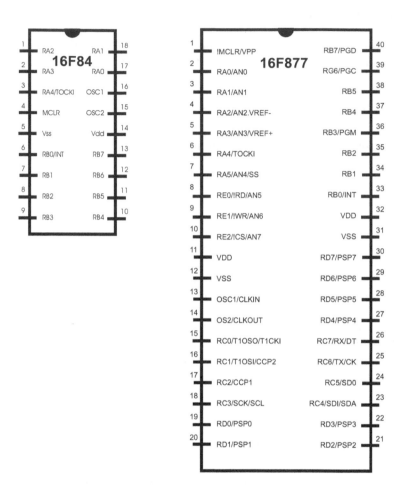

Figure 7-7 16F84A and 16F877 Pin Diagrams.

7.3.1 Ports

Ports provide PICs with access to the outside world and are mapped to physical pins on the device. In some mid-range PICs (see Figure 7-7), some pins used in I/O operations are multiplexed with alternate functions of peripheral modules. A pin ceases to be a port when a peripheral module function is activated for that pin.

General port pins are bi-directional; therefore they can be configured either as input or output. Each port has a corresponding TRIS register. The setting in this register determines if a port is designated as input or output. A value of 1 in the port's TRIS register makes the port an input and a value of 0 makes the mapped port an output. Input ports are used in communicating with input devices, such as switches, keypads, and input data lines from hardware devices. Output ports are used in communicating with output devices, such as LEDs, Seven-Segment displays, liquid-crystal displays (LCDs), and data output line to hardware devices such as motors.

Although port pins are bit mapped, they are read and written as a unit. For example, the PORTA register holds the status of the eight pins possibly mapped to port A. Writing to a specific port writes to the port latches. Write operations to ports are actually *read-modify-write* operations. The port pins are first read, then the value is modified, and finally written to the port's data latch. As previously mentioned, some of the port pins are multiplexed; for example, pin RA4 is multiplexed with the Timer0 module clock input; therefore, it is labeled RA4/T0CKI pin in the device's pinout. Other PORTA pins are multiplexed with analog inputs and with other peripheral functions. The device data sheets contain information regarding the functions assigned to each device pin.

7.3.2 Timers

Timer modules are available in all mid-range PICs; in fact, the TIMER0 module is present in all PICs of the mid-range family. The TIMER0 module has the following features:

- 8-bit timer/counter

- Readable and writable

- 8-bit software programmable prescaler

- Internal or external clock select

- Interrupt on overflow from FFh to 00h

- Edge select for external clock

Chapter 11 is devoted entirely to the architecture and programming of timers and counters.

7.3.3 Capture and Compare Module

Some mid-range devices contain one or more capture and compare modules, designated as Capture/Compare/PWM. In Figure 7-7 you can see that one of the functions multiplexed onto pin 17 of the 16F877 is labeled CCP1 (Capture and ComPare module number 1). The CCP2 module is multiplexed onto pin number 16. The principal function of the capture and compare modules is to enhance timer operations. Each module contains the following elements:

A 16-bit register that can operate as a:

A 16-bit Capture register

A 16-bit Compare register

A PWM Master/Slave Duty Cycle register

When more than one capture and compare module is implemented in a single device, they are all identical in operation. In the 16F877 the two available modules are designated CCP1 and CCP2. In each module a Capture/Compare/PWM Register1 (CCPR1) is comprised of two 8-bit registers: CCPR1L (low byte) and CCPR1H (high byte). The CCP1CON register controls the operation of CCP1.

The CCP modules find use in recording events such as measuring time periods, counting, generating pulses and periodic waveforms, and voltage averaging.

7.3.4 Master Synchronous Serial Port

Some mid-range PICs come equipped with hardware modules to implement serial protocols, including SPI and I2C. The module that provides these interfaces is called the *Master Synchronous Serial Port*, or MSSP. The MSSP module can operate in either the slave or the master mode, as well as in a free bus mode, also called the *multi-master function*. The MSSP module is useful for communicating with other peripheral or microcontroller devices. The peripheral devices can be serial EEPROMs, shift registers, display drivers, or A/D converters, etc. The MSSP module is discussed in Chapter 14 in the context of EEPROM data memory programming.

7.3.5 USART Module

The *Universal Synchronous Asynchronous Receiver Transmitter* (USART) module in the 16F87x family is also known as a *Serial Communications Interface*, or SCI. The USART module is used in communicating with devices and systems that support the RS-232 protocol, including computers and terminals. It can be configured as an asynchronous full-duplex device, as a synchronous half-duplex master, or as a synchronous half-duplex slave. In the synchronous mode, the USART is useful in communicating with analog-to-digital and digital-to-analog integrated circuits or for accessing serial EEPROMS. The USART is discussed extensively in the context of serial communications and programming EEPROMS. These topics are covered in Chapter 20 and Chapter 14, respectively.

7.3.6 A/D Module

Until recently A/D conversions required the use of dedicated devices, usually in the form of an integrated circuit component. In Microchip documentation, the *analog-to-digital conversions module* is referred to as ADC. Some mid-range PICs now come with on-board A/D hardware. One of the advantages of using on-board A/D converters is saving interface lines. Interfacing with a hardware analog-to-digital IC usually requires three to four lines. A similar function can be implemented with on-board A/C hardware with a single line. This saving in I/O lines can be significant in many PIC circuits.

Mid-range PICs equipped with A/D converters have either 8- or 10-bit resolution and can receive analog input in 2 to 16 different channels. For example, the 16F877 contains eight analog input channels at a 10-bit resolution. A/D converters use a sample-and-hold capacitor to store the analog charge and perform a successive approximation algorithm to produce the digital result. When the converter resolution is 10 bits these are stored in two 8-bit registers, one of them having only four significant bits. The ADC module on the 16F684 PIC provides a 1-bit binary output from a single analog signal. The presence of this module is sometimes the reason for preferring the 16F684 over its 16F84 and 16F84A predecessors.

The A/D module has high- and low-voltage reference inputs that are selected by software. The module can operate while the processor is in SLEEP mode, but only if the A/D clock pulse is derived from its internal RC oscillator.

7.4 Mid-Range PIC Core Features

Core features refer to those that apply to all members of the mid-range family. It includes the following elements:

- The device oscillator

- The reset mechanism

- CPU architecture and operation

- The ALU

- Data memory organization

- The interrupt system

- The instruction set

We have already discussed the architecture and general features of the CPU and the ALU, as well as to data memory organization. The remaining topics (oscillator, reset, interrupts, and instruction set) are covered in the following sub-sections.

7.4.1 Oscillator

Mid-range PICs require a clock signal for its operation. The clock signal source can be external to the PIC or an internal *oscillator*. Different mid-range PICs support different numbers of oscillator modes. For example, the 16F877 and the 16F684 can execute in any one of eight modes, but in the 16F84 only four oscillator modes are available. The 16F684 has eight possible clock modes, two of which originate in an internal oscillator.

The oscillator mode is selected at device programming time and cannot be changed at runtime. The *configuration bits*, which are nonvolatile flags set during device programming, determine which oscillator mode is used by the program, according to those supported by the device. Some mid-range PICs (like the 16F684) support unique oscillator modes that are not generally found in other members of the mid-range family. The following are the most common mid-range oscillator modes:

1. LP, low frequency crystal

2. XT, crystal resonator

3. HS, high speed crystal resonator

4. RC, external resistor/capacitor

5. EXTRC, external resistor/capacitor

6. EXTRC, external resistor/capacitor with CLKOUT

7. INTRC, internal 4-MHz resistor/capacitor

8. INTRC, internal 4-MHz resistor/capacitor with CLKOUT

The resistor/capacitor oscillator option is the cheapest to implement but also the least accurate one. This option should be used only in systems where clock accuracy and consistency are not issues. The *low-frequency crystal option* is the one with the lowest power consumption and is suitable for systems where power consumption is an important factor.

The first three oscillator modes (*LP*, *XT*, and *HS*) allow selecting different frequency ranges. The *HS* has the highest frequency range and consumes the most power. The *XT* option is based on a standard crystal resonator and has mid-range power consumption. The *LP* option has low gain and consumes the least power of the three crystal modes. The general rule is to use the oscillator with the lowest possible gain that still meets the circuit requirements. The *RC mode* with EXTRC and CLKOUT features have the same functionality as the straight RC oscillator option.

The XT option (*crystal resonator*) can be provided in a convenient ceramic package. This device, called a *ceramic resonator*, contains three pins. The ones on the extremes are connected to the corresponding oscillator input lines on the PIC, labeled OSC1 and OSC2. The center pin is connected to ground. Figure 7-8 shows the circuit diagram for an oscillator and a crystal resonator.

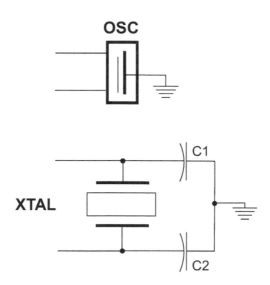

Figure 7-8 Circuit Diagrams for a Ceramic Resonator and a Crystal Oscillator.

Alternatively, the oscillator function can be provided by an integrated circuit (such as the ICS502) that can generate several different clock frequencies. Some circuits, especially in PIC demonstration boards, contain jumper pins that allow selecting among several clock rates.

7.4.2 System Reset

The *reset mechanism* is used to place the PIC in a known condition, to gain control of a run-away or hung-up program, as a forced interrupt in program execution, or to make the device ready at program load time. The processor's !MCLR pin produces the reset action when it reads logic zero. The exclamation sign preceding the pin's name (or alternatively, as a line over the text) indicates that the action is active-low. To prevent accidental resets the !MCLR pin must be connected to the positive voltage supply through a 5K or 10K resistor. When a resistor serves to place a logic 1 on a line, it is called a *pull-up resistor*.

The mid-range PICs are capable of several reset actions:

- Reset during power on (POR)
- !MCLR reset during normal operation
- Reset during SLEEP mode
- Watchdog timer reset (WDT)
- Brown-out reset (BOR)
- Parity error reset

The most common reset sources are the first two in the preceding list. POR reset serves to bring all PIC registers to an initial state, including the program counter register. This action is taken automatically by the microcontroller. The second source of a reset action takes place when the !MCLR line is intentionally brought down, usually by the action of a pushbutton reset switch. This switch is useful during program development because it provides a way of forcefully re-starting execution. Figure 7-9 shows a typical wiring of the !MCLR line to provide a reset activated by a pushbutton switch. The circuit includes an LED that lights up when the power is on.

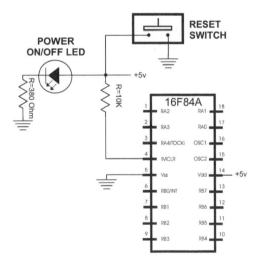

Figure 7-9 Typical Wiring of the Reset Switch.

User RAM memory is not affected by a reset. The PICs general-purpose registers (GPRs are in an unknown state during power up and are not changed by reset). SFR registers, on the other hand, are reset to an initial state. The initialization conditions for each of the SFRs can be found in the device data sheet. The most important of these is perhaps the *program counter* (PC), which is reset to zero. This action directs execution to the first instruction, effectively restarting the program.

During power-up, the processor itself initiates a reset when the power supply voltage goes from 1.2 to 1.8V. Several bits in various registers are related to the reset action, but these are not available in all mid-range devices. For example, some high-end devices in the mid-range group, such as the 16F87x, contain two bits in the PCON register that are reset related. One of them (named !POR) allows determining the power-on reset status. The other one (named !BOR) informs about the brown-out reset status. However, the PCON register does not exist in the 16F84 or 16F84A.

7.4.3 Interrupts

The *interrupt mechanism* provides a way of having the microcontroller respond to events as they occur, rather than being forced to *poll* devices in order to determine their state. The interrupt is said to work like a "tap on the shoulder" of the microcontroller, calling its attention to an event that requires an action or device that needs servicing. After responding to or ignoring the interrupt, the CPU resumes processing where it left off.

In computer technologies the interrupt mechanism is a complicated hardware/software system that often includes a programmable interrupt controller. Processors and microprocessors usually support hardware and software interrupts and maskable and nonmaskable interrupts, which can originate in practically any device connected to the system. In PICs, the interrupt mechanism is much simpler and varies considerably even among members of the same PIC family.

All PICs of the mid-range family support interrupts to some degree. The interrupt source usually originates in one of the hardware modules, although some sources can generate more than one interrupt. For example, the 16F684 supports eleven interrupt sources. The following are common interrupt sources in the mid-range family, although not all are supported by every PIC.

- INT pin interrupt (external interrupt)
- TMR0 overflow interrupt
- PORTX change interrupt
- Comparator change interrupt
- Parallel slave port interrupt
- USART interrupts
- Receive and transmit interrupt
- A/D conversion complete interrupt
- LCD interrupt
- Data EEPROM write complete interrupt
- Timer overflow interrupts
- CCP interrupt
- SSP interrupt

Several SFRs are related to the interrupt systems. The INTCON register provides interrupt enabling and control, and the PIE1, PIE2, PIR1, and PIR2 registers have specific device-related functions. Programming interrupts are discussed in the context of the corresponding operations later in this book.

7.5 Mid-Range Instruction Set

The mid-range PIC instruction set consists of 35 instructions, divided into three general groups:

- Byte-oriented and byte-wise file register operations
- Bit-oriented and bit-wise file register operations
- Literal and control instructions

Table 7.1 lists and briefly describes each instruction in the mid-range set.

Table 7.1

Mid-Range PIC Instruction Set

MNEMONIC	OPERAND	DESCRIPTION	CYCLES	BITS AFFECTED
BYTE-ORIENTED OPERATIONS:				
ADDWF	f,d	Add w and f	1	C,DC,Z
ANDWF	f,d	AND w with f	1	Z
CLRF	f	Clear f	1	Z
CLRW	-	Clear w	1	Z
COMF	f,d	Complement f	1	Z
DECF	f,d	Decrement f	1	Z
DECFSZ	f,d	Decrement, skip if 0	1(2)	-
INCF	f,d	Increment f	1	Z
INCFSZ	f,d	Increment, skip if 0	1(2)	-
IORWF	f,d	Inclusive OR w and f	1	Z
MOVF	f,d	Move f	1	Z
MOVWF	f	Move w to f	1	-
NOP	-	No operation	1	-
RLF	f,d	Rotate left through carry	1	C
RRF	f,d	Rotate right through carry	1	C
SUBWF	f,d	Subtract w from f	1	C,DC,Z
SWAPF	f,d	Swap nibbles in f	1	-
XORWF				
BIT-ORIENTED OPERATIONS				
BCF	f,b	Bit clear in f	1	-
BSF	f,b	Bit set in f	1	-
BTFSC	f,b	Bit test, skip if clear	1	-
BTFSS	f,b	Bit test, skip if set	1	-

(continues)

Table 7.1

Mid-Range PIC Instruction Set

MNEMONIC	OPERAND	DESCRIPTION	CYCLES	BITS AFFECTED
		LITERAL AND CONTROL OPERATIONS		
ADDLW	k	Add literal and w	1	C,DC,Z
ANDLW	k	AND literal and w	1	Z
CALL	k	Call procedure	2	-
CLRWDT	-	Clear Watchdog timer	1	TO,PD
GOTO	k	Go to address	2	-
IORLW	k	Inclusive OR literal with w	1	Z
MOVLW	k	Move literal to w	1	-
RETFIE	-	Return from interrupt	2	-
RETLWk	-	Return literal in w	2	-
RETURN	-	Return from procedure	2	-
SLEEP	-	Go into SLEEP mode	1	TO,PD
SUBLW	k	Subtract literal and w	1	C,DC,Z
XORLW	k	Exclusive OR literal with w	1	Z

Legend:
  f = file register
  d = destination: 0 = w register
                       1 = file register
  b = bit position
  k = 8-bit constant

7.5.1 STATUS and OPTION Registers

The STATUS register is one of the SFRs in mid-range PICs. The bits in this register reflect the arithmetic status of the ALU, the RESET status, and the bits that select which memory bank is currently being accessed. Because the bank selection bits are in the STATUS register, it must be present and at the same relative position in every bank. Figure 7-10 is a bitmap of the STATUS register.

The STATUS register can be the destination for any instruction. If it is, and the Z, DC, or C bits are affected, then the write operation to these bits is disabled. In addition, the TO and PD bits are not writable. Some instructions may have an unexpected action on the STATUS register bits; for example, the instruction

```
clrf     STATUS
```

clears the upper 3 bits, sets the Z bit, and leaves all other bits unchanged. For this reason it is recommended that only instructions that do not change the Z, C, and DC bits be used to alter the STATUS register. The only ones that qualify are BCF, BSF, SWAPF, and MOVWF.

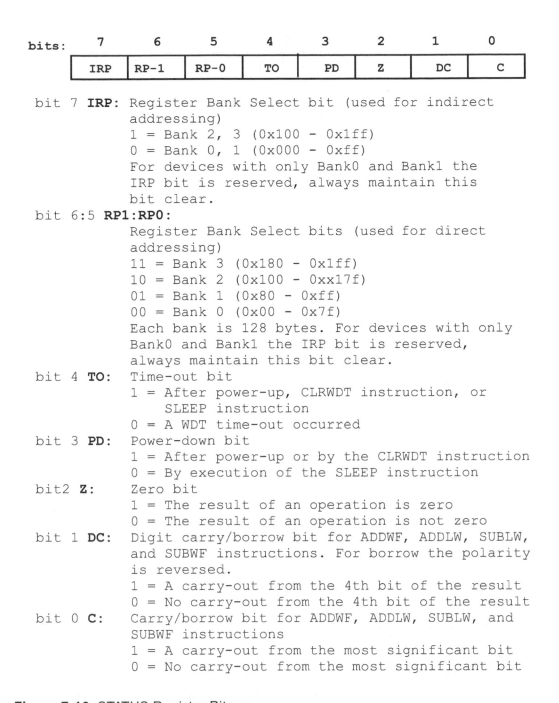

bits: 7   6   5   4   3   2   1   0

| IRP | RP-1 | RP-0 | TO | PD | Z | DC | C |

bit 7 **IRP:** Register Bank Select bit (used for indirect
addressing)
1 = Bank 2, 3 (0x100 - 0x1ff)
0 = Bank 0, 1 (0x000 - 0xff)
For devices with only Bank0 and Bank1 the
IRP bit is reserved, always maintain this
bit clear.

bit 6:5 **RP1:RP0:**
Register Bank Select bits (used for direct
addressing)
11 = Bank 3 (0x180 - 0x1ff)
10 = Bank 2 (0x100 - 0xx17f)
01 = Bank 1 (0x80 - 0xff)
00 = Bank 0 (0x00 - 0x7f)
Each bank is 128 bytes. For devices with only
Bank0 and Bank1 the IRP bit is reserved,
always maintain this bit clear.

bit 4 **TO:** Time-out bit
1 = After power-up, CLRWDT instruction, or
    SLEEP instruction
0 = A WDT time-out occurred

bit 3 **PD:** Power-down bit
1 = After power-up or by the CLRWDT instruction
0 = By execution of the SLEEP instruction

bit2 **Z:** Zero bit
1 = The result of an operation is zero
0 = The result of an operation is not zero

bit 1 **DC:** Digit carry/borrow bit for ADDWF, ADDLW, SUBLW,
and SUBWF instructions. For borrow the polarity
is reversed.
1 = A carry-out from the 4th bit of the result
0 = No carry-out from the 4th bit of the result

bit 0 **C:** Carry/borrow bit for ADDWF, ADDLW, SUBLW, and
SUBWF instructions
1 = A carry-out from the most significant bit
0 = No carry-out from the most significant bit

Figure 7-10 STATUS Register Bitmap.

The OPTION register is actually named the OPTION_REG to avoid a name clash
with the option instruction. The OPTION_REG register contains several bits related
to interrupts, the internal timers, and the Watchdog timer. Figure 7-11, on the following page, is a bitmap of the OPTION_REG register.

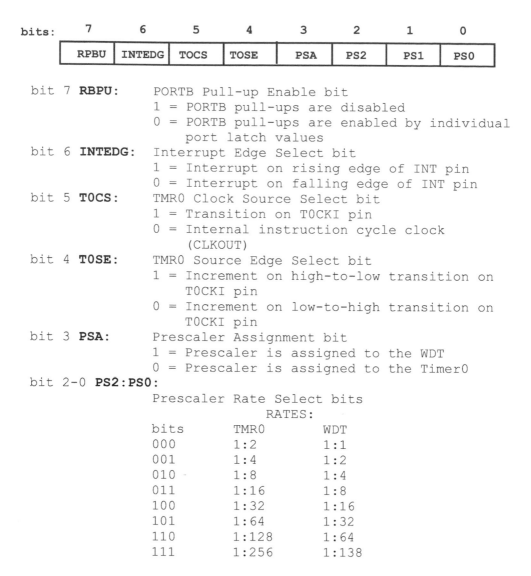

bits:	7	6	5	4	3	2	1	0
	RPBU	INTEDG	TOCS	TOSE	PSA	PS2	PS1	PS0

bit 7 **RBPU:** PORTB Pull-up Enable bit
1 = PORTB pull-ups are disabled
0 = PORTB pull-ups are enabled by individual
port latch values

bit 6 **INTEDG:** Interrupt Edge Select bit
1 = Interrupt on rising edge of INT pin
0 = Interrupt on falling edge of INT pin

bit 5 **T0CS:** TMR0 Clock Source Select bit
1 = Transition on T0CKI pin
0 = Internal instruction cycle clock
(CLKOUT)

bit 4 **T0SE:** TMR0 Source Edge Select bit
1 = Increment on high-to-low transition on
T0CKI pin
0 = Increment on low-to-high transition on
T0CKI pin

bit 3 **PSA:** Prescaler Assignment bit
1 = Prescaler is assigned to the WDT
0 = Prescaler is assigned to the Timer0

bit 2-0 **PS2:PS0:**
Prescaler Rate Select bits
RATES:

bits	TMR0	WDT
000	1:2	1:1
001	1:4	1:2
010	1:8	1:4
011	1:16	1:8
100	1:32	1:16
101	1:64	1:32
110	1:128	1:64
111	1:256	1:138

Figure 7-11 Bitmap of the OPTION_REG Register.

7.6 EEPROM Data Storage

EEPROM (pronounced double-e PROM or e squared PROM) stands for *Electrically-Erasable Programmable Read-Only Memory*. EEPROM memory is used in computers and digital devices as nonvolatile storage. EEPROM is not RAM, as RAM memory is volatile and EEPROM retains its data after power is removed. EEPROM memory is found in USB flash drives and in the nonvolatile storage of several microcontrollers, including many PICs.

The one advantage of EEPROM memory is that it can be erased and written electronically, without removing the chip. The predecessor technology, named EPROM, required that the chip be removed from the circuit and placed under ultraviolet light. EEPROM simplifies the erasing and rewriting process. In the PIC architecture, EEPROM data memory refers to both on-board EEPROM memory and to EEPROM memory ICs as separate circuit components. In general, EEPROM elements are classified according to their electrical interfaces into serial and parallel. Most EEPROM memories used in PICs are serial EEPROMs, also called SEEPROMs. The typical use of serial EEPROM on-board memory and EEPROM on ICs is in the storage of passwords, codes, configuration settings, and other information to be remembered after the system is turned off. For example, a PIC-based security system can use EEPROM memory to store the system password. Because EEPROM can be written, the user can change this password and the new one will also be remembered.

7.6.1 EEPROM in Mid-Range PICs

The mid-range PICs are equipped with EEPROM memory in three possible sizes: 64 bytes, 128 bytes, and 256 bytes. EEPROM memory allows read and write operations. This memory is not mapped into the processor's data or program area, but in a separate block that is addressed through some SFRs. The registers related to EEPROM operations are

- EECON1
- EECON2 (not a physically implemented register)
- EEDAT
- EEADR

EECON1 contains the control bits, and EECON2 is used to initiate the EEPROM read and write operations. The 8-bit data item to be written must first be stored in the EEDATA register, while the address of the location in EEPROM memory is stored in the EEADR register. The EEPROM address space always starts at 0x00 and extends linearly to maximum in the device.

When a write operation is performed, the contents of the EEPROM location is automatically erased. The EEPROM memory used in PICs is rated for high erase/write cycles. EEPROM programming is the topic of Chapter 14.

Chapter 8

Embedded Systems Programming

8.1 Assembly versus High-Level Languages

Some PIC microcontrollers can be programmed in high-level languages or in their native machine language. Machine language programming is facilitated by the use of an assembler program, and thus becomes assembly language programming. Although assembly language is the most used and popular way of PIC programming, some PIC microcontrollers can also be programmed in high-level languages such as C or BASIC.

The major argument in favor of high-level languages is their ease of use and their faster learning curve. The advantages of assembly language, on the other hand, are better control and greater efficiency. Another consideration is that only higher-level PICs can be programmed in high-level languages. The mid-range PICs, including the 16F877, 16F84, and 16F684 covered in this book can only be programmed in assembly language, which settles the programming language issue in this context.

8.1.2 Embedded Systems

At the heart of an embedded system is a microcontroller (such as a PIC), sometimes several of them. These devices are programmed to perform one or, at most, a few tasks. In the most typical case an embedded system also includes one or more "peripheral" circuits that are operated by dedicated ICs or use some functionality contained in the microcontroller itself. The term "embedded system" refers to the fact that the device is often found inside another one; for instance, the control circuit is embedded in a microwave oven. Furthermore, embedded systems do not have (in most cases) general-purpose devices such as hard disk drives, video controllers, printers, and network cards.

A typical embedded system is a control for a microwave oven. In this case the controller includes a timer, so that various operations can be clocked; a temperature sensor that provides information regarding the heat inside the oven; perhaps a motor to optionally rotate a table or tray; a sensor to detect when the oven door is open; and a set of pushbutton switches to select the various operational options. A program running on the embedded microcontroller reads the commands and parameters that are input through the keyboard, programs the timer and the rotating table, detects the state of the door, and turns the heating element on and off as required by the user's selection. Many other daily devices including automobiles, digital cameras, cell phones, and home appliances use embedded systems and many of them are PIC based.

The development process of an embedded system consists of the following steps:

- Define the system specifications. This step includes listing the functions that the system is to perform and determining the tests that will be used to validate their operations.

- Select the system components according to the specifications. This step includes locating the microcontroller that best suits the system as well as the other hardware components.

- Design the system hardware. This includes drawing the circuit diagrams.

- Implement a prototype of the system hardware by means of breadboards, wire wrapping, or another prototyping technology.

- Develop, load, and test the software. Loading software into a PIC is referred to as "blowing" the PIC.

- Implement the final system and test hardware and software.

8.2 Integrated Development Environment

The PIC assembly language development system provided provided by Microchip is named MPLAB. The package is furnished as an *integrated development environment* (IDE) and can be downloaded from the company's website at www.microchip.com. One limitation of the MPLAB package is that it is presently furnished only for the PC. If you are a Mac, Unix, or Linux user, you will not be able to use MPLAB. However, other development systems for Mac and Linux are available on the Web. The MPLAB IDE is intended for software development of embedded systems. The software in an embedded system is usually fixed and cannot be easily changed. For this reason it is called "firmware." The MPLAB development system, or integrated development environment, consists of a system of programs that runs on a PC. This software package is designed to help develop, edit, test, and debug PIC code.

Installing the MPLAB package is straightforward and simple. The package includes the following components:

- MPLAB editor. This tool allows creating and editing the assembly language source code. It behaves very much like any Windows editor and contains the standard editor functions including cut and paste, search and replace, and undo and redo functions.

- MPLAB assembler. The assembler reads the source file produced in the editor and generates either absolute or relocatable code. *Absolute code* executes directly in the PIC. *Relocatable code* can be linked with other separately assembled modules or with libraries.

- MPLAB linker. This component combines modules generated by the assembler with libraries or other object files, into a single executable file in .hex format.

- MPLAB debuggers. Several debuggers are compatible with the MPLAB development system. Debuggers are used to single-step through the code, breakpoint at critical places in the program, and watch variables and registers as the program executes. In addition to being a powerful tool for detecting and fixing program errors, debuggers prove an internal view of the processor, which is a valuable learning tool.

- MPLAB in-circuit emulators. These are development tools that allow performing basic debugging functions while the processor is installed in the circuit.

Figure 8-1 is a screen image of the MPLAB program. The application on the editor window is one of the programs developed later in this book.

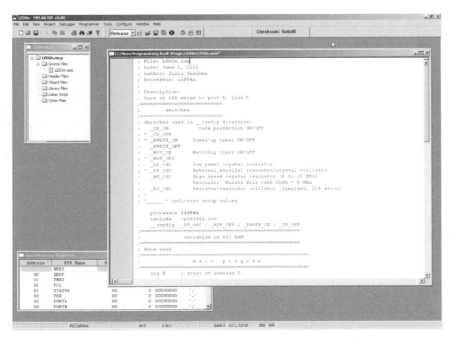

Figure 8-1 Screen Image of the MPLAB IDE.

8.2.1 Installing MPLAB

In the normal installation, the MPLAB executable will be placed in the following path:

C:\Program Files\Microchip\MPASM Suite

Once the development environment is installed, the software is executed by clicking the MPLAB IDE icon or double-clicking the executable file. It is usually a good idea to drag the icon onto the desktop so that the program can be easily activated.

Once the MPLAB software is installed, one should check that the applications were placed in the correct paths and folders. Failure to do so produces assembly-time failure errors with cryptic messages. To check the correct path for the software, open the MPLAB Project menu and select the *Select Language Toolsuite* command. Figure 8-2 shows the command screen in MPLAB version 8.0.

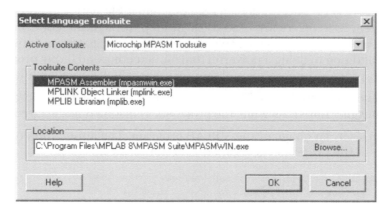

Figure 8-2 MPLAB Select Language Toolsuite Screen.

In the Select Language Toolsuite window, make sure that the file locations for all executables coincide with the actual installation path for the software. If in doubt, use the <Browse> button to navigate through the installation directories until the executable program is located. In Figure 8-2 the location of the mpasmwin.exe is shown. Follow the same process for all the executables in the Microchip MPASM Toolsuite window.

A more detailed control over the location of the various individual tools is provided by the *Set Language Tools Location* command, also in the Project menu. This command allows setting the installation path not only to the major suites, but also to the individual tools. Figure 8-3 shows the display screen of this command.

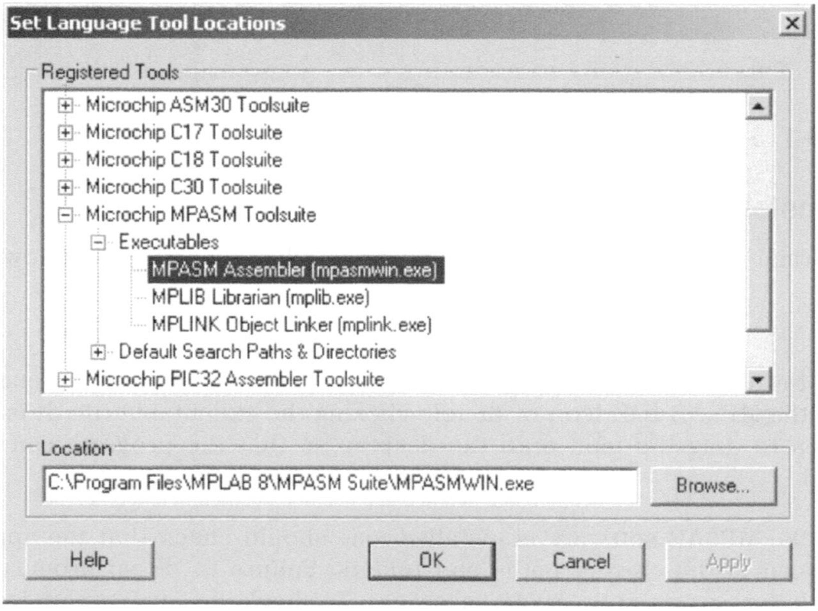

Figure 8-3 MPLAB Set Language Tools Location Screen.

8.2.2 MPLAB Project

In the MPLAB context, a *project* is a group of files generated or recognized by the IDE. Figure 8-4 shows the structure of an assembly language project.

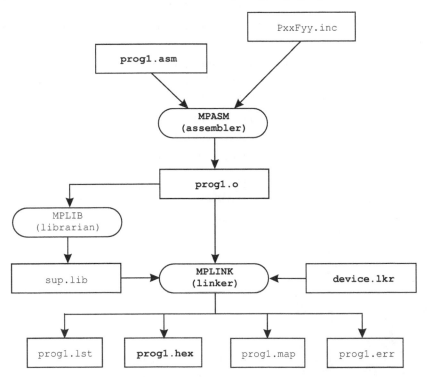

Figure 8-4 MPLAB Project Files.

In Figure 8-4 an assembly language source file (prog1.asm) and an optional processor-specific include file are used by the assembler program (MPASM) to produce an object file (prog1.o). Optionally, other sources and other include files may form part of the project. The resulting object file, as well as one or more optional libraries, and a device-specific script file (device.lkr), are then fed to the linker program (MPLINK), which generates a machine code file (prog1.hex) and several support files with listings, error reports, and map files. It is the .hex file that is used to blow the PIC.

In addition to the files in Figure 8-4, others may also be produced by the development environment according to the selected tools and options. For example, the assembler or the linker can generate a file with the extension .cod, which contains symbols and references used in debugging.

Projects can be created using the <New> command in the Project menu. The programmer then proceeds to configure the project manually and add to it the required files. An alternative option that simplifies creating a new project is to use the <Project Wizard> command in the Project menu. The Wizard will prompt for all the decisions and options that are required, as follows:

1. Device selection. Here the programmer selects the PIC hardware for the project, for example 16F84A.

2. Select Language Toolsuite. This screen is the same one shown in Figure 8-2. Its purpose is to make sure that the proper development tools and paths are active.

3. Next, the Wizard prompts the user for a project name and directory. It is possible to create a new directory at this time.

4. In the next step the user is given the option of adding existing files to the project and renaming these files if necessary. Because most projects reuse a template, an include file, or other preexisting resources, this can be a useful option.

5. Finally, the Wizard displays a summary of the project parameters. When the user clicks on the <Finish> button the project is created and programming can begin.

Figure 8-5 is the display of the final wizard screen.

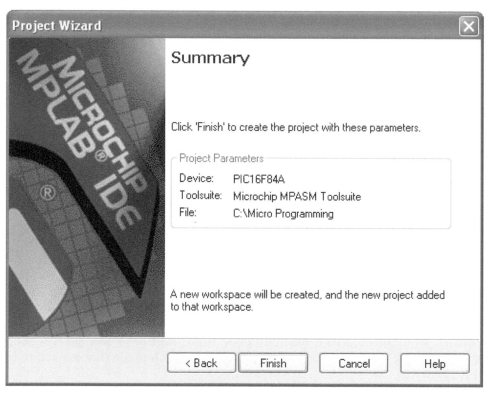

Figure 8-5 Final Screen of the Project Creation Wizard.

8.2.3 Project Build Options

The <Build Options: Project> command in the Project menu allows the user to customize the development environment. Of the tabs available on the Build Options screen, the MPASM Assembler one is probably the most used. The screen is shown in Figure 8-6.

Figure 8-6 MPASM Assembler Tab in the Build Options Screen.

The MPASM Assembler tab allows performing the following customizations:

1. Disable/enable case sensitivity. Normally the assembler is case sensitive. Enabling this option turns all variables and labels to uppercase.

2. Select the Default Radix. Numbers without formatting codes are assumed to be hex, decimal, or octal according to the selected option.

3. The Macro Definition window allows adding macro directives. Macros are discussed later in this chapter.

4. The Use Alternate Settings textbox is provided for command line commands in non-GUI environments.

5. The Restore Defaults box turns off all custom configurations.

8.2.4 Building the Project

Once all the options have been selected, the installation checked, and the assembly language source file written or imported, the development environment can be made to build the project. Building consists of calling the assembler, the linker, and any other support program in order to generate the files shown in Figure 8-4 and any others that may result from a particular project or IDE configuration.

The *build process* is initiated by selecting the <Build All> command in the Project menu. Once the building concludes, a screen labeled Output is displayed showing the results of the build operation. If the build succeeded, the last line of the Output screen will show this result. Figure 8-7 shows the output screen after a successful build.

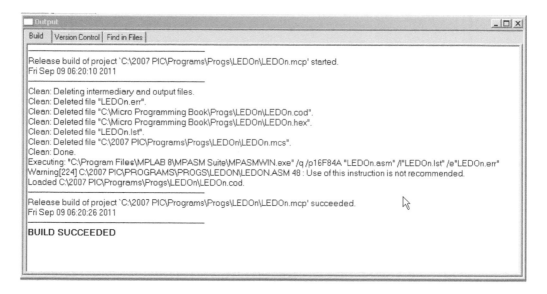

Figure 8-7 Output Window Showing the Build Command Result.

8.2.5 Quickbuild Option

Very often during prototype development we need to create a single executable file without having to deal with the complexity of a full-fledged project. For this case, MPLAB provides the Quickbuild option in the Project menu. The resulting .hex file can be used to blow a PIC in the conventional manner and contains all the necessary debugging information.

We have used the Quickbuild option for most of the programs developed for this book. The typical command sequence for using the Quickbuild option is as follows:

1. Make sure the correct PIC device is selected by using the Configure>Select Device command.

2. Make sure the Quickbuild option is active by selecting Project>Set Active Project>None. This places the environment in the Quickbuild mode.

3. Open an existing assembly language file in the editor window using the File>Open command sequence. If creating a new program from scratch, use the File>New command to open the editor window and type in the assembly code. Then use the File>Save option to save the file with the .asm extension. In any case, the file must have the .asm extension.

4. Select Project>Quickbuild file.asm to assemble your application, where file.asm is the name of your active assembly file.

5. Your application can now be used to blow a PIC or to debug.

8.3 Simulators and Debuggers

In the context of MPLAB documentation, the term *debugger* is reserved for hardware debuggers while the software versions are called *simulators*. Although this distinction is not always enforceable, we will abide by this terminology in order to avoid confusion. The reader should note that there are MPLAB functions in which the IDE considers a simulator as a debugger.

The MPLAB standard simulator is called *MPLAB SIM*. SIM is part of the Integrated Development Environment and can be selected at any time. The hardware debuggers offered by Microchip are named ICD 2, ICE 2000, and ICE 4000. A simulator, as the term implies, allows simulating the execution of a program one instruction at a time; also viewing file registers and symbols defined in the code. Debuggers, on the other hand, allow executing a program one step at a time or to a predefined breakpoint while the PIC is installed in the target system. This makes possible not only viewing the processor's internals, but also testing the circuit components in real-time.

8.3.1 MPLAB SIM

Microchip documentation describes the SIM program as a discrete-event simulator. SIM is part of the MPLAB IDE and is selected by clicking on the <Select Tool> command in the Debugger menu. The command offers several options, one of them being MPLAB SIM. Once the SIM program is selected, a special debug toolbar is displayed. The toolbar and its function is shown in Figure 8-8.

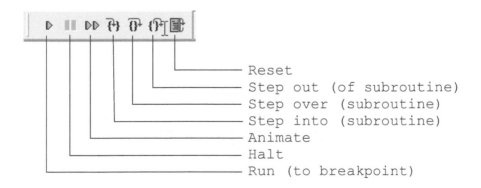

Figure 8-8 SIM Toolbar.

In order for the simulator to work, the program must first be successfully assembled. The most commonly used simulator method is single-stepping through the code and breakpoints. A *breakpoint* is a mark at a program line at which the simulator will stop and wait for user actions.

Breakpoints provide a way of inspecting program results at a particular place in the code. Single-stepping is executing the program one instruction at a time. The three buttons labeled <Step...> in Figure 8-8 are used in single-stepping. The first one allows breaking out of a subroutine or procedure. The second one is for bypassing a procedure or subroutine while in step mode. The third one single-steps into the program instruction that follows.

Breakpoints are set by double-clicking at the desired line while using the editor. The same action removes an existing breakpoint. Lines in which breakpoints have been placed are marked, on the left document margin, by a letter "B" enclosed in a red circle. Right-clicking while the cursor is on the program editor screen provides a context menu with several simulator-related commands. These include commands to set and clear breakpoints, to run to the cursor, and to set the program counter to the code location at the cursor.

The View menu contains several commands that provide useful features during program simulation and debugging. These include commands to program memory, file registers, EEPROM, and special function registers. One command in particular, named <Watch>, provides a way of inspecting the contents of FSRs and GPRs on the same screen. The <Watch> command displays a program window that contains reference to all file registers used by the program. The user then selects which registers to view and these are shown in the Watch window. The Watch window is shown in Figure 8-9.

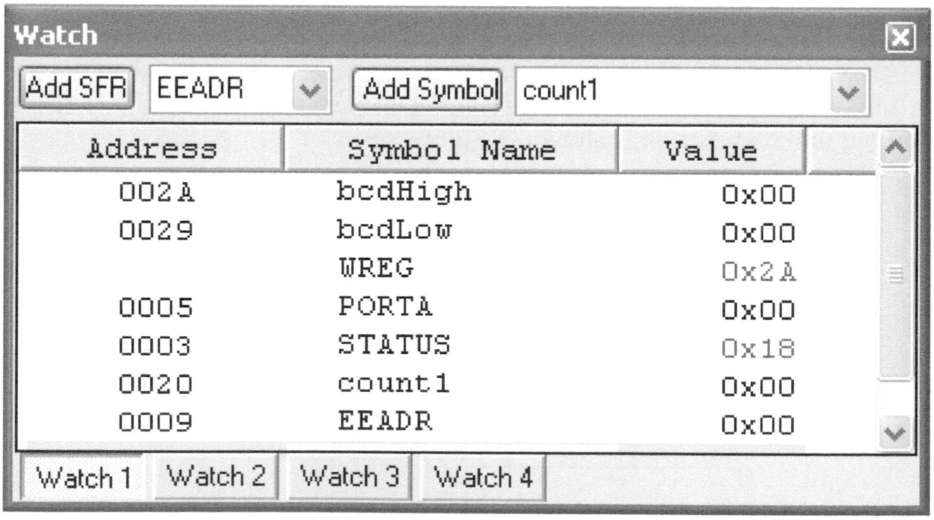

Figure 8-9 Use of Watch Window in MPLAB SIM.

When the program is in the single-step mode or breakpoint mode, the contents of the various registers can be observed in the Watch window. Those that have changed because the last step or breakpoint are displayed in red. The user can click on the corresponding arrows on the Watch window to display all the symbols or registers. The <Add Symbol> or <Add FSR> button is then used to display the item on the Watch screen. Four different Watch windows can be enabled, labeled Watch 1 to Watch 4, at the bottom of the screen in Figure 8-9.

Another valuable tool available from the View menu is the one labeled <Simulator Trace>. The Simulator Trace window provides a view of the current machine instruction combined with a window that displays the source code. The Simulator Trace window is shown in Figure 8-10.

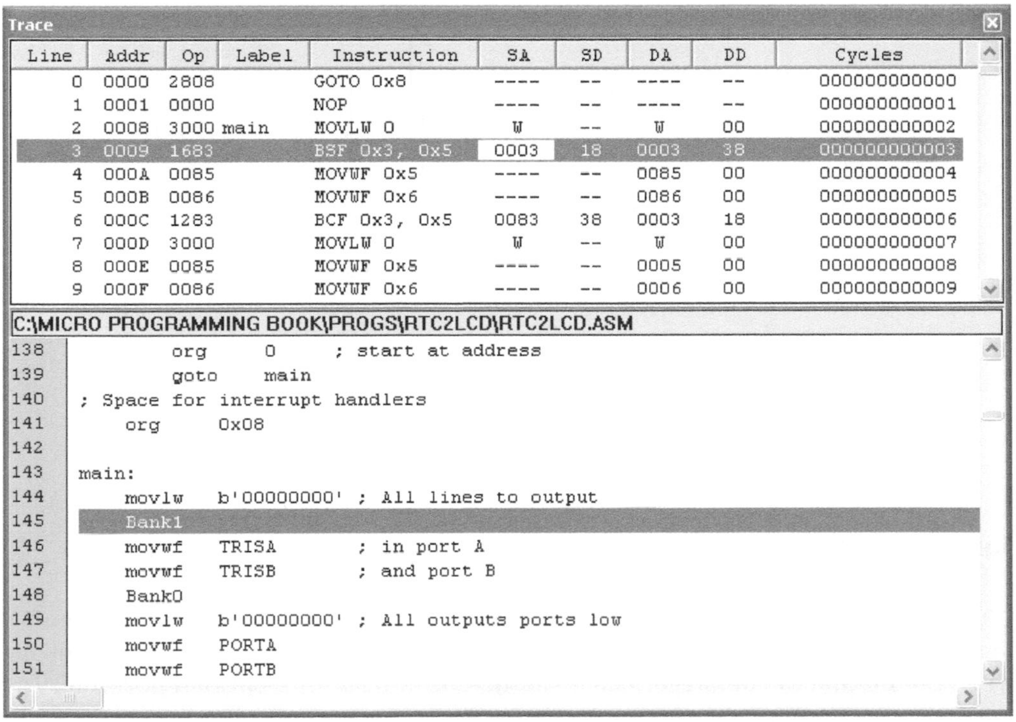

Trace ☒

Line	Addr	Op	Label	Instruction	SA	SD	DA	DD	Cycles
0	0000	2808		GOTO 0x8	----	--	----	--	000000000000
1	0001	0000		NOP	----	--	----	--	000000000001
2	0008	3000	main	MOVLW 0	W	--	W	00	000000000002
3	0009	1683		BSF 0x3, 0x5	0003	18	0003	38	000000000003
4	000A	0085		MOVWF 0x5	----	--	0085	00	000000000004
5	000B	0086		MOVWF 0x6	----	--	0086	00	000000000005
6	000C	1283		BCF 0x3, 0x5	0083	38	0003	18	000000000006
7	000D	3000		MOVLW 0	W	--	W	00	000000000007
8	000E	0085		MOVWF 0x5	----	--	0005	00	000000000008
9	000F	0086		MOVWF 0x6	----	--	0006	00	000000000009

C:\MICRO PROGRAMMING BOOK\PROGS\RTC2LCD\RTC2LCD.ASM

```
138          org      0      ; start at address
139          goto     main
140  ; Space for interrupt handlers
141      org      0x08
142
143  main:
144      movlw    b'00000000' ; All lines to output
145      Bank1
146      movwf    TRISA      ; in port A
147      movwf    TRISB      ; and port B
148      Bank0
149      movlw    b'00000000' ; All outputs ports low
150      movwf    PORTA
151      movwf    PORTB
```

Figure 8-10 MPLAB SIM Simulator Trace Window.

8.3.2 MPLAB Hardware Debuggers

A more powerful and versatile debugging tool is a hardware debugger, sometimes called an in-circuit debugger. Hardware debuggers allow tracing, breakpointing, and single-stepping through code while the PIC is installed in the target circuit. The typical in-circuit debugger requires several hardware components, as shown in Figure 8-11.

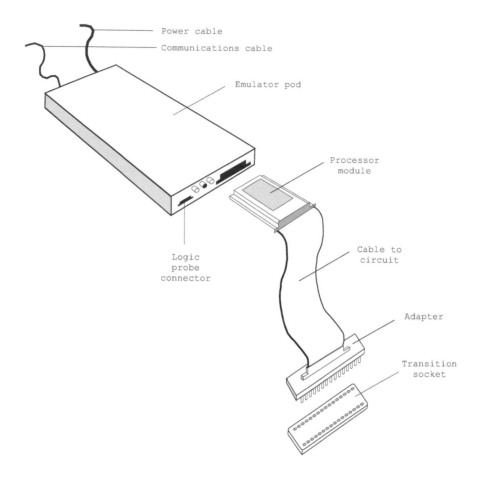

Figure 8-11 Components of a Typical Hardware Debugger.

The *emulator pod* with power supply and communications cable provides the basic communications and functionality of the debugger. The communications line between the PC and the debugger can be an RS-232, a USB, or a parallel port line. The *processor module* fits into a slot at the front of the pod module. The processor is device specific and provides these functions to the debugger. A *flex cable* connects the processor module to an interchangeable device adapter that allows connecting to the several PICs supported by the system. The *transition socket* allows connecting the device adapter to the target hardware. A separate socket allows connecting logic probes to the debugger.

Microchip provides two models of their in-circuit hardware debuggers, which they call *In Circuit Emulators*, or ICE. The ICE 2000 is designed to work with most PICs of the mid-range and lower series, while the ICE 4000 is for the PIC18x high-end family of PICs. Recently Microchip has released an in-circuit debugger designated as ICD 2 that offers many of the features of its full-fledged in-circuit emulators at much reduced prices. One of the disadvantages of the ICD 2 system is that it requires the exclusive use of some hardware and software resources in the target. Furthermore, the ICD requires that the system be fully functional. The ICEs, on the other hand, provide memory and clock so that the processor can run code even if it is not connected to the application board.

8.3.3 Improvised Debugger

The functionality of an actual hardware debugger can be replaced with a little ingenuity and a few lines of code. Most PICs are equipped with EEPROM memory. Programmers (covered in the following section) have the ability to read all the data stored in the PIC, including EEPROM. These two facts can be combined in order to obtain runtime information without resorting to a hardware debugger.

Suppose a defective application is suspect of not finding the expected value in a PIC port. The developer can write a few lines of code to store the port value on an EEPROM memory cell. An endless loop following this operation ensures that the stored value is not changed. Now the PIC is inserted in the circuit and the application executed. When the endless loop is reached, the PIC is removed from the circuit and placed back in the programmer. The value stored in EEPROM can now be inspected so as to determine the runtime state of the machine. In many cases this simple trick is less complicated and time consuming than setting up a hardware debugger, even if such a device is available.

8.4 Programmers

In the context of microcontroller technology, a *programmer* is a device that allows transferring the program onto the chip. The process is called "burning" a PIC, or more commonly "blowing" a PIC. Most programmers have three components:

1. A software package that runs on the PC

2. A cable connecting the PC to the programmer

3. A programmer device

Dozens of PIC programmers are available on the Internet. By releasing the programming specifications of the PIC to the public without requiring a nondisclosure agreement, Microchip originated a "cottage industry." The commercial programmers on the Internet range from a "no parts" PIC programmer that has been around because 1998, to sophisticated devices costing hundreds of dollars and providing many additional features and refinements. For the average new PIC user, a nice USB programmer with a ZIF socket and the required software can be purchased for about $50.00. Build-it-yourself versions are available for about half this amount.

An alternative programmer is made possible by the fact that some of the newer flash-based PICs can write to their own program memory. This allows placing a small *bootloader program* in PIC memory that will load an application over the RS-232 or USB lines.

Figure 8-12 is a screen capture of the driver software for a popular programmer from MicroPro.

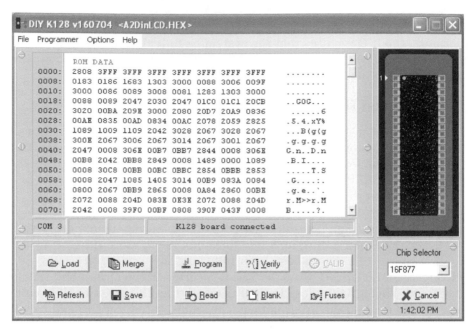

Figure 8-12 Control Program for the DIY MicroPro Programmer.

8.5 Engineering PIC Software

The program developer's main challenge is writing code that performs the task at hand. In this context this means writing a PIC assembly language program that assembles without errors (usually after some effort) and makes the circuit perform as intended. We have already reviewed the development environment (IDE) and the various hardware components and software tools. We now focus on the various elements that are used in developing the program itself.

8.5.1 Using Program Comments

One of the first realizations of novice programmers is how quickly we forget the reasoning and the logic that went into our code. It is common that a few weeks, even a few hours, after we coded a routine, we find that what was obvious then is now undecipherable and that the ideas that were clear in our minds a short time ago now evade our understanding. The only solution to the volatility of logic is to write good program comments that explain, not the elementary, but the trains of thought behind our code.

In PIC assembly language, the comment symbol is the semicolon (;). The presence of a semicolon indicates to the assembler that everything that follows, to the end of the line, must be ignored. Using comments judiciously and with good taste is the mark of the expert software engineer. Programs with few, cryptic, or confusing comments fall into the category of "spaghetti code." In the programming lingo, "spaghetti code" refers to a coding style that cannot be deciphered or understood. The worst that can be said about one's programming style is that one produces spaghetti code.

How we use comments to explain our code or even to decorate it is a matter of personal preference. However, there are certain common-sense rules that should always be considered:

- Do not use program comments to explain the programming language or reflect on the obvious.

- Abstain from humor in comments. Comedy has a place in the world but it is not in programs. By the same token, stay away from vulgarity, racist or sexist remarks, and anything that could be offensive. You can never anticipate who will read your code.

- Write short, clear, readable comments that explain how the program works. Decorate or embellish your code using ASCII graphics, according to your taste.

Program Header

Every program should have a commented header that contains, at least, the following information:

- Program name

- Programmer's or software company's name

- Copyright notice, if pertinent

- Target device or hardware

- Development environment

- Development dates

- Program description

Some of these elements allow various levels of detail. For example, the target device can be a simple reference to the PIC for which the program is written, a more-or-less detailed description of the target system, or a reference to a circuit diagram or board drawing. The development environment can also be described briefly or in abundant detail. The dates entry can be a single item that lists the first or the last program change, or a detailed description of all program changes, tests, and updates. The program description can be a short sentence or a mini-manual on using the application. In any case, the level of detail and the contents of each category are determined by the programmer's style, and the complexity and purpose of the application.

The following lines show the header of one of the programs developed for this book.

```
; File name: RTC2LCD.asm
; Last Update: June 6, 2011
; Authors: Sanchez and Canton
; Processor: 16F84A
;
; Dscription:
; Program to demonstrate use of the NJU6355 Real Time Clock
; IC. Program uses LCD to display results in hours, minutes,
; and seconds, as follows:
;
```

```
; Top LCD line:   H:xx M:yy S:zz
;
; Initialization values are in #define statements that start
; with i, such as iYear, iMonth, etc.
;
; For LCD display parameters see the LCDTest2 program.
;
; WARNING:
; Code assumes 4Mhz clock. Delay routines must be
; edited for faster clocke
```

Commented Banners

Very often we need to scroll through the code in search of a particular line or routine. Having banners that signal critical places in the program facilitate this search. Banners are created using comments and a framing symbol, as in the following code fragment:

```
;==================================
;   first text string procedure
;==================================
storeMS1:
; Procedure to store in PIC RAM buffer the message
; contained in the code area labeled msg1
; ON ENTRY:
;         variable pic_ad holds address of text buffer
;         in PIC RAM
;         w register hold offset into storage area
;         msg1 is routine that returns the string characters
;         an a zero terminator
;         index is local variable that hold offset into
;         text table. This variable is also used for
;         temporary storage of offset into buffer
; ON EXIT:
;         Text message stored in buffer
;
```

Sometimes the programmer needs to emphasize a program area with a large banner that extends from margin to margin, as follows:

```
;==============================================================
;==============================================================
;                 L O C A L    P R O C E D U R E S
;==============================================================
;==============================================================
;=========================
; init LCD for 4-bit mode
;=========================
initLCD:
; Initialization for Densitron LCD module as follows:
;    4-bit interface
;    2 display lines of 20 characters each
;    cursor on
;    left-to-right increment
;    cursor shift right
;    no display shift
```

Commented Bitmaps

It is also possible to use comments to signal the function of bit fields and individual bits of an operand, as in the following code fragment:

```
; OPTION_REG bitmap
;   7  6  5  4  3  2  1  0 <= OPTION bits
;   |  |  |  |  |  |  |__|__|_____ PS2-PS0 (prescaler bits)
;   |  |  |  |  |  |              Values for Timer0
;   |  |  |  |  |  |              000 = 1:2    001 = 1:4
;   |  |  |  |  |  |              010 = 1:8    011 = 1:16
;   |  |  |  |  |  |              100 = 1:32   101 = 1:64
;   |  |  |  |  |  |              110 = 1:128 *111 = 1:256
;   |  |  |  |  |  |_____ PSA (prescaler assign)
;   |  |  |  |  |               *1 = to WDT
;   |  |  |  |  |                0 = to Timer0
;   |  |  |  |  |_____ TOSE (Timer0 edge select)
;   |  |  |  |                  *0 = increment on low-to-high
;   |  |  |  |                   1 = increment in high-to-low
;   |  |  |_____ TOCS (TMR0 clock source)
;   |  |  |                 *0 = internal clock
;   |  |  |                  1 = RA4/TOCKI bit source
;   |  |_____ INTEDG (Edge select)
;   |                           *0 = falling edge
;   |_____ RBPU (Pullup enable)
;                                    *0 = enabled
;                                     1 = disabled
; * indicates options selected
        movlw           b'00001000' ; Value installed
        movwf           OPTION_REG
```

Clearly commented bitmaps, banners, and many other code embellishments do not add to the quality and functionality of the code. It is quite possible to write very sober and functional programs without using these gimmicks. The decision of how to comment and how much to decorate programs is one of style.

8.5.2 Defining Data Elements

Most programs will require the use of general-purpose file registers. These registers are allocated to memory addresses reserved for this purpose in the PIC architecture. Because the areas at these memory locations are already reserved for use as GPRs, the program can access the location either by address or by assigning a name to the address. The **equ** (equate) directory performs this function, as follows:

```
var1      equ      0x0c     ; Name var1 is assigned to location 0x0c
```

Actually the name (in this case var1) becomes an alias for the memory address to which it is linked. From that point on the program accesses the same variable. Program code can access the memory cell at address 0x0c as follows:

```
    movf      var1,w   ; Contents of var1 to w
or
    movf      0x0c,w   ; Same variable to w
```

In addition to the **equ** directive, PIC assembly language recognizes the C-like **#define** directive, so the name assignation could have been done as follows:

```
#define var1 0x0c
```

Although named variables are to be preferred over hard-coded addresses, there are times when we need to access an internal element of some multi-byte structure. In these cases the hard-coded form could be convenient, although not absolutely necessary.

cblock Directive

Another way of defining memory data is by using one of the data directives available in PIC assembly language. Although there are several of these, perhaps the most useful is the **cblock** directive. The **cblock** directive specifies an address for the first item and other items listed are allocated from this first address. The group ends with the **endc** directive. The following code fragment shows the use of the **cblock/endc** directives:

```
; Reserve 20 bytes for string buffer
    cblock     0x20
    strData
    endc

; Reserve three bytes for ASCII digits
    cblock     0x34
    asc100
    asc10
    asc1
    endc
```

The **cblock** directive actually defines a group of constants that are assigned consecutive addresses in RAM. In the previous code fragment the allocation of 20 bytes for the buffer named strData is illusory because no memory is actually reserved. The illusion works because the second **cblock** starts at address 0x34, which is 20 bytes after strData; and also because the programmer will abstain from allocating other variables in the buffer space.

8.5.3 Banking Techniques

Having to deal with memory banks is one of the aggravations of PIC programming. Banks are designated starting with bank 0. All PICs of the mid-range family have at least two banks, so bank shifting operations are virtually unavoidable. The issue is more how to switch banks designated because there are several possible techniques.

Bank selection is by means of bit RP0 and RP1 in the STATUS register. In mid-range PICs with four banks the various combinations are as shown in Table 8.1.

Table 8.1

STATUS Register Bank Selection Bits

RP1	RP0	BANK	ADDRESS RANGE
1	1	*Bank 3	0x180 - 0x1ff
1	0	*Bank 2	0x100 - 0xx17f
0	1	Bank 1	0x80 - 0xff
0	0	Bank 0	0x00 - 0x7f

* RP1 bit is not used in devices with two banks.

The most direct way to select the current bank is by clearing or setting the corresponding bits in the STATUS register. For example, to select bank 2 in a four-bank device, you could code:

```
bsf  STATUS,6 ; Set bit 6 in STATUS register
bcf  STATUS,5 ; Clear bit 5
```

banksel Directive

Alternatively, the application can use the **banksel** directive which selects the bank in which a particular register is located. For example, to select the bank in which the ADCON1 register is located, code could be as follows:

```
banksel   ADCON1
```

The banksel directive also works with registers defined by the user (GPRs).

Bank Selection Macros

An alternative way of performing bank selection is by coding *bank select macros*. A macro is an assembler structure that allows defining a series of instructions that is inserted in the code every time the macro is referenced. The PIC macro language defines the following format:

```
label macro [arg1, arg2... argn]
   .
   .
endm
```

In this example the ellipses serve as placeholders for PIC instructions, assembler directives, macro directives, or macro calls. Macros are usually defined at the beginning of the program because forward references to macros are not allowed. The optional arguments passed to the macro (arg1, arg2, etc.) are assigned values when the macro is invoked. For example, the following macros make the corresponding bank selections in a mid-range PIC with four banks:

```
; Macros to select the register banks
Bank0    MACRO              ; Select RAM bank 0
         bcf  STATUS,RP0
         bcf  STATUS,RP1
         ENDM

Bank1    MACRO              ; Select RAM bank 1
         bsf  STATUS,RP0
         bcf  STATUS,RP1
         ENDM

Bank2    MACRO              ; Select RAM bank 2
         bcf  STATUS,RP0
         bsf  STATUS,RP1
         ENDM

Bank3    MACRO              ; Select RAM bank 3
         bsf  STATUS,RP0
         bsf  STATUS,RP1
         ENDM
```

Once the bank switching macros have been defined, the application can change banks simply by calling the macro name; for example, if we know that the ADCON1 register is in bank 1 we can select the bank by calling

```
     Bank1
```

At this point in the code, the macro expansion will insert the corresponding operations to make the switch.

Which method to use when switching banks is a matter of personal preference and program constraints. Setting and clearing the RP1/RP0 bits is simple enough but can be error-prone. Using the banksel directive is convenient because we do not need to know in which bank the item is located. The objection to using banksel is that some unnecessary bank changes may take place. For example, if the program is already in bank 1 and the banksel directive appears with a register file in that same bank, then the bank switching code is generated anyway. The use of bank selection macros seems like a suitable method for most conditions. One advantage of the macro approach is that programs for different PICs can have their own banking macros. In this way code can be easily ported to different architectures.

Deprecated Banking Instructions

Several instructions in the mid-range instruction set have been deprecated and are no longer recommended by Microchip; these include **tris** and **option**. Microchip's reason for not recommending these instructions is to maintain compatibility with future mid-range products. From a programmer's viewpoint, it is difficult to see why using these instructions may be undesirable. In the unlikely case that our code will be ported to some other future device that does not support **tris** or **option**, then it will be easy enough to modify the code at that time.

The **tris** and **option** instructions are convenient because they allow loading the contents of the w register to the OPTION, TRISA, and TRISB registers directly, without bank concerns. For example, the following code fragment sets port line 1 to input and all others to output:

```
     movlw    b'00000010'    ; Line 1 is input
     tris     PORTA
```

We continue to use the deprecated instructions in programs in which there is no concern about future consequences. In programs in which portability is an issue, we use the banking macros discussed previously.

8.5.4 Processor and Configuration Controls

PIC programs must define the processor to be used by the development software. The **assembler** processor directive (and also the **list** directive) allow defining the PIC type. For example, a program for the 16F877 would contain the following line:

```
processor 16f877
```

Configuration Bits

The PIC microcontrollers contain a special register called the *configuration registe*r that allows customizing certain processor features to the needs of the application. In the mid-range PICs, the bits in the configuration register are mapped in program memory location 0x2007. This memory location can only be accessed during the programming mode, so the bits cannot be changed during normal program operation. The configuration bits cannot be read at runtime.

Microchip recommends that the configuration bit be set by means of the __**config** directive. The bits are mapped as follows:

```
CP1:CP0: Code Protection bits

    11 = Code protection off
    10 = See device data sheet
    01 = See device data sheet
    00 = All memory is code protected
```

Some devices use more or less bits to determine the level of code protection. Some use a single bit for this purpose. In this case the encoding is as follows:

```
    1 = Code protection off
    0 = Code protection on
```

DP: Data EEPROM Memory Code Protection bit

```
    1 = Code protection off
    0 = Data EEPROM Memory is code protected
```

BODEN: Brown-out Reset Enable bit

```
    1 = BOR enabled
    0 = BOR disabled
```

Enabling Brown-out Reset automatically enables the Power-up Timer (PWRT) regardless of the value of bit PWRTE. Ensure that the Power-up Timer is enabled anytime that Brown-out Reset is enabled.

PWRTE: Power-up Timer Enable bit

```
    1 = PWRT disabled
    0 = PWRT enabled
```

MCLRE: MCLR Pin Function Select bit

```
    1 = Pin's function is MCLR
    0 = Pin's function is as a digital I/O.
        MCLR is internally tied to VDD.
```

WDTE: Watchdog Timer Enable bit

```
    1 = WDT enabled
    0 = WDT disabled
```

FOSC1:FOSC0: Oscillator Selection bits

```
11 = RC oscillator
10 = HS oscillator
01 = XT oscillator
00 = LP oscillator
```

FOSC2:FOSC0: Oscillator Selection bits

```
111 = EXTRC oscillator, with CLKOUT
110 = EXTRC oscillator
101 = INTRC oscillator, with CLKOUT
100 = INTRC oscillator
011 = Reserved
010 = HS oscillator
001 = XT oscillator
000 = LP oscillator
```

The __**config** directive is used to embed configuration data in the source file. Alternatively, the configuration bits can be set at the time the PIC is blown. The following code fragment shows setting the configuration bits for a 16F877 PIC:

```
; Switches used in __config directive:
;   _CP_ON          Code protection ON/OFF
; * _CP_OFF
; * _PWRTE_ON       Power-up timer ON/OFF
;   _PWRTE_OFF
;   _BODEN_ON       Brown-out reset enable ON/OFF
; * _BODEN_OFF
; * _PWRTE_ON       Power-up timer enable ON/OFF
;   _PWRTE_OFF
;   _WDT_ON         Watchdog timer ON/OFF
; * _WDT_OFF
;   _LPV_ON         Low voltage IC programming enable ON/OFF
; * _LPV_OFF
;   _CPD_ON         Data EE memory code protection ON/OFF
; * _CPD_OFF
; OSCILLATOR CONFIGURATIONS:
;   _LP_OSC         Low power crystal occilator
;   _XT_OSC         External parallel resonator/crystal ocillator
; * _HS_OSC         High speed crystal resonator
;   _RC_OSC         Resistor/capacitor ocillator
; |                 (simplest, 20% error)
; |
; |_____ * indicates setup values presently selected
;
      __CONFIG _CP_OFF & _WDT_OFF & _BODEN_OFF & _PWRTE_ON & _HS_OSC &
_WDT_OFF & _LVP_OFF & _CPD_OFF
```

8.5.5 Naming Conventions

One of the style issues that the programmer must decide concerns the conventions followed for program labels and variable (register) names. The MPLAB assembler is case sensitive by default, so PORTB and portb can refer to different registers.

The programmer can define all the registers (SFRs and GPRs) used by an application using **equ** or **#define** directives. A safer approach is to import an include file (.inc extension) furnished in the MPALB package for each different PIC. The include files have the names of all SFRs and bits used by a particular device. The following code fragment is a listing of the MPLAB include file for the 16f84a:

```
        LIST
; P16F84A.INC  Standard Header File, Version 2.00
; Microchip Technology, Inc.
        NOLIST

; This header file defines configurations, registers, and other
; useful bits of information for the PIC16F84 microcontroller.
; These names are taken to match  the data sheets as closely as
; possible.
; Note that the processor must be selected before this file is
; included.  The processor may be selected the following ways:
;       1. Command line switch:
;               C:\ MPASM MYFILE.ASM /PIC16F84A
;       2. LIST directive in the source file
;               LIST   P=PIC16F84A
;       3. Processor Type entry in the MPASM full-screen interface
;==================================================================
;
;       Revision History
;
;==================================================================

;Rev:   Date:    Reason:

;1.00   2/15/99 Initial Release

;==================================================================
;
;       Verify Processor
;
;==================================================================

        IFNDEF __16F84A
            MESSG "Processor-header file mismatch.  Verify selected
 processor."
        ENDIF

;==================================================================
;
;       Register Definitions
;
;==================================================================

W               EQU     H'0000'
F               EQU     H'0001'

;- Register Files -

INDF            EQU     H'0000'
TMR0            EQU     H'0001'
PCL             EQU     H'0002'
STATUS          EQU     H'0003'
FSR             EQU     H'0004'
PORTA           EQU     H'0005'
PORTB           EQU     H'0006'
EEDATA          EQU     H'0008'
EEADR           EQU     H'0009'
```

```
        PCLATH          EQU     H'000A'
        INTCON          EQU     H'000B'

        OPTION_REG      EQU     H'0081'
        TRISA           EQU     H'0085'
        TRISB           EQU     H'0086'
        EECON1          EQU     H'0088'
        EECON2          EQU     H'0089'

;- STATUS Bits -

        IRP             EQU     H'0007'
        RP1             EQU     H'0006'
        RP0             EQU     H'0005'
        NOT_TO          EQU     H'0004'
        NOT_PD          EQU     H'0003'
        Z               EQU     H'0002'
        DC              EQU     H'0001'
        C               EQU     H'0000'


;- INTCON Bits -

        GIE             EQU     H'0007'
        EEIE            EQU     H'0006'
        T0IE            EQU     H'0005'
        INTE            EQU     H'0004'
        RBIE            EQU     H'0003'
        T0IF            EQU     H'0002'
        INTF            EQU     H'0001'
        RBIF            EQU     H'0000'

;- OPTION_REG Bits -

        NOT_RBPU        EQU     H'0007'
        INTEDG          EQU     H'0006'
        T0CS            EQU     H'0005'
        T0SE            EQU     H'0004'
        PSA             EQU     H'0003'
        PS2             EQU     H'0002'
        PS1             EQU     H'0001'
        PS0             EQU     H'0000'

;- EECON1 Bits -

        EEIF            EQU     H'0004'
        WRERR           EQU     H'0003'
        WREN            EQU     H'0002'
        WR              EQU     H'0001'
        RD              EQU     H'0000'

;=====================================================================
;
;       RAM Definition
;
;=====================================================================

        __MAXRAM H'CF'
        __BADRAM H'07', H'50'-H'7F', H'87'
```

```
;===================================================================
;
;          Configuration Bits
;
;===================================================================
_CP_ON          EQU     H'000F'
_CP_OFF         EQU     H'3FFF'
_PWRTE_ON       EQU     H'3FF7'
_PWRTE_OFF      EQU     H'3FFF'
_WDT_ON         EQU     H'3FFF'
_WDT_OFF        EQU     H'3FFB'
_LP_OSC         EQU     H'3FFC'
_XT_OSC         EQU     H'3FFD'
_HS_OSC         EQU     H'3FFE'
_RC_OSC         EQU     H'3FFF'
```

Note that all names in the include file are defined in all-capital letters. It is probably a good idea to adhere to this style instead of creating alternate names in lower case. The C-like **#include** directive is used to refer to the .inc files at assembly time; for example

```
#include <p16f84a.inc>
```

8.5.6 Errorlevel Directive

This directive allows controlling the warning and error messages produced at assembly and link times. One particular type of warning can be disturbing: those that refer to bank changes. Applications often turn off bank change-related warnings with the following line:

```
        errorlevel -302
```

8.6 Pseudo Instructions

It is sometimes disturbing to read in a code listing instructions that are not part of the standard set for the particular device. The reason this happens is that MPLAB includes a set of *pseudo instructions* for 12- and 14-bit devices. A list of all supported pseudo instructions can be found in the MPLAB documentation. In our programming we prefer not to use them because they tend to make code less readable. Incidentally, Microchip also recommends not using the pseudo instructions.

Chapter 9

I/O Circuits and Programs

9.1 Simple Input and Output

In this chapter we introduce embedded circuits and programming operations. We start with simple input and output devices that are controlled by one of the most basic PICs of the mid-range family (the 16F84A). Although using a PIC to control an LED or to read a switch is as elementary as it gets, these operations are by no means trivial to the beginner because they require building a working circuit and developing and installing the corresponding PIC program. Later in the chapter we progress to more complex input/output devices, including the Seven-Segment LED display and circuits with multiple switches. One of the circuits uses a bank of multiple LEDs to function as a binary output device and reads four toggle switches.

9.1.1 16F84A Programming Template

We have found that program development can be considerably simplified using code templates. A code template is a program devoid of functionality that serves to implement the most common and typical features of an application. The template not only saves the effort of re-doing the same tasks, but also reminds the programmer of program elements that could otherwise be forgotten. A professional developer will have collected many different templates over the years, for different types of applications, on various processors. The following template is for the 16F84A PIC:

```
;=============================================================
; File name:
; Date:
; Author:
; Processor:
; Reference circuit:
;=============================================================
; Copyright notice:
;=============================================================
; Program Description:
;
;===========================
; configuration switches
;===========================
```

```
; Switches used in __config directive:
;   _CP_ON          Code protection ON/OFF
; * _CP_OFF
; * _PWRTE_ON       Power-up timer ON/OFF
;   _PWRTE_OFF
;   _WDT_ON         Watchdog timer ON/OFF
; * _WDT_OFF
;   _LP_OSC         Low power crystal occilator
; * _XT_OSC         External parallel resonator
;   _HS_OSC         High speed crystal resonator (8 to 10 MHz)
;   _RC_OSC         Resistor/capacitor ocillator
;                   (simplest, 20% error)
; |
; |_____ * indicates setup values

;=========================
; setup and configuration
;=========================
      processor 16f84A
      include    <p16f84A.inc>
      __config   _XT_OSC & _WDT_OFF & _PWRTE_ON & _CP_OFF


;========================================================
;                   constant definitions
;========================================================
;========================================================
;                   PIC register equates
;========================================================
;========================================================
;                 variables in PIC RAM
;========================================================
      cblock    0x0c
      endc


;=========================================================
;                          program
;=========================================================
        org       0       ; start at address
        goto      main
; Space for interrupt handlers
      org       0x08

main:

;=========================================================
      end       ; END OF PROGRAM
;=========================================================
```

In addition to the template file, the program developer should keep at hand the necessary include files. In this case, p16f84a.inc.

9.2 Template Circuits

As the programmer can use a programming template for developing 16F84A code, the circuit designer can use a circuit template. Very often it is possible to build a new circuit by combining components and sections of existing templates. In the discussion of simple input and output circuits, we also introduce some basic circuit templates for the 16F84A PIC.

9.2.1 MCLR and Oscillator Template

Most 16F84 circuits require handling the MCLR (master clear) pin and the connections for an external oscillator. This basic circuit should also include a diagram of the PIC itself with the pin out, as well as the wiring of the standard components such as the power and ground. Figure 9-1 shows a circuit template for the 16F84A.

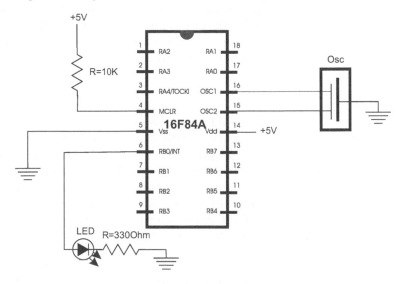

Figure 9-1 16F84A MCLR and Oscillator Circuit Template.

The circuit in Figure 9-1 assumes that the oscillator is an external parallel resonator. The circuit template in Figure 9-1 will not suit every possible circuit. Even the simplest components must sometimes be configured differently; for example, the reset line could be wired to a pushbutton switch, or a different oscillator may be used. In any case, it is always easier to make modifications to an existing diagram than to start from scratch every time.

9.2.2 Power Supplies

Every mid-range PIC-based circuit board requires a +5V power source. One possible source of power is one or more batteries. There is an enormous selection of battery types, sizes, and qualities. The most common ones for use in experimental circuits are listed in Table 9.1.

Table 9.1

Common Dry Cell Alkaline Battery Types

DESIGNATION	VOLTS	LENGTH MM.	DIAMETER MM.
D	1.5	61.5	34.2
C	1.5	50	26.2
AA	1.5	50	14.2
AAA	1.5	44.5	10.5
AAAA	1.5	42.5	8.3

All the batteries in Table 9.1 produce 1.5 volts. This means that for a PIC with a supply voltage from 2 to 6 volts, two to four batteries will be adequate. Note that in selecting the battery power source for a PIC-based circuit other elements, in addition to the microcontroller itself must be considered, such as the oscillator. Holders for several interconnected batteries are available at electronic supply sources.

Alternatively, the power supply can be a transformer with 120VAC input and 3 to 12VDC. These are usually called AC/DC or plug-in adapters or "wall wart." The most useful type for the experimenter are the ones with an ON/OFF switch and several selectable output voltages. Color-coded alligator clips at the output wires are also a convenience.

Voltage Regulator

A useful device for a typical PIC-based power source is a *voltage regulator IC*. The 7805 voltage regulator is ubiquitous in many PIC-based boards with AC/DC adapters. The IC is a three-pin device whose purpose is to ensure a stable voltage source that does not exceed the device rating. The 7805 is rated for 5V and will produce this output from any input source in the range 8 to 35V. Because the excess voltage is dissipated as heat, the 7805 is equipped with a metallic plate intended for attaching a heat sink. The heat sink is not required in a typical PIC application but it is a good idea to maintain the supply voltage closer to the device minimum rather than to its maximum.

The voltage regulator circuit also requires two capacitors: one electrolytic and the other one not. Figure 9-2 shows a power source circuit using the 7805.

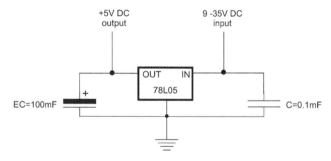

Figure 9-2 Voltage Stabilizer Circuit.

To simplify the schematics, the voltage regulator and power supply components are often not included in the circuit diagram. In this case the power input lines in the circuit are labeled +5V and the corresponding return lines with the conventional ground symbol, as shown in Figure 9.1.

9.3 Simple Circuits and Programs

In the following sub-sections we describe very simple PIC-based circuits that can be assembled with very few components using a breadboard. The corresponding programs exercise the circuit components. The beginner should not skip building these circuits and coding the programs because they demonstrate essential hardware and software elements.

As a learning experience it is a good idea to reverse-engineer the code in these sample programs. With the processor's instruction set at hand, listed in Appendix F, proceed to follow the code listing one instruction at a time until you can understand every processing detail. This is also a good opportunity to learn to use the MPLAB environment and the symbolic debugger (MPLAB SIM) discussed in Chapter 8. The fundamental exercises for using the debugger consist of inserting breakpoints in the code, running the program to the breakpoint, stepping through and stepping over subroutines, and inspecting variables and PIC registers.

9.3.1 Single LED Circuit

One of the simplest circuits consists of a single LED lamp wired to port B, line 0, of a 16F84A PIC, as shown in Figure 9-3.

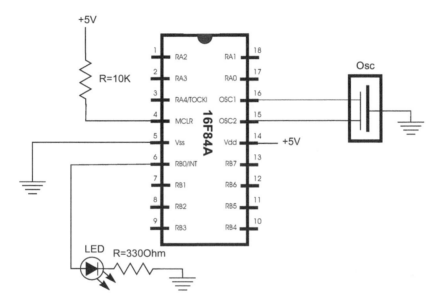

Figure 9-3 Simple LED Circuit.

Note that the power source for the circuit in Figure 9-3 is not shown in the circuit schematic. Typically, a battery source or an AC/DC converter and a voltage stabilizer circuit as the one in Figure 9-2 would be used.

A program to turn on the LED on port B, line 0, requires few but essential processing operations. Code must perform the following operations:

- Define and select processor (in this case, 16F84A).
- Link-in the corresponding include file (p16f84A.inc).
- Select the oscillator type (in this case, an external resonator, _XT type).
- Direct execution to the main label.
- Initialize port B for output.
- Set line 0 in port B high.

The entire program is as follows:

```
; File: LEDOn.asm
; Date: June 1, 2011
; Authors: Canton and Sanchez
; Processor: 16F84A
;
; Description:
; Turn on LED wired to port B, line 0
;==========================
;         switches
;==========================
; Switches used in __config directive:
;   _CP_ON          Code protection ON/OFF
; * _CP_OFF
; * _PWRTE_ON       Power-up timer ON/OFF
;   _PWRTE_OFF
;   _WDT_ON         Watchdog timer ON/OFF
; * _WDT_OFF
;   _LP_OSC         Low power crystal oscillator
; * _XT_OSC         External parallel resonator/crystal oscillator
;   _HS_OSC         High speed crystal resonator (8 to 10 MHz)
;                   Resonator: Murate Erie CSA8.00MG = 8 MHz
;   _RC_OSC         Resistor/capacitor oscillator (simplest)
; |
; |_____* indicates setup values
    processor 16f84A
    include   <p16f84A.inc>
    __config  _XT_OSC & _WDT_OFF & _PWRTE_ON & _CP_OFF
;=====================================================
;              variables in PIC RAM
;=====================================================
; None used
;=========================================================
;                m a i n   p r o g r a m
;=========================================================
    org       0       ; start at address 0
    goto      main
;==========================
; space for interrupt handler
;==========================
    org       0x04
;==========================
;      main program
;==========================
main:
; Initialize all lines in port B for output
    movlw     B'00000000'   ; w = 00000000 binary
    tris PORTB             ; Set up port B for output
; Turn on line 0 in port B. All others remain off
    movlw     B'00000001'
              ; -------|
              ; -------|____ Line 0 ON
              ;        |_____ All others off
    movwf     PORTB
; Endless loop intentionally hangs up program
wait:
    goto wait
        end
```

The preceding program, named LEDOn, can be found in the book's online software.

LED Flasher Program

A different program can be used on the same circuit to make the LED flash on and off. In this case the program must include a *delay loop* so as to keep the LED in either the ON or the OFF state for a short time period. The delay loop can be easily implemented using a file register as a counter. The program logic turns on the LED and counts down to zero; then resets the counter, turns the LED off, and counts down again.

The counter routine can also be used to demonstrate the creation of a subroutine, called a *procedure* in PIC programming. The procedure is nothing more than a routine marked by a label at its entry point and terminated with a return statement. The procedure is executed by a call statement to its initial label. If the procedure is named "delay", the procedure call is as follows:

```
call delay          ; Call procedure
    .
    .
    .
```

Elsewhere in the program,

```
delay:
    ; procedure instructions go here
    return          ; End of procedure
```

The simplest delay loop operates by wasting processor time. Because each instruction takes four clock cycles, the delay can be accurately calculated by multiplying the number of instructions in the loop by the device's clock speed, divided by 4. The details of delay loops are discussed in Chapter 11 on timers and counters.

Implementing a timer loop usually requires two counters. Because the maximum value that can be stored in a register file is 255, and a delay of 255 machine cycles is a very short one. In the code sample that follows we get around this limitation by creating a double-loop counter: the inner loop counts down 200 cycles and an outer loop repeats the inner-loop 200 times. The result is that the routine repeats 200 times 200 times, 40,000 iterations, which is sufficient for the purpose at hand. Code is as follows:

```
delay:
    movlw    .200       ; w = 200 decimal
    movwf    j          ; j = w
jloop:
    movwf    k          ; k = w
kloop:
    decfsz   k,f        ; k = k-1, skip next if zero
    goto     kloop
    decfsz   j,f        ; j = j-1, skip next if zero
    goto     jloop
    return
```

Code assumes that two variables were created in the processor's GPR space, as follows:

```
; Declare variables at 2 memory locations
j     equ  0x0c
k     equ  0x0d
```

The listing for the entire LEDFlash program, contained in this book's software resource is as follows:

```
; File: LEDFlash.asm
; Date: June 2, 2011
; Authors: Canton and Sanchez
; Processor: 16F84A
;
; Description:
; Turn on and off LED wired to port B, line 0
;===========================
;          switches
;===========================
; Switches used in __config directive:
;   _CP_ON          Code protection ON/OFF
; * _CP_OFF
; * _PWRTE_ON       Power-up timer ON/OFF
;   _PWRTE_OFF
;   _WDT_ON         Watchdog timer ON/OFF
; * _WDT_OFF
;   _LP_OSC         Low power crystal oscillator
; * _XT_OSC         External parallel resonator/crystal oscillator
;   _HS_OSC         High speed crystal resonator (8 to 10 MHz)
;                   Resonator: Murate Erie CSA8.00MG = 8 MHz
;   _RC_OSC         Resistor/capacitor oscillator (simplest, 20% error)
; |
; |_____ * indicates setup values

      processor 16f84A
      include   <p16f84A.inc>
      __config  _XT_OSC & _WDT_OFF & _PWRTE_ON & _CP_OFF
;=======================================================
;                 variables in PIC RAM
;=======================================================
; Declare variables at 2 memory locations
j     equ       0x0c
k     equ       0x0d
;=========================================================
;               m a i n   p r o g r a m
;=========================================================
      org       0       ; start at address 0
      goto      main
;=============================
; space for interrupt handler
;=============================
      org       0x04
;=============================
;      main program
;=============================
main:
; Initialize all lines in port B for output
      movlw     B'00000000'   ; w = 00000000 binary
```

```
        tris      PORTB             ; Set up port B for output
;
; Program loop to turn LED on and off
LEDonoff:
; Turn on line 0 in port B. All others remain off
        movlw     B'00000001'   ; LED ON
        movwf     PORTB
        call      delay             ; Local delay routine
; Turn off line 0 in port B.
        movlw     B'00000000'   ; LED OFF
        movwf     PORTB
        call      delay
        goto      LEDonoff
;================================
;       delay sub-routine
;================================
delay:
        movlw     .200              ; w = 200 decimal
        movwf     j                 ; j = w
jloop:
        movwf     k                 ; k = w
kloop:
        decfsz    k,f               ; k = k-1, skip next if zero
        goto      kloop
        decfsz    j,f               ; j = j-1, skip next if zero
        goto      jloop
        return
        end
```

9.3.2 LED/Pushbutton Circuit

A slightly more complex circuit contains a pushbutton switch. In this case the program can monitor the state of the pushbutton and turn on the LED when the switch is closed. Figure 9-4 shows one possible wiring for the LED/pushbutton circuit.

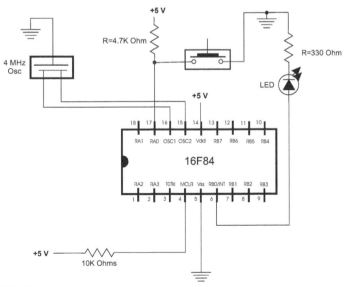

Figure 9-4 LED/Pushbutton Circuit.

If a switch reports a zero bit when active it is described as *active-low*. A switch that reports a one bit when pressed is said to be *active-high*. The pushbutton switch in Figure 9-4 is wired active-low. In the same manner, an output device can be wired so that it is turned on with a logic 0 and off with logic 1 on the port pin. A device turned on by the port current it is said to *source the current*. When the device is turned on when the port reports logic 0, the line is said to *sink the current*. PICs and other CMOS devices operate better sinking than sourcing current. Table 9.2 shows the maximum sink and source currents for the 16F84 ports.

Table 9.2

Sink and Source Current for 16F84 Ports

SOURCE	ANY I/O PIN	PORT A	PORT B
sink current	25 mA	80 mA	150 mA
source current	20 mA	50 mA	100 mA

The 4.7K-Ohm resistor in the circuit of Figure 9-4 keeps RA0 high until the switch is pressed. This switch action determines that RA0 reads a binary 1 when the switch is released and a binary 0 (low) when the switch is pressed (active).

To test if the switch in the circuit of Figure 9-4 is closed, the application can read port pin RA0. If the value in the port is 1, then the switch is open (released); if it is 0, then the switch is closed. The following program, called LEDandPb, exercises the circuit in Figure 9-4.

```
; File: LEDandPb.asm
; Date: June 2, 2011
; Authors: Canton and Sanchez
; Processor: 16F84A
;
; Description:
; Circuit with LED wired to RB0 and pushbutton switch,
; active low, wired to RA0. Pushbutton action turns LED
; OFF when pressed and ON when released.
;============================
;        switches
;============================
; Switches used in __config directive:
;   _CP_ON        Code protection ON/OFF
; * _CP_OFF
; * _PWRTE_ON     Power-up timer ON/OFF
;   _PWRTE_OFF
;   _WDT_ON       Watchdog timer ON/OFF
; * _WDT_OFF
;   _LP_OSC       Low power crystal oscillator
; * _XT_OSC       External parallel resonator/crystal oscillator
;   _HS_OSC       High speed crystal resonator (8 to 10 MHz)
;                 Resonator: Murate Erie CSA8.00MG = 8 MHz
;   _RC_OSC       Resistor/capacitor oscillator (simplest, 20% error)
; |
; |_____ * indicates setup values

        processor 16f84A
        include    <p16f84A.inc>
```

```
        __config   _XT_OSC & _WDT_OFF & _PWRTE_ON & _CP_OFF
;========================================================
;                   variables in PIC RAM
;========================================================
; Not used in this program
;
;=========================================================
;                   m a i n   p r o g r a m
;=========================================================
        org         0        ; start at address 0
        goto        main
;
;==============================
; space for interrupt handler
;==============================
        org         0x04
;
;==============================
;       main program
;==============================
main:
; Initialize all lines in port B for output
        movlw       B'00000000'   ; w = 00000000 binary
        tris        PORTB         ; Set up port B for output
; Initialize port A, line 0, for input
        movlw       B'00000001'   ; w = 00000001 binary
        tris        PORTA         ; Set up RA0 for input
; Program loop to test state of pushbutton switch
;
;==============================
;    read PB switch state
;==============================
LEDctrl:
; Push button switch on demo board is wired to port A bit 0
; Switch logic is active low
        btfss       PORTA,0       ; Test. Skip next line if
                                  ; bit is set
        goto        turnOFF       ; Turn LED off routine
; At this point port A bit 0 is not set
; Switch is pressed (active low action)
; Turn ON line 0 in port B
        bsf         PORTB,0       ; RB0 high
        goto        LEDctrl
turnOFF:
; Routine to turn OFF LED
        bcf         PORTB,0       ; RB0 low
        goto        LEDctrl

        end
```

The electronic file for the LEDandPb program previously listed can be found in this book's online software.

9.3.3 Multiple LED Circuit

The circuit in Figure 9-5 introduces a few more circuit and programming complications because it contains a battery of eight LEDs, all wired to port B.

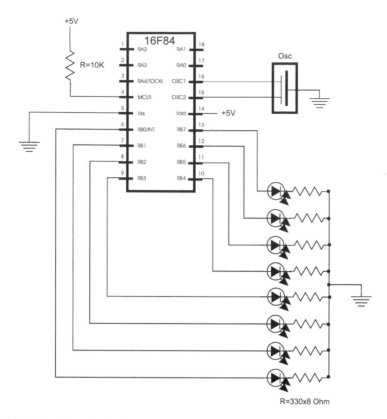

Figure 9-5 Multiple LED Circuit.

The circuit in Figure 9-5 can be programmed to do different functions. For example, the eight LEDs can be visualized as representing an 8-bit binary number, and the circuit can be made to count in binary from 0 to 255. Because the eight LEDs are all wired to port B, the binary count can be stored directly in the port register (PORTB). In this case the LEDs wired to port B lines that hold a binary 1 will be turned on and those wired to lines that hold binary 0 will be turned off. The following program, called LEDCount, performs this operation.

```
; File: LEDCount.asm
; Date: June 3, 2011
; Authors: Canton and Sanchez
; Processor: 16F84A
; Description:
; Circuit with eight LEDs wired to RB0 to RB7.
; Program displays a binary count from 0 to 255 on
; LEDs.
;============================
;         switches
;============================
; Switches used in __config directive:
;   _CP_ON          Code protection ON/OFF
; * _CP_OFF
; * _PWRTE_ON       Power-up timer ON/OFF
;   _PWRTE_OFF
;   _WDT_ON         Watchdog timer ON/OFF
```

```
;   *  _WDT_OFF
;      _LP_OSC         Low power crystal oscillator
;   *  _XT_OSC         External parallel resonator/crystal oscillator
;      _HS_OSC         High speed crystal resonator (8 to 10 MHz)
;                      Resonator: Murate Erie CSA8.00MG = 8 MHz
;      _RC_OSC         Resistor/capacitor oscillator (simplest, 20% error)
;   |
;   |_____  * indicates setup values

        processor 16f84A
        include   <p16f84A.inc>
        __config  _XT_OSC & _WDT_OFF & _PWRTE_ON & _CP_OFF
;========================================================
;                   variables in PIC RAM
;========================================================
; Declare variables at 2 memory locations
j       equ       0x0c
k       equ       0x0d
;========================================================
;                m a i n   p r o g r a m
;========================================================
        org       0       ; start at address 0
        goto      main
;==============================
; space for interrupt handler
;==============================
        org       0x04
;==============================
;        main program
;==============================
main:
; Initialize all lines in port B for output
        movlw     B'00000000'   ; w = 00000000 binary
        tris      PORTB         ; Set up port B for output
; Set port B bit 0 ON
        movlw     B'00000000'   ; w := 0 binary
        movwf     PORTB         ; port B itself := w
; Clear the carry bit
        bcf       STATUS,C
mloop:
        incf      PORTB,f       ; Add 1 to register value
        call      delay
        goto      mloop
;================================
;        delay sub-routine
;================================
delay:
        movlw     .200 ; w = 200 decimal
        movwf     j          ; j = w
jloop:
        movwf     k          ; k = w
kloop:
        decfsz    k,f       ; k = k-1, skip next if zero
        goto      kloop
        decfsz    j,f       ; j = j-1, skip next if zero
        goto      jloop
        return

        end
```

9.4 Seven-Segment LED

A Seven-Segment display can be connected to output ports on the PIC and used to display numbers and some digits. The circuit in Figure 9-6 shows one possible wiring scheme.

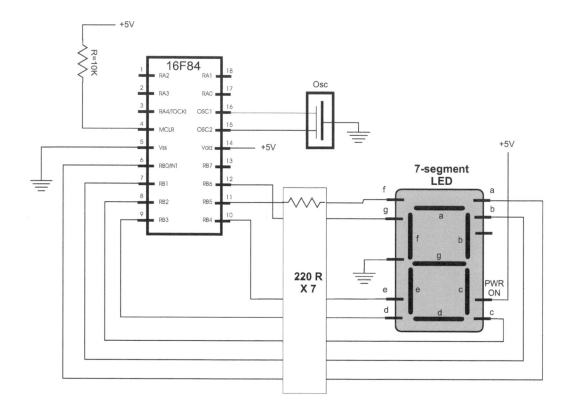

Figure 9-6 Seven-Segment LED Circuit.

As the name indicates, the Seven-Segment display has seven linear LEDs that allow forming all the decimal and hex digits and some symbols and letters. Once the mapping of the individual bars of the display to the PIC ports has been established, digits and letters can be shown by selecting which port lines are set and which are not. For example, in the Seven-Segment LED of Figure 9-6, the digit 2 can be displayed by setting segments a, b, g, e, and d. In this particular wiring, these segments correspond to port B lines 0, 1, 6, 4, and 5.

Conversion of the individual digits to port display codes can be accomplished by means of a *lookup table*. The processing depends on three special features of PIC assembly language:

1. The program counter file register (labeled PC and located at offset 0x02) holds the address in memory of the current instruction. Because each PIC instruction takes

up a single byte (except for those that modify the PC), one can jump to consecutive entries in a table by adding an integer value to the program counter.

2. The addwf instruction can be used to add a value in the w register to the program counter.

3. The retlw instruction returns to the caller a literal value stored in the w register. In the case of retlw, the literal value is the instruction operand.

If the lookup table is located at a subroutine called getcode, then the processing can be implemented as follows:

```
getcode:
    addwf     PC,f        ; Add value in w register to program  counter
    retlw     0x3f        ; code for number 0
    retlw     0x06        ; code for number 1
    retlw     0x5b        ; code for number 2
    ...
    retlw     0x6f        ; code for number 9
```

The calling routine places in the w register the numeric value whose code is desired, and then calls the table lookup as follows:

```
    movlw     0x03        ; Code for number 3 desired
    call      getcode
    movwf     PORTB       ; Display 3 in 7-segment display
```

A program and circuit to demonstrate programming the Seven-Segment LED is offered in the following section.

9.5 I/O Demo Board

Demonstration (or demo) *boards* are a useful tool in mastering PIC programming. Many are available commercially and, like programmers, a cottage industry of PIC demo boards has appeared on the Internet. Constructing your own demo boards is not a difficult task and serves to acquire circuit prototyping skills. Here again, the components can be placed on a breadboard, wire-wrapped onto a special circuit board, a printed circuit board can be home-made, or ordered through the Internet. These options have been previously discussed, and Appendix G contains instructions on how to build your own PCB for the demonstration boards used in the book.

Figure 9-7 shows the schematics of a simple 16F84-based demo board for experimenting with simple input and output devices, including a Seven-Segment LED, a bank of eight LEDs, buzzer, pushbutton switch, and a bank of four toggle switches.

Note that the Demo Board A in Figure 9-7 contains three jumper switches, labeled J1, J2, and J3. The jumper switches are wired to ground and are used to de-select the corresponding component. For example, if the J1 switch is closed and J2 is open, then port B is displayed on the Seven-Segment LED but not in the eight-LED bank. By the same token, jumper J3 de-selects the buzzer.

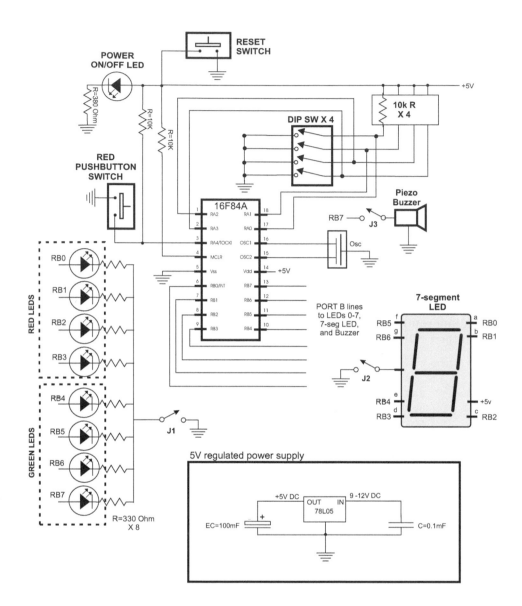

Figure 9-7 PIC 16F84 Demo Board.

9.5.1 TestDemo1 Program

The following program exercises some of the experiments that can be implemented on the demo board in Figure 9-7. The program reads the state of the pushbutton switch wired to port RA4. If the switch is closed, the buzzer is activated. Otherwise, the buzzer is turned off. The program also reads the state of the four toggle switches wired to the four low-order lines of port B. The resulting value is used to display the decimal digits 0 to 9 and the hexadecimal letters A through F on the Seven-Segment LED. The corresponding code is obtained from a local lookup table as described earlier in this chapter.

```
; File: TestDemo1.asm
; Date: June 2, 2011
; Authors: Canton and Sanchez
; Processor: 16F84A
;
; Description:
; Program to exercise the demonstration circuit and board
; in Figure 9-7
;============================
;          switches
;============================
; Switches used in __config directive:
;    _CP_ON          Code protection ON/OFF
; * _CP_OFF
; * _PWRTE_ON        Power-up timer ON/OFF
;    _PWRTE_OFF
;    _WDT_ON         Watchdog timer ON/OFF
; * _WDT_OFF
;    _LP_OSC         Low power crystal oscillator
; * _XT_OSC          External parallel resonator/crystal
;                    oscillator
;    _HS_OSC         High speed crystal resonator (8 to 10 MHz)
;                    Resonator: Murate Erie CSA8.00MG = 8 MHz
;    _RC_OSC         Resistor/capacitor oscillator
; |
; |_____ * indicates setup values

     processor 16f84A
     include   <p16f84A.inc>
     __config  _XT_OSC & _WDT_OFF & _PWRTE_ON & _CP_OFF
;===================================================
;                variables in PIC RAM
;===================================================
     cblock    0x0c ; Start of block
     count1         ; Counter # 1
     j              ; counter J
     k              ; counter K
     endc
;=====================================================
;                     P R O G R A M
;=====================================================
     org       0     ; start at address 0
     goto      main
;
; Space for interrupt handlers
     org       0x08
main:
; Port A (5 lob) for input
     movlw     B'00011111'   ; w := 00001111 binary
     tris      PORTA         ; port A (lines 0 to 4) to input
; Port bit (8 lines) for output
     movlw     B'00000000'   ; w := 00000000 binary
     tris      PORTB         ; port B to output
;==============================
; Pushbutton switch processing
;==============================
pbutton:
; Pushbutton switch on demo board is wired to RA4
; Switch logic is active low
     btfss     PORTA,4       ; Test and skip if bit is set
```

```
        goto      buzzit          ; Buz if switch ON
; At this point port A bit 4 is set (switch is off)
        call      buzoff          ; Buzzer off
        goto      readdip         ; Read DIP switches
buzzit:
        call      buzon           ; Turn on buzzer
        goto      pbutton
;==============================
;     DIP switch processing
;==============================
; Read all bits of port A
readdip:
        movf      PORTA,w         ; Port A bits to w
; If board uses active low then all switch bits must be negated
; This is done by XORing with 1-bits
        xorlw     b'11111111'     ; Invert all bits in w
; Eliminate all 4 high order bits
        andlw     b'00001111'     ; And with mask
; Get digit into w
        call      segment         ; get digit code
        movwf     PORTB           ; Display digit
        call      delay           ; Give time
; Updade digit and loop counter
        goto      pbutton


;*******************************
;   7-segment table of hex codes
;*******************************
segment:
        addwf     PCL,f      ; PCL is program counter latch
        retlw     0x3f       ; 0 code
        retlw     0x06       ; 1
        retlw     0x5b       ; 2
        retlw     0x4f       ; 3
        retlw     0x66       ; 4
        retlw     0x6d       ; 5
        retlw     0x7d       ; 6
        retlw     0x07       ; 7
        retlw     0x7f       ; 8
        retlw     0x6f       ; 9
        retlw     0x77       ; A
        retlw     0x7c       ; B
        retlw     0x39       ; C
        retlw     0x5b       ; D
        retlw     0x79       ; E
        retlw     0x71       ; F
        retlw     0x7f       ; Just in case all on


;**************************
;   piezo buzzer ON
;**************************
; Routine to turn on piezo buzzer on port B bit 7
buzon:
        bsf       PORTB,7         ; Tune on bit 7, port B
        return


;**************************
;   piezo buzzer OFF
;**************************
; Routine to turn off piezo buzzer on port B bit 7
```

```
buzoff:
    bcf         PORTB,7         ; Bit 7 port b clear
    return
;=================================
;        delay sub-routine
;=================================
delay:
    movlw       .200        ; w = 200 decimal
    movwf       j           ; j = w
jloop:
    movwf       k           ; k = w
kloop:
    decfsz      k,f         ; k = k-1, skip next if zero
    goto        kloop
    decfsz      j,f         ; j = j-1, skip next if zero
    goto        jloop
    return

    end
```

9.6 Comparisons in PIC Programming

The power and usefulness of programs is due, in great measure, to their decision-making ability, and decisions are based on comparison. In a *comparison*, code is able to make decisions based on the relative values of two operands. For example, compare the values a and b. If a is greater than b, execute a certain coded routine; if b is greater than a, execute another one; and if both operands have the same value, then proceed to a third code branch.

CISC and even some RISC microprocessors contain a compare operator in their instruction set. However, the compare can be substituted, with some inconvenience, by a subtraction. Because there is no compare operation in the mid-range IC instruction set, the programmer is forced to simulate the comparison by subtracting the w register from a literal value or from a file register. The sublw and subwf instructions can be used for this. After the subtraction takes place, code can make decisions based on the state of the zero and the carry flags. For example, the following code fragment compares the value in the two registers, labeled OP1 and OP2 respectively, and directs execution to three possible routines:

```
; Declare variables at 2 memory locations
OP1         equ 0x0c    ; First operand
OP2         equ 0x0d    ; second operand
.
.
.
main:
    movlw       0x30        ; First operand
    movwf       OP1         ; to OP1 register
    movlw       0x50        ; Second operand
    movwf       OP2         ; To OP2 register
    movf        OP2,w       ; OP2 to w register (not really necessary)
    subwf       OP1,w       ; Subtract w (OP2) from OP1
    btfsc       STATUS,2    ; 2 is zero bit. Test zero flag.
                            ; Skip next instruction if Z bit = 0,
                            ; that is if both numbers are not the
                            ; same
    goto ops_are_eq    ; OP2 = w routine
```

```
; At this point the zero flag is not set. Therefore the two operand
; are not equal
; Now test the carry flag for OP1 < OP2, in this case C = 1
     btfss     STATUS,0  ; 0 is carry bit. Test carry flag
                         ; and skip next instruction if
                         ; C bit = 1
     goto op2big          ; OP2 > w routine
; Processing for the case OP1 > OP2
     nop
     goto      done
ops_are_eq:
;   Processing for the case OP1 = OP2l
     nop
     nop
     goto      done
op2big:
; Processing for the case OP1 < OP2
     nop
     nop
done:
     goto      done
     end
```

9.6.1 PIC Carry Flag

In mid-range PIC microcontrollers it is somewhat unusual that the effects on the *carry flag* are different in addition than in subtraction. During addition (addwf and addlw), the carry flag indicates a *carry out* of the most significant bit of the result. In this case, C = 1 if there was a carry out, and C = 0 otherwise. However, in subtraction, the carry flag is described in the Microchip documentation as behaving as an *inverted borrow*. When two numbers are subtracted and the result is too big to fit in the destination operand, then the carry flag is clear. What this amounts to is that in PIC subtraction (sublw and subwf operations), the carry bit is set if there is no carry out of the high-order bit. This unusual behavior is shown in the preceding code fragment.

Chapter 10

PIC Interrupt System

10.1 Interrupts

An interrupt is an asynchronous signal calling for processor attention. Interrupts can originate in hardware or in software. The interrupt mechanism is a way to avoid wasting processor time, because without interrupts code has to poll hardware devices in ineffective, closed-loops. With interrupts the processor can continue to do its work because the interrupt mechanism ensures that the CPU will receive a signal whenever an event occurs that requires its attention. PIC microcontrollers provide varying levels of support for interrupts. We focus on interrupts on the 16F84. Other members of the mid-range PIC family support interrupts with minor variations.

10.1.1 16F84 Interrupts

Four different sources of interrupts are available in the 16F84. These are discussed in the following section. One instruction named RETFIE (for return from interrupt) is specifically related to interrupt processing. Its purpose is to return to the program counter the address of the instruction that follows the location in code where the interrupt took place. It does so by loading into the program counter register the 13-bit address saved at the top of the stack. In addition, RETFIE sets the *Global Interrupt Enable bit* in the INTCON register (discussed in the following section) an action that automatically reenables interrupts.

In addition to the RETFIE instruction, two PIC hardware elements relate directly to interrupts: the OPTION register and the INTCON (interrupt control) register. Both registers are readable and writeable and contain bits that allow setting up, controlling, and detecting the various interrupts. The *Interrupt Control Register* (INTCON) records individual interrupt requests in flag bits. It also contains the individual and global interrupt enable bits. The *OPTION register* has several bits that must be accessed in order to initialize interrupts.

10.1.2 Interrupt Control Register

The Interrupt Control (INTCON) register is a readable and writeable register located at offset 0x08 in bank 0. The INTCON register contains two classes of bits: bits to enable and disable the various interrupt sources, and flag bits that allow detecting the occurrence of the various interrupts. The bits to enable and disable interrupts have names that end with the letter E, while the interrupt flag bit names end with the letter F. This has suggested calling them the INTCON E and the F bits. Figure 10-1 is a bitmap of the INTCON register.

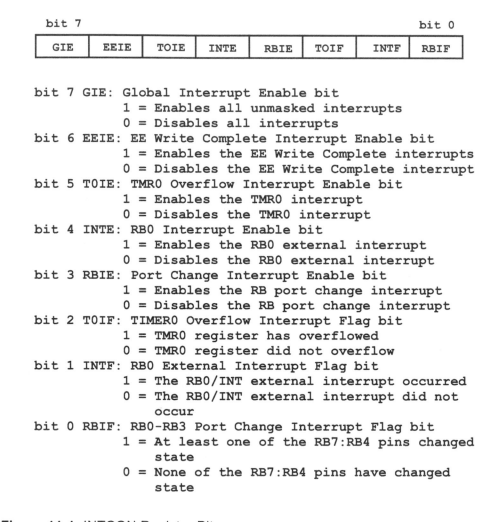

bit 7   bit 0

| GIE | EEIE | TOIE | INTE | RBIE | TOIF | INTF | RBIF |

```
bit 7 GIE: Global Interrupt Enable bit
            1 = Enables all unmasked interrupts
            0 = Disables all interrupts
bit 6 EEIE: EE Write Complete Interrupt Enable bit
            1 = Enables the EE Write Complete interrupts
            0 = Disables the EE Write Complete interrupt
bit 5 T0IE: TMR0 Overflow Interrupt Enable bit
            1 = Enables the TMR0 interrupt
            0 = Disables the TMR0 interrupt
bit 4 INTE: RB0 Interrupt Enable bit
            1 = Enables the RB0 external interrupt
            0 = Disables the RB0 external interrupt
bit 3 RBIE: Port Change Interrupt Enable bit
            1 = Enables the RB port change interrupt
            0 = Disables the RB port change interrupt
bit 2 T0IF: TIMER0 Overflow Interrupt Flag bit
            1 = TMR0 register has overflowed
            0 = TMR0 register did not overflow
bit 1 INTF: RB0 External Interrupt Flag bit
            1 = The RB0/INT external interrupt occurred
            0 = The RB0/INT external interrupt did not
                occur
bit 0 RBIF: RB0-RB3 Port Change Interrupt Flag bit
            1 = At least one of the RB7:RB4 pins changed
                state
            0 = None of the RB7:RB4 pins have changed
                state
```

Figure 11-1 INTCON Register Bitmap.

10.1.3 OPTION Register

The OPTION register is a readable and writeable register that contains controls for configuring the prescaler bits and assigning them to either Timer0 or the Watchdog Timer; for selecting the increment mode on the RA4/TOCKI pin, the Timer0 source clock, the raising or falling edge in the RB0 interrupt; and for enabling and disabling the internal Port B pull-up resistors. The OPTION register is located in Bank1, at address 0x81. Although this register is not directly related to interrupts, several of its bits are related to the various interrupts. Figure 10-2 is a bitmap of the OPTION register.

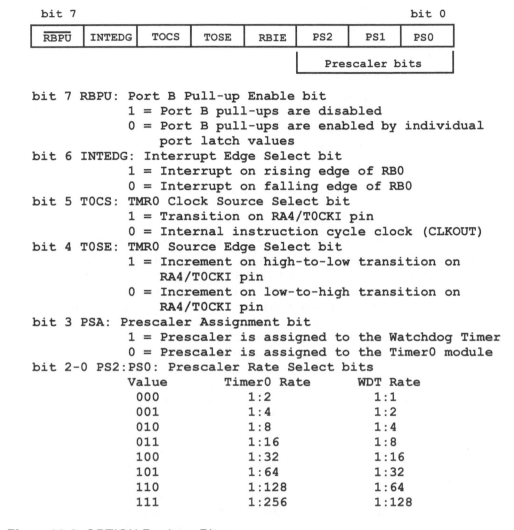

bit 7   bit 0

$\overline{\text{RBPU}}$	INTEDG	TOCS	TOSE	RBIE	PS2	PS1	PS0

Prescaler bits

```
bit 7 RBPU: Port B Pull-up Enable bit
            1 = Port B pull-ups are disabled
            0 = Port B pull-ups are enabled by individual
                port latch values
bit 6 INTEDG: Interrupt Edge Select bit
            1 = Interrupt on rising edge of RB0
            0 = Interrupt on falling edge of RB0
bit 5 T0CS: TMR0 Clock Source Select bit
            1 = Transition on RA4/T0CKI pin
            0 = Internal instruction cycle clock (CLKOUT)
bit 4 T0SE: TMR0 Source Edge Select bit
            1 = Increment on high-to-low transition on
                RA4/T0CKI pin
            0 = Increment on low-to-high transition on
                RA4/T0CKI pin
bit 3 PSA: Prescaler Assignment bit
            1 = Prescaler is assigned to the Watchdog Timer
            0 = Prescaler is assigned to the Timer0 module
bit 2-0 PS2:PS0: Prescaler Rate Select bits
            Value       Timer0 Rate       WDT Rate
             000          1:2               1:1
             001          1:4               1:2
             010          1:8               1:4
             011          1:16              1:8
             100          1:32              1:16
             101          1:64              1:32
             110          1:128             1:64
             111          1:256             1:128
```

Figure10-2 OPTION Register Bitmap.

10.2 Interrupt Sources

The 16F84 supports four different sources of interrupt:

1. External interrupt detected by line 0 of port B

2. Interrupts that originate in the timer (TMR0 overflow interrupt)

3. Interrupts originated in changes of lines RB7 to RB4 in port B

4. EEPROM complete data write interrupt

Other members of the mid-range PIC family support additional interrupts; for example, the 16F684 has ten sources of interrupts, as follows:

1. External interrupt RA2/INT

2. Timer0 overflow interrupt

3. PORTA change interrupts

4/5. Two comparator interrupts

6. A/D interrupt

7. Timer1 overflow interrupt

8. Timer2 match interrupt

9. EEPROM data write interrupt

10. Fail-safe clock monitor interrupt

11. Enhanced CCP interrupt

In the sections that follow we discuss the 16F84 interrupt sources.

10.2.1 Port B External Interrupt

This external interrupt is triggered by either the rising or falling signal edge on port B, line 0. Whether the interrupt takes place on the rising or falling edge of the signal depends on the setting of the INTEDG bit of the OPTION register. The port B interrupt is useful in detecting and responding to external events, for example, in measuring the frequency of a signal or in responding with some PIC action to a change in the state of a hardware device. This interrupt can be disabled by clearing the corresponding bit in the INTCON register. If enabled, once the interrupt takes place, code must clear the corresponding flag bit before reenabling the interrupt.

Suppose there is a circuit containing an emergency switch that the user will activate on the occurrence of some critical event. One possible approach is to check the state of the switch by continuously polling the port to which it is wired. But in a complex program, code would have to ensure that the switch polling routine is called with sufficient frequency so that an emergency event is detected immediately. A more effective solution is to connect the emergency switch to line number 0 of port B and set up the port B external interrupt source. In this case, whenever the emergency switch is activated, the program immediately responds via the interrupt mechanism. Furthermore, once the interrupt code has been developed and debugged, it will continue to function correctly no matter what changes are made to the rest of the program.

10.2.2 Timer0 Interrupt

The 16F84 is equipped with a special timer module, named Timer0, which can serve both as a timer and as a counter. The Timer0 module, discussed in greater detail in the next chapter, consists of an 8-bit readable register operated by an internal or external clock and attached to an 8-bit programmable *prescaler*. The prescaler is used to delay the timer by dividing the clock signal. The Timer0 module can be setup to interrupt on overflow. In this case an interrupt is generated whenever the counter goes from 0xff to 0x00.

The Timer0 counter interrupt can be used to measure events and to respond to elapsed periods. For example, the timer is used to measure events by determining the number of timer interrupts that have taken place because an event occurred, because the timer of each interrupt can be determined from the processor clock speed and the prescaler setup. In this manner the event time is calculated by multiplying the time of each interrupt by the number of interrupts that have occurred. In this case the interrupt routine would increment a counter register, which is accessible to code anywhere in the program. So the actual count can be reset from inside or outside the service routine.

In responding to an elapsed period, the Timer0 interrupt service routine not only keeps track of the time elapsed because the event, but also tests for a certain counter value that represents the desired time limit. Once the timer counter reaches this preset limit, the service routine responds directly to the required action.

One powerful and common application of a Timer0 interrupt is in implementing serial communications. In this case the timer interrupt is set up to take place at the baud rate at which the serial line must be polled for data or at which individual data bits are sent. The sample program LapseTmrInt, developed later in this book, demonstrates this use of the timer interrupt.

10.2.3 Port B Line Change Interrupt

The third 16F84 interrupt source relates to a change in the value stored in port B lines 4 to 7. When this interrupt is enabled, any change in status in any of the four port B pins labeled RB7, RB6, RB5, and RB4 will trigger an interrupt. The interrupt can be set up to take place when the pin status changes from logic 1 to logic0, or vice versa. For this interrupt to take place, port B pins 4 to 7 must be defined as input. Otherwise, the interrupt does not take place.

The port B line change interrupt provides a mechanism for monitoring up to four different interrupt sources, typically originating in hardware devices. When the interrupt is enabled, the current state of the port B lines is constantly compared to the old values. If there is a change in state in any of the four lines, the interrupt is generated.

Implementation of the line change interrupt is not without complications. The circuit and software designer must take into account the characteristics of the external signal because only then can code be developed that will correctly handle the various possible sources. Two pieces of information that are necessary in this case are

1. The signal's rising edge and falling edges

2. The pulse width of the interrupt trigger

The need to determine whether the signal is on a rising or falling edge is to ensure that the service routine is entered only for the desired edge. For example, if the device is an active-low pushbutton switch, an interrupt will typically be desired on the signal's falling edge, that is, when it goes from high to low.

Knowledge about the signal's width determines the processing required by the service routine. This is due to the fact that both the rising and the falling edge of the signal can trigger the interrupt. So, if the triggering signal has a small pulse width compared to the time of execution of the interrupt handler, then the interrupt line will have returned to the inactive state before the service routine completes and a possible false interrupt on the signal's falling edge is not possible. On the other hand, if the pulse width of the interrupt signal is large and the service routine completes before the signal returns to the inactive state, then the signal's falling edge can trigger a false interrupt. Figure 10-3 shows both situations.

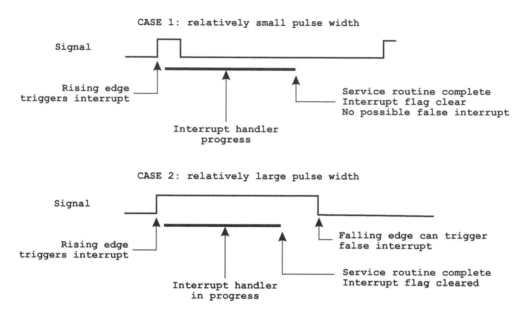

Figure 10-3 Signal Pulse Width and Interrupt Latency.

In the context of Figure 11-3 the period between the edge that triggers the interrupt and the termination of the interrupt handler is sometimes called the *mismatch period*. The mismatch period terminates when the service routine completes and the corresponding interrupt is reenabled. If this happens after the interrupt signal is reset, no possible false interrupt can take place and no special provision is required in the handler. In fact, the interrupt handler will run correctly as long as the service routine takes longer to execute than the interrupt frequency. However, if the handler terminates before the signal returns to its original state, then the handler must make special provisions to handle a possible false interrupt. In order to do this, the handler must first determine if the interrupt took place on the rising or the falling signal edge, which can be done by examining the corresponding port B line. For example, if the interrupt is to take place on the rising edge only, and the line is low, then it can be ignored because it took place on the falling edge.

When an interrupt can take place on either the rising or the falling edge of the triggering signal, the interrupt source must have a minimum pulse width in order to ensure that both edges are detected. In this case, the minimum pulse width is the maximum time from the edge that triggered the interrupt to the moment when the interrupt flag is cleared. Otherwise, the interrupt will be lost because the interrupt mechanism will be disabled at the time it takes place.

The preceding discussion leads directly to the possibility of an interrupt taking place while the service routine of a previous interrupt is still in progress. These are called *reentrant* or *nested interrupts*. Several things must happen in order to allow reentrant interrupts. One of them is that interrupts must be reenabled before the handler terminates. In addition, the service routine must be able to create different instances of the variables in use, usually allocated in the stack. The lack of a program-accessible stack and the PIC interrupt mechanism itself, forces the conclusion that reentrant interrupts should not be attempted in PIC programs.

Multiple External Interrupts

One of the practical applications of the port B line change interrupt is in handling several different interrupt sources. For example, a circuit can contain four pushbutton switches that activate four different circuit responses. If the switches are wired to the corresponding pins in port B (RB4 to RB7) and the line change interrupt is enabled, the interrupt will take place when any one of the four switches changes level, that is, when any one of the interrupt lines goes from high to low or from low to high. The interrupt handler software can determine which of the switches changed state and if the change took place on the signal's rising or falling edge. The corresponding software routines will then handle each case.

Later in this chapter we develop a sample program that uses the port B line change interrupt to respond to action on four pushbutton switches.

10.2.4 EEPROM Data Write Interrupt

The origin of this interrupt relates to the relative slowness of the EEPROM data write operation, which is 10 ms on the 16F84. The interrupt serves no other function than to allow the microcontroller to continue execution while the data write operation is in progress. The interrupt service routine informs the microcontroller when writing has ended through the EEIF bit located in the EECON1 register. The use of this interrupt is considered in Chapter 14, in the context of EEPROM data memory access and programming.

10.3 Developing the Interrupt Handler

The *interrupt handler*, also called the *interrupt service routine* or the ISR, is the code that receives control upon the occurrence of the interrupt. Most of the programming that goes into the service routine is specific to the application; however, there are certain housekeeping operations that should be included almost universally. The following list describes the structure of an interrupt service routine for the mid-range PICs:

- Preserve the value of the w register.
- Preserve the value of the STATUS register.
- Execute the application-specific operations.
- Restore the value of the STATUS register at the time of the interrupt.
- Restore the value of the w register at the time of the interrupt.
- Issue the RETFIE instruction to end the interrupt handler.

In the PIC 16F84, the interrupt service routine must be located at offset 0x004 in code memory. A simple org directive takes care of ensuring this location, as in the following code fragment:

```
    org       0x000          ; Beginning of code area
    goto      start          ; Jump to program start
    org       0x004          ; Start of Service routine
    .
    .                        ; SERVICE ROUTINE GOES HERE
    .
    retfie                   ; End of ISR

start:                       ; Program starts here
```

Alternatively, code can place a jump at offset 0x004 and locate the Service Routine elsewhere in the code. In this case it is important to remember not to call the Service Routine, but to access it with a goto instruction. The reason is that the call opcode places a return address in the stack, which will make the retfie instruction fail.

10.3.1 Context Saving Operations

The only value automatically preserved by the interrupt mechanism is the Program Counter (PC), which is stored in the stack. Applications often need to restore the hardware registers in the CPU to the same state as when the interrupt took place, so the first operation of most interrupt handlers is saving the processor context. This context usually includes the W and the STATUS registers. Occasionally at this time the application may save other user-defined registers.

Saving W and STATUS Registers

Saving the W and the STATUS registers requires using register variables, but the process must take some other elements into account. Saving the w register is simple enough: its value at the start of the Service Routine is stored in a local variable from which it is restored at termination. But saving the STATUS register cannot be done with MOVF instruction because this instruction changes the zero flag. The solution is to use the SWAPF instruction which does not affect any of the flags. Of course, SWAPF inverts the nibbles in the operand, so the process must be repeated in order to restore the original state. The following code fragment assumes that file register variables named old_w and old_status were previously created:

```
save_cntx:
    movwf     old_w          ; Save w register
    swapf     STATUS,w       ; STATUS to w
    movwf     old_status     ; Save STATUS
;
; Interrupt handler operations go here
;
    swapf     old_status,w   ; Saved status to w
    movfw     STATUS         ; To STATUS register
; At this point all operations that change the
; STATUS register must be avoided, but swapf does not.
    swapf     old_w,f        ; Swap file register in itself
    swapf     old_w,w        ; re-swap back to w
    retfie
```

10.4 Interrupt Programming

In the sections that follow we discuss programming interrupts that originate in port B, line 0, and those that originate in changes of port B lines RB4 to RB7. Interrupts that relate to the Timer0 overflow or to EEPROM data write operations are covered in the chapter on EEPROM Data Operations.

10.4.1 Programming the External Interrupt

Interrupts detected on port B, line 0, are referred to as the *External Interrupt source*. The name is not the most adequate because other interrupts can also have external sources. One of the important uses of this interrupt source is to wake the processor from the SLEEP mode. This allows developing applications that can run on a small power source (such as batteries) because the program uses almost no power until some action associated with the interrupt source wakes up the PIC. A sample program using the RB0 interrupt is developed later in this chapter. Our first sample program is a simple demonstration of the installation and action of the interrupt. The program is based on the circuit in Figure 10-4.

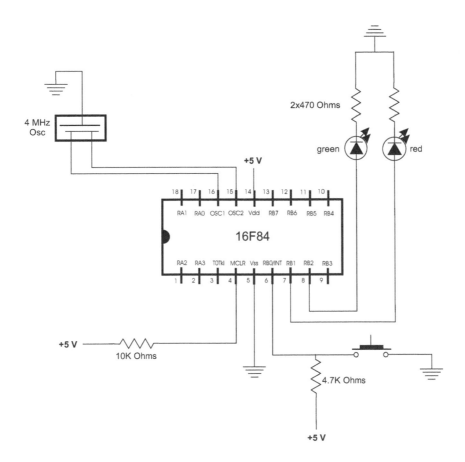

Figure 10-4 Circuit for RB0 Interrupt Demonstration.

In the circuit of Figure 10-4, a pushbutton switch is wired to the RB0 port. This switch produces the interrupt when pressed. A red LED is wired to port RB1 and a green LED to port RB2. The main program flashes the green LED on and off at a rate of approximately one-half second. The red LED is toggled on and off when the pushbutton switch is pressed. The switch contains a 4.7K-Ohm resistor that keeps the port high until the contact is made and sent to ground. This makes the switch active low, and the interrupt is programmed on the falling edge of the signal, which takes place when the contact is made.

RB0 Interrupt Initialization

In order to initialize the RB0 interrupt the following operations must take place:

- Port B, line 0, must be initialized for input.
- The interrupt source must be set to take place either on the falling or the rising edge of the signal.
- The external interrupt flag (INTF in the INTCON register) must be initially cleared.
- Global interrupts must be enabled by setting the GIE bit in the INTCON register.
- The External Interrupt on RB0 must be enabled by setting the INTE bit in the INTCON register.

The following code fragment, from the program RB0Int in this book's software package, performs these operations:

```
      org  0x00
      goto main
;==============================
;       interrupt handler
;==============================
      org 0x04
      goto IntServ
;==============================
;       main program
;==============================
main:
; Set up interrupt on falling edge
; by clearing OPTION register bit 6
      movlw      b'10111111'
      option
      movlw      b'11111111'          ; Set port A for input
      tris       porta                ; (not necessary for this program)
      movlw      b'00000001'          ; Port B bit 0 is input
      tris       portb                ; all others are output
      clrf       portb                ; All port B to 0
; Initially turn on LED
      bsf        portb,0              ; Set line 0 bit
;==========================
;      setup interrupts
;==========================
; Clear external interrupt flag (intf = bit 1)
      bcf        INTCON,intf          ; Clear flag
; Enable global interrupts (gie = bit 7)
; Enable RB0 interrupt (inte = bit 4)
      bsf        INTCON,gie           ; Enable global int (bit 7)
      bsf        INTCON,inte          ; Enable RB0 int (bit 4)
;==============================
```

```
;           flash LED
; ==============================
; Program flashes LED wired to port B, line 2
lights:
     movlw     b'00000010'          ; Mask with bit 1 set
     xorwf     portb,f              ; Complement bit 1
     call      long_delay           ; Local delay routine
     call      long_delay
     call      long_delay
     goto      lights
```

RB0 ISR

The service routine for the RB0 interrupt will depend on the specific application. Nevertheless, the following processing steps should be considered:

- Determine if the source is an RB0 interrupt.

- Clear the RB0 interrupt flag (INTF bit) in the INTCON register.

- Save the context. Which registers and variables need to be saved depends on the specific application.

- Perform the interrupt action.

- Restore the context.

- Return from the interrupt with the retfie instruction.

In addition, the interrupt handler may have to perform operations that are specific to the application; for example, debounce a switch or initialize local variables. The following Interrupt Service routine is from the program RB0Int in this book's software resource:

```
; =========================================================
;                  Interrupt Service Routine
; =========================================================
; Service routine receives control when there is
; action on pushbutton switch wired to port B, line 0
IntServ:
; First test if source is an RB0 interrupt
     btfss      INTCON,INTF    ; INTF flag is RB0 interrupt
     goto       notRB0         ; Go if not RB0 origin
; Save context
     movwf      old_w          ; Save w register
     swapf      STATUS,w       ; STATUS to w
     movwf      old_status     ; Save STATUS
; =========================
;    interrupt action
; =========================
; Debounce switch
;      Logic:
;      Debounce algorithm consists in waiting until the
; same level is repeated on a number of samplings of the
; switch. At this point the RB0 line is clear because the
; interrupt takes place on the falling edge. The routine
; waits until the low value is read several times.
     movlw      D'10'          ; Number of repetitions
     movwf      count2         ; To counter
wait:
```

```
; Check to see that port B bit 0 is still 0
; If not, wait until it changes
     btfsc      portb,0         ; Is bit set?
     goto       exitISR         ; Go if bit not 0
; At this point RB0 bit is clear
     decfsz     count2,f        ; Count this iteration
     goto       wait            ; Continue if not zero
; Interrupt action consists of toggling bit 2 of
; port B to turn LED on and off
     movlw      b'00000100'     ; Xoring with a 1-bit produces
                                ; the complement
     xorwf            portb,f   ; Complement bit 2, port B
;=========================
;          exit ISR
;=========================
exitISR:
; Restore context
     swapf      old_status,w    ; Saved status to w
     movfw      STATUS          ; To STATUS register
     swapf      old_w,f         ; Swap file register in itself
     swapf      old_w,w         ; re-swap back to w
notRB0:
; Reset interrupt
     bcf        INTCON,intf     ; Clear INTCON bit 1
     retfie
```

Note that the interrupt handler listed previously contains a debouncing routine in order to clean the switch's signal. In this particular implementation, the detection of a signal of the wrong value determines that the interrupt is aborted. For the particular switch used in the test circuit, this approach seemed to work better. Alternatively, the routine can be designed so that if a wrong edge is detected, execution continues in the wait loop. In any case, the entire complication of software debouncing can be avoided by debouncing the switch in hardware.

10.4.2 Wake-Up from SLEEP Using the RB0 Interrupt

The PIC microcontroller *sleep mode* provides a mechanism for saving power that is particularly useful in battery-operated devices. The sleep mode is activated by executing the SLEEP instruction, which suspends all normal operations and switches off the clock oscillator. The sleep mode is suitable for applications that are not required to run continuously. For example, a device that records temperature at daybreak can be designed so that a light-sensitive switch generates an interrupt that turns the device on each morning. Once the data is recorded, the device goes into the sleep mode until the next daybreak.

Several events can force the device to wake-up from the sleep mode:

- A reset signal on the !MCLR pin
- Watchdog timer wake-up signal, if WDT is enabled
- Interrupt on RB0 line
- Port change interrupt on RB4 to RB7 lines
- EEPROM write complete interrupt

In the sleep mode the device is placed in a power-down state that generates the lowest power consumption. The system clock is turned off in the sleep mode so signals that depend on the clock cannot be used to terminate the sleep. If enabled, the Watchdog Timer is cleared by the sleep instruction but keeps running. The PD bit in the STATUS register is also cleared and the TO bit is set. The ports maintain the status they had before the SLEEP instruction was executed.

The TO and PD bits in the STATUS register can be used to determine the cause of wake-up, as the TO bit is cleared if a Watchdog Timer wake-up took place. The corresponding interrupt enable bit must be set for the device to wake due to an interrupt. Wake-up takes place regardless of the state of the *General Interrupt Enable* (GIE) bit. If the bit is clear, the device continues execution at the instruction following SLEEP. Otherwise, the device executes the instruction after the SLEEP instruction and then branches to the interrupt address. If the execution of the instruction following SLEEP is undesirable, the program should contain a NOP instruction after the SLEEP instruction.

SleepDemo Program

The program called SleepDemo in this book's online software package is a trivial demonstration of using the RB0 interrupt to wakeup the processor from the sleep mode. The program can be tested using the circuit in Figure 10-4. SleepDemo flashes the green LED at one-half second intervals during twenty iterations and then goes into the sleep mode. Pressing the pushbutton switch on line RB0 generates an interrupt that wakes the processor from the sleep mode. The following code fragment shows the coding of the main loop in the program:

```
;=============================
;    flash LED 20 times
;=============================
wakeUp:
; Program flashes LED wired to port B, line 2
; 20 times before entering the sleep state
      movlw    D'20'            ; Number of iterations
      movwf    count2           ; To counter
lights:
      movlw    b'00000010'      ; Mask with bit 1 set
      xorwf    portb,f          ; Complement bit 1
      call     long_delay
      call     long_delay
      call     long_delay
      decfsz   count2           ; Decrement counter
      goto     lights
; 20 iterations have taken place
      clrwdt                    ; Clear WDT
      sleep
      nop                       ; Recommended!
      goto     wakeUp           ; Resume execution
```

In the SleepDemo program, the Interrupt Service Routine does nothing. Its coding is as follows:

```
;=========================================================
;            Interrupt Service Routine
;=========================================================
```

```
; The interrupt service routine performs no operation
IntServ:
    bcf        INTCON,INTF        ; Clear flag
    retfie
```

The initialization of the RB0 interrupt is identical to the one in the RB0Int program previously listed.

10.4.3 Port B Bits 4-7 Status Change Interrupt

In the PIC 16F84 microcontroller, a change of input signal on port B, lines 4 to 7, generates an interrupt. This interrupt will set the RBIF bit in the INTCON register to indicate that at least one of the ports has changed value. The port change takes place when the port's previous value changes from logic 1 to logic 0, or vice versa. In order for port pins to recognize this interrupt, they must have been defined as input. If any one of the port pins (4 to 7) is defined as output, the interrupt will take place. The status change of the ports is in reference to the last time port B was read.

The principal application of this interrupt source is in detecting several different interrupt sources. Its principal disadvantage is that it forces the declaration of four port B lines as input, although during processing not all lines need be recognized as interrupt sources. The conclusion is that applications that only need a single external interrupt source should use the RB0 interrupt described in previous sections. Only applications that require more than one external interrupt should use the port B lines 4 to 7 interrupt on change source.

Because the interrupt takes place on any status change (high-to-low or low-to-high) the service routine executes on both signal edges. If interrupt processing is required on only one edge, that is, either when the port goes high or low, then the filtering must be performed in software. The circuit in Figure 10-5 allows testing the Port B Status Change Interrupt.

In the circuit of Figure 10-5, a pushbutton switch is wired to the RB7 port and another one to RB4. Both of these switches produce the interrupt when pressed. A red LED is wired to port RA0 and a green LED to port RA1. The red and green LEDs are toggled on and off when the corresponding pushbutton switches are pressed. The switches contain a 4.7K-Ohm resistor that keeps the port high until the contact is made and sent to ground. This makes both switches active low, and the interrupt is programmed on the falling edge of the signal.

RB4-7 Interrupt Initialization

In order to initialize the RB4-7 change interrupt, the following operations must take place:

- Port B lines 4 to 7 must be initialized for input.

- The interrupt source must be set to take place either on the falling or the rising edge of the signal.

- The RB port change interrupt flag (RBIF in the INTCON Register) must be initially cleared.

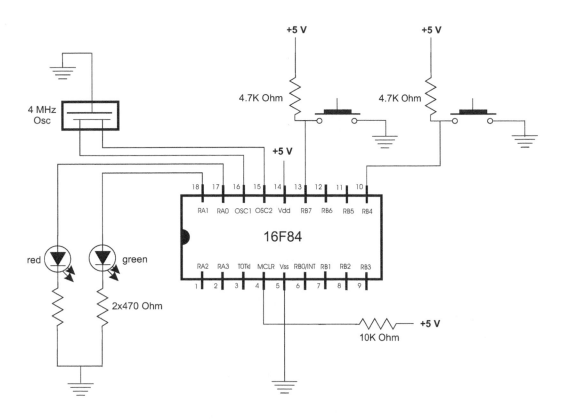

Figure 10-5 Circuit for Testing the Port B Status Change Interrupt.

- Global interrupts must be enabled by setting the GIE bit in the INTCON register.
- The RB port change interrupt must be enabled by setting the RBIE bit in the INTCON register.
- Internal pull-ups on port B should be disabled in the OPTION register.

The following code fragment from the program RB4to7Int in this book's online software package shows the required processing:

```
;==============================
;         main program
;==============================
main:
; Disable port B internal pullups
; Interrupts on falling edge of pushbutton action
    Movlw          b'10111111'
    option
; Wiring:
;       7  6  5  4  3  2  1  0  <= port B
;       |        |_____ red pushbutton
;       |_____ black pushbutton
;
```

```
;      7  6  5  4  3  2  1  0   <= port A
;                        |  |_____ red LED
;                        |_____ green LED
;
    movlw     b'00000000'    ; Set port A for ouput
    tris      porta
    movlw     b'11110000'    ; Port B bit 0-3 are output
                             ; bits 4-7 are input
    tris      portb          ; all others are output
    clrf      portb          ; All port B to 0
    movlw     b'00000000'    ; Zero to w
    movwf     bitsB47        ; Store in local variable
; Initially turn on LEDs
    bsf       porta,0        ; Set LEDs on line 0
    bsf       porta,1        ; and on line 1
;===========================
;      setup interrupts
;===========================
; Clear external interrupt flag (intf = bit 1)
    bcf       INTCON,rbif    ; Clear flag
; Enable global interrupts (gie = bit 7)
; Enable RB0 interrupt (inte = bit 4)
    bsf       INTCON,gie     ; Enable global int (bit 7)
    bsf       INTCON,rbie    ; Enable RB0 int (bit 3)
```

RB4-7 Change ISR

The Service Routine for the RB4-7 change interrupt will depend on the specific application. Nevertheless, the following processing steps should be considered:

- Determine if the source is an RB4-7 change interrupt.

- Clear the RBIF interrupt flag in the INTCON register.

- Save the context. Which registers and variables need to be saved depends on the specific application.

- Perform the interrupt action.

- Restore the context.

- Return from the interrupt with the retfie instruction.

Here again, the interrupt handler may have to perform operations that are specific to the application: for example, debounce a switch or initialize local variables. The following Interrupt Service routine is from the program RB4to7Int in the book's online software:

```
;========================================================
;               Interrupt Service Routine
;========================================================
; Service routine receives control whenever any of
; port B lines 4 to 7 change state
IntServ:
; First test: make sure source is an RB4-7 interrupt
    btfss     INTCON,rbif       ; RBIF flag is interrupt
    goto      notRBIF           ; Go if not RBIF origin
; Save context
    movwf     old_w             ; Save w register
```

```
      swapf     STATUS,w              ; STATUS to w
      movwf     old_status           ; Save STATUS
;=========================
;   interrupt action
;=========================
; The interrupt occurs when any of port B bits 4 to 7
; have changed status.
      movf      portb,w        ; Read port B bits
      movwf     temp           ; Save reading
      xorwf     bitsB47,f      ; Xor with old bits, result in f
; Test each meaningful bit (4 and 7 in this example)
      btfsc     bitsB47,4      ; Test bit 4
      goto      bit4Chng ; Routine for changed bit 4
; At this point bit 4 did not change
      btfsc     bitsB47,7      ; Test bit 7
      goto      bit7Chng       ; Routine for changed bit 7
; Invalid port line change. Exit
      goto      pbRelease
;=========================
; bit 4 change routine
;=========================
; Check for signal falling edge, ignore if not
bit4Chng:
      btfsc     portb,4        ; Is bit 4 high
      goto      pbRelease ; Bit is high. Ignore
; Toggling bit 1 of port A turns LED on and off
      movlw     b'00000010'    ; Xoring with a 1-bit produces
                               ; the complement
      xorwf     porta,f        ; Complement bit 1, port A
      goto      pbRelease
;=========================
; bit 7 change routine
;=========================
; Check for signal falling edge, ignore if not
bit7Chng:
      btfsc     portb,7        ; Is bit 7 high
      goto      exitISR        ; Bit is high. Ignore
; Toggling bit 0 of port A turns LED on and off
      movlw     b'00000001'    ; Xoring with a 1-bit produces
                               ; the complement
      xorwf     porta,f        ; Complement bit 1, port A
;
pbRelease:
      call      delay          ; Debounce switch
      movf      portb,w        ; Read port B into w
      andlw     b'10010000'    ; Eliminate unused bits
      btfsc     STATUS,z       ; Check for zero
      goto      pbRelease      ; Wait
; At this point all port B pushbuttons are released
;=========================
;       exit ISR
;=========================
exitISR:
; Store new value of port B
      movf      temp,w         ; This port B value to w
      movwf     bitsB47        ; Store
; Restore context
      swapf     old_status,w   ; Saved status to w
      movfw     STATUS         ; To STATUS register
      swapf     old_w,f        ; Swap file register in itself
```

```
     swapf     old_w,w          ; re-swap back to w
; Reset,interrupt
notRBIF:
     bcf             INTCON,rbif    ; Clear INTCON bit 0
     retfie
```

The processing by the Interrupt Service routine is straightforward. The code first determines which line caused the interrupt and takes the corresponding action in each case. In either case, the handler waits until all pushbuttons have been released before returning from the interrupt. This serves to debounce the switches.

10.5 Sample Programs

Three sample programs in this book's software package demonstrate the interrupt programming discussed in this chapter. The programs can be executed in the Interrupts Demo Board labeled Demo Board I, shown in Figure 10-6.

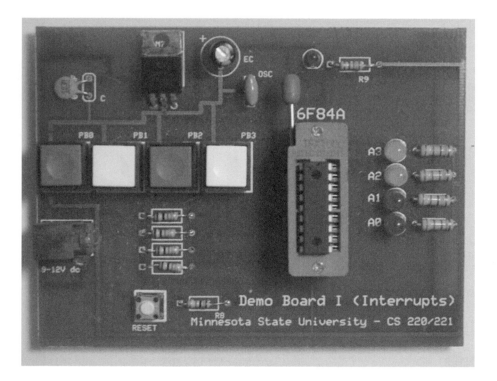

Figure 10-6 A Demo Board for Interrupts.

This book's online package contains support files and instructions for building the demo board in Figure 10.6. The sample programs to demonstrate interrupt programming are also found the in this package. Their function is as follows:

- The RB0Int program tests the interrupt on port RB0. A pushbutton switch, connected to this port, toggles an LED wired to port B, line 2. Another LED on port B, line 1, flashes on and off at one-half second intervals.

- The SleepDemo program uses the External Interrupt on port RB0 to terminate the power-down state caused by the SLEEP instruction. A pushbutton switch is connected to port RB0. This pushbutton generates the interrupt that ends the SLEEP condition. An LED on port B, line 1, flashes on and off at one-half second intervals for twenty iterations. At that time the program enters the SLEEP condition. Pressing the pushbutton switch on line RB0 gnerates the interrupt that ends the SLEEP.

- The RB4to7Int program tests the STATUS change interrupt. Pushbutton switches are connected to port B lines 4 and 7. A red LED is wired to port RA0 and a green LED to port RA1. The pushbuttons generate interrupts that toggle the LEDs on and off.

10.6 Demonstration Programs

The following programs demonstrate the programming discussed in this chapter.

10.6.1 RB0Int Program

```
; File: RB0Int.ASM
; Date: April 2, 2011
; Authors: Sanchez and Canton
; Processor: 16F84A
;
; Description:
; Program to test interrupt on port RB0
; A pushbutton switch is connected to port RB0.
; The pushbutton toggles an LED on port-B, line 2
; Another LED on port-B, line 1, flashes on and off
; at one-half second intervals
;===========================
;         switches
;===========================
; Switches used in __config directive:
;   _CP_ON        Code protection ON/OFF
; * _CP_OFF
; * _PWRTE_ON     Power-up timer ON/OFF
;   _PWRTE_OFF
;   _WDT_ON       Watchdog timer ON/OFF
; * _WDT_OFF
;   _LP_OSC       Low power crystal oscillator
; * _XT_OSC       External parallel resonator/crystal oscillator
;
;   _HS_OSC       High speed crystal resonator (8 to 10 MHz)
;                 Resonator: Murate Erie CSA8.00MG = 8 MHz
;   _RC_OSC       Resistor/capacitor oscillator (simplest, 20% ;
;                  error)
; |
; |_____ * indicates setup values


;===========================
```

```
; setup and configuration
;==========================
        processor 16f84A
        include    <p16f84A.inc>
        __config  _XT_OSC & _WDT_OFF & _PWRTE_ON & _CP_OFF
;========================================================
;                variables in PIC RAM
;========================================================
; Local variables
        cblock  0x0d              ; Start of block
        J                 ; counter J
        K                 ; counter K
        count1            ; Auxiliary counter
        count2            ; ISR counter
        old_w             ; Context saving
        old_STATUS        ;
        endc


;===========================================================
;                m a i n   p r o g r a m
;===========================================================
        org     0         ; start at address 0
        goto    main
;
;==============================
;       interrupt handler
;==============================
        org             0x04
        goto    IntServ
;==============================
;         main program
;==============================
main:
; Set up interrupt on falling edge
; by clearing OPTION register bit 6
        movlw   b'10111111'
        option
        movlw   b'11111111'       ; Set port a for input
        tris    PORTA
        movlw   b'00000001'       ; Port-B bit 0 is input
        tris    PORTB             ; all others are output
        clrf    PORTB             ; All port-B to 0
; Initially turn on LED
        bsf     PORTB,0           ; Set line 0 bit
;==========================
;     setup interrupts
;==========================
; Clear external interrupt flag (INTF = bit 1)
```

```
        bcf       INTCON,INTF      ; Clear flag
; Enable global interrupts (GIE = bit 7)
; Enable RB0 interrupt (INTE = bit 4)
        bsf       INTCON,GIE       ; Enable global int (bit 7)
        bsf       INTCON,INTE      ; Enable RB0 int (bit 4)
;============================
;          flash LED
;============================
; Program flashes LED wired to Port-B, line 2
lights:
        movlw     b'00000010'      ; Mask with bit 1 set
        xorwf     PORTB,f          ; Complement bit 1
        call      long_delay
        call      long_delay
        call      long_delay
        goto      lights
;========================================================
;             Interrupt Service Routine
;========================================================
; Service routine receives control when there is
; action on pushbutton switch wired to Port-B, line 0
IntServ:
; First test if source is an RB0 interrupt
        btfss     INTCON,INTF      ; INTF flag is RB0 interrupt
        goto      notRB0           ; Go if not RB0 origin
; Save context
        movwf     old_w            ; Save w register
        swapf     STATUS,w ; STATUS to w
        movwf     old_STATUS       ; Save STATUS
; Make sure that interrupt occurred on the falling edge
; of the signal. If not, abort handler
        btfsc     PORTB,0          ; Is bit set?
        goto      exitISR          ; Go if clear
;=========================
;   interrupt action
;=========================
; Debounce switch
;      Logic:
;       Debounce algorithm consists of waiting until the
; same level is repeated on a number of samplings of the
; switch. At this point the RB0 line is clear as the
; interrupt takes place on the falling edge. An initial
; short delay makes sure that spikes are ignored.
        movlw     D'10'            ; Number of repetitions
        movwf     count2           ; To counter
wait:
; Check to see that port-B bit 0 is still 0
; If not, wait until it changes
```

```
        btfsc    PORTB,0          ; Is bit set?
        goto     exitISR          ; Go if bit not 0
; At this point RB0 bit is clear
        decfsz   count2,f ; Count this iteration
        goto     wait             ; Continue if not zero
; Interrupt action consists of toggling bit 2 of
; port-B to turn LED on and off
        movlw    b'00000100'      ; Xoring with a 1-bit produces
                                  ; the complement
        xorwf    PORTB,f          ; Complement bit 2, port-B
;=========================
;         exit ISR
;=========================
exitISR:
; Restore context
        swapf    old_STATUS,w     ; Saved STATUS to w
        movfw    STATUS           ; To STATUS register
        swapf    old_w,f          ; Swap file register in itself
        swapf    old_w,w          ; re-swap back to w
; Reset,interrupt
notRB0:
        bcf      INTCON,INTF      ; Clear INTCON bit 1
        retfie
;=======================
;  Procedure to delay
;    10 machine cycles
;=======================
delay:
        movlw    D'4'             ; Repeat 12 machine cycles
        movwf    count1           ; Store value in counter
repeat
        decfsz   count1,f         ; Decrement counter
        goto     repeat           ; Continue if not 0
        return
;=============================
;   long delay sub-routine
;      (for debugging)
;=============================
long_delay
        movlw    D'200'   ; w = 200 decimal
        movwf    J                ; J = w
jloop:  movwf    K                ; K = w
kloop:  decfsz   K,f              ; K = K-1, skip next if zero
        goto     kloop
        decfsz   J,f              ; J = J-1, skip next if zero
        goto     jloop
        return
        end
```

10.6.2 SleepDemo Program

```
; File: SleepDemo
; Date: April 4, 2011
; Author: Canton and Sanchez
; Processor: 16F84A
;
; Description:
; Program to use the External Interrupt on port RB0
; to terminate the power-down state caused by the
; SLEEP instruction. A pushbutton switch is connected to
; port RB0. The pushbutton generates the interrupt that
; ends the SLEEP conditions.
; Demonstration:
; An LED on port-B, line 1, flashes on and off at 1/2
; second intervals for 20 iterations. At that time the
; program enters the SLEEP condition. Pressing the
; pushbutton switch on line RB0 generates the interrupt
; that ends the SLEEP.
;===========================
;          switches
;===========================
; Switches used in __config directive:
;   _CP_ON        Code protection ON/OFF
; * _CP_OFF
; * _PWRTE_ON     Power-up timer ON/OFF
;   _PWRTE_OFF
;   _WDT_ON       Watchdog timer ON/OFF
; * _WDT_OFF
;   _LP_OSC       Low power crystal oscillator
; * _XT_OSC       External parallel resonator/crystal oscillator

;   _HS_OSC       High speed crystal resonator (8 to 10 MHz)
;                 Resonator: Murate Erie CSA8.00MG = 8 MHz
;   _RC_OSC       Resistor/capacitor oscillator (simplest, 20%
;                 error)
; |
; |_____ * indicates setup values

;==========================
; setup and configuration
;==========================
        processor 16f84A
        include   <p16f84A.inc>
        __config _XT_OSC & _WDT_OFF & _PWRTE_ON & _CP_OFF


;=========================================================
```

```
;                 variables in PIC RAM
;========================================================
; Local variables
        cblock    0x0d              ; Start of block
        J                           ; counter J
        K                           ; counter K
        count1                      ; Auxiliary counter
        count2                      ; Second auxiliary counter
        old_w                       ; Context saving
        old_STATUS                  ;
        endc


;==========================================================
;                 m a i n   p r o g r a m
;==========================================================
        org     0              ; start at address 0
        goto    main
;
;==============================
;       interrupt handler
;==============================
        org              0x04
        goto    IntServ
;==============================
;       main program
;==============================
main:
; Set up interrupt on falling edge
; by clearing OPTION register bit 6
        movlw   b'10111111'
        option
        movlw   b'11111111'       ; Set port a for input
        tris    PORTA
        movlw   b'00000001'       ; Port-B bit 0 is input
        tris    PORTB             ; all others are output
        clrf    PORTB             ; All port-B to 0
;==============================
;       setup interrupts
;==============================
; Clear external interrupt flag (INTF = bit 1)
        bcf     INTCON,INTF       ; Clear flag
; Enable global interrupts (GIE = bit 7)
; Enable RB0 interrupt (INTE = bit 4)
        bsf     INTCON,GIE        ; Enable global int (bit 7)
        bsf     INTCON,INTE       ; Enable RB0 int (bit 4)
;==============================
;    flash LED 20 times
;==============================
```

```
wakeUp:
; Program flashes LED wired to port-B, line 2
; 20 times before entering the sleep state
        movlw   D'20'              ; Number of iterations
        movwf   count2             ; To counter
lights:
        movlw   b'00000010'        ; Mask with bit 1 set
        xorwf   PORTB,f            ; Complement bit 1
        call    long_delay
        call    long_delay
        call    long_delay
        decfsz  count2,f ; Decrement counter
        goto    lights
; 20 iterations have taken place
        sleep
        nop                        ; Recommended!
        goto    wakeUp             ; Resume execution


;===========================================================
;               Interrupt Service Routine
;===========================================================
; The interrupt service routine performs no operation
IntServ:
        bcf     INTCON,INTF               ; Clear flag
        retfie


;==============================
;   long delay sub-routine
;==============================
long_delay
        movlw   D'200'   ; w = 200 decimal
        movwf   J                 ; J = w
jloop:
        movwf   K                 ; K = w
kloop:
        decfsz  K,f               ; K = K-1, skip next if zero
        goto    kloop
        decfsz  J,f               ; J = J-1, skip next if zero
        goto    jloop
        return

        end
```

10.6.3 RB4to7Int Program

```
; File: RB4to7Int.ASM
; Date: May 26, 2011
```

```
; Authors: Sanchez and Canton
; Processor: 16F84A
;
; Description:
; Program to test the port-B, bits 4 to 7, STATUS
; change interrupt. Pushbutton switches are connected
; to port-B lines 4 and 7. A red LED is wired to port
; RA1 and a green LED to port RA0. The pushbuttons
; generate interrupts that toggle the LEDs on and off.
;===========================
;         switches
;===========================
; Switches used in __config directive:
;   _CP_ON          Code protection ON/OFF
; * _CP_OFF
; * _PWRTE_ON       Power-up timer ON/OFF
;   _PWRTE_OFF
;   _WDT_ON         Watchdog timer ON/OFF
; * _WDT_OFF
;   _LP_OSC         Low power crystal oscillator
; * _XT_OSC         External parallel resonator/crystal oscillator
;
;   _HS_OSC         High speed crystal resonator (8 to 10 MHz)
;                   Resonator: Murate Erie CSA8.00MG = 8 MHz
;   _RC_OSC         Resistor/capacitor oscillator (simplest, 20%
;                   error)
; |
; |_____ * indicates setup values

;=========================
; setup and configuration
;=========================
        processor 16f84A
        include   <p16f84A.inc>
        __config  _XT_OSC & _WDT_OFF & _PWRTE_ON & _CP_OFF
;=====================================================
;              variables in PIC RAM
;=====================================================
; Local variables
        cblock  0x0d      ; Start of block
        J                 ; counter J
        K                 ; counter K
        count1            ; Auxiliary counter
        count2            ; ISR counter
        old_w             ; Context saving
        old_STATUS        ;
        bitsB47           ; Storage for previous value
                                ; in port-B bits 4-7
```

```
        temp                    ; Temporary storage
        endc

;============================================================
;                   m a i n   p r o g r a m
;============================================================
        org     0               ; start at address 0
        goto    main
;
;=============================
;       interrupt handler
;=============================
        org             0x04
        goto    IntServ
;=============================
;       main program
;=============================
main:
; Disable port-B internal pullups
; Interrupts on falling edge of pushbutton action
        movlw   b'10111111'
        option
; Wiring:
;       7  6  5  4  3  2  1  0  <= port-B
;       |           |_____ red pushbutton
;       |_____ black pushbutton
;
;       7  6  5  4  3  2  1  0  <= Port-A
;                       |  |_____ red LED
;                       |_____ green LED
;
        movlw   b'00000000'     ; Set Port-A for ouput
        tris    PORTA
        movlw   b'11110000'     ; Port-B bit 0-3 are output
                                ; bits 4-7 are input
        tris    PORTB           ; all others are output
        clrf    PORTB           ; All Port-B to 0
        movlw   b'00000000'     ; Zero to w
        movwf   bitsB47         ; Store in local variable
; Initially turn on LEDs
        bsf     PORTA,0         ; Set LEDs on line 0
        bsf     PORTA,1         ; and on line 1
;=============================
;       set up interrupts
;=============================
; Clear external interrupt flag (intf = bit 1)
        bcf     INTCON,RBIF     ; Clear flag
; Enable global interrupts (GIE = bit 7)
```

```
; Enable RB0 interrupt (inte = bit 4)
        bsf     INTCON,GIE      ; Enable global int (bit 7)
        bsf     INTCON,RBIE     ; Enable RB0 int (bit 3)
;=============================
;          flash LED
;=============================
; Main program does nothing. All action takes place in
; Interrupt Service Routine
lights:
        nop
        goto    lights
;=======================================================
;                  Interrupt Service Routine
;=======================================================
; Service routine receives control whenever any of
; port-B lines 4 to 7 change state
IntServ:
; First test: make sure source is an RB4-7 interrupt
        btfss   INTCON,RBIF     ; RBIF flag is interrupt
        goto    notRBIF         ; Go if not RBIF origin
; Save context
        movwf   old_w           ; Save w register
        swapf   STATUS,w ; STATUS to w
        movwf   old_STATUS      ; Save STATUS
;==========================
;   interrupt action
;==========================
; The interrupt occurs when any of Port-B bits 4 to 7
; have changed STATUS.
        movf    PORTB,w         ; Read Port-B bits
        movwf   temp            ; Save reading
        xorwf   bitsB47,f       ; Xor with old bits,
                                ; result in f
; Test each meaningful bit (4 and 7 in this example)
        btfsc   bitsB47,4       ; Test bit 4
        goto    bit4Chng ; Routine for changed bit 4
; At this point bit 4 did not change
        btfsc   bitsB47,7       ; Test bit 7
        goto    bit7Chng ; Routine for changed bit 7
; Invalid port line change. Exit
        goto    pbRelease
;=======================
; bit 4 change routine
;=======================
; Check for signal falling edge, ignore if not
bit4Chng:
        btfsc   PORTB,4         ; Is bit 4 high
        goto    pbRelease       ; Bit is high. Ignore
```

```
; Toggling bit 1 of Port-A turns LED on and off
        movlw   b'00000010'       ; Xoring with a 1-bit produces
                                   ; the complement
        xorwf   PORTA,f           ; Complement bit 1, Port-A
        goto    pbRelease
;========================
; bit 7 change routine
;========================
; Check for signal falling edge, ignore if not
bit7Chng:
        btfsc   PORTB,7           ; Is bit 7 high
        goto    exitISR           ; Bit is high. Ignore
; Toggling bit 0 of Port-A turns LED on and off
        movlw   b'00000001'       ; Xoring with a 1-bit produces
                                   ; the complement
        xorwf   PORTA,f           ; Complement bit 1, Port-A
;
pbRelease:
        call    delay             ; Debounce switch
        movf    PORTB,w           ; Read port-B into w
        andlw   b'10010000' ; Eliminate unused bits
        btfsc   STATUS,Z ; Check for zero
        goto    pbRelease         ; Wait
; At this point all port-B pushbuttons are released
;==========================
;         exit ISR
;==========================
exitISR:
; Store new value of port-B
        movf    temp,w            ; This port-B value to w
        movwf   bitsB47           ; Store
; Restore context
        swapf   old_STATUS,w  ; Saved STATUS to w
        movfw   STATUS            ; To STATUS register
        swapf   old_w,f           ; Swap file register in itself
        swapf   old_w,w           ; re-swap back to w
; Reset,interrupt
notRBIF:
        bcf     INTCON,RBIF       ; Clear INTCON bit 0
        retfie


;======================
;  Procedure to delay
;   10 machine cycles
;======================
delay:
        movlw   D'6'              ; Repeat 18 machine cycles
        movwf   count1            ; Store value in counter
```

```
repeat:
        decfsz    count1,f          ; Decrement counter
        goto      repeat            ; Continue if not 0
        return
;==============================
;   long delay sub-routine
;      (for debugging)
;==============================
long_delay
        movlw     D'200'    ; w = 200 decimal
        movwf     J         ; J = w
jloop:
        movwf     K         ; K = w
kloop:
        decfsz    K,f       ; K = K-1, skip next if zero
        goto      kloop
        decfsz    J,f       ; J = J-1, skip next if zero
        goto      jloop
        return

        end
```

Chapter 11

Timers and Counters

11.1 Controlling the Time Lapse

The mid-range PICs contain facilities and devices for controlling and manipulating *time lapses* in a program; these are usually in the form of timing and counting operations. It is difficult to find a PIC application that does not require some form of counting or timing. In the programs previously developed we have used a simple iteration counter that produces a delay by means of a loop that implements a delay by wasting a series of machine cycles. In this chapter we expand the theory and the use of delay loops, and explore the use of the *built-in timing and counting circuits* on the 16F84. The material presented relates to Chapter 10 because timing and counting operations can be set up to generate interrupts. The interrupt-based sample programs developed in this chapter can be executed and tested in Demo Board I described in Chapter 10.

11.1.1 16F84 Timer0 Module

The basic timer facility on the 16F84 PIC is known as the Timer0 module, as the free-running timer, as the timer/counter, or simply as TMR0. Timer0 is an internal 8-bit register that increments automatically with every PIC instruction cycle until the count overflows the timer capacity, which takes place when the timer count goes from 0xff to 0x00. At that time the timer restarts the count.

   The Timer0 module is in fact a peripheral device that adds a specific functionality to the microcontroller. Learning to program the Timer0 module serves as an introduction to programming PIC peripheral devices. There are few such devices in the 16F84 but other mid-range PICs (such as the 16F877 and the 16F684) contain other peripherals. These include several timers, capture/compare modules, various serial and parallel interfaces, USART and other communications hardware, comparators, and converters.

   The 16F84 Timer0 module has the following characteristics:

- A timer register that is readable and writeable by software
- Can be powered by an external or internal clock
- Timing edge for external clock can be selected
- 8-bit software programmable prescaler
- Interrupt capability
- Can be used as a timer or as a counter

Figure 11-1 is a simplified block diagram of the Timer0 hardware.

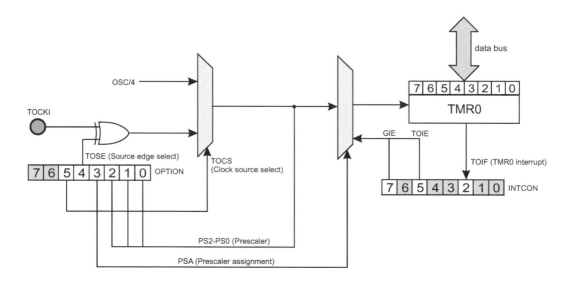

Figure 11-1 Timer0 Block Diagram.

11.1.2 Timer0 Operation

Block diagrams, such as the one in Figure 11-1, provide valuable and compact circuit information. In this example the block diagram, which is read left to right, shows that timer operation can be assigned to the internal clock (OSC/4 in the diagram) or to the PICs TOCKI pin that is shared with port RA4. Bit 5 of the OPTION register (labeled TOCS in the diagram) performs the selection. If TOCS is set, then the timer is linked to the RA4/TOCKI pin. In this mode the timer is used as a counter. If TOCS is reset, then the timer uses the PIC's instruction cycle clock signal. The shaded trapezoids in the illustration depict multiplexer functions, sometimes called *mux* (see Section 5.5.1). Bits inside the multiplexer symbol show which input is selected by the TOCS bit.

If an external source is selected by setting the TOCS bit, then bit 4 of the OPTION register (labeled TOSE) allows selecting whether the timer increments on the high-to-low or low-to-high transition of the signal on the RA4/TOCKI pin. The XOR symbol with inputs from the TOCKI pin and bit 4 of the OPTION register indicate this action. If the TOSE bit is set, then increment takes place on the high-to-low transition of the TOCKI pin, as it corresponds with the logical XOR operation. Otherwise the increment takes place on the low-to-high transition of the TOCKI pin. The gray shading of bits 6 and 7 of the OPTION register (and of other register bits in the diagram) indicate that these bits are not used in configuring Timer0 operations.

The rightmost multiplexer symbol depicts the control of the prescaler function. The prescaler serves as a frequency divider for the signal and allows slowing down the clock action on Timer0. As shown in Figure 11-1, the control bit for the prescaler function is bit 3 of the OPTION register, labeled PSA. If this bit is clear, the prescaler is assigned to the Timer0 module. If the PSA bit is set, then the prescaler is assigned to the Watchdog timer. Bits 0, 1, and 2 of the OPTION register define eight possible prescaler settings.

The Timer0 module can be visualized as a register that increments with every instruction cycle at one-fourth the clock rate, that is, as a timer. In a PIC equipped with a 4-Mhz oscillator, the timer register increments at a rate of one pulse per millisecond without the prescaler. Because there are eight bits in the counter register, the value stored is in the range 0 to 255 decimal (shown as TMR0 in the block diagram of Figure 11-1). When the counter overflows, the register is reset. The prescaler allows slowing down this rate. The various settings of the prescaler bits of the OPTION register (bits 0, 1, and 2) allow selecting one of eight possible rate divisors (2, 4, 8, 16, 32, 64, 128, or 256)

Timer0 Interrupt

PIC applications sometimes read the timer register (TMR0) directly. Alternatively, Timer0 can be set up to generate an interrupt at every transition from 0xff to 0x00, that is, every time the counter overflows. The timer register can be accessed in bank 0, offset 0x01. The timer interrupt is enabled by setting bit 5 (labeled TOIE) of the INTCON register. In this case the *Global Interrupt Enable bit* (labeled GIE) of the INTCON register must also be set. Once the timer interrupt is enabled, the *Timer Interrupt Flag*, assigned to bit 2 of the INTCON register (labeled TOIF) will be set on every overflow of the timer register. At that time an interrupt will take place. The TOIF bit (also called the Timer0 flag) must be cleared by the interrupt handler so that the timer interrupt can take place again. Later in this chapter we develop a sample program that uses Timer0 as an interrupt source.

Timer0 Prescaler

We have seen that the counter prescaler consists of the three low-order bits in the OPTION register. These bits allow selecting eight possible values that serve as a divisor for the counter rate. When the prescaler is disabled, the counter rate is one-fourth the processor's clock speed. If the prescaler is set to the maximum value (255) then one of 255 clock signals actually reach the timer. Table 11.1 shows the prescaler settings and their action on the rate of the Timer0 module.

Note that the prescaler can be assigned to either Timer0 or the Watchdog timer, but not to both. If bit 3 of the OPTION register is set, then the prescaler is assigned to the Watchdog timer. If it is clear it, is assigned to the Timer0 module.

Table 11.1

Prescaled Bits Selected Rates

BIT VALUE		TMR0 RATE
BINARY	HEX	
000	0x00	1:2
001	0x01	1:4
010	0x02	1:8
011	0x03	1:16
100	0x04	1:32
101	0x05	1:64
110	0x06	1:128
111	0x07	1:256

11.2 Delays Using Timer0

The simplest application of the Timer0 module is to count instruction cycles in a simple delay loop. Applications in which the Timer0 register is polled directly are said to use a *free running timer.* There are two advantage in free running timers over conventional delay loops: the prescaler provides a way of slowing down the count, and the delay is independent of the number of machine cycles in the loop body. In most cases this means that it is easier to implement an accurate time delay using the Timer0 module than by counting instruction cycles. The drawback is that Timer0 must first be set up for this mode of operation.

Calculating the time taken by each counter iteration requires that we first divide the clock speed by four to obtain the instruction speed. We know that a PIC running on a 4-MHz oscillator clock increments the counter every 1 MHz. If the prescaler is not used, the counter register is incremented at a rate of 1 μs, which is the same as saying that the Timer0 clock beats at a rate of 1,000,000 cycles per second. If the prescaler is set to the maximum divisor value (256), then each increment of the timer takes place at a rate of 1,000,000/256 μs, which is approximately 3,906 μs or 3.906 ms. Because this is the slowest possible rate of the timer in a machine running at 4 MHz, it is often necessary to employ supplementary counters in order to achieve longer delays.

The fact that the timer register (TMR0) is both readable and writeable makes possible some interesting timing techniques. For example, an application can set the timer register to an initial value and then count up until the timer overflows or a predetermined limit is reached. If the difference between the limit and the initial value is, let's say, 100, then the routine will count 100 times the timer rate per beat instead of 256.

Two simple elements can be used in designing a timing routine with a specific delay: the number of timer iterations counted by the Timer0 hardware and the prescaler assignment. Suppose there is a routine that allows Timer0 to start from zero and count up. In this case when the count reaches the maximum value (0xff), the routine would have introduced a delay of 256 times the Timer0 clock rate. In this case a prescaler assignment can be used to further reduce the number of clock beats read by the timer. If we select the maximum value for the prescaler (assuming a machine clock of 4 MHz) then each timer beat will take place at a rate of 1,000,000/256, or approximately 3,906 timer beats per second. Because in this example the routine reads the maximum number of iterations in the counter register (256), the delay can be calculated by dividing the number of beats per second (3,906) by the number of counts in the delay loop. In this case 3,906/256 results in a delay of approximately 15.26 iterations per second.

A general formula for calculating the number of timer beats per second is as follows:

$$T = \frac{C}{4PR}$$

where T is the number of clock beats per second, C is the system clock speed in Hz, P is the value stored in the prescaler, and R is the number of iteration counted in the TMR0 register. The range of both P and R in this formula is from 1 to 256. Also notice that the reciprocal of T ($1/T$) gives the time delay, in seconds, per iteration of the delay routine.

11.2.1 Long Delay Loops

In the previous section we saw that even using the largest possible prescaler and counting the maximum number of timer beats, the longest possible timer delay in a 4-MHz system is approximately of 1/15 of a second. Add to this that applications must sometimes devote the prescaler to the Watchdog timer, which impedes its use in Timer0. Without the prescaler the maximum delay is approximately 3,906 timer beats per second. Consequently, applications that must measure time in seconds (or in minutes, hours, or days) must find ways of keeping count of large number of repetitions of the timer beat.

In implementing counters for larger delays we must be careful not to introduce round-off errors. For instance, in the previous example, a timer cycles at the rate of 15.26 times per second. The closest integer to 15.25 is 15, so if we now set up a one-second counter by counting fifteen iterations, we would introduce an error of approximately 2%. Considering that (in the previous example) each iteration of the timer contains 256 individual beats, there are 3,906.25 individual timer beats per second at the maximum prescaled rate. If we were to implement a timer by keeping track of the individual prescaled beats (instead of timer iterations), the count would proceed from 0 to 3,906 instead of from 0 to 15. Approximating 3,906.25 by the closest integer, 3,906, introduces a much smaller round-off error than what results from approximating 15.26 with 15.

Finally, in this same example, we could eliminate the prescaler so that the timer beats at the clock rate, that is, at 1,000,000 beats per second. In this option a counter that counts from 0 to 1,000,000 would have no intrinsic error due to round-off.

Which solution is more adequate depends on the accuracy required by the application and the resulting complexity of the code. A timer counter in the range 0 to 15 can be implemented in a single 8-bit register. A counter in the range 0 to 3,906 requires two bytes. One to count from 0 to 1,000,000 requires three bytes. Because arithmetic operations in the 16F84 are 8-bits, manipulating multiple-register counters requires more complicated processing.

How Accurate Is the Delay?

The actual implementation of a delay routine based on multi-byte counters presents some difficulties. If the timer register (TMR0) is used to keep track of timer beats, then detecting the end of the count presents a subtle problem. Intuitively thinking, our program could detect timer overflow by reading the value in TMR0 and testing the zero flag in the status register. Because the **movf** instruction affects the zero flag, one could be tempted to code:

```
wait:
      movf      tmr0,w         ; Timer value into w
      btfss     status,z       ; Was it zero?
      goto      wait
; If this point is reached tmr0 has overflowed
```

But there is a problem: the timer ticks as each instruction executes. Because the **goto** instruction takes two machine cycles, it is possible that the timer overflows while the **goto** instruction is in progress; therefore the overflow condition would not be detected. A solution to this problem proposed in the Microchip documentation is to check for less than a nominal value by testing the carry flag, as follows:

```
wait1:
      movlw     0x03           ; 3 to w
      subwf     tmr0,w         ; Subtract w - tmr0
      btfsc     status,c       ; Test carry
      goto      wait1
```

One adjustment that is sometimes necessary in free running timers results from the fact that when the TMR0 register is written, the count is inhibited for the following two instruction cycles. Software can usually compensate for this by writing an adjusted value to the timer register. If the prescaler is assigned to Timer0, then a write operation to the timer register determines that the timer will not increment for four clock cycles.

A more elegant and accurate solution has been described by Roman Black in a Web article titled "Zero-error One Second Timer". Black credits Bob Ammerman with the suggestion of using *Bresenham's algorithm* for creating accurate PIC timer periods. In the Black-Ammerman method the counter works in the background, either by polling or Interrupt-Driven. So the program can continue executing while the counter runs. In either case the timer count value is stored in a 3-byte register that is decremented by the software. The Internet article offers through coverage of this algorithm.

11.3 Timer0 as a Counter

Timer0 operations can be assigned to the PIC RA4/TOCKI pin by setting bit 5 of the OPTION register (labeled TOCS). This mode is referred to as the *counter mode*. When the timer is setup to work as a counter, then bit 4 of the OPTION register (labeled

TOSE) allows selecting whether the counter increments on the high-to-low or low-to-high transition of the signal.

When an external clock input is present in the RA4/TOCKI pin, the code must meet certain requirements in order to ensure that the external source can be synchronized with the internal clock. When no prescaler is used, the external clock input must be high and low for at least twice the internal clock rate. In addition, there must be a resistor-capacitor induced delay of 20 ns on both the high and low cycles. When a prescaler is used, the external clock input must be high and low for at least four times the rate of the internal clock rate. In addition, there must be a resistor-capacitor induced delay of 40 ns on both the high and low cycles.

Once the counter mode is enabled, any pulse on pin RA4/TOCKI is automatically counted in the TMR0 register. The mechanism can be compared to an *automatic interrupt* because no program action is required to keep track of the number of pulses. The routine can be coded so that when the timer count overflows an interrupt is generated. The interrupt handler can then increment a supplementary counter so that events that exceed 256 pulses can be recorded.

The program named TMR0Counter developed later in this chapter and contained in this book's online software is an example of using the counter function of the Timer0 module.

11.4 Timer0 Programming

Software routines that use the Timer0 module range in complexity from simple delay loops to configurable, Interrupt-Driven counters that must meet high timing accuracy requirements. When the time period to be measured does not exceed the one that can be obtained with the prescaler and the timer register count, then the coding is straightforward and the processing is uncomplicated. But often this is not the case. The following elements should be examined before attempting to design and code a Timer0-based routine:

1. What is the required accuracy of the timer delay?
2. Can the prescaler be used, or is the prescaler devoted to the Watchdog timer?
3. Does the program suspend execution while the delay is in progress, or does the application continue executing in the foreground?
4. Can the timer be Interrupt-Driven, or must it be polled?
5. Will the delay be the same on all calls to the timer routine, or must the routine provide delays of different magnitudes?
6. How long must the delay last?

In this section we explore several timer routines of different complexity and meeting various constraints. The first one uses the Timer0 module as a counter, as described in Section 11.3. Later we develop a simple delay loop that uses the Timer0 register instead of an instruction count. We conclude with an Interrupt-Driven timer routine that can be changed to implement different delays.

11.4.1 Programming a Counter

The 16F84 can be programmed so that port RA4/TOCKI is used to count events or pulses. This is accomplished by initializing the Timer0 module as a counter. If interrupts are not used, the process requires the following preparatory steps:

1. Port A, line 4, (RA4/TOCKI) is defined for input.

2. The Timer0 register (TMR0) is cleared.

3. The Watchdog timer internal register is cleared by means of the CLRWDT instruction.

4. The OPTION register bits PSA and PS0:PS2 are initialized if the prescaler is to be used.

5. The OPTION register bit TOSE is set so as to increment the count on the high-to-low transition of the port pin if the port source is active low. Otherwise the bit is cleared.

6. The OPTION register bit TOCS is set to select action on the RA4/TOCKI pin.

Once the timer is set up as a counter, any pulse received on the RA4/TOCKI pin (as long as it meets the restrictions mentioned in Section 11.3) is counted in the TMR0 register. Software can read and write to the TMR0 register, located at address 0x01 in bank 0, in order to obtain or change the event count. If the timer interrupt is enabled when the timer is defined as a counter, the interrupt takes place every time the counter overflows, that is, when the count cycles from 0xff to 0x00.

Timer/Counter Test Circuit

The circuit shown in Figure 11-2 contains a pushbutton switch wired to port RA4/TOCKI and a Seven-Segment LED display wired to port B lines 0 to 6.

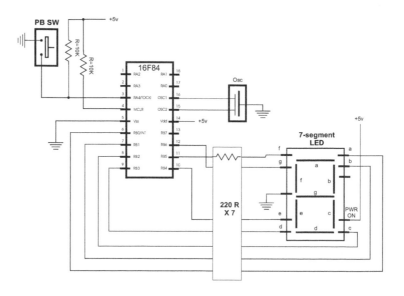

Figure 11-2 Test Circuit for Timer/Counter Program.

Note that the circuit in Figure 11-2 is a section of the one used by Virtual Board A (in Figure 9.7). Programs that execute in the circuit of Figure 11-2 will also run in Virtual Board A.

TimerCounter Program

The program named TimerCounter in this book's online software package uses the circuit in Figure 11-2 (and in Virtual Board A) to demonstrate the programming of the Timer0 module in the counter mode. The program detects and counts action on the pushbutton switch wired to port RA4/TOCKI. The count (in hex digits in the range 0x00 to 0x0f) is displayed in the Seven-Segment LED connected to port B. When using Virtual Board A to demonstrate the program, make sure that toggle switch J1 is closed at the time the program executes. The programming of Seven-Segment LEDs was discussed in Chapter 9.

Code Details

Code first clears the TMR0 register and the Watchdog timer, as follows:

```
main:
; Clear the timer and the Watchdog
    clrf        TMR0
    clrwdt
```

Next, the OPTION register must be set up as follows:

- Disable the prescaler (accomplished by setting the PSA bit that assigns it to the Watchdog timer).

- Set the Timer0 edge select bit (TOSE) to increment in the high-to-low signal transition.

- Select the RA4/TOCKI pin as a beat source.

The bits of the OPTION register that are not meaningful for this mode of Timer0 operation are set to their default state, as follows:

```
; Set up the OPTION register. First move value into w
    movlw       b'10111000'
; Note that OPTION register is in bank 1 and its register
; name is OPTION_REG
    bsf         STATUS,RP0      ; RP0 is bank select bit
    movwf       OPTION_REG      ; Copy w to OPTION
    bcf         STATUS,RP0      ; Bank 0
```

Now port B must be "trissed" for output and port A for input. Port B is also cleared:

```
; Setup ports
    movlw       0x00            ; Set port B to output
    tris        PORTB
    clrf        PORTB           ; All port B to 0
; Port A. Five low-order lines set for input
    movlw       B'00011111'     ; w = 00011111 binary
    tris        PORTA           ; port A (lines 0 to 4) to input
```

At this point, every time the pushbutton switch wired to port A, line RA4/TOCKI, is pressed, one is added to the value in TMR0. Because the value to be displayed in the Seven-Segment LED goes from 0 to 0xf, the value in TMR0 must be scaled to this range. This is easily accomplished by masking off the four high-order bits of TMR0 at display time. Note that a binary counter is reminiscent of the odometer in an automobile, that is, the count proceeds from the lower digits to the higher ones. Therefore masking off a series of higher-order digits does not affect the count kept in the lower-order digits. More clearly, if you were to place a piece of tape over the three high-order digits of a car's odometer, the value in the uncovered digits would continue to be correct. Code reads the value in TMR0 into w, masks off the four high-order bits, displays the resulting value, and loops back endlessly, as follows:

```
checkTmr0:
      movf       TMR0,w          ; Timer register to w
; Eliminate four high order bits
      andlw      b'00001111'     ; Mask off high bits
; At this point the w register contains a 4-bit value
; in the range 0 to 0xf. Use this value (in w) to
; obtain Seven-Segment display code
      call       segment
      movwf      PORTB           ; Display switch bits
      goto       checkTmr0       ; Endless loop
```

11.4.2 Timer0 as a Delay Timer

Another simple use of the Timer0 module is to implement a delay loop. In this application the Timer0 module is initialized to use the internal clock by setting the TOSE bit of the OPTION register. If the prescaler is to be used, as is often the case, the PSA bit is cleared and the desired pre-scaling is entered in bits PS2 to PS0 of the OPTION register.

Delay Timer Circuit

The circuit in Figure 11-3 allows testing several timer-related programs developed in this chapter. Notice that this circuit is a section of the one for Virtual Board A in Figure 9.7; therefore applications developed for this circuit will also run in the Virtual Board A program.

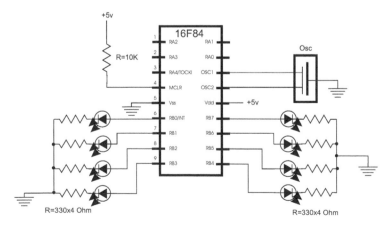

Figure 11-3 Circuit for Testing Several Timer Programs.

11.4.3 DelayTimer Program

The program named DelayTimer, in this book's online software package, uses a timer-based delay loop to flash eight LEDs displaying a binary value from 0x00 to 0xff. The delay routine executes in the foreground, so that processing is suspended while the count is in progress.

Code Details

Code first clears the Watchdog timer:

```
; Clear the Watchdog timer and reset prescaler
    clrwdt
```

The prescaler is assigned to Timer0 by clearing the PSA bit, and bits PS2 to PS0 are set to assign a 1:256 prescaler to the timer. The bits of the OPTION register that are not meaningful for this mode of Timer0 operation are set to their default state. Code is as follows:

```
; Set up the OPTION register
    movlw     b'11010111'
;   7  6  5  4  3  2  1  0 <= OPTION bits
;   |  |  |  |  |  |  |__|__|_____ PS2-PS0 (prescaler bits)
;   |  |  |  |  |  |          Values for Timer0
;   |  |  |  |  |  |          000 = 1:2    001 = 1:4
;   |  |  |  |  |  |          010 = 1:8    011 = 1:16
;   |  |  |  |  |  |          100 = 1:32   101 = 1:64
;   |  |  |  |  |  |          110 = 1:128 *111 = 1:256
;   |  |  |  |  |_____ PSA (prescaler assign)
;   |  |  |  |                 1 = to WDT
;   |  |  |  |                *0 = to Timer0
;   |  |  |  |_____ TOSE (Timer0 edge select)
;   |  |  |                    0 = increment on low-to-high
;   |  |  |                   *1 = increment on high-to-low
;   |  |  |_____ TOCS (TMR0 clock source)
;   |  |                       *0 = internal clock
;   |  |                        1 = RA4/TOCKI bit source
;   |  |_____ INTEDG (Edge select)
;   |                          *0 = falling edge
;   |_____ RBPU (Pullup enable)
;                              0 = enabled
;                             *1 = disabled
; Note that OPTION register is in bank 1 and its register
; name is OPTION_REG
    bsf       STATUS,RP0        ; RP0 is bank select bit
    movwf     OPTION_REG        ; Copy w to OPTION
    bcf       STATUS,RP0        ; Bank 0
```

Port B is then "trissed" for output and cleared. Because the program does not use port A, it is ignored by the code.

```
; Setup ports
    movlw     0x00            ; Set port B to output
    tris      portb
    clrf      portb           ; All port B to 0
```

The program's main loop consists of adding one to the value in port B so as to bump the count. The delay routine named TM0delay is then called, followed by a goto instruction to the loop start.

```
mloop:
    incf       portb,f          ; Add 1 to register value
    call       TM0delay
    goto       mloop
```

The delay procedure named TM0delay provides the necessary time lapse between successive increments in the count displayed. First the TMR0 register is cleared to make sure the timing starts correctly. Then the value in TMR0 is read into the w register and the constant 0xff is subtracted from W. When the count reaches 0xff, this subtraction sets the zero flag. Until that happens, code loops to a label named *cycle*. The routine returns to the caller when the time lapses.

```
TM0delay:
; Initialize the timer register
    clrf       tmr0             ; Clear SFR for Timer0
; Routine tests the value in the tmr0 register by
; subtracting 0xff from the value in tmr0. The zero flag
; is set if tmr0 = 0xff
cycle:
    movf       tmr0,w           ; Timer to w
; w has tmr0 register value
    sublw      0xff             ; Subtract max value
; Zero flag is set if value in tmr0 = 0xff
    btfss      status,z         ; Test for zero
    goto       cycle            ; Repeat
    return
```

11.4.4 Variable Time Lapse

A time-lapse routine can be designed so it can be modified (or reconfigured at call time) to produce a specific delay. This adjustable time-lapse procedure can be a useful tool in any programmer's library. In previous sections we have developed delay routines that do so by counting timer pulses. This same idea can be used to develop routines that can be adjusted so as to produce accurate delays within a range by modifying the value of the count.

The routine itself can be implemented to varying degrees of sophistication. One variation is a procedure that receives the desired time lapse as a parameter. Another one is a procedure that reads the desired time lapse from program constants, which can be edited at assembly time. A third option, used by the program VariLapse in this book's software package, uses three variables to hold the number of machine cycles in the desired wait period. By using machine cycles instead of time units (such as microseconds or milliseconds), the procedure becomes adaptable to devices running at different clock speeds. Because each instruction (or Timer0 iteration) requires four clock cycles, the device's clock speed in Hz is divided by four in order to determine the number of machine cycles per time unit.

For example, a processor equipped with an 8-MHz clock executes at a rate of 8,000,000/4 machine cycles per second, that is, 2,000,000 instruction cycles per second. To produce a one-quarter second delay requires a wait period of 2,000,000/4 or 500,000 instruction cycles. By the same token, a 16F84 running at 4 MHz executes 1,000,000 instructions per second. In this case, a one-quarter second delay would require waiting 250,000 instruction cycles.

11.4.5 Variable Lapse Timer Program

The program titled VariLapse in this book's software package uses Timer0 to produce a *variable-lapse delay*. The delay is calculated based on the number of machine cycles necessary for the desired wait period, as described in the preceding paragraph. The program uses the Black-Ammerman methods described earlier in this chapter, which require a prescaler of 1:2 so that each timer iteration takes place at one-half the clock rate.

Code Details

The program first initializes the OPTION register in order to achieve the following setup:

- The prescaler is assigned to Timer0 (PSA bit) and a 1:2 rate is selected in OPTION_REG bits 0 to 2.

- The internal clock is selected as a timing source (TOCS bit).

- Timing count is set to increment in the high-to-low signal transition (TOSE bit).

- OPTION register bits not used by the program are set to their default values.

- Next, port B is selected for output and cleared. Port A is ignored by the code because it is not used by the program.

The VariLapse program requires a one-half second delay and assumes a 16F84 running at 4 MHz. The delay procedure must wait for 500,000 clock beats to take place before returning to the caller. The value 500,000 requires three storage bytes. This is easily seen by converting 500,000 to hexadecimal: 500,000 = 0x07a120. At this point a simple decimal-to-hex calculator would be convenient. Several screen-based hex calculators (including a screen version of the HP 16C) are available free on the Internet. The program defines the hex equivalent of 500,000 in three constants:

```
; Constants for a 1/2 second delay on a 4 MHz machine
; 500,000 decimal = 0x07a1x20
;
;              0x07 0xa1 0x20
;              ---- ---- ----
;                |    |    |___ low_byte
;                |    |_____ mid_byte
;                |_____ high_byte
;
low_byte   equ  0x20
mid_byte   equ  0xa1
high_byte equ  0x07
```

The advantage of using **equ** directives in defining these values is that they can easily be edited in order to accommodate a different processor speed or to change the delay. In the sample program we coded a procedure called setDelay, which initializes three counter variables using the values contained in the constants low_byte, mid_byte, and high_byte. The setDelay procedure is coded as follows:

```
; Procedure to initialize local variables for a delay defined
; in program constants. For a one-half second on a 16F84 at
; 4 MHz initialization values are as follows:
;
```

```
; 500,000 = 0x07 0xa1 0x20
;           ---- ---- ----
;            |    |    |___ low_byte
;            |    |_____ mid_byte
;            |_____ high_byte
setDelay:
    movlw       high_byte
    movwf       countH
    movlw       mid_byte
    movwf       countM
    movlw       low_byte
    movwf       countL
    return
```

The logic for a 3-byte counter consists of decrementing the low-order counter until it expires, then the mid-order counter, and when it goes to zero the high-order counter is decremented. If the high-order counter expires, the delay concludes. The flowchart in Figure 11-4 shows this sequence of operations.

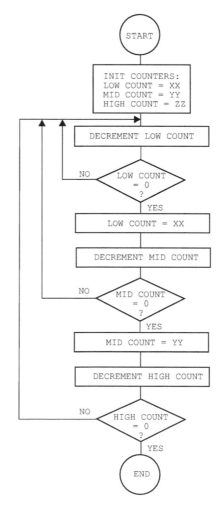

Figure 11-4 Multiple Counter Flowchart.

Roman Black points out in his Web article (mentioned earlier in this chapter) that one of the difficulties of timing routines based on counters is that the exact count is lost by the manipulations necessary to detect the end of each counter iteration. One possible solution is to use interrupts to detect the end of the count. Another one (proposed by Black) is based on introducing a 1:2 delay in the timer by means of the prescaler. Because the prescaled timer beats at one-half the instruction rate, 128 timer cycles will be required for one complete iteration at the full instruction rate. This allows us to test the state of the counter's high-order bit to detect when the count has expired. A new count can be started by clearing this high-order bit without affecting the validity of the count held in the remaining seven bits. The result is a timer-counter that does not lose step while detecting the end of each iteration level.

The delay routine uses the Timer0 register to provide the low-order level of the count. Because the counter counts up from zero, in order to ensure that the initial low-level delay count is correct, the value 128 − (xx/2) must be calculated, where xx is the value in the original countL register. The program performs the division by 2 by shifting bits to the right one position. The resulting value is subtracted from 128 and the result stored in TMR0, as follows:

```
; First calculate xx/2 by bit shifting
     bcf        STATUS,C        ; Clear carry flag
     rrf        countL,f        ; Divide by 2
; now subtract 128 - (xx/2)
     movf       countL,w        ; w holds low-order byte
     sublw      .128
; Now w has adjusted result. Store in TMR0
     movwf      TMR0
```

The delay routine detects timer overflow by testing bit 7 of the TMR0 register. If the bit is set, then 256 time cycles have elapsed (except in the first iteration) and the mid-order counter register is decremented. If the mid-order register overflows when it is decremented, then the high-order register is decremented. If it overflows, the counter has gone to zero and the delay routine ends, as follows:

```
; Routine tests timer overflow by testing bit 7 of
; the TMR0 register.
cycle:
     btfss      TMR0,7          ; Is bit 7 set?
     goto       cycle           ; Wait if not set
; At this point TMR0 bit 7 is set
; Clear the bit
     bcf        TMR0,7          ; All other bits are preserved
; Subtract 256 from beat counter by decrementing the
; mid-order byte
     decfsz     countM,f
     goto       cycle           ; Continue if mid-byte not zero
; At this point the mid-order byte has overflowed.
; High-order byte must be decremented.
     decfsz     countH,f
     goto       cycle
; At this point the time cycle has elapsed
     return
```

The circuit in Figure 11-3 and the Virtual Board A application can be used to test the VariLapse program.

11.4.6 Interrupt-Driven Timer

Interrupt-driven timers and counters have two major advantages over polled routines:

1. The time lapse counting take place in the background so that the application can continue to do other work in the foreground.

2. Processing can take place without the prescaler, which can be used for the Watchdog timer.

Developing a timer routine that is Interrupt-driven is quite similar to developing any other interrupt routines. The initialization consists of configuring the OPTION and the INTCON register bits for the task at hand. In the particular case of an Interrupt-Driven timer, the following are necessary:

1. The external interrupt flag (INTF in the INTCON Register) must be initially cleared.

2. Global interrupts must be enabled by setting the GIE bit in the INTCON register.

3. The Timer0 overflow interrupt must be enabled by setting the TOIE bit in the INTCON register.

11.4.7 TimerInt Program

The TimerInt program is an Interrupt-Driven version of the VariLapse program developed earlier. Here again, code uses Timer0 to produce a variable-lapse delay. The delay of one-fourth second is calculated based on the number of machine cycles necessary for the desired wait period.

Code Details

In the TimerInt program the prescaler is not used with the timer, so the initialization code sets the PSA bit in the OPTION register so that the prescaler is assigned to the Watchdog timer. The following code fragment is from the TimerInt program:

```
; Clear the Watchdog timer and reset prescaler
    clrf TMR0
    clrwdt
; Set up the OPTION register
    movlw    b'11011000'
;   7  6  5  4  3  2  1  0 <= OPTION bits
;   |  |  |  |  |  |  |__|__|_____ PS2-PS0 (prescaler bits)
;   |  |  |  |  |  |            Values for Timer0
;   |  |  |  |  |  |            000 = 1:2    001 = 1:4
;   |  |  |  |  |  |            010 = 1:8    011 = 1:16
;   |  |  |  |  |  |            100 = 1:32   101 = 1:64
;   |  |  |  |  |  |            110 = 1:128 *111 = 1:256
;   |  |  |  |  |_____ PSA (prescaler assign)
;   |  |  |  |                  *1 = to WDT
;   |  |  |  |                   0 = to Timer0
;   |  |  |  |_____ TOSE (Timer0 edge select)
;   |  |  |                     0 = increment on low-to-high
;   |  |  |                    *1 = increment on high-to-low
;   |  |  |_____ TOCS (TMR0 clock source)
;   |  |                        *0 = internal clock
```

```
;    |   |                              1 = RA4/TOCKI bit source
;    |   |_____ INTEDG (Edge select)
;    |                               *0 = falling edge
;    |_____ RBPU (Pullup enable)
;                                    0 = enabled
;                                   *1 = disabled
; Note that OPTION register is in bank 1 and its register
; name is OPTION_REG
      bsf        STATUS,RP0    ; RP0 is bank select bit
      movwf      OPTION_REG    ; Copy w to OPTION
      bcf        STATUS,RP0    ; Bank 0
; Setup ports
      movlw      0x00          ; Set port B to output
      tris       PORTB
      clrf       PORTB         ; All port B to 0
; Port A is not used in this program
```

Interrupts are set up clearing the INTF flag in the INTCON register and setting the GIE and T0IE bits.

```
;=============================
;      setup interrupts
;=============================
; Clear external interrupt flag (INTF = bit 1)
      bcf        INTCON,INTF   ; Clear flag
; Enable global interrupts (GIE = bit 7)
; Enable RB0 interrupt (inte = bit 4)
      bsf        INTCON,GIE    ; Enable global int (bit 7)
      bsf        INTCON,T0IE   ; Enable TMR0 overflow interrupt
```

As in the program VariLapse, developed previously in this chapter, the timer operates by decrementing a 3-byte counter that holds the number of timer beats required for the delay. In the case of the TimerInt program, the routine that initializes the register variables from the local constants (called setDelay) also makes the correction so that the initial value loaded into the TMR0 register is correctly adjusted. The code is as follows:

```
;===============================
;   Initialize delay counters
;===============================
; Procedure to initialize local variables for a delay defined
; in program constants. For a one-half second on a 16F84 at
; 4 MHz initialization values are as follows:
; 500,000 = 0x07 0xa1 0x20
;           ---- ---- ----
;            |    |    |___ low_byte
;            |    |_____ mid_byte
;            |_____ high_byte
setDelay:
      movlw      high_byte
      movwf      countH
      movlw      mid_byte
      movwf      countM
      movlw      low_byte
      movwf      countL
; The Timer0 register provides the low-order level
; of the count. Because the counter counts up from zero,
; in order to ensure that the initial low-level delay
; count is correct the value 256 - xx must be calculated
```

```
; where xx is the value in the original countL register.
    movf       countL,w        ; w holds low-order byte
    sublw      .255
; Now w has adjusted result. Store in TMR0
    movwf      TMR0
    return
```

The interrupt service routine in the TimerInt program receives control when the TMR0 register underflows, that is, when the count goes from 0xff to 0x00. The service routine then proceeds to decrement the mid-range counter register and adjust, if necessary, the high-order counter. If the count goes to zero, the handler toggles the LED on port B, line 0, and reinitializes the counter variables by calling the setDelay procedure. The interrupt handler is coded as follows:

```
;===========================================================
;                Interrupt Service Routine
;===========================================================
; Service routine receives control when the timer
; register TMR0 overflows, that is, when 256 timer beats
; have elapsed
IntServ:
; First test if source is a Timer0 interrupt
    btfss      INTCON,T0IF     ; T0IF is Timer0 interrupt
    goto       notT0IF         ; Go if not RB0 origin
; If so clear the timer interrupt flag so that count continues
    bcf        INTCON,T0IF     ; Clear interrupt flag
; Save context
    movwf      old_w           ; Save w register
    swapf      STATUS,w        ; STATUS to w
    movwf      old_STATUS      ; Save STATUS
;=========================
;   interrupt action
;=========================
; Subtract 256 from beat counter by decrementing the
; mid-order byte
    decfsz     countM,f
    goto       exitISR         ; Continue if mid-byte not zero
; At this point the mid-order byte has overflowed.
; High-order byte must be decremented.
    decfsz     countH,f
    goto       exitISR
; At this point count has expired so the programmed time
; has elapsed. Service routine turns the LED on line 0,
; port B on and off at every conclusion of the count.
; This is done by XORing a mask with a one-bit at the
; port B line 0 position
    movlw      b'00000001'     ; Xoring with a 1-bit produces
                               ; the complement
    xorwf      PORTB,f         ; Complement bit 2, port B
; Reset delay constants
    call       setDelay
;=========================
;        exit ISR
;=========================
exitISR:
; Restore context
    swapf      old_STATUS,w    ; Saved STATUS to w
    movfw      STATUS          ; To STATUS register
    swapf      old_w,f         ; Swap file register in itself
```

```
     swapf     old_w,w          ; re-swap back to w
; Reset,interrupt
notTOIF:
          retfie
```

Note that one of the initial operations of the service routine is to clear the TOIF bit in the INTCON register. This action reenables the timer interrupt and prevents counting cycles from being lost. Because the interrupt is generated every 256 beats of the timer, there is no risk that by enabling the timer interrupt flag a re-entrant interrupt will take place.

The interrupt-based timer program named TimerInt can be tested on the circuit shown in Figure 11-3 or on Virtual Board A. Recall that Virtual Board I is used for interrupts that are based on port B input lines, which is not the case with the TimerInt program.

11.5 Watchdog Timer

The 16F84 contains an independent timer with its own clock source called the WWatchdog timer or WDT. The purpose of the Watchdog timer is to provide a way for the processor to recover from a software error that impedes program continuation, such as an endless loop. The Watchdog timer is not designed to recover from hardware faults, such as a *brown-out*.

The Watchdog timer hardware is independent of the PIC's internal clock. It has a time-out period from approximately 18 milliseconds to 2.3 seconds, depending on whether the prescaler is used. It is also not very accurate because it is sensitive to temperature. According to Microchip documentation, under worst case conditions its time-out period can take up to several seconds. The following program elements relate to Watchdog timer operation:

1. Configuration bit 2, labeled WDTE, enables and disables the Watchdog timer during system configuration. The WDT cannot be set or reset at runtime. It is enabled and disabled during programming.

2. The PSA bit in the OPTION register selects whether the prescaler is assigned to the Watchdog timer or to the Timer0 module.

3. Bits PS2 to PS0 in the OPTION register allow assigning eight rates to the Watchdog timer, from 1:1 to 1:128.

4. Bit 4 of the STATUS register, named the TO bit, is cleared when a time-out condition occurrs that originated in the WDT.

5. The power-down bit (PD) in the STATUS register is set after the execution of the **clrwdt** instruction.

6. The **clrwdt** instruction clears the Watchdog timer. It also clears the prescaler count (if the prescaler is assigned to the Watchdog timer) and sets STATUS bits TO and PD.

The purpose of the WDT is to provide a recovery mechanism for software errors. When the WDT timesout, the TO flag in the STATUS register is cleared and the program counter is reset to 0x000 so that the program restarts. Applications can prevent the reset by issuing the clrwdt instruction before the time-out period ends. When clrwdt executes, the WDT time-out period restarts.

11.5.1 Watchdog Timer Programming

Not much information is available regarding details of the operation or practical uses of the Watchdog timer in the 16F84. However, we can see that using the WDT in applications is not just a simple matter of restarting the counter with the clrwdt instruction. The timer is designed to detect software errors that can hangup a program, but how does it detect these errors and which conditions trigger the WDT operation is not clear from the information currently available. For example, an application that contains a long delay loop may determine that the Watchdog timer forces an untimely break out of the loop. The Watchdog timer provides a powerful error-recovery mechanism but its use requires careful consideration of program conditions that could make the timer malfunction.

11.6 Demonstration Programs

The following programs demonstrate the programming discussed in this chapter:

11.6.1 Tmr0Counter program

```
; File name: Tmr0Counter.asm
; Date: April 30, 2011
; Authors: Canton and Sanchez
; Processor: 16F84A
;
; Reference: SevenSeg Circuit and Board
;
; Description:
; Test program for the Timer0 counter. The program counts
; the number of presses of the pushbutton switch on port
; RA4/TOCKI and display the count on a seven-segment LED.
; Switch is wired active low.
;
; Switches used in __config directive:
;   _CP_ON         Code protection ON/OFF
; * _CP_OFF
; * _PWRTE_ON      Power-up timer ON/OFF
;   _PWRTE_OFF
;   _WDT_ON        Watchdog Timer ON/OFF
; * _WDT_OFF
;   _LP_OSC        Low power crystal oscillator
; * _XT_OSC        External parallel resonator/crystal ocillator
;
;   _HS_OSC        High speed crystal resonator (8 to 10 MHz)
;                  Resonator: Murate Erie CSA8.00MG = 8 MHz
;   _RC_OSC        Resistor/capacitor oscillator
; |
; |_____ * indicates set up values
```

```
;
;==========================
; set up and configuration
;==========================
          processor 16f84A
          include   <p16f84A.inc>
          __config  _XT_OSC & _WDT_OFF & _PWRTE_ON & _CP_OFF
;
;======================================================
;                   constant definitions
;                 (per circuit wiring diagram)
;======================================================
#define Pb_sw  4 ; Port-A line 4 to pushbutton switch
;
;============================
;       local variables
;============================
          cblock   0x0c      ; Start of block
          J                  ; counter J
          K                  ; counter K
          endc
;=============================================================
;                             program
;=============================================================
          org      0         ; start at address 0
          goto     main
;
; Space for interrupt handlers
          org      0x08

main:
; Clear the timer and the Watchdog
          clrf     TMR0
          clrwdt
; Set up the OPTION regiser bit map
          movlw    b'10111000'
;   7 6  5  4  3  2  1  0 <= OPTION bits
;   | |  |  |  |  |  |__|__|_____ PS2-PS0 (prescaler bits)
;   | |  |  |  |  |              Values for Timer0
;   | |  |  |  |  |              *000 = 1:2    001 = 1:4
;   | |  |  |  |  |              010 = 1:8    011 = 1:16
;   | |  |  |  |  |              100 = 1:32   101 = 1:64
;   | |  |  |  |  |              110 = 1:128  *111 = 1:256
;   | |  |  |  |  |_____ PSA (prescaler assign)
;   | |  |  |  |                 *1 = to WDT
;   | |  |  |  |                  0 = to Timer0
;   | |  |  |  |_____ TOSE (Timer0 edge select)
;   | |  |  |                     0 = increment on low-to-high
```

```
;   |   |   |                            *1 = increment on high-to-low
;   |   |   |_____ TOCS (TMR0 clock source)
;   |   |                            0 = internal clock
;   |   |                           *1 = RA4/TOCKI bit source
;   |   |_____ INTEDG (Edge select)
;   |                               *0 = falling edge
;   |_____ RBPU (Pullup enable)
;                                    0 = enabled
;                                   *1 = disabled
            option
; Set up ports
            movlw    0x00      ; Set Port-B to output
            tris     PORTB
            clrf     PORTB    ; All Port-B to 0
; Port-A. Five low-order lines set for for input
            movlw    B'00011111'        ; w = 00011111 binary
            tris     PORTA    ; Port-A (lines 0 to 4) to
                                ; input
;===================================
; Check value in TMR0 and display
;===================================
; Every press of the pushbutton switch connected to line
; RA4/TOCKI adds one to the value in the TMR0 register.
; Loop checks this value, adjusts to the range 0 to 15
; and displays the result in the Seven-Segment LED on
; Port-B
checkTmr0:
            movf     TMR0,w              ; Timer register to w
; Eliminate four high order bits
            andlw    b'00001111' ; Mask off high bits
; At this point the w register contains a 4-bit value
; in the range 0 to 0xf. Use this value (in w) to
; obtain Seven-Segment display code
            call     segment
            movwf    PORTB               ; Display switch bits
            goto     checkTmr0           ; Endless loop
;
;===================================
;   routine to returns 7-segment
;             codes
;===================================
segment:
            addwf    PCL,f    ; PCL is program counter latch
            retlw    0x3f     ; 0 code
            retlw    0x06     ; 1
            retlw    0x5b     ; 2
            retlw    0x4f     ; 3
            retlw    0x66     ; 4
```

```
        retlw     0x6d      ; 5
        retlw     0x7d      ; 6
        retlw     0x07      ; 7
        retlw     0x7f      ; 8
        retlw     0x6f      ; 9
        retlw     0x77      ; A
        retlw     0x7c      ; B
        retlw     0x39      ; C
        retlw     0x5b      ; D
        retlw     0x79      ; E
        retlw     0x71      ; F
        retlw     0x7f      ; Just in case all on
        end
```

11.6.2 Timer0 Program

```
; File: Timer0.ASM
; Date: April 7, 2011
; Authors: Canton and Sanchez
; Processor: a6F84A
;
; Description:
; Program to demonstrate programming of the 16F84A
; Timer0 module. Program flashes eight LEDs in sequence
; counting from 0 to 0xff. Timer0 is used to delay
; the count.
;============================
;         switches
;============================
; Switches used in __config directive:
;   _CP_ON          Code protection ON/OFF
; * _CP_OFF
; * _PWRTE_ON       Power-up timer ON/OFF
;   _PWRTE_OFF
;   _WDT_ON         Watchdog Timer ON/OFF
; * _WDT_OFF
;   _LP_OSC         Low power crystal oscillator
; * _XT_OSC         External parallel resonator/crystal oscillator
;
;   _HS_OSC         High speed crystal resonator (8 to 10 MHz)
;                   Resonator: Murate Erie CSA8.00MG = 8 MHz
;   _RC_OSC         Resistor/capacitor oscillator (simplest, 20%
;                   error)
; |
; |_____ * indicates set up values

        processor 16f84A
        include   <p16f84A.inc>
```

```
        __config  _XT_OSC & _WDT_OFF & _PWRTE_ON & _CP_OFF
;=========================================================
;                variables in PIC RAM
;=========================================================
; None in this application
;
;=========================================================
;                m a i n   p r o g r a m
;=========================================================
        org     0              ; start at address 0
        goto    main
;
;==============================
;      interrupt handler
;==============================
        org              0x08
;==============================
;       main program
;==============================
main:
; Clear the Watchdog Timer and reset prescaler
        clrwdt
; Set up the OPTION regiser bit map
        movlw   b'11010111'
;   7  6  5  4  3  2  1  0 <= OPTION bits
;   |  |  |  |  |  |__|__|_____ PS2-PS0 (prescaler bits)
;   |  |  |  |  |  |            Values for Timer0
;   |  |  |  |  |  |            000 = 1:2    001 = 1:4
;   |  |  |  |  |  |            010 = 1:8    011 = 1:16
;   |  |  |  |  |  |            100 = 1:32   101 = 1:64
;   |  |  |  |  |  |            110 = 1:128 *111 = 1:256
;   |  |  |  |  |_____ PSA (prescaler assign)
;   |  |  |  |             1 = to WDT
;   |  |  |  |             *0 = to Timer0
;   |  |  |  |_____ TOSE (Timer0 edge select)
;   |  |  |               0 = increment on low-to-high
;   |  |  |               *1 = increment on high-to-low
;   |  |  |_____ TOCS (TMR0 clock source)
;   |  |                 *0 = internal clock
;   |  |                 1 = RA4/TOCKI bit source
;   |  |_____ INTEDG (Edge select)
;   |                   *0 = falling edge
;   |_____ RBPU (Pullup enable)
;                      0 = enabled
;                      *1 = disabled
        option
; Set up ports
        movlw   0x00                    ; Set Port-B to output
```

```
             tris      PORTB
             clrf      PORTB              ; All Port-B to 0
; Port-A is not used in this program
mloop:
             incf      PORTB,f            ; Add 1 to register value
             call      TM0delay
             goto      mloop
;*****************************
;      delay sub-routine
;          uses Timer0
;*****************************
TM0delay:
; Initialize the timer register
             clrf      TMR0      ; Clear SFR for Timer0
; Routine tests the value in the TMR0 register by
; subtracting 0xff from the value in TMR0. The zero flag
; is set if TMR0 = 0xff
cycle:
             movf      TMR0,w             ; Timer to w
; w has TMR0 register value
             sublw     0xff               ; Subtract max value
; Zero flag is set if value in TMR0 = 0xff
             btfss     STATUS,Z ; Test for zero
             goto      cycle              ; Repeat
             return

             end
```

11.6.3 LapseTimer Program

```
; File: LapseTimer.ASM
; Date: May 1, 2011
; Authors: Canton and Sanchez
; Processor: 16F84A
;
; Description:
; Using Timer0 to produce a variable-lapse delay.
; The delay is calculated based on the number of machine
; cycles necessary for the desired wait period. For
; example, a machine running at a 4 MHz clock rate
; executes 1,000,000 instructions per second. In this
; case a 1/2 second delay requires 500,000 instructions.
; The wait period is passed to the delay routine in three
; program registers which hold the high-, middle-, and
; low-order bytes of the counter.
;===========================
;      switches
;===========================
```

```
; Switches used in __config directive:
;   _CP_ON        Code protection ON/OFF
; * _CP_OFF
; * _PWRTE_ON     Power-up timer ON/OFF
;   _PWRTE_OFF
;   _WDT_ON       Watchdog Timer ON/OFF
; * _WDT_OFF
;   _LP_OSC       Low power crystal oscillator
; * _XT_OSC       External parallel resonator/crystal oscillator
;
;   _HS_OSC       High speed crystal resonator (8 to 10 MHz)
;                 Resonator: Murate Erie CSA8.00MG = 8 MHz
;   _RC_OSC       Resistor/capacitor oscillator
; |
; |_____ * indicates set up values

        processor 16f84A
        include   <p16f84A.inc>
        __config  _XT_OSC & _WDT_OFF & _PWRTE_ON & _CP_OFF
;=======================================================
;                 variables in PIC RAM
;=======================================================
; Local variables
        cblock    0x0d    ; Start of block
                          ; 3-byte auxiliary counter for delay.
        countH            ; High-order byte
        countM            ; Medium-order byte
        countL            ; Low-order byte
        endc
;=========================================================
;                 m a i n   p r o g r a m
;=========================================================
        org       0       ; start at address 0
        goto      main
;
;=============================
;       interrupt handler
;=============================
        org       0x04
;       goto      IntServ
;=============================
;       main program
;=============================
main:
; Clear the Watchdog Timer and reset prescaler
        clrf      TMR0
        clrwdt
; Set up the OPTION regiser bitmap
```

```
              movlw    b'11010000'
;     7   6   5   4   3   2   1   0 <= OPTION bits
;     |   |   |   |   |   |   |__|__|_____ PS2-PS0 (prescaler bits)
;     |   |   |   |   |   |                Values for Timer0
;     |   |   |   |   |   |                *000 = 1:2    001 = 1:4
;     |   |   |   |   |   |                 010 = 1:8    011 = 1:16
;     |   |   |   |   |   |                 100 = 1:32   101 = 1:64
;     |   |   |   |   |   |                 110 = 1:128 *111 = 1:256
;     |   |   |   |   |   |_____ PSA (prescaler assign)
;     |   |   |   |   |                     1 = to WDT
;     |   |   |   |   |                    *0 = to Timer0
;     |   |   |   |   |_____ TOSE (Timer0 edge select)
;     |   |   |   |                         0 = increment on low-to-high
;     |   |   |   |                        *1 = increment on high-to-low
;     |   |   |   |_____ TOCS (TMR0 clock source)
;     |   |   |                            *0 = internal clock
;     |   |   |                             1 = RA4/TOCKI bit source
;     |   |   |_____ INTEDG (Edge select)
;     |   |                               *0 = falling edge
;     |   |_____ RBPU (Pullup enable)
;     |                                0 = enabled
;     |                               *1 = disabled
              option
; Set up ports
              movlw    0x00              ; Set Port-B to output
              tris     PORTB
              clrf     PORTB             ; All Port-B to 0
; Port-A is not used in this program
;============================
;      display loop
;============================
mloop:
; Turn on LED
          bsf      PORTB,0
; Initialize counters and delay
              call     onehalfSec
              call     TM0delay
; Turn off LED
          bcf      PORTB,0
; Re-initialize counter and delay
              call     onehalfSec
              call     TM0delay
              goto     mloop
;==================================
;  variable-lapse delay procedure
;          using Timer0
;==================================
; ON ENTRY:
```

```
;            Variables countL, countM, and countH hold
;            the low-, middle-, and high-order bytes
;            of the delay period, in timer units
; Routine logic:
; The prescaler is assigned to Timer0 and set up so
; that the timer runs at 1:2 rate. This means that
; every time the counter reaches 128 (0x80) a total
; of 256 machine cycles have elapsed. The value 0x80
; is detected by testing bit 7 of the counter
; register.
; Note:
;     The Timer0 register provides the low-order level
; of the count. Because the counter counts up from zero,
; in order to ensure that the initial low-level delay
; count is correct the value 128 - (xx/2) must be calculated
; where xx is the value in the original countL register.
; First calculate xx/2 by bit shifting
TM0delay:
        bcf       STATUS,C ; Clear carry flag
        rrf       countL,f ; Divide by 2
; now subtract 128 - (xx/2)
        movf      countL,w ; w holds low-order byte
        sublw     d'128'
; Now w has adjusted result. Store in TMR0
        movwf     TMR0
; Routine tests timer overflow by testing bit 7 of
; the TMR0 register.
cycle:
        btfss     TMR0,7             ; Is bit 7 set?
        goto      cycle              ; Wait if not set
; At this point TMR0 bit 7 is set
; Clear the bit
        bcf       TMR0,7             ; All other bits are preserved
; Subtract 256 from beat counter by decrementing the
; mid-order byte
        decfsz    countM,f
        goto      cycle              ; Continue if mid-byte not
zero
; At this point the mid-order byte has overflowed.
; High-order byte must be decremented.
        decfsz    countH,f
        goto      cycle
; At this point the time cycle has elapsed
        return
;===============================
;  set register variables for
;     one-half second delay
;===============================
```

```
; Procedure to initialize local variables for a
; delay of one-half second on a 16F84 at 4 MHz.
; Timer is set up for 500,000 clock beats as
; follows: 500,000 = 0x07 0xa1 0x20
; 500,000 = 0x07 0xa1 0x20
;               ---- ---- ----
;                |    |    |___ countL)
;                |    |_____ countM
;                |_____ countH
onehalfSec:
        movlw   0x07
        movwf   countH
        movlw   0xa1
        movwf   countM
        movlw   0x20
        movwf   countL
        return

        end
```

11.6.4 LapseTmrInt Program

```
; File: LapseTmrInt.ASM
; Date: May 1, 2011
; Authors: Canton and Sanchez
; Processor: 16F84A
;
; Description:
; Interrupt-Driven version of the LapseTimer program.
; Using Timer0 to produce a variable-lapse delay.
; The delay is calculated based on the number of machine
; cycles necessary for the desired wait period. For
; example, a machine running at a 4 MHz clock rate
; executes 1,000,000 instructions per second. In this
; case a 1/2 second delay requires 500,000 instructions.
; The wait period is passed to the delay routine in three
; register variables which hold the high-, middle-, and
; low-order bytes of the counter.
;==========================
;          switches
;==========================
; Switches used in __config directive:
;    _CP_ON        Code protection ON/OFF
; * _CP_OFF
; * _PWRTE_ON      Power-up timer ON/OFF
;    _PWRTE_OFF
;    _WDT_ON       Watchdog Timer ON/OFF
; * _WDT_OFF
```

```
;   _LP_OSC         Low power crystal oscillator
; * _XT_OSC         External parallel resonator/crystal oscillator
;
;   _HS_OSC         High speed crystal resonator (8 to 10 MHz)
;                   Resonator: Murate Erie CSA8.00MG = 8 MHz
;   _RC_OSC         Resistor/capacitor oscillator
; |
; |_____ * indicates set up values

        processor 16f84A
        include   <p16f84A.inc>
        __config _XT_OSC & _WDT_OFF & _PWRTE_ON & _CP_OFF


;======================================================
;               variables in PIC RAM
;======================================================
; Local variables
        cblock   0x0d     ; Start of block
                          ; 3-byte auxiliary counter for delay.
        countH            ; High-order byte
        countM            ; Medium-order byte
        countL            ; Low-order byte
        old_w             ; Context saving
        old_STATUS        ; Idem
        endc
;==========================================================
;                 m a i n   p r o g r a m
;==========================================================
        org      0        ; start at address 0
        goto     main
;
;==============================
;      interrupt handler
;==============================
        org      0x04
        goto     IntServ
;==============================
;      main program
;==============================
main:
; Clear the Watchdog Timer and reset prescaler
        clrf     TMR0
        clrwdt
; Set up the OPTION regiser bit map
        movlw    b'11011000'
;   7  6  5  4  3  2  1  0 <= OPTION bits
;   |  |  |  |  |  |__|__|_____ PS2-PS0 (prescaler bits)
;   |  |  |  |  |                  Values for Timer0
```

```
;    |   |   |   |   |                  000 = 1:2    001 = 1:4
;    |   |   |   |   |                  010 = 1:8    011 = 1:16
;    |   |   |   |   |                  100 = 1:32   101 = 1:64
;    |   |   |   |   |                  110 = 1:128 *111 = 1:256
;    |   |   |   |   |_____ PSA (prescaler assign)
;    |   |   |   |                     *1 = to WDT
;    |   |   |   |                      0 = to Timer0
;    |   |   |   |_____ TOSE (Timer0 edge select)
;    |   |   |                         0 = increment on low-to-high
;    |   |   |                        *1 = increment on high-to-low
;    |   |   |_____ TOCS (TMR0 clock source)
;    |   |                            *0 = internal clock
;    |   |                             1 = RA4/TOCKI bit source
;    |   |_____ INTEDG (Edge select)
;    |                               *0 = falling edge
;    |_____ RBPU (Pullup enable)
;                                     0 = enabled
;                                    *1 = disabled
          option
; Set up ports
          movlw   0x00              ; Set Port-B to output
          tris    PORTB
          clrf    PORTB             ; All Port-B to 0
; Port-A is not used in this program
;=============================
;      set up interrupts
;=============================
; Clear external interrupt flag (INTF = bit 1)
          bcf     INTCON,INTF       ; Clear flag
; Enable global interrupts (GIE = bit 7)
; Enable RB0 interrupt (inte = bit 4)
          bsf     INTCON,GIE        ; Enable global int (bit 7)
          bsf     INTCON,T0IE       ; Enable TMR0 overflow
                                    ; interrupt
; Init count
          call    onehalfSec
;=============================
;     do-nothing loop
;=============================
; All work is performed by the interrupt handler
mloop:
          goto    mloop
;=============================
;  set register variables for
;     one-half second delay
;=============================
; Procedure to initialize local variables for a
; delay of one-half second on a 16F84 at 4 MHz.
```

```
; Timer is set up for a 500,000 clock beats as
; follows: 500,000 = 0x07 0xa1 0x20
; 500,000 = 0x07 0xa1 0x20
;            ---- ---- ----
;              |    |    |___ countL)
;              |    |_____ countM
;              |_____ countH
onehalfSec:
        movlw   0x07
        movwf   countH
        movlw   0xa1
        movwf   countM
        movlw   0x20
        movwf   countL
; The Timer0 register provides the low-order level
; of the count. Because the counter counts up from zero,
; in order to ensure that the initial low-level delay
; count is correct the value 256 - xx must be calculated
; where xx is the value in the original countL register.
        movf    countL,w ; w holds low-order byte
        sublw   d'255'
; Now w has adjusted result. Store in TMR0
        movwf   TMR0
        return
;========================================================
;             Interrupt Service Routine
;========================================================
; Service routine receives control when there the timer
; register TMR0 overflows, that is, when 256 timer beats
; have elapsed
IntServ:
; First test if source is a Timer0 interrupt
        btfss   INTCON,T0IF      ; T0IF is Timer0 interrupt
        goto    notT0IF      ; Go if not RB0 origin
; If so clear the timer interrupt flag so that count continues
        bcf             INTCON,T0IF      ; Clear interrupt flag
; Save context
        movwf   old_w             ; Save w register
        swapf   STATUS,w ; STATUS to w
        movwf   old_STATUS        ; Save STATUS
;==========================
;    interrupt action
;==========================
; Subtract 256 from beat counter by decrementing the
; mid-order byte
        decfsz  countM,f
        goto    exitISR    ; Continue if mid-byte not zero
; At this point the mid-order byte has overflowed.
```

```
; High-order byte must be decremented.
        decfsz  countH,f
        goto    exitISR
; At this point count has expired so the programmed time
; has elapsed. Service routine turns the LED on line 0,
; Port-B on and off at every conclusion of the count.
; This is done by xoring a mask with a one-bit at the
; Port-B line 0 position
        movlw   b'00000001'         ; Xoring with a 1-bit produces
                                    ; the complement
        xorwf   PORTB,f             ; Complement bit 2, Port-B
; Reset one-half second counter
        call    onehalfSec
;=========================
;          exit ISR
;=========================
exitISR:
; Restore context
        swapf   old_STATUS,w        ; Saved STATUS to w
        movfw   STATUS              ; To STATUS register
        swapf   old_w,f             ; Swap file register in itself
        swapf   old_w,w             ; re-swap back to w
; Reset,interrupt
notTOIF:
        retfie

        end
```

Chapter 12

LCD Hardware and Programming

12.1 Liquid Crystal Display

Liquid crystal displays (LCDs) are devices frequently used for alphanumeric output in microcontroller-based embedded systems. Their advantages are their reduced size, moderate cost, and the convenience of mounting the LCD directly on the circuit board. According to their interface, LCDs are classified into serial and parallel. Serial LCDs require less I/O resources but execute slower than their parallel counterparts. Although serial LCDs require less control lines, they are considerably more expensive than the parallel type. In this chapter we discuss parallel-driven LCD devices based on the Hitachi HD44780 character-based controller, which is by far the most popular controller for PIC-driven LCDs. LCD technology was discussed in Chapter 4.

12.1.1 LCD Features and Architecture

The HD44780 is a *dot-matrix* liquid crystal display controller and driver. The device displays ASCII alphanumeric characters, Japanese kana characters, and some symbols. A single HD44780 can display up to two 28-character lines. An available extension driver makes possible addressing up to 80 characters.

The HD44780U contains a 9,920-bit character-generator ROM that produces a total of 240 characters: 208 characters with a 5 x 8 dot resolution and 32 characters at a 5 x 10 dot resolution. The device is capable of storing 64 times 8-bit character data in its character generator RAM. This corresponds to eight custom characters in 5 times 8-dot resolution or four characters in 5 times 10-dot resolution.

The controller is programmable to three different duty cycles: 1/8 for one line of 5 × 8 dots with cursor, 1/11 for one line of 5 × 10 dots with cursor, and 1/16 for two lines of 5 × 8 dots with cursor. The built-in commands include clearing the display, homing the cursor, turning the display on and off, turning the cursor on and off, setting display characters to blink, shifting the cursor and the display left-to-right or right-to-left, and reading and writing data to the character generator and to display data ROM.

12.1.2 LCD Functions and Components

The following hardware elements form part of the HD44780 controller: two internal registers labeled the data register and the instruction register, a busy flag, an address counter, a RAM area of display data (DDRAM), a character generator ROM, a character generator RAM, a timing generation circuit, a liquid crystal display driver circuit, and a cursor and blink control circuit. The controller itself is often referred to as the *MPU* in the Hitachi literature.

Internal Registers

The HD44780 contains an IR (*instruction register*) and a DR (*data register*). The IR is used to store instruction codes, such as those to clear the display, define an address, or store a bitmap in character generator RAM. The IR is written only from the controller.

The *data register, DR*, is used to temporarily store data to be written into DDRAM or CGRAM as well as temporarily store data read from DDRAM or CGRAM. Data placed in the data register is automatically written into DDRAM or CGRAM by an internal operation.

Busy Flag

When *BF* (the *busy flag*) is 1, the HD44780U is in the internal operation mode, and the next instruction not accepted. The busy flag is mapped to data bit 7. Software must ensure that the busy flag is reset (BF = 0) before the next instruction is entered.

Address Counter

AC (the *address counter*) stores the current address used in operations that access DDRAM or CGRAM. When an instruction contains address information, the address is stored in the address counter. The RAM area accessed—DDRAM or CGRAM—is also determined by the instruction that stores the address in the AC.

The AC is automatically incremented or decremented after each instruction that writes or reads DDRAM or CGRAM data. The variations and options in operations that change the AC are described later in this chapter.

Display Data RAM (DDRAM)

DDRAM (the *display data RAM area*) is used to store the 8-bit bitmaps that represent the display characters and graphics. Display data is represented in 8-bit character codes. When equipped with the extension, its capacity is 80 times 8 bits, or 80 characters. The area not used for storing display character can be used by software for storing any other 8-bit data. The mapping of DDRAM locations to the LCD display is discussed in Section 12.1.3.

Character Generator ROM (CGROM)

The character generator is a ROM that has the bitmaps for 208 characters in 5 times 8 dot resolution or 32 characters in 5 times 10 dot resolution. Figure 12-1 shows the standard character set in the HD44780.

Figure 12-1 HD44780 Character Set.

With a few exceptions, the characters in the range 0x20 to 0x7f correspond to those of the ASCII character set. The remaining characters are Japanese kana characters and special symbols. The characters in the range 0x0 to 0x1f, ASCII control characters, do not function as such in the HD44780. Sending a backspace (0x08), a bell (0x07), or a carriage return (0x0d) code to the controller has no effect.

Character Generator RAM (CGRAM)

CGRAM (the *character generator RAM*) allows the creation of customized characters by defining the corresponding 5 x 8 bitmaps. Eight custom characters can be stored in the 5 x 8 dot resolution and four in the 5 x 10 resolution. The creation and use of custom characters is addressed later in this chapter.

Timing Generation Circuit

This circuit produces the timing signals for the operation of internal components circuits such as DDRAM, CGROM, and CGRAM. The *timing generation circuit* is not accessible to the program.

Liquid Crystal Display Driver Circuit

The *liquid crystal display driver circuit* consists of sixteen common signal drivers and forty segment signal drivers. The circuit responds to the number of lines and the character font selected. Once this is done, the circuit performs automatically and is not otherwise accessible to the program.

Cursor/Blink Control Circuit

The *cursor and blink control circuit* generates both the cursor and the character blinking. The cursor or the character blinking is applied to the character located in the data RAM address referenced in the address counter (AC).

12.1.3 Connectivity and Pin-Out

LCDs are powerful yet complex devices. Fortunately, the programmer does not have to deal with all the complexities of LCD displays because these devices are usually furnished in a module that includes the LCD controller. Furthermore, most LCDs used in microcontroller circuits are equipped with the same controller: the Hitachi HD44780. This controller provides a relatively simple interface between a microcontroller and the LCD.

But the fact that the HD44780 has become almost ubiquitous in LCD controller technology does not mean that these devices are without complications. The first difficulty confronted by the circuit designer is selecting the most appropriate LCD for the application among dozens (perhaps hundreds) of available configurations, each one with its own resolution, interface technology, size, graphics options, pin patterns, and other individual features. In this sense, it may be better to experiment with a simple LCD in a breadboard circuit before attempting a final circuit with hardware.

Two common connectors used with 44780-based LCDs have either fourteen pins in a single row, each pin spaced 0.100" apart, or two rows of eight pins each, also spaced 0.100" apart. In both cases, the pins are labeled in the LCD board. The two common connectors are shown in Figure 12-2.

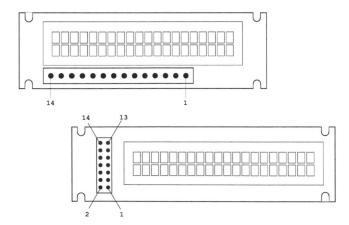

Figure 12-2 Typical HD44780 Connector Pin-Outs.

In LCDs with a backlight option, sometimes the connectors have two extra pins, usually numbered 15 and 16. Pin number 15 is connected to a 5V source for the backlight and pin number 16 to ground. Typical LCD wiring is shown in Table 12.1.

Table 12.1

Hitachi HD44780 LCD Controller Pin-Out (80 characters or less)

PIN NUMBER	SYMBOL	DESCRIPTION
1	Vss	Ground
2	Vcc	Vcc (Power supply +5V)
3	Vee	Contrast control
4	RS	Set/reset
		0 = instruction input
		1 = data input
5	R/W	R/W (read/write select)
		0 = write to LCD
		1 = read LCD data
6	E	Enable. Clock signal to
		initiate data transfer
7	DB0	Data bus line 0
8	DB1	Data bus line 1
9	DB2	Data bus line 2
10	DB3	Data bus line 3
11	DB4	Data bus line 4
12	DB5	Data bus line 5
13	DB6	Data bus line 6
14	DB7	Data bus line 7

The pin-out in Table 12-1 refers to controllers that address no more than eighty characters. In addition, some LCDs with LED backlighting contain two additional pins, usually numbered 15 and 16. In these cases, pin number 15 is a +5VDC source for the backlight and pin 16 is the backlight ground.

From the pin-out in Table 12-1, it is evident that the interface to the LCD uses eight parallel lines (lines 7 to 14). However, it is also possible to drive the LCD using just four lines, saving connections on limited circuits.

The reader should be aware that LCDs are often furnished in custom boards that may or may not have other auxiliary components. These boards are often wired differently from the examples shown in Figure 12-2. In all cases, the device's documentation and the corresponding data sheets should provide the appropriate wiring information.

12.2 Interfacing with the HD44780

The Hitachi 44780 controller allows parallel interfacing using 4- or 8-bit data paths. In the 4-bit mode, each data byte must be divided into a high-order and a low-order nibble and are transmitted sequentially, the high nibble first. In the 8-bit parallel mode, each data byte is transmitted from the PIC to the controller as a unit. The advantage of using the 4-bit mode is greater economy of I/O lines on the PIC side. The disadvantages are slightly more complicated programming and minimally slower execution speed. Our first example and circuit uses the 8-bit data mode so as to avoid complications. Once the main processing routines are developed, make the necessary modifications so as to make possible the 4-bit data mode.

In addition to the *data transmission mode*, there are other circuit options to consider. Two control lines between the microcontroller and the HD44780-driven LCD are necessary in all cases: one to the RS line to select between data and instruction input modes, and another one to the E line to provide the pulse that initiates the data transfer. The *R/W control line*, which selects between the read and the write mode of the LCD controller, can be connected or grounded. If the R/W line is not connected to a microcontroller port, then the HD44780 operates only in the write data mode and all read operations are unavailable.

12.2.1 Busy Flag or Timed Delay Options

Because many applications do not read text data from controller memory, the write-only mode is often an attractive option, especially considering that microcontroller I/O ports are often in short supply and that this option saves one port for other duties. However, there is a less apparent drawback to not being able to read LCD data, which is that the application is not able to monitor the *busy* flag. This flag, which indicates that the controller has concluded its operation, is mapped to bit 7. Because testing the BF requires reading this bit, not connecting the R/W line has the effect that applications cannot use the busy flag and must rely on timing routines to ensure that each operation completes before the next one begins. The timing requirements for each instruction are discussed in detail later in this chapter.

For the circuit designer, to read or not to read controller data is a decision with several trade-offs. Using time delay routines to ensure that each controller operation has concluded is a viable option that saves one interface line. On the other hand, code that relies on timing routines is externally dependent on the clocks and timer hardware. If code that relies on timing routines is ported to another circuit with different microcontroller, clocks, or timer hardware, the delays may change and the routines could fail. Furthermore, the use of delay routines often is not efficient, as controller operations can terminate before the timed delay has expired.

On the other hand, code that reads the busy flag to determine the termination of a controller operation is not without pitfalls. If the controller or the circuit fails, then the program can hang up in an endless loop, waiting for the busy flag to clear. To be absolutely safe, the code would have to contain an external wait loop when testing the busy flag, so that if the external loop expires, then the processing can assume that there is a hardware problem and break out of the flag test loop. The programmer must decide whether this safety mechanism for reading the busy flag is necessary because its implementation requires a somewhat complicated exception response.

In the code samples developed in this chapter, we implement both ways of ensuring operation completion. The code also furnishes a software switch that allows selecting the preferred option.

12.2.2 Contrast Control

In addition to the control lines that require processor interface, the HD44780 contains other control lines. One such line is used for the LCD contrast. The *contrast control line* (usually labeled *Vee*) is connected to pin number 3 (see Table 12.1). The actual implementation of the contrast control function varies according to the manufacturer. In general, for an LCD with a normal temperature range, the contrast control line is wired as shown in Figure 12-3.

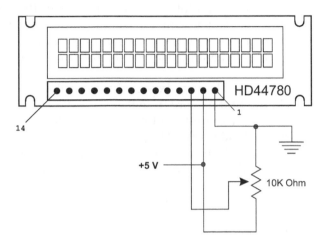

Figure 12-3 Typical Contrast Adjustment Circuit.

12.2.3 Display Backlight

Some LCDs are equipped with an LED backlight so as to make the displayed characters more visible. In different LCDs, backlight is implemented in different ways. Some manufacturers wire the backlight directly to the LCD power supply, while others provide additional pins that allow turning the backlight on or off independently of the LCD display. Backlit displays with fourteen pins belong to the first type, while those with sixteen pins have independent backlight control. If the backlight pins are adjacent to the other display pins, then they are numbered 15 and 16. In this case, pin number 15 is wired, through a current limiting resistor, to the +5V source and pin 16 to ground. Sometimes the current-limiting resistor is built into the display. This information is available in the device's data sheet.

Note that some four-line displays use pins 15 and 16 for other purposes. In these systems, backlight control, if available, is provided by separate pins.

12.2.4 Display Memory Mapping

The Hitachi HD44780 is a memory-mapped system in which characters are displayed by storing their ASCII codes in the corresponding memory address associated with each digit-display area. The area of controller RAM mapped to character-display memory has a capacity of eighty characters. This area is known as *display data RAM* or *DDRAM*.

In order to save circuitry, the common lines of the controller outputs to the liquid crystal display hardware are multiplexed. In this context, the *duty ratio* of a system is the number of multiplexed common lines. The most common duty ratio is 1/16, although 1/8 and 1/11 are found in some systems. Because the duty ratio measures the number of multiplexed lines, it also determines the display mapping. For example, in a single-line-by-sixteen character display with a 1/16 duty ratio the first eight characters are mapped to one set of consecutive memory addresses and the second eight characters to another set of addresses. The reason is that in every display line, sixteen common access lines are multiplexed, instead of eight. By the same token, a two-line-by-sixteen-character display with a 1/16 duty ratio requires sixteen common lines. In this case, the address of the second lines is not a continuation of the address of the first line, but is in another address set not contiguous to the first one.

For example, in a typical two-line-by-sixteen-character display, the addresses of the sixteen characters in the first line are from 0x00 to 0x0F, while the addresses of the characters in the second line are from 0x40 to 0x4F. Because there are eighty memory locations in the controller's DDRAM, each line contains storage for a total of forty characters. The range of the entire first line is from 0x00 to 0x27 (forty characters total) but of these, only 16 are actually displayed. The same applies to the second line of 16 characters. In this case, the storage area is in the range 0x28 to 0x4f, but only 16 characters are displayed. In the single-line-by-sixteen-character display mentioned first the addresses of the first eight characters would be a set from 0x00 to 0x07 and the addresses of the second eight characters in the line are from 0x40 to 0x47. Table 12.2 lists the memory address mapping of some common LCD configurations.

Table 12.2

7-bit DDRAM Address Mapping for Common LCDs

CHARACTERS/ ROW	LINE NUMBER	CHARACTER NUMBER	FIRST IN GROUP	NEXT IN GROUP	LAST IN GROUP
8/1	1	1	0x00	0x01	0x07
8/2	1	1	0x00	0x01	0x07
	2	1	0x40	0x41	0x47
16/1	1	1	0x00	0x01	0x07
	1	9	0x40	0x41	0x47
16/2	1	1	0x00	0x01	0x0f
	2	1	0x40	0x41	0x4f
20/2	1	1	0x00	0x01	0x13
	2	1	0x40	0x41	0x53
24/2	1	1	0x00	0x01	0x17
	2	1	0x40	0x41	0x57
16/4	1	1	0x00	0X01	0x0f
	2	1	0x40	0x41	0x4f
	3	1	0x10	0x11	0x1f
	4	1	0x50	0x51	0x5f
20/4	1	1	0x00	0x01	0x13
	2	1	0x40	0x41	0x53
	3	1	0x14	0x15	0x27
	4	1	0x54	0x55	0x67

Note that systems that exceed a total of 80 characters require two or more HD44780 controllers. Although the information provided in Table 12.2 corresponds to the mapping in most LCDs, it is a good idea to consult the data sheet of the specific hardware in order to corroborate the address mapping in a particular device.

Table 12.3 contains the seven low-order bits of DDRAM addresses. HD44780 commands to set the DDRAM address for read or write operations require that the high-order bit (bit number 7) be set. Therefore, to write to DDRAM memory address 0x07, code uses the value 0x87; and to write to DDRAM address 0x43, code uses 0xc3 as the instruction operand.

12.3 HD44780 Instruction Set

The HD44780 instruction set includes operators to initialize the system and set operational modes, clear the display, manipulate the cursor, set, reset, and control automatic display address shift, set and reset the interface parameters, poll the busy flag, and read and write to CGRAM and DDRAM memory.

12.3.1 Instruction Set Overview

Pin number 4 in Table 12.1 selects two modes of operation on the HD44780 controller: instruction and data input. When the instruction mode is enabled (RS pin is set low), the controller receives commands that set up the hardware and determine its configuration and mode of operation. These commands are part of the HD44780 instruction set shown in Table 12.3.

Table 12.3

HD44780 Instruction Set

INSTRUCTION	RS	R/W	B7	B6	B5	B4	B3	B2	B1	B0	TIME
Clear Display	0	0	0	0	0	0	0	0	0	1	1.64
Return home	0	0	0	0	0	0	0	0	1	#	1.64
Entry mode set	0	0	0	0	0	0	0	1	I/D	S	37
Display/Cursor ON/OFF	0	0	0	0	0	0	1	D	C	B	37
Cursor/display shift	0	0	0	0	0	1	S/C	R/L	#	#	37
Function set	0	0	0	0	1	DL	N	F	#	#	37
Set CGRAM address	0	0	0	1	---------- address ----------------						37
Set DDRAM address	0	0	1	----------------- address ------------------							37
Read busy flag and Address register	0	1	BF	--------------- address ------------------							0
Write data	1	0	------------------------ data -------------------								37
Read data	0	1	------------------------ data -------------------								37

Note: Bits labeled # have no effect.

Clearing the Display

Clearing the display fills the display with blanks by writing the code 0x20 into all DDRAM addresses. It also returns the cursor to the *home position* (top-left display corner) and sets address 0 in the DDRAM address counter. After this command executes, the display disappears and the cursor goes to the left edge of the display.

Return Home

Return home returns the cursor to home position at the upper left position of the first character line. It sets DDRAM address 0 in the address counter and sets the display to its default status if it was shifted. DDRAM contents remain unchanged.

Entry Mode Set

Entry mode set sets the direction of cursor movement and the display shift mode. If B1 (I/D) bit is set, cursor handling is set to the increment mode; that is, left-to-right. If this bit is clear, then cursor movement is set to the decrement mode, that is, right-to-left.

If B0 (S) bit is set, *display shift* is enabled. In the display shift mode, it appears as if the display moves instead of the cursor; otherwise display shift is disabled. Operations that read or write to CGRAM and operations that read DDRAM do not shift the display.

Display and Cursor ON/OFF

If B2 (D) bit is set, display is turned on. Otherwise, it is turned off. When the display is turned off, data in DDRAM is not changed.

If B1 (C) bit is set, the cursor is turned on. Otherwise, it is turned off. Operations that change the current address in the DDRAM Address register, like those to automatically increment or decrement the address, are not affected by turning off the cursor. The cursor is displayed at the eighth line in the 5 times 8 character matrix.

If B0 (B) bit is set, the character at the current cursor position blinks. Otherwise, the character does not blink. Note that character blinking and cursor are independent operations and that both can be set to work simultaneously.

Cursor/Display Shift

Cursor/display shift moves the cursor or shifts the display according to the selected mode. The operation does not change the DDRAM content. Because the cursor position always coincides with the value in the Address register, the instruction provides software with a mechanism for making DDRAM corrections or to retrieve display data at specific DDRAM locations. Table 12.4 lists the four available options:

Table 12.4

Cursor/Display Shift Options

BITS		
S/C	R/L	OPERATION
0	0	Cursor position is shifted left. Address counter is decremented by one
0	1	Cursor position is shifted right. Address counter is incremented by one
1	0	Cursor and display are shifted left
1	1	Cursor and display are shifted right

Function set

Function set sets the parallel interface data length, the number of display lines, and the character font. If B4 (DL) bit is set, then the interface is set to eight bits. Otherwise it is set to four bits. If B3 (N) bit is zero, the display is initialized for 1/8 or 1/11 duty cycle. When the N bit is set, the display is set to 1/16 duty cycle. Displays with multiple lines typically use the 1/16 duty cycle. The 1/16 duty cycle on a one-line display appears as if it were a two-line display, that is, the line consists of two separate address groups (see Table 12.3).

If B2 (F) bit is set, then the display resolution is 5 times 10 pixels. Otherwise the resolution is 5 times 8 pixels. This bit is not significant when the 1/16 duty cycle is selected; that is, when the N bit is set.

The function set instruction should be issued during controller initialization. No other instruction can be executed before this one, except for changing the interface data length.

Set CGRAM address

Set CGRAM address sets the *CGRAM (character generator RAM)* address to which data is sent or received after this operation. The CGRAM address is a 6-bit field in the range 0 to 64 decimal. Once a value is entered in the CGRAM Address register, data can be read or written from CGRAM.

Set DDRAM address

Set DDRAM address sets the *DDRAM (display data RAM)* address to which data is sent or received after this operation. The DDRAM address is a 7-bit field in the range 0 to 127 decimal. Once a value is entered in the DDRAM Address register, data can be read or written from CGRAM. DDRAM address mapping is discussed in Section 12.1.4.

Read busy flag and Address register

Read busy flag and Address register reads the busy flag to determine if an internal operation is in progress and reads the address counter content. The value in the Address register is reported in bits 0 to 6. Bit 7 (BF) is the busy flag bit. This bit is read only. The address counter is incremented or decremented by 1 (according to the mode set) after the execution of a data write or read instruction.

Write data

Write data writes eight data bits to CGRAM or DDRAM. Before data is written to either controller RAM area, software must first issue a set DDRAM address or set CGRAM address instruction (described previously). These two instructions not only set the next valid address in the Address register, but also select either CGRAM or DDRAM for writing operations. What other actions take place as data is written to the controller depends on the settings selected by the *entry mode set* instruction. If the direction of cursor movement or data shift is in the increment mode, then the data write operation adds one to the value in the Address register. If the cursor movement is enabled, then the cursor is moved accordingly after data write takes place. If the display shift mode is active, then the displayed characters are shifted either right or left.

Read data

Read data reads eight data bits to CGRAM or DDRAM. Before data is read from either controller RAM area, software must first issue a set DDRAM address or set CGRAM address instruction. These instructions not only set the next valid address in the Address register, but also select either CGRAM or DDRAM for writing operations. Failing to set the corresponding RAM area results in reading invalid data.

What other actions take place as data is read from the controller RAM depends on the settings selected by the entry mode set instruction. If the direction of cursor movement or data shift is in the increment mode, then the data read operation adds one to the value in the Address register. However, display is not shifted by a read operation even if the display shift is active.

The cursor shift instruction has the effect of changing the content of the Address register. So if a cursor shift precedes a data read instruction, there is no need to reset the address by means of an address set command.

12.3.2 16F84 8-Bit Data Mode Circuit

The first circuit presented in this chapter is experimental. Its purpose is to exercise LCD display functions in the simplest forms; therefore, the circuit uses 8-bit *parallel data transmission interfacing* with a 16F84 microcontroller. The circuit is shown in Figure 12-4.

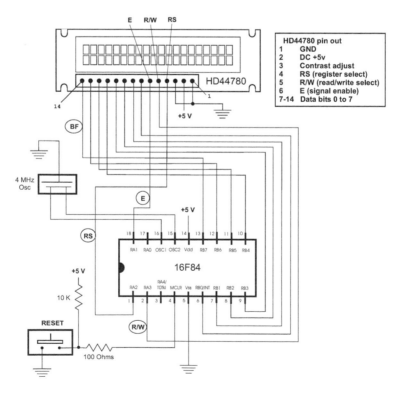

Figure 12-4 16F84 to LCD 8-Bit Mode Circuit.

In the circuit of Figure 12-4, three control lines are wired between the microcontroller and the LCD. The line designations are shown inside ovals. The R/W line is not necessary, because it is possible to devise a system that does not read LCD data. In the diagram the R/W line is included because it allows reading the busy flag in synchronizing operations. Table 12.5 shows the control and data connections for the circuit in Figure 12-4.

Table 12.5

Connections for 16F84/LCD 8-bit Data Mode Circuit

16F84		LCD	LINE	
PIN	PORTBIT	PIN	NAME	FUNCTION
1	A2	4	RS	Select instruction/ data register
2	A3	5	R/W	Read/write select
18	A1	6	E	Enable signal
13	B7	14	before	Busy flag
6-13	B0-B7	7-14	Data	Data lines

12.4 LCD Programming

LCD programming is usually device specific. Before attempting to write code, the programmer should become familiar with the circuit wiring diagram, the setup parameters, and the specific hardware requirements. It is risky to make assumptions that a specific device conforms exactly to the HD44780 interface because often a style sheet contains specifications that are not in strict conformance with the standard. In addition to the PIC setup and initialization functions, code to display a simple text message on the LCD screen consists of the following display-related functions:

1. Define the required constants, variables, and buffers.

2. Set up and initialize ports used by the LCD.

3. Initialize the LCD to circuit and software specifications.

4. Store text in PIC text buffer.

5. Select DDRAM start address on LCD.

6. Display text by transferring characters in PIC text buffer to LCD DDRAM.

If the LCD display consists of multiple lines, then the previous steps 4, 5, and 6 are repeated for each line. LCD initialization and display operations vary according to whether the interface is 4- or 8-bits and whether the code uses delay loops or busy flag monitoring to synchronize operations. All of these variations are considered in the examples in this chapter.

12.4.1 Defining Constants and Variables

In any program, defining and documenting constants and fixed parameters should be done centrally, rather than hard-coded through the code. Centralizing the elements that are variable under different circumstances makes it possible to adapt code to circuit and hardware changes.

Two common ways are available for defining constants: the C-like **#define** directive and the **equ** (equate) directive. In most cases, it is a matter of personal preference which is used, but a general guideline is to use the **#define** statement to create literal constants; that is, constants that are not associated with program registers or variables. The **equ** directive is then used to define registers, flags, and local variables.

According to this scheme, an LCD display driver program could use **#define** statements to create literals that are related to the wiring diagram or the specific LCD values obtained form the data sheet, such as the DDRAM addresses for each display line, as in the following code fragment:

```
;=======================================================
;                       constant definitions
;   for PIC-to-LCD pin wiring and LCD line addresses
;=======================================================
#define E_line 1            ; |
#define RS_line 2           ; |  => from wiring diagram
#define RW_line 3           ; |
; LCD line addresses (from LCD data sheet)
#define LCD_1 0x80          ; First LCD line constant
#define LCD_2 0xc0          ; Second LCD line constant
```

By the same token, the values associated with PIC register addresses and bit flags are defined using equ, as follows:

```
;=======================================================
;                       PIC register equates
;=======================================================
porta      equ       0x05
portb      equ       0x06
fsr        equ       0x04
status     equ       0x03
indf       equ       0x00
z          equ       2
```

One advantage of this scheme is that constants are easier to locate, because they are grouped by device. Those for the LCD are in the **#define** directives area and those for the PIC hardware in an area of **equ** directives.

There are also drawbacks to this approach, as symbols created in **#define** directives are not available for viewing in the MPLAB debuggers. However, if the use of the **#define** directive is restriced to literal constants, then their viewing during a debugging session is not essential.

MPLAB also supports the **constant** directive for creating a constant symbol. Its use is identical to the **equ** directive but the latter is more commonly found in code.

Using MPLAB Data Directives

Often a program needs to define a block of sequential symbols and assign to each one a corresponding name. In the PIC 16f84, the address space allocated to general-purpose registers allocated by the user is of 68 bytes, starting at address 0x0c. One possible way of allocating user-defined registers is to use the **equ** directive to assign addresses in the PIC SRAM space:

```
Var1      equ      0x0c
Var2      equ      0x0d
Var3      equ      0x0e
Buf1      equ      0x0f      ; 10-byte buffer space
Var4      equ      0x19      ; Next variable
```

Although this method is functional, it depends on the programmer calculating the location of each variable in the PIC's available SRAM space. Alternatively, MPLAP provides a **cblock** directive that allows defining a group of consecutive sequential symbols while referring only to the address of the first element in the group. If no address is entered in **cblock**, then the assembler assigns the address. This address is one higher than the final address in the previous **cblock**. Each **cblock** ends with the **endc** directive. The following code fragment showing the use of the **cblock** directive is from one of the sample programs for this chapter:

```
;=======================================================
;                 variables in PIC RAM
;=======================================================
; Reserve 16 bytes for string buffer
        cblock   0x0c
                 strData
        endc
; Leave 16 bytes and continue with local variables
        cblock   0x1d      ; Start of block
        count1             ; Counter # 1
        count2             ; Counter # 2
        count3             ; Counter # 3
        pic_ad             ; Storage for start of text area
                           ; (labeled strData) in PIC RAM
        J                  ; counter J
        K                  ; counter K
        index              ; Index into text table
        endc
```

Note in the preceding code fragment that the allocation for the 16-byte buffer space named strData is ensured by entering the corresponding start address in the second **cblock**. The PIC microcontrollers do not contain a directive for reserving memory areas inside **cblock**, although the **res** directive can be used to reserve memory for individual variables.

12.4.2 LCD Initialization

LCD initialization depends on the specific hardware in use and on the circuit wiring. Information about the specific LCD can be obtained from the device's data sheet. Sometimes, the data sheet includes examples of initialization values for different conditions and even code listings. The information is usually sufficient to ensure correct initialization.

A word of warning: The popular LCD literature available online often contains initialization "myths" for specific components requiring that a certain mystery code be used for no documented reason, or that a certain function be repeated a given number of times. The programmer should make sure that the code is rational and that every operation is actually required and documented.

Before the LCD initialization commands are used, it is necessary to set the communications lines correctly. The E line should be low, the RS line should be low for command, and the R/W line (if connected) should be low for *write mode*. After the lines are set accordingly, there should be a 125-ms delay. Note that, at this point, the LCD busy flag is not yet reliable. The following code fragment shows the processing:

```
bcf     porta,E_line      ; E line low
bcf     porta,RS_line     ; RS line low for command
bcf     porta,RW_line     ; Write mode
call    delay_125         ; delay 125 microseconds
```

The procedure delay_125 in the previous code fragment is described later in this chapter.

Function Set Command

Function set is the first initialization command sent to the LCD. The command determines whether the display font consists of 5 times 10 or 5 times 7 pixels. The latter is by far the more common. It determines the duty cycle, which is typically 1/8 or 1/11 for single-line displays and 1/16 for multiple lines. The interface width is also determined in the FUNCTION SET command. It is 4 bits or 8 bits. The following code fragment shows the commented code for the Function Set command:

```
;********************|
;    Function Set    |
;********************|
        movlw   0x38     ; 0 0 1 1 1 0 0 0 (FUNCTION SET)
                         ;     | | | |__ font select:
                         ;     | | |     1 = 5x10 in 1/8 or 1/11
                         ;     | | |     0 = 1/16 dc
                         ;     | | |___ Duty cycle select
                         ;     | |       0 = 1/8 or 1/11
                         ;     | |       1 = 1/16
                         ;     | |___ Interface width
                         ;     |       0 = 4 bits
                         ;     |       1 = 8 bits
```

```
                            ;          |___ FUNCTION SET COMMAND
        movwf    portb
        call     pulseE    ;pulse E line to force LCD command
```

In the preceding code fragment, the LCD is initialized to multiple lines, 5 times 7 font, and 8-bit interface, as in the program LCDTest1 found in this book's online software package.

The procedure named pulseE sets the E line bit off and on to force command recognition by the LCD. The procedure is listed and described later in this chapter.

Display Off

Some initialization routines in LCD documentation and data sheets require that the display be turned off following the Function Set command. If so, the Display Off command is executed as follows:

```
;***********************|
;      Display Off      |
;***********************|
        movlw    0x08     ; 0 0 0 0 1 0 0 0 (DISPLAY ON/OFF)
                          ;         | | | |___ Blink character at
                          ;         | | | |    Cursor
                          ;         | | |      1 = on, 0 = off
                          ;         | | |___ Cursor on/off
                          ;         | |       1 = on, 0 = off
                          ;         | |____ Display on/off
                          ;         |        1 = on, 0 = off
                          ;         |____ COMMAND BIT
        movwf    portb
        call     pulseE   ; pulse E line to force LCD command
```

Display and Cursor On

Whether or not the display is turned off, it must be turned on first. Also, code must select if the cursor is on or off, and whether the character at the cursor position is to blink. The following command sets the cursor and the display on, and the character blink off:

```
;***********************|
; Display and Cursor On |
;***********************|
        movlw    0x0e     ; 0 0 0 0 1 1 1 0 (DISPLAY ON/OFF)
                          ;         | | | |___ Blink character at
                          ;         | | | |    cursor
                          ;         | | |      1 = on, 0 = off
                          ;         | | |___ Cursor on/off
                          ;         | |       1 = on, 0 = off
                          ;         | |____ Display on/off
                          ;         |        1 = on, 0 = off
```

```
               ;              |____ COMMAND BIT
     movwf   portb
     call    pulseE    ; pulse E line to force LCD command
```

Set Entry Mode

The Entry Mode command sets the direction of cursor movement or *display shift mode*. Normally, the display is set to the *increment mode* when writing in Western European languages. The Entry Mode command controls *display shift*. If enabled, the displayed characters appear to scroll. This mode is used to simulate an electronic billboard effect by storing more than one line of characters in DDRAM and the then scrolling the characters left-to-right. The following code sets entry mode to increment mode and no shift:

```
;***********************|
;   Set Entry Mode      |
;***********************|
     movlw   0x06     ; 0 0 0 0 0 1 1 0 (ENTRY MODE SET)
                      ;             | | |___ display shift
                      ;             | |      1 = shift
                      ;             | |      0 = no shift
                      ;             | |____ cursor increment
                      ;             |        mode
                      ;             |        1 = left-to-right
                      ;             |        0 = right-to-left
                      ;             |___ COMMAND BIT
     movwf   portb   ;00000110
     call    pulseE
```

Operations that read or write to CGRAM and operations that read DDRAM do not shift the display.

Cursor and Display Shift

These commands determine whether the cursor or the display shift according to the selected mode. Shifting the cursor or the display provides a software mechanism for making DDRAM corrections or for retrieving display data at specific DDRAM locations. The four available options appear in Table 12.4 previously in this chapter. The following instructions set the cursor to *shift right* and disable *display shift*:

```
;***********************|
; Cursor/Display Shift  |
;***********************|
     movlw   0x14     ; 0 0 0 1 0 1 0 0 (CURSOR/DISPLAY
                      ;         | | | | |  SHIFT)
                      ;         | | | |_|___ don't care
                      ;         | |_|__ cursor/display shift
                      ;         |       00 = cursor shift left
                      ;         |       01 = cursor shift right
                      ;         |       10 = cursor and display
```

```
                        ;          |               shifted left
                        ;          |       11 = cursor and display
                        ;          |               shifted right
                        ;          |___ COMMAND BIT
        movwf    portb  ;0001 1111
        call     pulseE
```

Clear Display

The final initialization command is usually one to clear the display. It is entered as follows:

```
;**********************|
;    Clear Display     |
;**********************|
        movlw    0x01   ; 0 0 0 0 0 0 0 1 (CLEAR DISPLAY)
                        ;               |___ COMMAND BIT
        movwf    portb  ;0000 0001
        call     pulseE
        call     delay_5 ;delay 5 milliseconds after init
```

Note that the last command is followed by a 5ms delay. The delay procedure delay_5 is listed and described later in this chapter.

12.4.3 Auxiliary Operations

Several support routines are required for effective text display in LCD devices. These include time delay routines for timed access, a routine to pulse the E line in order to force the LCD to execute a command or to read or write text data, routines to read the busy flag when this is the method used for processor/LCD synchronization, and routines to merge data with port bits so as to preserve the status of port lines not being addressed by code.

Time Delay Routine

There are several ways of producing time delays in PIC microcontrollers. A book by David Benson is devoted almost entirely to timing and counting routines. The present concern is quite simple: to develop a software routine that ensures the time delay that must take place in LCD programming, as shown in Table 12.3.

One mechanism for producing time delays in PIC programming is by means of the Timer0 module, a built-in 8-bit timer counter. Once enabled, Port-A pin 4, labeled the TOCKI bit and associated with file register 01 (TMR0), is used to time processor operations. In the particular case of LCD timing routines, using the Timer0 module seems somewhat of an overkill, in addition to the fact that it requires the use of a Port-A line, which is often required for other purposes.

Alternatively, timing routines that serve the purpose at hand can be developed using simple delay loops. In this case, no port line is sacrificed and coding is considerably simplified. These routines are generically labeled *software timers*, in contrast to the hardware timers that depend on the PIC timer/counter device described previously. Software timers provide the necessary delay by means of program loops; that is, by wasting time. The length of delay provided by the routine depends on the execution time of each instruction and on the number of repeated instructions.

Instructions on the PIC 16f84 consume four clock cycles. If the processor clock is running at 4 MHz, then one fourth of 4 MHz is the execution time for each instruction, which is 1 µs. So if each instruction requires 1 µs, repeating 1,000 instructions produces a delay of 1 ms. The following routines provide convenient delays for LCD interfacing:

```
;=======================
;   Procedure to delay
;    125 microseconds
;=======================
delay_125mics:
        movlw    D'42'              ; Repeat 42 machine cycles
        movwf    count1             ; Store value in counter
repeat:
        decfsz   count1,f           ; Decrement counter (1 cycle)
        goto     repeat             ; Continue if not 0 (2 cycles)
                                    ; 42 * 3 = 126
        return                      ; End of delay
;=======================
;   Procedure to delay
;    5 milliseconds
;=======================
delay_5ms:
        movlw    D'41'              ; Counter = 41
        movwf    count2             ; Store in variable
delay:
        call     delay_125mics      ; Delay 41 microseconds
        decfsz   count2,f           ; 41 times 125 = 5125 ms
                                    ; or approximately 5 ms
        goto     delay
        return                      ; End of delay
```

Actually, the delay loop of the procedure named delay_5ms is not exactly the product of 41 iterations times 125 µs, because the instructions to decrement the counter and the **goto** to the label delay are also inside the loop. Three instruction cycles must be added to those consumed by the delay_125mics procedure. This results in a total of 41 * 3 or 123 instruction cycles that must be added to the 5,125 consumed by delay_125mics. In fact, there are several other minor delays by the instructions to initialize the counters that are not included in the calculation. In reality, the delay loops required for LCD interfacing need not be exact, as long as they are not shorter than the recommended minima.

For calculating software delays in the 16f84, the instruction execution time is determined by an external clock either in the form of an oscillator crystal, a resonator, or an RC oscillator furnished in the circuit. The PIC 16f84A is available in various processor speeds, from 4 MHz to 20 MHz. These speeds describe the maximum capacity of the PIC hardware. The actual instruction speed is determined by the clocking device, so a 20-MHz 16f84A using a 4-MHz oscillator effectively runs at 4 MHz.

Pulsing the E Line

The LCD hardware does not recognize data as it is placed in the input lines. When the various control and data pins of the LCD are connected to ports in the PIC and data is placed in the port bits, no action takes place in the LCD controller. In order for the controller to respond to commands or to perform read or write operations, it must be activated by *pulsating* (or *strobing*) the E line. The pulsing or strobing mechanism requires that the E line be kept low and then raised momentarily. The LCD checks the state of its lines on the rising edge of the E line. Once the command has completed, the E line is brought low again. The following code fragment pulses the E line in the manner described:

```
;========================
;      pulse E line
;========================
pulseE
        bsf       porta,E_line      ; pulse E line
        bcf       porta,E_line
        call      delay_125mics     ; delay 125 microseconds
        return
```

Note that the listed routine includes a 125 μs delay following the pulsing operation. This delay is not part of the pulse function but is required by most LCD hardware. Some pulse functions in the popular PIC literature include a *no operation opcode* (**nop**) between the commands to set and clear the E line. In most cases this short delay does not hurt, but some LCDs require a minimum time lapse during the pulse and will not function correctly if the **nop** is inserted in the code.

Reading the Busy Flag

Synchronization between LCD commands and between data access operations is based on time delay loops or on reading the LCD *busy flag*. The busy flag, which is in the same pin as the bit 7 data line, is read clear when the LCD is ready to receive the next command, read, or write operation, and is set if the device is not ready. By reading the state of the busy flag, code can accomplish more effective synchronization than with time delay loops. The sample program named LCDTest2, in this book's online software package, performs LCD display using the busy flag method. The following procedure shows busy flag synchronization:

```
;========================
; busy flag test routine
;========================
; Procedure to test the HD44780 busy flag
; Execution returns when flag is clear
busyTest:
        movlw     b'11111111'       ; All lines to input
        tris      portb                        ; in port B
```

```
        bcf       porta,RS_line     ; RS line low for control
        bsf       porta,RW_line     ; Read mode
        bsf       porta,E_line      ; E line high
        movf      portb,w           ; Read port B into W
                                    ; Port B bit 7 is busy flag
        bcf       porta,E_line      ; E line low
        andlw     0x80              ; Test bit 7, high is busy
        btfss     status,z          ; Test zero bit in STATUS
        goto      busyTest          ; Repeat if set
; At this point busy flag is clear
; Reset R/W line and port B to output
        bcf       porta,RW_line     ; Clear R/W line
        movlw     b'00000000'       ; All lines to output
        tris      portb                         ; in port B
        return
```

Note that testing the busy flag requires setting the LCD to read mode, which in turn requires implementing a connection between a PIC port and the R/W line; also that the listed procedure contains no safety mechanism for detecting a hardware error condition in which the busy flag never clears. If such were the case, the program would hang in a forever loop. To detect and recover from this error, the routine would have to include an external timing loop or some other means of recovering a possible hardware error.

Bit Merging Operations

Often, PIC/LCD circuits do not use all the lines in an individual port. In this case, the routines that manipulate PIC/LCD port access should not change the settings of other port bits. This situation is not exclusive to LCD interfacing; the discussion that follows has many other applications in PIC programming.

A processing routine can change one or more port lines without affecting the remaining ones. For example, an application that uses a 4-bit interface between the PIC and the LCD typically leaves four unused lines in the access port, or uses some of these lines for interface connections. In this case, the programming problem can be described as merging bits of the data byte to be written to the port and some existing port bits. One operand is the access port value, and the other one is the new value to write to this port. If the operation at hand uses the four high-order port bits, then its four low-order bits must be preserved. The logic required is simple: AND the corresponding operands with masks that clear the unneeded bits and preserve the significant ones, then OR the two operands. The following procedure shows the required processing:

```
;=================
;  merge bits
;=================
; Routine to merge the 4 high-order bits of the
; value to send with the contents of port B
; so as to preserve the 4 low-bits in port B
; Logic:
```

```
;          AND value with 1111 0000 mask
;          AND port B with 0000 1111 mask
;          At this point low nibble in value and high
;          nibble in port B are al 0 bits:
;                       value = vvvv 0000
;                       port B = 0000 bbbb
;          OR value and port B resulting in:
;                       vvvv bbbb
; ON ENTRY:
;       w contain value bits
; ON EXIT:
;       w contains merged bits
merge4:
          andlw    b'11110000'       ; ANDing with 0 clears the
                                     ; bit. ANDing with 1 preserves
                                     ; the original value
          movwf    store2            ; Save result in variable
          movf     portb,w           ; port B to w register
          andlw    b'00001111'       ; Clear high nibble in port b
                                     ; and preserve low nibble
          iorwf    store2,w          ; OR two operands in w
          return
```

Note that this particular example refers to merging two operand nibbles. The code can be adapted to merge other size bit-fields by modifying the corresponding masks. For example, the following routine merges the high-order bit of one operand with the seven low-order bits of the second one:

```
; Routine to merge the high-order bit of the first operand with
; the seven low-order bits of the second operand
; ON ENTRY:
;          w contains value bits of first operand
;          port b is the second operand
merge1:
          andlw    b'10000000'       ; ANDing with 0 clears the
                                     ; bit. ANDing with 1 preserves
                                     ; the original value
          movwf    store2            ; Save result in variable
          movf     portb,w           ; port B to w register
          andlw    b'01111111'       ; Clear high-order  bit in
                                     ; port b and preserve the
                                     ; seven low order bits
          iorwf    store2,w          ; OR two operands in w
          return
```

Popular PIC literature describes routines to merge bit-fields by assuming certain conditions in the destination operand and then testing the first operand bit to determine if the assumed condition should be preserved or changed. This type of operation is sometimes called "bit flipping," for example:

```
flipBit7:
; Code fragment to test the high-order bit in the variable named
; oprnd1 and preserve its status in the register variable portb
        bcf         portb,7             ; Assume oprnd1 bit is reset
        btfsc       oprnd1,7            ; Test operand bit and skip if
                                        ; clear (assumption valid)
        bsf         portb,7             ; Set bit if necessary
        return
```

The logic in bit-flipping routines contains one critical flaw: If the assumed condition is false then the second operand is changed improperly, even if for only a few microseconds. However, the incorrect value can produce errors in execution if it is used by another device during this period. Because there is no such objection to the merge routines based on *masking*, the programmer should always prefer them.

12.4.4 Text Data Storage and Display

Text display operations require some way of generating the ASCII characters that are to be stored in DDRAM memory. Although the PIC Assembler contains several operators to generate ASCII data in program memory, there is no convenient way of storing a string in the general-purpose register area. Even if this was possible, SRAM is typically in short supply and text strings gobble up considerable data space.

Several possible approaches are available. The most suitable one depends on the total string length to be generated or stored, whether the strings are reused in the code, and other program-related circumstances. In this sense, short text-strings can be produced character-by-character and sent sequentially to DDRAM memory by placing the characters in the corresponding port and pulsing the E line.

The following code fragment consecutively displays the characters in the word "Hello." Code assumes that the command to set the Address register has been entered previously:

```
; Generate characters and send directly to DDRAM
        movlw       'H'                 ; ASCII for H in w
        movwf       portb               ; Store code in port B
        call        pulseE      ; Pulse E line
        movlw       'e'                 ; Continues
        movwf       portb
        call        pulseE
        movlw       'l'
        movwf       portb
        call        pulseE
        movlw       'l'
        movwf       portb
        call        pulseE
        movlw       'o'
        movwf       portb
        call        pulseE
        call        delay_5
```

Note in the preceding fragment that the code assumes that the LCD has been initialized to automatically increment the Address register left-to-right. For this reason, the Address register is bumped to the next address with each port access.

Generating and Storing a Text String

An alternative approach suitable for generating and displaying longer strings consists of storing the string data in a local variable (sometimes called a *buffer*) and then transferring the characters, one by one, from the buffer to DDRAM. This kind of processing has the advantage of allowing the reuse of the same string and the disadvantage of using up scarce data memory. The logic for one possible routine consists of first generating and storing in PIC RAM the character string, then retrieving the characters from the PIC RAM buffer and displaying them. The character generation and storage logic is shown in Figure 12-5.

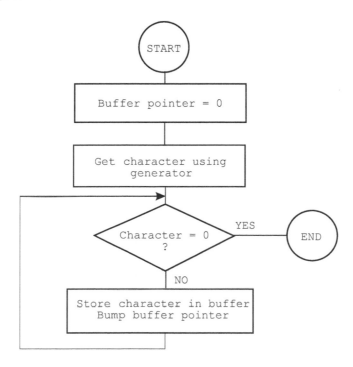

Figure 12-5 Flowchart for String Generation Logic.

The processing is demonstrated in the following procedure.

```
;===================================
;  first text string procedure
;===================================
storeMN:
; Procedure to store in PIC RAM buffer the message
; contained in the code area labeled msg1
; ON ENTRY:
;          variable pic_ad holds address of text buffer
;          in PIC RAM
```

```
;           w register holds offset into storage area
;           msg1 is routine that returns the string characters
;           and a zero terminator
;           index is local variable that holds offset into
;           text table. This variable is also used for
;           temporary storage of offset into buffer
; ON EXIT:
;           Text message stored in buffer
;
; Store offset into text buffer (passed in the w register)
; in temporary variable
          movwf     index               ; Store w in index
; Store base address of text buffer in fsr
          movf      pic_ad,w ; first display RAM address to W
          addwf     index,w  ; Add offset to address
          movwf     fsr      ; W to FSR
; Initialize index for text string access
          movlw     0        ; Start at 0
          movwf     index    ; Store index in variable
; w still = 0
get_msg_char:
          call      msg1     ; Get character from table
; Test for zero terminator
          andlw     0x0ff
          btfsc     status,z ; Test zero flag
          goto      endstr1  ; End of string
; ASSERT: valid string character in w
;         store character in text buffer (by fsr)
          movwf     indf     ; store in buffer by fsr
          incf      fsr,f    ; increment buffer pointer
; Restore table character counter from variable
          movf      index,w  ; Get value into w
          addlw     1        ; Bump to next character
          movwf     index    ; Store table index in variable
          goto      get_msg_char      ; Continue
endstr1:
          return
; Routine for returning message stored in program area
msg1:
          addwf     PCL,f               ; Access table
          retlw     'M'
          retlw     'i'
          retlw     'n'
          retlw     'n'
          retlw     'e'
          retlw     's'
          retlw     'o'
          retlw     't'
```

```
        retlw    'a'
        retlw    0                    ; terminator character
```

The auxiliary procedure named msg1, listed in the preceding code fragment, performs the character-generator function by producing each of the ASCII characters in the message string. Because a **retlw** instruction is necessary for each character, one instruction space in program memory is used for each character generated, plus a final binary zero for the string terminator.

Displaying the Text String

Once the string is stored in a local buffer, it is displayed by moving each ASCII code from the buffer into LCD DDRAM. Here again, we assume that the LCD has previously been set to the auto increment mode and that the Address register has been properly initialized with the corresponding DDRAM address. The following procedure demonstrates initialization of the DDRAM Address register to the value defined in the constant named LCD_1:

```
;=========================
; Set Address register
;    to LCD line 1
;=========================
; ON ENTRY:
;         Address of LCD line 1 in constant LCD_1
line1:
        bcf      porta,E_line       ; E line low
        bcf      porta,RS_line      ; RS line low, set up for
                                    ; control
        call     delay_125          ; delay 125 microseconds
; Set to second display line
        Movlw    LCD_1              ; Address and command bit
        movwf    portb
        call     pulseE             ; Pulse and delay
; Set RS line for data
        bsf      porta,RS_line      ; Set up for data
        call     delay_125mics      ; Delay
        return
```

Once the Address register has been set up, the display operation consists of transferring characters from the PIC RAM buffer into LCD DDRAM. The following procedure can be used for this:

```
;===============================
;   LCD display procedure
;===============================
; Sends 16 characters from PIC buffer, with address stored
; in variable pic_ad, to LCD line previously selected
display16:
; Set up for data
        bcf      porta,E_line       ; E line low
```

```
        bsf      porta,RS_line    ; RS line low for control
        call     delay_125                 ; Delay
; Set up counter for 16 characters
        movlw    D'16'            ; Counter = 16
        movwf    count3
; Get display address from local variable pic_ad
        movf     pic_ad,w ; First display RAM address to W
        movwf    fsr              ; W to FSR
getchar:
        movf     indf,w   ; get character from display RAM
                          ; location pointed to by file select
                          ; register
        movwf    portb
        call     pulseE   ;send data to display
; Test for 16 characters displayed
        decfsz   count3,f ; Decrement counter
        goto     nextchar ; Skipped if done
        return
nextchar:
        incf     fsr,f                 ; Bump pointer
        goto     getchar
```

Note that the procedure display16, previously listed, assumes that the address of the local buffer is stored in a variable called pic_ad. This allows reusing the procedure to display text stored at other locations in PIC RAM.

The previously listed procedures demonstrate just one of many possible variations on this technique. Another approach is to store the characters directly in DDRAM memory as they are produced by the message-returning routine, thus avoiding the display procedure entirely. In this last case, the programming saves some data memory space at the expense of having to generate the message characters each time they are needed. The most suitable approach depends on the application.

12.4.5 Data Compression Techniques

Circuits based on the parallel data transfer of eight data bits require eight devoted port lines. Assuming that three other lines are required for LCD commands and interfacing (RS, E, and R/W lines), then 11 PIC-to-LCD lines are needed, leaving two free port lines at the most, on an 16f84 microcontroller. Not many useful devices can make do with just two port lines. Several possible solutions allow compressing the data transfer function. The most obvious one is to use the 4-bit data transfer mode to free four port lines. Other solutions are based on dedicating logic components to the LCD function.

4-Bit Data Transfer Mode

One possible solution is to use the capability of the Hitachi 44780 controller that allows a parallel interface using just four data paths instead of eight. The objections are that programming in 4-bit mode is slightly more convoluted and there is a very minor performance penalty. In 4-bit mode, data must be sent one nibble at a time, so execution is slower. Because the delay is required only after the second nibble, the execution time penalty for 4-bit transfers is not very large.

Many of the previously developed routines for 8-bit data mode can be reused without modification in the 4-bit mode. Others require minor changes, and there is one specific display procedure that must be developed ad hoc. The first required change is in the LCD initialization because bit 4 in the Function Set command must be clear for a 4-bit interface. The remaining initialization commands should require no further change, although it is a good idea to consult the data sheet for the LCD hardware in use.

Displaying data using a 4-bit interface consists of sending the high-order nibble followed by the low-order nibble, through the LCD 4-high-order data lines, usually labeled DB5 to DB7. The pulsing of line E follows the last nibble sent. It is usually the case in the 16f84 PIC that circuit wiring in the 4-bit mode uses four of five lines in port A, or four of eight lines in port B. Software must provide a way of reading and writing to the appropriate port lines, the ones used in the data transfer, without altering the value stored in the port bits dedicated to other uses. *Bit merging* routines, discussed in Section 12.3, are quite suitable for the purpose at hand.

The following procedures are designed to send the two nibbles of a data byte through the four high-order lines in port B. The auxiliary procedure named merge4 performs the bit-merging operation while the procedure named send8 does the actual write operation:

```
;========================
;   send 2 nibbles in
;      4-bit mode
;========================
; Procedure to send two 4-bit values to port B lines
; 7, 6, 5, and 4. High-order nibble is sent first
; ON ENTRY:
;         w register holds 8-bit value to send
send8:
        movwf   store1          ; Save original value
        call    merge4          ; Merge with port B
; Now w has merged byte
        movwf   portb           ; w to port B
        call    pulseE          ; Send data to LCD
; High nibble is sent
        movf    store1,w        ; Recover byte into w
        swapf   store1,w        ; Swap nibbles in w
        call    merge4
        movwf   portb
        call    pulseE          ; Send data to LCD
        call    delay_125
        return
;==================
;   merge bits
```

```
;==================
; Routine to merge the 4 high-order bits of the
; value to send with the contents of port B
; so as to preserve the 4 low-order bits in port B
; Logic:
;       AND value with 1111 0000 mask
;       AND port B with 0000 1111 mask
;       Now low nibble in value and high nibble in
;       port B are all 0 bits:
;                   value = vvvv 0000
;                   port B = 0000 bbbb
;       OR value and port B resulting in:
;                   vvvv bbbb
; ON ENTRY:
;       w contain value bits
; ON EXIT:
;       w contains merged bits
merge4:
        andlw    b'11110000'      ; ANDing with 0 clears the
                                  ; bit. ANDing with 1 preserves
                                  ; the original value
        movwf    store2           ; Save result in variable
        movf     portb,w          ; port B to w register
        andlw    b'00001111'      ; Clear high nibble in port b
                                  ; and preserve low nibble
        iorwf    store2,w         ; OR two operands in w
        return
```

The program named LCDTest3 in this book's online software package is a demonstration using the 4-bit interface mode. Figure 12-6 shows a PIC/LCD circuit that is wired for the 4-bit data transfer mode.

Note in the circuit of Figure 12-6 that a total of six port lines remain unused. Two of these lines are in port-A and four in port-B.

Master/Slave Systems

To this point we have assumed that driving the LCD is one of the functions performed by the PIC microcontroller, which also executes the other circuit functions. In practice, such a scheme is rarely viable for two reasons: the number of interface lines required and the amount of PIC code space used up by the LCD driver routines. A more efficient approach is to dedicate a PIC exclusively to controlling the LCD hardware, while one or more other microcontrollers perform the main circuit functions. In this scheme, the PIC devoted to the LCD function is referred to as a *slave* while the one that sends the display commands is called the *master*.

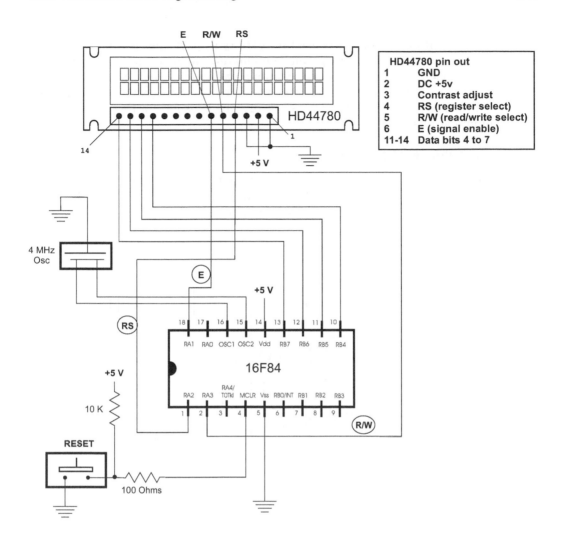

Figure 12-6 PIC/LCD Circuit for 4-Bit Data Mode.

When sufficient numbers of interface lines are available, the connection between master and slave can be simplified using a *parallel interface*. For example, if four port lines are used to interconnect the two PICs, then sixteen different command codes can be sent to the slave. The slave reads the communications lines much like it would read a multiple toggle switch. A simple protocol can be devised so that the slave uses these same interface lines to provide feedback to the master. For example, the slave sets all four lines low to indicate that it is ready for the next command, and sets them high to indicate that command execution is in progress and that no new commands can be received. The master, in turn, reads the communications lines to determine when it can send another command to the slave.

But using parallel communications between master and slave can be a self-defeating proposition, because it requires at least seven interface lines to be able to send ASCII characters. Because the scarcity of port lines is the original reason for using a master/slave setup, parallel communications may not be a good solution in many cases. On the other hand, communications between master and slave can take place serially, using a single interface line. The discussion of using *serial interface* between a master and an LCD slave driver PIC is left for the chapter on serial communications.

12.5 Sample Programs

Three sample programs in this book's software package demonstrate the LCD programming discussed in this chapter. The programs can be executed in the LCD and Real-Time clock demo board. Schematics for this board are shown in Figure 12-7.

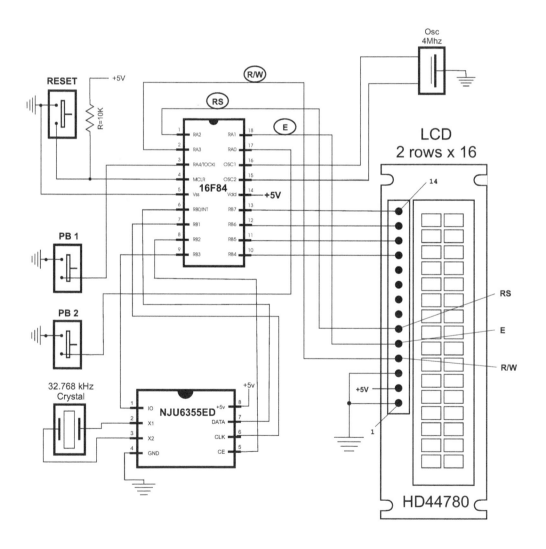

Figure 12-7 Demo Board B (LCD and Real-Time Clock).

In this book's software resource you will find support files and instructions for building the demo board in Figure 12-7, as well as sample programs to demonstrate LCD programming. All three programs assume that the LCD is driven by the Hitachi HD44780 controller and that the display supports two lines, each one with sixteen characters. The wiring and base address of each display line is stored in **#define** statements. These statements can be edited to accommodate a different setup. Their function is as follows:

- The LCDTest1 program exercises the 8-bit PIC-to-LCD interface using delay loops for interface timing.

- The LCDTest2 program exercises the 8-bit PIC-to-LCD interface using the busy flag to synchronize processor access.

- The LCDTest3 program exercises the 4-bit PIC-to-LCD interface. Program uses delay loops for interface timing.

Chapter 13

Analog-to-Digital and Real-Time Clocks

13.1 Clocks and the Digital Revolution

Digits are a human invention; nature does not count or measure using numbers. We measure natural forces and phenomena using digital representations, but the forces and phenomena themselves are continuous. Time, pressure, voltage, current, temperature, humidity, gravitational attraction; all exist as continuous entities that we measure in volts, pounds, hours, amperes, or degrees, so as to better understand them and to be able to perform numerical calculations.

In this sense, natural phenomena occur in analog quantities. Sometimes they are digitized so as to facilitate measurements and manipulations. For example, a potentiometer in an electrical circuit allows reducing the voltage level from the circuit maximum to ground, or zero level. In order to measure and control the action of the potentiometer, we need to quantify its action by producing a digital value within the physical range of the circuit; that is, we need to convert an *analog* quantity that varies continuously between 0 and 5 volts, to a discrete *digital* value range. If, in this case, the voltage range of the potentiometer is from 5 to 0 volts, we can digitize its action into a numeric range of 0 to 500 units, or measure the angle or rotation of the potentiometer disk in degrees from 0 to 180. The device that performs either conversion is called an A/D or *analog-to-digital converter*. The reverse process, digital-to-analog, is also necessary, although not as often as A/D. In this chapter we explore A/D conversions in PIC software and hardware.

The second topic of this chapter is the measurement of time in discrete (albeit, digital) units. In this context we speak of "real-time" as years, days, hours, minutes, and so on. So a real-time clock measures time in hours, minutes, and seconds, and a real-time calendar measures it in years, months, weeks, and days. Not all time units are in proportional relationship with one another. There are 60 seconds in a minute and 60 minutes in an hour, but 24 hours in a day, and 28, 29, 30, or 31 days in a month. Furthermore, the months and the days of the week have traditional names. Finally, the *Gregorian calendar* requires adding a twenty-ninth day to February on any year that is evenly divisible by 4. The device or software to perform all of these time calculations is referred to as a realtime clock. In this chapter we discuss the use of real-time clocks in PIC circuits.

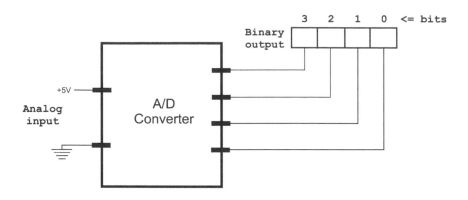

Figure 13-1 A/D Converter Block Diagram.

13.2 A/D Converters

In electronics, the typical *A/D* or *ADC converter* is a device that takes a voltage input and returns a binary digital number. Figure 13-1 is a block diagram of an A/D converter.

The electronic A/C converter requires an input in the form of an electrical voltage. Nonelectric quantities must be changed into a voltage level before the conversion can be performed. The device that performs this conversion is called a *transducer*. For example, a digital barometer must be equipped with a transducer that converts the measurement into voltage levels. The voltage levels can then be fed into an A/D converter and the result output in digital form.

13.2.1 Converter Resolution

An ideal A/D converter outputs into an infinite number of discrete steps that exactly represent the analog quantity. Needless to say, such a device cannot exist, and a real A/D converter must be limited to a numeric range. For example, the device in Figure 16-1 outputs a voltage range of 0 to +5 volts in four binary digits that represent values between 0 and 15. Another A/D converter may produce output in eight binary digits, and another in sixteen binary digits. The number of discrete values in the conversion is called the *resolution*. The converter's resolution is usually expressed in bits. Figure 16-2 represents an A/C converter with a voltage range of 0 to +5 volts and a resolution of three bits.

Suppose that a value of 2.5 volts were input into the A/D converter in Figure 13-2. Because the output has a resolution in the range 0 to 7, the converter's output would be either 4 or 5. The nonlinear characteristic of the output determines a *quantization error* that increases as the converter resolution decreases. Converters used in PIC circuits have a resolution of either 8, 10, or 12 bits. In each case the *output range*, or *quantization level*, is 0 to 255, 0 to 1023, or 0 to 4095. The voltage resolution of the converter is its maximum voltage range divided by the number of quantization levels. A device with a voltage range of 5 volts and a range of 255 levels has a voltage resolution of

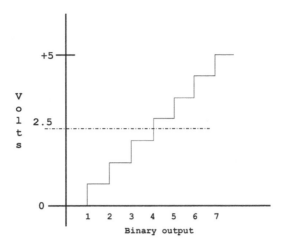

Figure 13-2 Converter Quantization Error.

$$Voltage\ resolution = \frac{5}{255} = 0.01960\ volts = 19.60\ mV$$

13.2.2 ADC Implementation

The analog-to-digital converter performs accurately only if the input voltage is within the converter's valid range. This range is usually selected by setting high and low voltage references on converter pins. For example, if +4 volts is input into the converter's *positive reference pin* and +2 volts into the *negative reference pin*, then the converter's voltage range lies between these values. In many PIC applications the converter range is selected as the system's supply voltage and ground, that is, +5 and 0 volts. When a different range is externally referenced, there is a general restriction that the range cannot exceed the system's positive and negative limits (Vdd and Vss). Also, a minimum difference is required between the high and low voltage references.

The output of the ADC is a digital representation of the original analog signal. In this context, the term *quantization* refers to subdividing a range into small but measurable increments. The quantization process can introduce a quantization error, which is similar to a rounding error.

The time required for the holding capacitor on the ADC to charge is called the *acquisition time*. The holding capacitor on the ADC must be given sufficient time to settle to the analog input voltage level before the actual conversion is initiated. Otherwise, the conversion is not accurate. The acquisition time is determined by the impedance of the internal multiplexer and that of the analog source. The exact acquisition time can be determined from the device's data sheet, although 10K Ohms is the maximum recommended source impedance for 8- and 10-bit converters and 2.5K Ohms for 12-bit converters.

Most analog-to-digital converters in PIC applications, either internal or external, are of the successive approximation type. The *successive approximation algorithm* performs a conversion on one bit at a time, beginning with the most significant bit and ending with the least-significant bit. To determine each bit in the range, the value of the input signal is tested to see if it is in the upper or lower portion of this range. If in the upper portion, the conversion bit is a 1, otherwise it is a 0. The next most-significant bit is then tested in the lower half of the remaining range. The process is continued until the least-significant bit has been determined.

13.3 A/D Integrated Circuits

Several popular integrated circuits are used to perform as A/D converters, among them the ADC0831, the LTC1298, and the MAX 190 and MAX 191. The variations consist in the resolution and interfacing of the different ICs. Of these, the ADC0831, from National Semiconductor, is an 8-bit resolution, serial interface A/D quite suited to applications for small, mid-range PICs such as the 16F84. The input range of the 0831 is 0 to 5 volts, which matches the TTL voltage levels used in PIC circuits. The 0831 pin diagram is shown in Figure 13-3.

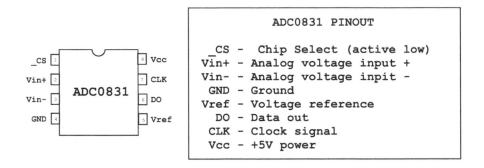

Figure 13-3 ADC0831 Pin Diagram.

The ADC0831 uses three control lines, labeled *DO* (*data out*), *CLK* (*clock*), and *_CS* (*chip select*) in Figure 13-3. Interfacing the ADC0831 requires three I/O lines. Of these, two can be multiplexed with other functions or with other ADC0831. Actually, only the chip-select (CS) pin requires a dedicated line. This allows for several ADCs to be multiplexed on the CLK and DO lines as long as each one has its own CS connection to the microcontroller. In this case, the controller determines which device is being read by the port to which its CS line is connected.

The input voltage range of the ADC0831 is determined by the *Vref* (*positive voltage reference* line) and *Vin-* (*negative voltage reference line*) pins. Vref is used to set the maximum level and Vin- the minimum. Because the ADC0831 has an 8-bit range, the voltage reading that matches the Vref value is read as 255 and the one that matches the Vin- value is read as 0. The minimum difference between the voltage limits is 1 volt.

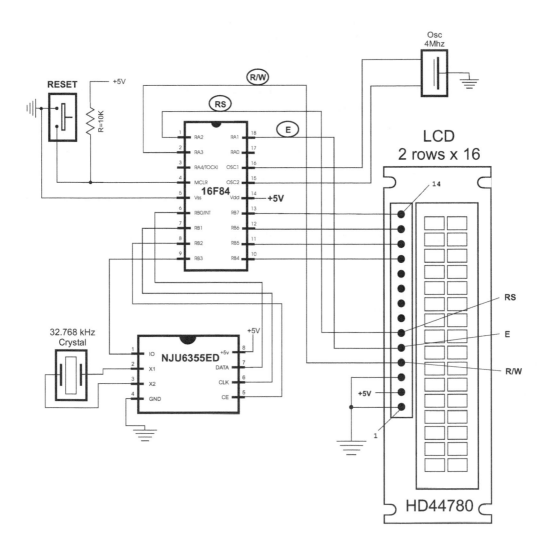

Figure 13-4 ADC0831 Demonstration Circuit.

13.3.1 ADC0331 Sample Circuit and Program

A simple circuit to illustrate the action of an analog-to-digital converter consists of connecting a potentiometer with the positive voltage reference line, as sown in Figure 13-4. In the circuit the potentiometer was selected so as to produce a voltage range between 0 and +5 volts. Vref was wired to the circuit's +5V source and Vin- was wired to ground. The potentiometer variable line was connected to the ADC0831 Vin+ line and the other ADC lines to the corresponding 16F84 port B pins.

The sample program is called ADF84, and can be found in this book's online software. The ADF84 program uses the ADC0831 to convert the analog voltage from the potentiometer, in the range +5 to 0 volts, into a digital value in the range 0 to 255. The value read is then displayed on the LCD. The initialization routine defines port-B, line 0 as input because this is the one connected to the DO line. The remaining lines in ports A and B are defined as output. ADC0831 processing consists of a single procedure that reads the analog line and returns an 8-bit digital value. The processing required is performed in the following steps:

1. The *data return register* (named rcvdata) is cleared and the bit counter register is initialized to count 8 bits.

2. The ADC0831 is prepared by bringing the CS line low and pulsing the CLK line.

3. The CLK line is pulsed and one bit is read from the low-order bit (DO line) of port B.

4. The bit is shifted into the data return register and the bit counter is decremented.

5. If the bit counter is exhausted, execution ends and the ADC is turned off. Otherwise processing continues at Step 3.

The following procedure, from the ADF84 program, reads digital data from the ADC0831:

```
;==============================
;   procedure to read and
;     convert analog line
;==============================
; ON ENTRY:
; Code assumes that the ADC0831 DO line is initialized for
; input, while CLK and CS lines are output
; from ADC0831 wiring diagram. All lines in port B
;       DO        =       RB0    ==> INPUT
;       CLK       =       RB1    <== OUTPUT
;       CS        =       RB2    <== OUTPUT
; ON EXIT:
; Returns 8-bit digital value in the register rcvdata
;
ana2dig:
; Clear data register and init counter for 8 bits
        clrf      rcvdata ; Clear register
        movlw     0x08             ; Initialize counter
        movwf     bitCount
; Prepare to read analog line
        bcf       PORTB,CS ; CS pin low to enable ADC
        nop                        ; Delay for 4 MHz clock
        bsf       PORTB,CLK        ; Set CLK high
        nop
        bcf       PORTB,CLK ; Reset CLK to start conversion
        nop
nextB:
; Pulse CLK line to read bit from ADC
        bsf       PORTB,CLK         ; CLK high
        nop       bcf               PORTB,CLK          ; CLK low
        nop
; Read analog line and store data, bit by bit
```

```
        movf      PORTB,w              ; Read all Port-B bits
        movwf     store1               ; Store value for later
        rrf       store1,f             ; Rotate bit into carry flag
        rlf       rcvdata,f            ; Rotate carry flag into
                                       ; result register
        decfsz    bitCount,f           ; Bump counter, skip next
                                       ; if counter zero
        goto      nextB
; Value read is stored in rcvdata register
        bsf       PORTB,CLK            ; Final clock pulse
        Nop
        bcf       PORTB,CLK
        nop
        bsf       PORTB,CS ; Turn off ADC
        call      long_delay           ; Time to settle
        Return
```

13.4 PIC Onboard A/D Hardware

A few years ago, A/D conversions always required the use of devices such as the ones described in the previous sections. Nowadays, many PIC microcontrollers come with onboard A/D hardware. One of the advantages of using onboard A/D converters is saving interface lines. The circuit shown in Figure13-4 requires devoting three lines to the interface between the ADC0831 and the PIC 16F84. On the other hand, a similar circuit can be implemented in a PIC with internal A/C conversion by simply connecting the analog device to the corresponding PIC port. In the PIC world, where I/O lines are often in short supply, this advantage is not insignificant.

At the time of writing, PICs equipped with A/D converters have either 8- or 10-bit resolution and can receive analog input in two to sixteen different channels. The 16F877 with eight analog input channels at a 10-bit resolution is discussed. Nowadays, these PICs are easy to obtain. On the other hand, if the resolution required exceeds 10 bits, then the designer has to resort to an independent A/D IC, such as the LTC1298, which has a 12-bit resolution, or to others with even higher numbers of output bits.

13.4.1 A/D Module on the 16F87x

The PICs of the 16F87x family are equipped with an analog-to-digital converter module. The number of lines depends on the specific version of the device: 28-pin devices have five A/D lines and all others have eight lines. The converter *uses a sample and hold* capacitor to store the analog charge and performs a successive approximation algorithm to produce the digital result. The converter resolution is 10 bits, which are stored in two 8-bit registers. One of the registers has only four significant bits.

The A/D module has high- and low-voltage reference inputs that are selected by software. The module can operate while the processor is in SLEEP mode, but only if the A/D clock pulse is derived from its internal RC oscillator. The module contains four registers accessible to the application:

1. ADRESH - Result High Register

2. ADRESL - Result Low Register

3. ADCON0 - Control Register 0

4. ADCON1 - Control Register 1

Of these, it is the ADCON0 register that controls most of the operations of the A/C module. Port A pins RA0 to RA5 and port E pins RE0 to RE2 are multiplexed as analog input pins into the A/C module. In the 28-pin versions of the 16F87x, port pins RA0 to RA5 provide the five input channels. In all other implementations of the 16F87X, port E pins RE0 to RE2 provide the three additional channels.

Figure 13-5 shows the registers associated with A/D module operations.

REGISTER NAME	7	6	5	4	3	2	1	0	bits
INTCON	GIE	PEIE							
PIR1		ADIF							
PIE1		ADIE							
ADRESH	A/D Result Register High Byte								
ADRESL	A/D Result Register Low Byte								
ADCON0	ADSC1	ADSC0	CHS2	CHS1	CHS0	GO/DONE		ADON	
ADCON1	ADFM				PCFG3	PCFG2	PCFG1	PCFG0	

Figure 13-5 Registers Related to A/C Module Operations.

ADCON0 Register

The ADCON0 register is located in bank 0, at address 0x1f. Seven of the eight bits are meaningful in A/D control and status operations. Figure 13-6 is a bitmap of the ADCON0 register.

In Figure 13-6, bits 7 and 6, labeled ASCC1 and ADSC0, are the selection bits for the *A/D conversion clock*. The conversion time per bit is defined as *TAD* in PIC documentation. A/D conversion requires a minimum of 12 TAD in a 10-bit ADC. The source of the A/D conversion clock is software selected. The four possible options for TAD are:

1. Fosc/2

2. Fosc/8

3. Fosc/32

4. Internal A/D module RC oscillator (varies between 2 and 6 µs)

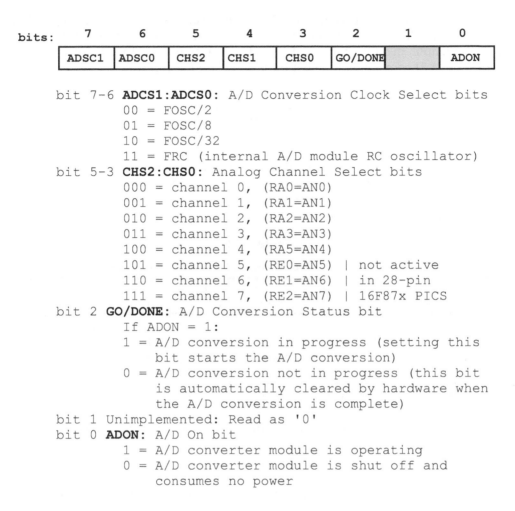

bits:	7	6	5	4	3	2	1	0
	ADSC1	ADSC0	CHS2	CHS1	CHS0	GO/DONE		ADON

bit 7-6 **ADCS1:ADCS0:** A/D Conversion Clock Select bits
       00 = FOSC/2
       01 = FOSC/8
       10 = FOSC/32
       11 = FRC (internal A/D module RC oscillator)
bit 5-3 **CHS2:CHS0:** Analog Channel Select bits
       000 = channel 0, (RA0=AN0)
       001 = channel 1, (RA1=AN1)
       010 = channel 2, (RA2=AN2)
       011 = channel 3, (RA3=AN3)
       100 = channel 4, (RA5=AN4)
       101 = channel 5, (RE0=AN5) | not active
       110 = channel 6, (RE1=AN6) | in 28-pin
       111 = channel 7, (RE2=AN7) | 16F87x PICS
bit 2 **GO/DONE:** A/D Conversion Status bit
       If ADON = 1:
       1 = A/D conversion in progress (setting this
          bit starts the A/D conversion)
       0 = A/D conversion not in progress (this bit
          is automatically cleared by hardware when
          the A/D conversion is complete)
bit 1 Unimplemented: Read as '0'
bit 0 **ADON:** A/D On bit
       1 = A/D converter module is operating
       0 = A/D converter module is shut off and
          consumes no power

Figure 13-6 ADCON0 Register Bitmap.

The conversion time is the analog-to-digital clock period multiplied by the number of bits of resolution in the converter, plus the two to three additional clock periods for settling time, as specified in the data sheet of the specific device. The various sources for the analog-to-digital converter clock represent the main oscillator frequency divided by 2, 8, or 32. The third choice is the use of a dedicated internal RC clock that has a typical period of 2 to 6 μs. Because the conversion time is determined by the system clock, a faster clock results in a faster conversion time.

The A/D conversion clock must be selected to ensure a minimum Tad time of 1.6 μs. The formula for converting processor speed (in MHz) into Tad microseconds is as follows:

$$Tad = \frac{1}{\dfrac{Tosc}{Tdiv}}$$

Where *Tad* is A/D conversion time, *Tosc* is the oscillator clock frequency in MHz, and *Tdiv* is the divisor determined by bits ADSC1 and ADSC0 of the ADCON0 register. For example, in a PIC running at 10 MHz, if we select the Tosc/8 option (divisor equal 8) the A/D conversion time per bit is calculated as follows:

$$Tad = \frac{1}{\dfrac{5\ MHz}{8}} = 1.6$$

In this case, the minimum recommended conversion speed of 1.6 μs is achieved. However, in a PIC with an oscillator speed of 10 MHz, this option produces a conversion speed of 0.8 μs, less than the recommended minimum. In this case we would have to select the divisor 32 option, giving a conversion speed of 3.2 μs.

Table 13.1

A/C Converter Tad at various Oscillator Speeds

OPERATION	ADCS1:ADCS0	TAD IN MICROSECONDS			
		20MHZ	10MHZ	5MHZ	1.25MHZ
Fosc/2	00	0.1	0.2	0.4	**1.6**
Fosc/8	01	0.4	0.8	**1.6**	**6.4**
Fosc/32	10	**1.6**	**3.2**	**6.4**	25.6
RC	11	**2-6**	**2-6**	**2-6**	**2-6**

Note: values in bold are within the recommended limits

In Table 13.1, converter speeds of less than 1.6 μs or higher than 10 μs are not recommended. Recall that the Tad speed of the converter is calculated per bit, so the total conversion time in a 10-bit device (such as the 16F87x) is approximately the Tad speed multiplied by 10 bits, plus three additional cycles; therefore, a device operating at a Tad speed of 1.6 μs requires 1.6 μs * 13, or 20.8 μs, for the entire conversion.

Bits CHS2 to CHS0 in the ADCON0 register (see Figure 13-6) determine which of the analog channels is selected. This is required, because there are several channels for analog input but only one A/2 converter circuitry. So the setting of this bit field determines which of six or eight possible channels is currently read by the A/C converter. An application can change the setting of these bits in order to read several analog inputs in succession.

Bit 2 of the ADCON0 register, labeled GO/DONE, is both a control and a status bit. Setting the GO/DONE bit starts A/D conversion. Once conversion has stared, the bit indicates if it is still in progress. Code can test the status of the GO/DONE bit in order to determine if conversion has concluded.

Bit 0 of the ADCON0 register turns the A/D module on and off. The initialization routine of an A/D-enabled application turns on this bit. Programs that do not use the A/D conversion module leave the bit off to conserve power.

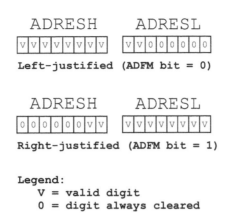

Figure 13-7 Left- and Right-Justification of A/D Result.

ADCON1 Register

The ADCON1 register also plays an important role in programming the A/D module. Bit 7 of the ADCON1 register is used to determine the *bit justification* of the digital result. This is possible because the 10-bit result is returned in two 8-bit registers; therefore, the six unused bits can be placed either on the left- or the right-hand side of the 16-bit result. If ADCON1 bit 7 is set then the result is right-justified, otherwise it is left-justified. Figure 13-7 shows the location of the significant bits.

One common use of right justification is to reduce the number of significant bits in the conversion result. For example, an application on the 16F877 that uses the A/D conversion module requires only 8-bit accuracy in the result. In this case, code can left-justify the conversion result, read the ADRESH register, and ignore the low-order bits in the ADRESL register. By ignoring the two low-order bits, the 10-bit accuracy of the A/D hardware is reduced to eight bits and the converter performs as an 8-bit accuracy unit.

The bit field labeled PCFG3 to PCFG0 in the ADCON1 register determines port configuration as analog or digital and the mapping of the positive and negative voltage reference pins. The number of possible combinations is limited by the four bits allocated to this field, so the programmer and circuit designer must select the option that is most suited to the application when the ideal one is not available. Table 13-2 (in the following page) shows the port configuration options.

For example, there is a circuit that calls for two analog inputs, wired to ports RA0 and RA1, with no reference voltages. In Table 13-2 we can find two options that select ports RA0 and RA1 and are analog inputs: these are the ones selected with PCFG bits 0100 and 0101. The first option also selects port RA3 as analog input, even though not required in this case. The second one also selects port RA3 as a positive voltage reference, also not required.

Either option works in this case; however, any pin configured for analog input produces incorrect results if used as a digital source. Therefore, a channel configured for analog input cannot be used for non-analog purposes. On the other hand, a

Table 13.2

A/D Converter Port Configuration Options

PCFG3: PCFG0	An7 Re2	An6 Re1	An5 Re0	An4 Ra5	An3 Ra3	An2 Ra2	An1 Ra1	An0 Ra0	Vref+	Vref-	CHAN/ Refs
0000	A	A	A	A	A	A	A	A	VDD	VSS	8/0
0001	A	A	A	A	Vre+	A	A	A	RA3	VSS	7/1
0010	D	D	D	A	A	A	A	A	VDD	VSS	5/0
0011	D	D	D	A	Vre+	A	A	A	RA3	VSS	4/1
0100	D	D	D	D	A	D	A	A	VDD	VSS	3/0
0101	D	D	D	D	Vre+	D	A	A	RA3	VSS	2/1
011x	D	D	D	D	D	D	D	D	VDD	VSS	0/0
1000	A	A	A	A	Vre+	Vre-	A	A	RA3	RA2	6/2
1001	D	D	A	A	A	A	A	A	VDD	VSS	6/0
1010	D	D	A	A	Vre+	A	A	A	RA3	VSS	5/1
1011	D	D	A	A	Vre+	Vre-	A	A	RA3	RA2	4/2
1100	D	D	D	A	Vre+	Vre-	A	A	RA3	RA2	3/2
1101	D	D	D	D	Vre+	Vre-	A	A	RA3	RA2	2/2
1110	D	D	D	D	D	D	D	A	VDD	VSS	1/0
1111	D	D	D	D	Vre+	Vre-	D	A	RA3	RA2	1/2

Legend:
   D = digital input
   A = analog input
   CHAN/Refs = analog channels/voltage reference inputs

channel configured for digital input should not be used for analog data because extra current is consumed by the hardware. Finally, channels to be used for analog-to-digital conversion must be configured for input in the corresponding TRIS register.

SLEEP Mode Operation

The A/D module can be made to operate in SLEEP mode. As mentioned previously, SLEEP mode operation requires that the A/D clock source be set to RC by setting both ADCS bits in the ADCON0 register. When the RC clock source is selected, the A/D module waits one instruction cycle before starting the conversion. During this period, the SLEEP instruction is executed, thus eliminating all *digital switching noise* from the conversion. The completion of the conversion is detected by testing the GO/DONE bit. If a different clock source is selected, then a SLEEP instruction causes the conversion-in-progress to be aborted and the A/D module to be turned off.

13.4.2 A/D Module Sample Circuit and Program

The circuit in Figure 13-8 is designed to demonstrate the use of the A/D converter module in PICs of the 16F87x family.

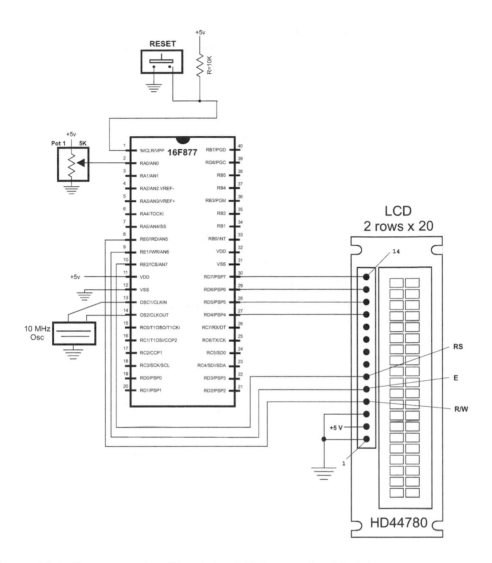

Figure 13-8 Demonstration Circuit for A/D Conversion Module.

Comparing Figure 13-8 with Figure 13-4, which uses the ADC0831 IC, we notice the economy of resources that results from selecting a PIC with an onboard A/D module. In the circuit of Figure 13-4, three microcontroller I/O ports must be used to connect the converter IC to the PIC. In the circuit of Figure 13-8, the potentiometer is connected directly to a single PIC port, saving two I/O lines. Considering the number of different PIC architectures that are equipped with onboard A/D converters, the circuit designer should explore this possibility before deciding on using a separate converter IC. At the same time, recall that two of the three input lines used by converter ICs can be shared. In a design with more than one converter IC, the use of input lines is not a 3 to 1 ratio.

The circuit in Figure 13-8 consists of a 5K potentiometer wired to analog port RA0 of a 16F877 PIC. The LCD display is used to show three digits, in the range 0 to 255, that represent the relative position of the potentiometer's disk. The program called A2DinLCD, in this book's online software, uses the built-in A/D module.

Programming the A/D module consists of the following steps:

1. Configure the PIC I/O lines to be used in the conversion. All analog lines are initialized as input in the corresponding TRIS registers.

2. Select the ports to be used in the conversion by setting the PCFGx bits in the ADCON1 register. Selects right- or left-justification.

3. Select the analog channels, select the A/D conversion clock, and enable the A/D module.

4. Wait the acquisition time.

5. Initiate the conversion by setting the GO/DONE bit in the ADCON0 register.

6. Wait for the conversion to complete.

7. Read and store the digital result.

The following procedure from the A2DinLCD program initialized the A/D module for the required processing:

```
;==============================
;      init A/D module
;==============================
; 1. Procedure to initialize the A/D module, as follows:
;     Configure the PIC I/O lines. Init analog lines as input
; 2. Select ports to be used by setting the PCFGx bits in the
;     ADCON1 register. Selects right- or left-justification.
; 3. Select the analog channels, select the A/D conversion
;     clock, and enable the A/D module.
; 4. Wait the acquisition time.
; 5. Initiate the conversion by setting the GO/DONE bit in the
;     ADCON0 register.
; 6. Wait for the conversion to complete.
; 7. Read and store the digital result.
InitA2D:
        Bank1                  ; Select bank for TRISA register
        movlw    b'00000001'
        movwf    TRISA    ; Set Port-A, line 0, as input
; Select the format and A/D port configuration bits in
; the ADCON1 register
; Format is left-justified so that ADRESH bits are the
; most significant
;  0   x   x   x   1   1   1   0   <== value installed in ADCON1
;  7   6   5   4   3   2   1   0   <== ADCON1 bits
;  |               |__|__|__|____ RA0 is analog.
;  |                               Vref+ = Vdd
;  |                               Vref- = Vss
;  |_____ 0 = left-justified
; ADCON1 is in bank 1
```

```
        movlw   b'00001110'
        movwf   ADCON1  ; RA0 is analog. All others digital
                        ; Vref+ = Vdd
; Select D/A options in ADCON0 register
; For a 10-MHz clock the Fosc32 option produces a conversion
; speed of 1/(10/32) = 3.2 microseconds, which is within the
; recommended range of 1.6 to 10 microseconds.
;  1  0  0  0  0  0  0  1  <== value installed in ADCON0
;  7  6  5  4  3  2  1  0  <== ADCON0 bits
;  |  |  |  |  |  |     |____ A/D function select
;  |  |  |  |  |  |          1 = A/D ON
;  |  |  |  |  |  |_____ A/D status bit
;  |  |  |__|__|_____ Analog Channel Select
;  |  |                     000 = Channel 0 (RA0)
;  |__|_____ A/D Clock Select
;                            10 = Fosc/32
; ADCON0 is in bank 0
        Bank0
        movlw   b'10000001'
        movwf   ADCON0  ; Channel 0, Fosc/32, A/D enabled
; Delay for selection to complete
        call    delayAD ; Local procedure
        return
```

Once the module is initialized, the analog line is read by the following procedure:

```
;=============================
;         read A/D line
;=============================
; Procedure to read the value in the A/D line and convert
; to digital
ReadA2D:
; Initiate conversion
        Bank0                   ; Bank for ADCON0 register
        bsf     ADCON0,GO       ; Set the GO/DONE bit
; GO/DONE bit is cleared automatically when conversion ends
convWait:
        btfsc   ADCON0,GO       ; Test bit
        goto    convWait ; Wait if not clear
; At this point, conversion has concluded
; ADRESH register (bank 0) holds 8 MSBs of result
; ADRESL register (bank 1) holds 4 LSBs.
; In this application value is left-justified. Only the
; MSBs are read
        movf    ADRESH,W ; Digital value to w register
        return
```

The delay routine required in this case is coded as follows:

```
;========================
;    delay procedure
;========================
; For a 10 MHz clock the Fosc32 option produces a conversion
; speed of 1/(10/32) = 3.2 microseconds. At 3.2 ms per bit,
; 13 bits require approximately 41 ms. The instruction time
; at 10 MHz is 10 ms. 4/10 = 0.4 ms per insctruction. To delay
; 41 ms, a 10 MHz PIC must execute 11 instructions. Add one
; more for safety.
delayAD:
        movlw   .12              ; Repeat 12 machine cycles
        movwf   count1   ; Store value in counter
repeat11:
        decfsz  count1,f   ; Decrement counter
        goto    repeat11 ; Continue if not 0
        return
```

13.5 Real-Time Clocks

In the context of microcontrollers and embedded systems, *real-time clocks* (also called *RTCs*) are integrated circuits designed to keep track of time in conventional hours, that is, in years, days, hours, minutes, and seconds. Many real-time clock ICs are available with various characteristics, data formats, modes of operation, and interfaces. Most of the ones used in PIC circuits have a serial interface in order to save access ports. Most RTC chips provide a battery connection so that time can be kept when the system is turned off.

In the sections that follow, we discuss one popular RTC chip: the NJU6355, but this is by no means the only option for embedded systems.

13.5.1 NJU6355 Real-Time Clock

The NJU6355 series is a serial I/O real-time clock used in microcontroller-based embedded systems. The IC includes its own quartz crystal oscillator, counter, shift register, voltage regulator, and interface controller. The PIC interface requires four lines. Operating voltage is TTL level so it can be wired directly on the typical PIC circuit. The output data includes year, month, day-of-week, hour, minutes, and seconds. Figure 13-9 is the pin diagram for the NJU6355.

NJU6355 output is in packed BCD format, that is, each decimal digit is represented by a 4-bit binary number. The chip's logic correctly calculates the number of days in each month as well as the leap years. All unused bits are reported as binary 0. Figure 13-10 is a bitmap of the formatted timer data.

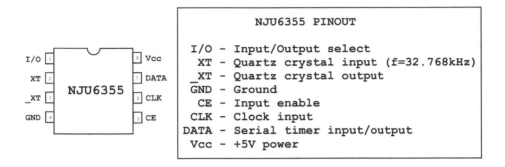

Figure 13-9 NJU6355 Pin Diagram.

Timer data is read when the I/O line is low and the CE line is high. Output from the 6355 is LSB first. A total of fifty-two significant bits are read in bottom-up order for data as shown in Figure 13-10. That is, the first bit received is the least-significant bit of the year, then the month, then the day, and so forth. All date items are eight bits, except the day of week, which is four bits. Nonsignificant bits in each field are reported as zero; this means that the value for the tenth month (October) is encoded as binary digits 00001010. Reporting unused digits as zero simplifies the conversion into BCD and ASCII.

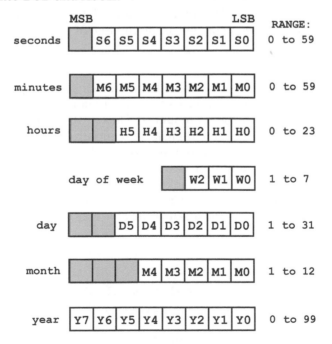

Figure 13-10 NJU6355 Timer Data Format.

The NJU6355 does not report valid time data until after it has been initialized, even if there are power and clock signals into the chip. Initialization requires writing data into the 6355 registers. In order to write to the IC, code must set the I/O and the CE lines high. At this moment, all clock updates stop and the RTC goes into the write mode. Input data is latched in LSB first, starting with the year and concluding with the minutes. There is no provision for writing seconds into the RTC, so the total number of bits written is 44.

The 6355 contains a mechanism for detecting conditions that could compromise the clock's operation, such as low power. In this case, the special value 0xee is written into each digit of the internal registers to inform processing routines that the timer has been compromised.

The NJU6355 requires the installation of an external crystal oscillator. The crystal must have a frequency of 32.768 KHz. The time-keeping accuracy of the RTC is determined by the quartz oscillator. The capacity of the oscillator must match that of the RTC and of the circuit. A standard crystal with a capacitance of 12.5 pF works well for applications that do not demand high clock accuracy. For more exacting applications, the 6355 can be programmed to check the clock frequency and determine its error. The chip's frequency-checking mode is described in an NJU6355 Application Note available from New Japan Radio Co., Ltd.

13.5.2 RTC Demonstration Circuit and Program

The circuit shown in Figure 13-11 is a simple application of the 6355 RTC. The circuit uses a NJU6355 in conjunction with a 16F86 PIC and an LCD. The demonstration program, named RTC2LCD, sets up RTC and reads clock data in an endless loop. The hours, minutes, and seconds are displayed at the top line of the LCD as follows:

        H:xx M:xx S:xx

where xx represents the two BCD digits read from the clock and converted to ASCII decimal for display. The program initializes the 6355 to some arbitrary values contained in the corresponding **#define** statements. These values are copied into program variables by a local procedure and then used to initialize the RTC registers. Two procedures relate to RTC operation: one to initialize the clock hardware and the other one to read the current time. In addition, two auxiliary procedures are implemented: one to read clock data and one to write clock data. Because clock data can be in 8- or 4-bit formats, each procedure contains a separate entry point to handle the 4-bit option. The procedure to initialize and the one to write clock data are coded as follows:

```
;==============================
;          init RTC
;==============================
; Procedure to initialize the real time clock chip. If chip
; is not initialized it will not operate and the values
; read will be invalid.
; because the 6355 operates in BCD format the stored values must
; be converted to packed BCD.
; According to wiring diagram
```

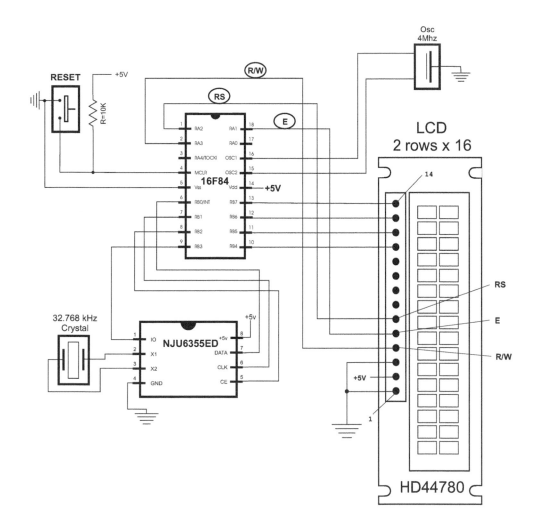

Figure 13-11 Real-Time Clock Demonstration Circuit.

```
; NJU6355 Interface for setting time:
; DAT           PORTB,0             Output
; CLK           PORTB,1             Output
; CE            PORTB,2             Output
; IO            PORTB,3             Output
setRTC:
        Bank1
        movlw   b'00000000'         ; All lines are output
        movlw   TRISB
        Bank0
; Writing to the 6355 requires that the CLK bit be held
; low while the IO and CE lines are high
        bcf     PORTB,CLK           ; CLK low
        call    delay_5
```

```
          bsf       PORTB,IO ; IO high
          call      delay_5
          bsf       PORTB,CE ; CE high
; Data is stored in RTC as follows:
;  year            8 bits (0 to 99)
;  month           8 bits (1 to 12)
;  day             8 bits (1 to 31)
;  dayOfWeek       4 bits (1 to 7)
;  hour            8 bits (0 to 23)
;  minutes         8 bits (0 to 59)
;                    ======
;   Total             44 bits
; Seconds cannot be written to RTC. RTC seconds register
; is automatically initialized to zero
          movf      year,w   ; Get item from storage
          call      bin2bcd  ; Convert to BCD
          movwf     temp1
          call      writeRTC

          movf      month,w
          call      bin2bcd
          movwf     temp1
          call      writeRTC

          movf      day,w
          call      bin2bcd
          movwf     temp1
          call      writeRTC

          movf      dayOfWeek,w        ; day of week is 4-bits
          call      bin2bcd
          movwf     temp1
          call      write4RTC

          movf      hour,w
          call      bin2bcd
          movwf     temp1
          call      writeRTC

          movf      minutes,w
          call      bin2bcd
          movwf     temp1
          call      writeRTC
; Done
          bcf       PORTB,CLK          ; Hold CLK line low
          call      delay_5
          bcf       PORTB,CE           ; and the CE line
                                       ; to the RTC
```

```
          call      delay_5
          bcf       PORTB,IO          ; RTC in output mode
          return
;=============================
;    write 4/8 bits to RTC
;=============================
; Procedure to write 4 or 8 bits to the RTC registers
; ON ENTRY:
;    temp1 register holds value to be written
; ON EXIT:
;    nothing
write4RTC
          movlw     .4                ; Init for 4 bits
          goto      allBits
writeRTC:
          movlw     .8                ; Init for 8 bits
allBits:
          movwf     counter           ; Store in bit counter
writeBits:
          bcf       PORTB,CLK         ; Clear the CLK line
          call      delay_5           ; Wait
          bsf       PORTB,DAT         ; Set the data line to RTC
          btfss     temp1,0           ; Send LSB
          bcf       PORTB,DAT         ; Clear data line
          call      delay_5           ; Wait for operation to
complete
          bsf       PORTB,CLK         ; Bring CLK line high to
validate
          rrf       temp1,f           ; Rotate bits in storage
          decfsz    counter,1         ; Decrement bit counter
          goto      writeBits         ; Continue if not last bit
          return
```

The following procedures are used by the RTC2LCD program to read the data in the RTC registers:

```
;=============================
;       read RTC data
;=============================
; Procedure to read the current time from the RTC and store
; data (in packed BCD format) in local time registers.
; According to wiring diagram
; NJU6355 Interface for read operations:
; DAT              PORTB,0           Input
; CLK              PORTB,1           Output
; CE               PORTB,2           Output
; IO               PORTB,3           Output
Get_Time
; Clear Port-B
```

```
        movlw    b'00000000'
        movwf    PORTB
; Make data line input
        Bank1
        movlw    b'00000001'
        movwf    TRISB
        Bank0
; Reading RTC data requires that the IO line be low and the
; CE line be high. CLK line is held low
        bcf      PORTB,CLK          ; CLK low
        call     delay_125
        bcf      PORTB,IO ; IO line low
        call     delay_125
        bsf      PORTB,CE ; and CE line high
; Data is read from RTC as follows:
;   year            8 bits (0 to 99)
;   month           8 bits (1 to 12)
;   day             8 bits (1 to 31)
;   dayOfWeek       4 bits (1 to 7)
;   hour            8 bits (0 to 23)
;   minutes         8 bits (0 to 59)
;   seconds         8 bits (0 to 59)
;                            ======
;   Total              52 bits
;
        call     readRTC
        movwf    year
        call     delay_125

        call     readRTC
        movwf    month
        call     delay_125

        call     readRTC
        movwf    day
        call     delay_125

; day of week is a 4-bit value
        call     read4RTC
        movwf    dayOfWeek
        call     delay_125

        call     readRTC
        movwf    hour
        call     delay_125

        call     readRTC
        movwf    minutes
```

```
            call        delay_125

            call        readRTC
            movwf       seconds
            bcf         PORTB,CE ; CE line low to end output
            return

;===============================
;   read 4/8 bits from RTC
;===============================
; Procedure to read 4/8 bits stored in 6355 registers
; Value returned in w register
read4RTC
            movlw       .4                  ; 4 bit read
            goto        anyBits
readRTC:
            movlw       .8                  ; 8 bits read
anyBits:
            movwf       counter
; Read 6355 read operation requires the IO line be set low
; and the CE line high. Data is read in the following order:
; year, month, day, day-of-week, hour, minutes, seconds
readBits:
            bsf         PORTB,CLK; Set CLK high to validate data
            bsf         STATUS,C ; Set the carry flag (bit = 1)
; Operation:
;   If data line is high, then bit read is a 1-bit
;   otherwise bit read is a 0-bit
            btfss       PORTB,DAT           ; Is data line high?
                                 ; Leave carry set (1 bit) if high
            bcf         STATUS,C ; Clear the carry bit (make bit 0)
; At this point the carry bit matches the data line
            bcf         PORTB,CLK           ; Set CLK low to end read
; The carry bit is now rotated into the temp1 register
            rrf         temp1,1
            decfsz      counter,1           ; Decrement the bit counter
            goto        readBits ; Continue if not last bit
; At this point all bits have been read (8 or 4)
            movf        temp1,0             ; Result to w
            return
```

BCD Conversion Procedures

In addition to the RTC procedures to initialize the clock registers and to read clock data, the application requires auxiliary procedures to manipulate and display data in BCD format. BCD encodings are a way of representing decimal digits in binary form. Two common BCD formats are used: packed and unpacked. In the unpacked format, each byte encodes a single BCD value. In packed form, two BCD digits are encoded per byte. The 6355 uses the packed BCD format.

Because program data is usually in binary form, it is useful to have a routine to convert binary data into BCD form. A simple algorithm for converting binary to BCD is as follows:

1. The value 10 is subtracted from the source operand until the remainder is less than 0 (carry cleared). The number of subtractions is the high-order BCD digit.

2. The value 10 is then added back to the subtrahend to compensate for the last subtraction.

3. The final remainder is the low-order BCD digit.

The binary to BCD conversion procedure is coded as follows:

```
;=============================
;  binary to BCD conversion
;=============================
; Convert a binary number into two packed BCD digits
; ON ENTRY:
;        w register has binary value in range 0 to 99
; ON EXIT:
;        output variables bcdLow and bcdHigh contain two
;        unpacked BCD digits
;        w contains two packed BCD digits
; Routine logic:
;   The value 10 is subtracted from the source operand
;   until the remainder is < 0 (carry cleared). The number
;   of subtractions is the high-order BCD digit. 10 is
;   then added back to the subtrahend to compensate
;   for the last subtraction. The final remainder is the
;   low-order BCD digit
; Variables:
;     inNum       storage for source operand
;     bcdHigh     storage for high-order nibble
;     bcdLow      storage for low-order nibble
;     thisDig     Digit counter
bin2bcd:
        movwf     inNum             ; Save copy of source value
        clrf      bcdHigh ; Clear storage
        clrf      bcdLow
        clrf      thisDig
min10:
        movlw     .10
        subwf     inNum,f           ; Subtract 10
        btfsc     STATUS,C          ; Did subtract overflow?
        goto      sum10             ; No. Count subtraction
        goto      fin10
sum10:
```

```
        incf     thisDig,f          ; Increment digit counter
        goto     min10
; Store 10th digit
fin10:
        movlw    .10
        addwf    inNum,f            ; Adjust
        movf     thisDig,w          ; Get digit counter contents
        movwf    bcdHigh            ; Store it
; Calculate and store low-order BCD digit
        movf     inNum,w            ; Store units value
        movwf    bcdLow             ; Store digit
; Combine both digits
        swapf    bcdHigh,w          ; High nibble to HOBs
        iorwf    bcdLow,w           ; ORin low nibble
        return
```

Because the program requires displaying values encoded in BCD format, a routine is necessary to convert two packed BCD digits into two ASCII decimal digits. The conversion logic is quite simple, as the BCD digit is converted to ASCII by adding 0x30 to its value. All that is necessary is to shift bits in the packed BCD operand so as to isolate each digit and then add 0x30 to each one. The routine's code is as follows:

```
;===============================
;     BCD to ASCII decimal
;          conversion
;===============================
; ON ENTRY:
;         w register has two packed BCD digits
; ON EXIT:
;         output variables asc10 and asc1 have
;         two ASCII decimal digits
; Routine logic:
;   The low order nibble is isolated and the value 30H
;   added to convert to ASCII. The result is stored in
;   the variable asc1. Then the same is done to the
;   high-order nibble and the result is stored in the
;   variable asc10

Bcd2asc:
        movwf    store1   ; Save input
        andlw    b'00001111'      ; Clear high nibble
        addlw    0x30             ; Convert to ASCII
        movwf    asc1             ; Store result
        swapf    store1,w ; Recover input and swap digits
        andlw    b'00001111'      ; Clear high nibble
        addlw    0x30             ; Convert to ASCII
        movwf    asc10            ; Store result
        return
```

13.6 Demonstration Programs

The following subsections contain the sample programs discussed in this chapter.

13.6.1 ADF84 Program

```
; File name: ADCF84.asm
; Last Update: June 8, 2011
; Authors: Sanchez and Canton
; Processor: 16F84A
;
; Description:
; Program to demonstrate use of the ADC0831 Analog to
; Digital converter with the 16F84 PIC. Program reads the
; value of a potentionmeter connected to Port-A, line 0
; and displays resistance in the range 0 to 255 on the
; attached LCD.
; Circuit:
;    ADC0831              16F84              CIRCUIT
; PIN  LINE
;  6    DO ------------- RB0
;  7    CLK ------------- RB1
;  1    CS ------------- RB2
;  2  Vin+ ----------------------------- POT2
;  3  Vin- ----------------------------- GND
;  5  Vref ----------------------------- +5v
;  8   Vcc ----------------------------- +5v
;
; For LCD display parameters see the LCDTest2 program.
; WARNING:
; Code assumes 4 MHz clock. Delay routines must be
; edited for faster clock
;
;===========================
;        switches
;===========================
; Switches used in __config directive:
;   _CP_ON       Code protection ON/OFF
; * _CP_OFF
; * _PWRTE_ON    Power-up timer ON/OFF
;   _PWRTE_OFF
;   _WDT_ON      Watchdog timer ON/OFF
; * _WDT_OFF
;   _LP_OSC      Low power crystal oscillator
; * _XT_OSC      External parallel resonator/crystal oscillator
;
;   _HS_OSC      High speed crystal resonator (8 to 10 MHz)
;                Resonator: Murate Erie CSA8.00MG = 8 MHz
;   _RC_OSC      Resistor/capacitor oscillator
```

```
;  |                   (simplest, 20% error)
;  |
;  |_____ * indicates setup values presently selected


;==========================
; setup and configuration
;==========================
        processor 16f84A
        include   <p16f84A.inc>
        __config  _XT_OSC & _WDT_OFF & _PWRTE_ON & _CP_OFF

        errorlevel -302
; Suppress bank-related warning
;=============================================================
;                         M A C R O S
;=============================================================
; Macros to select the register banks in 16F84
Bank0   MACRO               ; Select RAM bank 0
        bcf       STATUS,RP0
        ENDM

Bank1   MACRO                       ; Select RAM bank 1
        bsf       STATUS,RP0
        ENDM
;=====================================================
;               constant definitions
;   for PIC-to-LCD pin wiring and LCD line addresses
;=====================================================
#define E_line 1            ; |
#define RS_line 2           ; | => from circuit wiring diagram
#define RW_line 3           ; |
; LCD line addresses (from LCD data sheet)
#define LCD_1 0x80          ; First LCD line constant
#define LCD_2 0xc0          ; Second LCD line constant
; Note: The constants that define LCD display line
;       addresses have the high-order bit set in
;       order to facilitate the controller command

; Defines from ADC0831 wiring diagram
; all lines in Port-A
#define  DO                 0   ; |
#define  CLK                1   ; | => from circuit wiring diagram
#define  CS                 2   ; |
;
;=====================================================
;               variables in PIC RAM
;=====================================================
; Reserve 16 bytes for string buffer
```

```
            cblock    0x0c
            strData
            endc
; Reserve three bytes for ASCII digits
            cblock    0x1d
            asc100
            asc10
            asc1
            endc
; Continue with local variables
            cblock    0x20              ; Start of block
            count1                ; Counter # 1
            count2                ; Counter # 2
            count3                ; Counter # 3
            pic_ad                ; Storage for start of text area
            J                     ; Counter J
            K                     ; Counter K
            index                 ; Index into text table (also used
                                  ; for auxiliary storage)
            store1                ; Local temporary storage
            store2                ; Storage # 2
            rcvdata               ; Received data
            bitCount
; Storage for ASCII decimal conversion and digits
            inNum                 ; Source operand
            thisDig               ; Digit counter
            endc


;============================================================
;                           program
;============================================================
            org       0          ; start at address
            goto      main
; Space for interrupt handlers
            org       0x08
main:
            Bank1
            movlw     b'00000000'     ; All lines to output
            movwf     TRISA           ; in Port-A
            movlw     b'00000001'     ; B line 0 to input
            movwf     TRISB
            Bank0
            movlw     b'00000000'     ; All outputs ports low
            movwf     PORTA
            movwf     PORTB
; Wait and initialize HD44780
            call      delay_5         ; Allow LCD time to initialize
                                      ; itself
```

```
        call      delay_5
        call      initLCD           ; Then do forced
initialization
        call      delay_5           ; Wait again
; Store base address of text buffer in PIC RAM
        movlw     0x0c              ; Start address for buffer
        movwf     pic_ad            ; to local variable
;======================
;   first LCD line
;======================
; Store 16 blanks in PIC RAM, starting at address stored
; in variable pic_ad
        call      blank16
; Call procedure to store ASCII characters for message
; in text buffer
        movlw     d'0'              ; Offset into buffer
        call      storeMS1 ; Store message text in buffer
; Initialize ADC0831
nextAna:
        call      ana2dig           ; Read analog line
        call      delay_125
; Display result
        movf      rcvdata,w
        call      bin2asc           ; Conversion routine
; At this point three ASCII digits are stored in local
; variables. Move digits to display area
        movf      asc1,w            ; Unit digit
        movwf     .26               ; Store in buffer
        movf      asc10,w           ; same with other digits
        movwf     .25
        movf      asc100,w
        movwf     .24
; Display line
; Set DDRAM address to start of first line
        call      line1
; Call procedure to display 16 characters in LCD
        call      display16
        call      long_delay
        goto      nextAna


;================================================================
;               initialize LCD for 4-bit mode
;================================================================
initLCD:
; Initialization for Densitron LCD module as follows:
;       4-bit interface
;    2 display lines of 16 characters each
;    cursor on
```

```
;     left-to-right increment
;     cursor shift right
;     no display shift
;======================|
;   set command mode    |
;======================|
         bcf        PORTA,E_line        ; E line low
         bcf        PORTA,RS_line       ; RS line low
         bcf        PORTA,RW_line       ; Write mode
         call       delay_125           ; delay 125 microseconds
;*********************|
;     FUNCTION SET     |
;*********************|
         movlw      0x28     ; 0 0 1 0 1 0 0 0 (FUNCTION SET)
         call       send8    ; 4-bit send routine

; Set 4-bit mode command must be repeated
         movlw      0x28
         call       send8


;*********************|
; DISPLAY AND CURSOR ON |
;*********************|
         movlw      0x0e     ; 0 0 0 0 1 1 1 0 (DISPLAY ON/OFF)
         call       send8
;*********************|
;   set entry mode     |
;*********************|
         movlw      0x06     ; 0 0 0 0 0 1 1 0 (ENTRY MODE SET)
         call       send8
;
;*********************|
; cursor/display shift |
;*********************|
         movlw      0x14     ; 0 0 0 1 0 1 0 0 (CURSOR/DISPLAY
                             ; SHIFT)
         call       send8
;*********************|
;   clear display      |
;*********************|
         movlw      0x01     ; 0 0 0 0 0 0 0 1 (CLEAR DISPLAY)
         call       send8
; Per documentation
         call       delay_5 ; Test for busy
         return
;
;======================
;  Procedure to delay
```

```
;    42 microseconds
;=======================
delay_125:
        movlw   D'42'           ; Repeat 42 machine cycles
        movwf   count1          ; Store value in counter
repeat:
        decfsz  count1,f        ; Decrement counter
        goto    repeat          ; Continue if not 0
        return                  ; End of delay


;=======================
;   Procedure to delay
;    5 milliseconds
;=======================
delay_5:
        movlw   D'41'           ; Counter = 41
        movwf   count2          ; Store in variable
delay:
        call    delay_125       ; Delay
        decfsz  count2,f        ; 40 times = 5 milliseconds
        goto    delay
        return                  ; End of delay
;=======================
;     pulse E line
;=======================
pulseE:
        bsf     PORTA,E_line    ; Pulse E line
        nop
        bcf     PORTA,E_line
        return


;==============================
;   long delay sub-routine
;       (for debugging)
;==============================
long_delay:
        movlw   D'200'          ; w = 200 decimal
        movwf   J               ; J = w
jloop:
        movwf   K               ; K = w
kloop:
        decfsz  K,f             ; K = K-1, skip next if zero
        goto    kloop
        decfsz  J,f             ; J = J-1, skip next if zero
        goto    jloop
        return
;==============================
;   LCD display procedure
```

```
;===============================
; Sends 16 characters from PIC buffer with address stored
; in variable pic_ad to LCD line previously selected
display16
        call    delay_5             ; Make sure not busy
; Set up for data
        bcf     PORTA,E_line        ; E line low
        bsf     PORTA,RS_line       ; RS line high for data
; Set up counter for 16 characters
        movlw   D'16'               ; Counter = 16
        movwf   count3
; Get display address from local variable pic_ad
        movf    pic_ad,w ; First display RAM address to W
        movwf   FSR                 ; W to FSR
getchar:
        movf    INDF,w   ; get character from display RAM
                        ; location pointed to by file select
                        ; register
        call    send8    ; 4-bit interface routine
; Test for 16 characters displayed
        decfsz  count3,f            ; Decrement counter
        goto    nextchar            ; Skipped if done
        return
nextchar:
        incf    FSR,f               ; Bump pointer
        goto    getchar


;========================
;   send 2 nibbles in
;     4-bit mode
;========================
; Procedure to send two 4-bit values to Port-B lines
; 7, 6, 5, and 4. High-order nibble is sent first
; ON ENTRY:
;       w register holds 8-bit value to send
send8:
        movwf   store1              ; Save original value
        call    merge4              ; Merge with Port-B
; Now w has merged byte
        movwf   PORTB               ; w to Port-B
        call    pulseE              ; Send data to LCD
; High nibble is sent
        movf    store1,w            ; Recover byte into w
        swapf   store1,w            ; Swap nibbles in w
        call    merge4
        movwf   PORTB
        call    pulseE              ; Send data to LCD
        call    delay_125
```

```
        return
;==================
;   merge bits
;==================
; Routine to merge the 4 high-order bits of the
; value to send with the contents of Port-B
; so as to preserve the 4 low-bits in Port-B
; Logic:
;       AND value with 1111 0000 mask
;       AND Port-B with 0000 1111 mask
;       Now low nibble in value and high nibble in
;       Port-B are all 0 bits:
;           value = vvvv 0000
;          Port-B = 0000 bbbb
;       OR value and Port-B resulting in:
;                    vvvv bbbb
; ON ENTRY:
;       w contain value bits
; ON EXIT:
;       w contains merged bits
merge4:
        andlw   b'11110000'       ; ANDing with 0 clears the
                                  ; bit. ANDing with 1 preserves
                                  ; the original value
        movwf   store2            ; Save result in variable
        movf    PORTB,w           ; Port-B to w register
        andlw   b'00001111'       ; Clear high nibble in Port-b
                                  ; and preserve low nibble
        iorwf   store2,w          ; OR two operands in w
        return


;========================
;     blank buffer
;========================
; Procedure to store 16 blank characters in PIC RAM
; buffer starting at address stored in the variable
; pic_ad
blank16:
        movlw   D'16'             ; Setup counter
        movwf   count1
        movf    pic_ad,w ; First PIC RAM address
        movwf   FSR               ; Indexed addressing
        movlw   0x20              ; ASCII space character
storeit:
        movwf   INDF              ; Store blank character in PIC
RAM
                                  ; buffer using FSR register
        decfsz  count1,f          ; Done?
```

```
        goto     incfsr              ; no
        return                       ; yes
incfsr:
        incf     FSR,f               ; Bump FSR to next buffer
                           ; space
        goto     storeit


;========================
; Set address register
;    to LCD line 1
;========================
; ON ENTRY:
;          Address of LCD line 1 in constant LCD_1
line1:
        bcf      PORTA,E_line        ; E line low
        bcf      PORTA,RS_line       ; RS line low, setup for
                                     ; control
        call     delay_5             ; busy?
; Set to second display line
        movlw    LCD_1               ; Address and command bit
        call     send8               ; 4-bit routine
; Set RS line for data
        bsf      PORTA,RS_line       ; Setup for data
        call     delay_5             ; Busy?
        return


;===============================
;  first text string procedure
;===============================
storeMS1:
; Procedure to store in PIC RAM buffer the message
; contained in the code area labeled msg1
; ON ENTRY:
;          variable pic_ad holds address of text buffer
;          in PIC RAM
;          w register holds offset into storage area
;          msg1 is routine that returns the string characters
;          and a zero terminator
;          index is local variable that holds offset into
;          text table. This variable is also used for
;          temporary storage of offset into buffer
; ON EXIT:
;          Text message stored in buffer
;
; Store offset into text buffer (passed in the w register)
; in temporary variable
        movwf    index               ; Store w in index
; Store base address of text buffer in FSR
```

```
          movf      pic_ad,w ; first display RAM address to W
          addwf     index,w            ; Add offset to address
          movwf     FSR                ; W to FSR
; Initialize index for text string access
          movlw     0                  ; Start at 0
          movwf     index              ; Store index in variable
; w still = 0
get_msg_char:
          call      msg1               ; Get character from table
; Test for zero terminator
          andlw     0x0ff
          btfsc     STATUS,Z           ; Test zero flag
          goto      endstr1            ; End of string
; ASSERT: valid string character in w
;         store character in text buffer (by FSR)
          movwf     INDF               ; store in buffer by FSR
          incf      FSR,f              ; increment buffer pointer
; Restore table character counter from variable
          movf      index,w            ; Get value into w
          addlw     1                  ; Bump to next character
          movwf     index              ; Store table index in
variable
          goto      get_msg_char       ; Continue
endstr1:
          return

; Routine for returning message stored in program area
; Message has 10 characters
msg1:
          addwf     PCL,f              ; Access table
          retlw     'P'
          retlw     'o'
          retlw     't'
          retlw     ' '
          retlw     'R'
          retlw     'e'
          retlw     's'
          retlw     'i'
          retlw     's'
          retlw     't'
          retlw     ':'
          retlw     0
;==============================
;   binary to ASCII decimal
;         conversion
;==============================
; ON ENTRY:
;         w register has binary value in range 0 to 255
```

```
; ON EXIT:
;          output variables asc100, asc10, and asc1 have
;          three ASCII decimal digits
; Routine logic:
;   The value 100 is subtracted from the source operand
;   until the remainder is < 0 (carry cleared.) The number
;   of subtractions is the decimal hundreds result. 100 is
;   then added back to the subtrahend to compensate
;   for the last subtraction. Now 10 is subtracted in the
;   same manner to determine the decimal tenths result.
;   The final remainder is the decimal units result.
; Variables:
;     inNum      storage for source operand
;     asc100     storage for hundreds position result
;     asc10      storage for tenth position result
;     asc1       storage for unit position result
;     thisDig    Digit counter
bin2asc:
        movwf     inNum      ; Save copy of source value
        clrf      asc100     ; Clear hundreds storage
        clrf      asc10      ; Tens
        clrf      asc1       ; Units
        clrf      thisDig
sub100:
        movlw     .100
        subwf     inNum,f         ; Subtract 100
        btfsc     STATUS,C        ; Did subtract overflow?
        goto      bump100         ; No. Count subtraction
        goto      end100
bump100:
        incf      thisDig,f       ;increment digit counter
        goto      sub100
; Store 100th digit
end100:
        movf      thisDig,w       ; Adjusted digit counter
        addlw     0x30            ; Convert to ASCII
        movwf     asc100          ; Store it
; Calculate tenth position value
        clrf      thisDig
; Adjust minuend
        movlw     .100            ; Minuend
        addwf     inNum,f         ; Add value to minuend to
                                  ; compensate for last
operation
sub10:
        movlw     .10
        subwf     inNum,f         ; Subtract 10
        btfsc     STATUS,C        ; Did subtract overflow?
```

```
          goto      bump10              ; No. Count subtraction
          goto      end10
bump10:
          incf      thisDig,f           ;increment digit counter
          goto      sub10
; Store 10th digit
end10:
          movlw     .10
          addwf     inNum,f             ; Adjust for last subratct
          movf      thisDig,w           ; get digit counter contents
          addlw     0x30                ; Convert to ASCII
          movwf     asc10               ; Store it
; Calculate and store units digit
          movf      inNum,w     ;       Store units value
          addlw     0x30                ; Convert to ASCII
          movwf     asc1                ; Store digit
          return


;============================================================
;                    ADC0831 procedures
;============================================================
;=============================
;   procedure to read and
;    convert analog line
;=============================
; ON ENTRY:
; Code assumes that the ADC0831 DO line is initialized for
; input, while CLK and CS lines are output
; From ADC0831 wiring diagram. All lines in Port-B
;       DO        =         RB0     ==> INPUT
;       CLK       =         RB1     <== OUTPUT
;       CS        =         RB2     <== OUTPUT
; ON EXIT:
; Returns 8-bit digital value in the register rcvdata
;
ana2dig:
; Clear data register and init counter for 8 bits
          clrf      rcvdata             ; Clear register
          movlw     0x08                ; Initialize counter
          movwf     bitCount
; Prepare to read analog line
          bcf       PORTB,CS ; CS pin low to enable ADC
          nop                           ; Delay for 4 MHz clock
          bsf       PORTB,CLK           ; Set CLK high
          nop
          bcf       PORTB,CLK           ; Reset CLK to start
                                        ; conversion
          nop
```

```
nextB:
; Pulse CLK line to read bit from ADC
        bsf       PORTB,CLK        ; CLK high
        nop
        bcf       PORTB,CLK        ; CLK low
        nop
; Read analog line and store data, bit by bit
        movf      PORTB,w ; Read all Port-B bits
        movwf     store1           ; Store value for later
        rrf       store1,f ; Rotate bit into carry flag
        rlf       rcvdata,f        ; Rotate carry flag into
result
                                   ; register
        decfsz    bitCount,f       ; Bump counter, skip next
                                   ; if counter zero
        goto      nextB
; Value read is stored in rcvdata register
        bsf       PORTB,CLK        ; Final clock pulse
        nop
        bcf       PORTB,CLK
        nop
        bsf       PORTB,CS ; Turn off ADC
        call      long_delay  ; Time to settle
        return

        end
```

13.6.2 A2DinLCD Program

```
; File name: A2DinLCD.asm
; Last revision: June 2, 2011
; Authors: Sanchez and Canton
; Processor: 16F877
;
; Description:
; Program to demonstrate use of the Analog to Digital
; Converter (A/D) module on the 16F877. Program reads the
; value of a potentionmeter connected to Port-A, line 0
; and displays resistance in the range 0 to 255 on the
; attached LCD.
;
; WARNING:
; Code assumes 10 MHz clock. Delay routines must be
; edited for faster clock. Clock speed is also used to
; set up the A/D converter clock.
;
;===========================
;       16F877 switches
```

```
;============================
; Switches used in __config directive:
;   _CP_ON          Code protection ON/OFF
; * _CP_OFF
; * _PWRTE_ON      Power-up timer ON/OFF
;   _PWRTE_OFF
;   _BODEN_ON      Brown-out reset enable ON/OFF
; * _BODEN_OFF
; * _PWRTE_ON      Power-up timer enable ON/OFF
;   _PWRTE_OFF
;   _WDT_ON        Watchdog timer ON/OFF
; * _WDT_OFF
;   _LPV_ON         Low voltage IC programming enable ON/OFF
; * _LPV_OFF
;   _CPD_ON         Data EE memory code protection ON/OFF
; * _CPD_OFF
; OSCILLATOR CONFIGURATIONS:
;   _LP_OSC         Low power crystal oscillator
;   _XT_OSC         External parallel resonator/crystal oscillator

; * _HS_OSC         High speed crystal resonator
;   _RC_OSC         Resistor/capacitor oscillator
; |                 (simplest, 20% error)
; |
; |_____  * indicates setup values presently selected

        processor       16f877              ; Define processor
        #include <p16f877.inc>
        __CONFIG _CP_OFF & _WDT_OFF & _BODEN_OFF & _PWRTE_ON &
_HS_OSC & _WDT_OFF & _LVP_OFF & _CPD_OFF
; __CONFIG directive is used to embed configuration data
; within the source file. The labels following the directive
; are located in the corresponding .inc file.


        errorlevel -302
; Suppress bank-related warning
;=============================================================
;                       M A C R O S
;=============================================================
; Macros to select the register banks
Bank0   MACRO               ; Select RAM bank 0
        bcf     STATUS,RP0
        bcf     STATUS,RP1
        ENDM

Bank1   MACRO               ; Select RAM bank 1
        bsf     STATUS,RP0
        bcf     STATUS,RP1
```

```
        ENDM

Bank2   MACRO               ; Select RAM bank 2
        bcf       STATUS,RP0
        bsf       STATUS,RP1
        ENDM

Bank3   MACRO               ; Select RAM bank 3
        bsf       STATUS,RP0
        bsf       STATUS,RP1
        ENDM
;========================================================
;                   constant definitions
;   for PIC-to-LCD pin wiring and LCD line addresses
;========================================================
#define E_line 1        ;|
#define RS_line 0       ;| => from wiring diagram
#define RW_line 2       ;|
; LCD line addresses (from LCD data sheet)
#define LCD_1 0x80          ; First LCD line constant
#define LCD_2 0xc0          ; Second LCD line constant
#define LCDlimit .20; Number of characters per line
#define  spbrgVal .64; For 2400 baud on 10-MHz clock
; Note: The constants that define the LCD display
;       line addresses have the high-order bit set
;       so as to meet the requirements of controller
;       commands.
;========================================================
;                 variables in PIC RAM
;========================================================
; Reserve 20 bytes for string buffer
        cblock    0x20
        strData
        endc

; Reserve three bytes for ASCII digits
        cblock    0x34
        asc100
        asc10
        asc1
        endc

; Data
        cblock    0x37                  ; Start of block
        count1              ; Counter # 1
        count2              ; Counter # 2
        count3              ; Counter # 3
        pic_ad
```

```
        J                       ; counter J
        K                       ; counter K
        index
        store1                  ; Local storage
        store2
; For LCDscroll procedure
        LCDcount ; Counter for characters per line
        LCDline                 ; Current display line (0 or 1)
        endc

; Common RAM area for most critical variables
        cblock  0x70
; Storage for ASCII decimal conversion and digits
        inNum                   ; Source operand
        thisDig                 ; Digit counter
        endc


;================================================================
;                       P R O G R A M
;================================================================
        org     0               ; start at address
        goto    main
; Space for interrupt handlers
        org     0x08
main:
; Wiring:
;       LCD data to Port-D, lines 0 to 7
;       E line -> Port-E, 1
;       RW line -> Port-E, 2
;       RS line -> Port-E, 0
; Set PORTE D and E for output
; First, initialize Port-B by clearing latches
        clrf    STATUS
        clrf    PORTB
; Select bank 1 to tris Port-D for output
        Bank1
; Tris Port-D for output. Port-D lines 4 to 7 are wired
; to LCD data lines. Port-D lines 0 to 4 are wired to LEDs.
        movlw   B'00000000'
        movwf   TRISD           ; and Port-D
; By default Port-A lines are analog. To configure them
; as digital code must set bits 1 and 2 of the ADCON1
; register (in bank 1)
        movlw   0x06                    ; binary 0000 0110  is code to
                                        ; make all Port-A lines
digital
        movwf   ADCON1
; Port-B, lines are not used by this application. Init
```

```
; to output
        movlw    b'00000000'
        movwf    TRISB
; Tris Port-E for output. LCD lines are in Port-E
        movwf    TRISE              ; Tris Port-E
; Enable Port-B pullups for switches in OPTION register
;    7  6  5  4  3  2  1  0 <= OPTION bits
;    |  |  |  |  |  |__|__|_____ PS2-PS0 (prescaler bits)
;    |  |  |  |  |  |                Values for Timer0
;    |  |  |  |  |  |                000 = 1:2    001 = 1:4
;    |  |  |  |  |  |                010 = 1:8    011 = 1:16
;    |  |  |  |  |  |                100 = 1:32  101 = 1:64
;    |  |  |  |  |  |                110 = 1:128 *111 = 1:256
;    |  |  |  |  |  |_____ PSA (prescaler assign)
;    |  |  |  |  |                *1 = to WDT
;    |  |  |  |  |                0 = to Timer0
;    |  |  |  |  |_____ TOSE (Timer0 edge select)
;    |  |  |  |                   *0 = increment on low-to-high
;    |  |  |  |                    1 = increment on high-to-low
;    |  |  |_____ TOCS (TMR0 clock source)
;    |  |  |                    *0 = internal clock
;    |  |  |                     1 = RA4/TOCKI bit source
;    |  |_____ INTEDG (Edge select)
;    |  |                       *0 = falling edge
;    |_____ RBPU (Pullup enable)
;                              *0 = enabled
;                               1 = disabled
        movlw    b'00001000'
        movwf    OPTION_REG
; Back to bank 0
        Bank0
; Clear all output lines
        movlw    b'00000000'
        movwf    PORTD
        movwf    PORTE
; Wait and initialize HD44780
        call     delay_5            ; Allow LCD time to initialize
itself
        call     initLCD            ; Then do forced
initialization
        call     delay_5            ; (Wait probably not
necessary)
; Clear character counter and line counter variables
        clrf     LCDcount
        clrf     LCDline
; Initialize A/D conversion lines
        call     InitA2D            ; Local procedure
; Store base address of text buffer in PIC RAM
```

```
            movlw    0x20              ; Start address for buffer
            movwf    pic_ad            ; to local variable
; Store 20 blanks in PIC RAM, starting at address stored
; in variable pic_ad
            call     blank20
; Call procedure to store ASCII characters for message
; in text buffer
            movlw    d'0'              ; Offset into buffer
            call     storeMS1
;=============================
;          read POT digital value
;=============================
readPOT:
            call     ReadA2D           ; Local procedure
; w has digital value read from analog line RA0
; Display result
            call     bin2asc           ; Conversion routine
; At this point three ASCII digits are stored in local
; variables. Move digits to display area
            movf     asc1,w            ; Unit digit
            movwf    0x2e              ; Store in buffer
            movf     asc10,w           ; same with other digits
            movwf    0x2d
            movf     asc100,w
            movwf    0x2c
; Display line
; Set DDRAM address to start of first line
showLine:
            call     line1
; Call procedure to display 16 characters in LCD
            call     display20
            goto     readPOT


;================================================================
;================================================================
;               L O C A L     P R O C E D U R E S
;================================================================
;================================================================
;==========================
; init LCD for 4-bit mode
;==========================
initLCD:
; Initialization for Densitron LCD module as follows:
;    4-bit interface
;    2 display lines of 20 characters each
;    cursor on
;    left-to-right increment
;    cursor shift right
```

```
;   no display shift
;=====================|
;   set command mode  |
;=====================|
        bcf       PORTE,E_line      ; E line low
        bcf       PORTE,RS_line     ; RS line low
        bcf       PORTE,RW_line     ; Write mode
        call      delay_125         ; delay 125 microseconds
;*********************|
;     FUNCTION SET    |
;*********************|
        movlw     0x28     ; 0 0 1 0 1 0 0 0 (FUNCTION SET)
        call      send8    ; 4-bit send routine

; Set 4-bit mode command must be repeated
        movlw     0x28
        call      send8
;*********************|
; DISPLAY AND CURSOR ON |
;*********************|
        movlw     0x0e     ; 0 0 0 0 1 1 1 0 (DISPLAY ON/OFF)
        call      send8
;*********************|
;   set entry mode    |
;*********************|
        movlw     0x06     ; 0 0 0 0 0 1 1 0 (ENTRY MODE SET)
        call      send8
;*********************|
; cursor/display shift |
;*********************|
        movlw     0x14     ; 0 0 0 1 0 1 0 0 (CURSOR/DISPLAY
SHIFT)
        call      send8
;*********************|
;    clear display    |
;*********************|
        movlw     0x01     ; 0 0 0 0 0 0 0 1 (CLEAR DISPLAY)
        call      send8
; Per documentation
        call      delay_5 ; Test for busy
        return

;=====================
;  Procedure to delay
;   125ms. at 10 MHz
;=====================
delay_125:
        movlw     .110                ; Repeat 110 machine cycles
```

```
        movwf    count1          ; Store value in counter
repeat:
        decfsz   count1,f        ; Decrement counter
        goto     repeat          ; Continue if not 0
        return                   ; End of delay


;=======================
;   Procedure to delay
;    5 milliseconds
;=======================
delay_5:
        movlw    .110            ; Counter = 110
        movwf    count2          ; Store in variable
delay:
        call     delay_125       ; Delay
        decfsz   count2,f        ; 40 times = 5 milliseconds
        goto     delay
        return                   ; End of delay
;=======================
;      pulse E line
;=======================
pulseE:
        bsf      PORTE,E_line    ; Pulse E line
        nop
        bcf      PORTE,E_line
        return


;=============================
;    long delay sub-routine
;=============================
long_delay:
        movlw    .200            ; w delay count
        movwf    J               ; J = w
jloop:
        movwf    K               ; K = w
kloop:
        decfsz   K,f             ; K = K-1, skip next if zero
        goto     kloop
        decfsz   J,f             ; J = J-1, skip next if zero
        goto     jloop
        return


;=============================
;    display buffer on LCD
;=============================
; Sends 20 characters from PIC buffer with address stored
; in variable pic_ad to LCD line previously selected
display20:
```

```
        call    delay_5                      ; Make sure not busy
; Set up for data
        bcf     PORTA,E_line      ; E line low
        bsf     PORTA,RS_line     ; RS line high for data
; Set up counter for 20 characters
        movlw   D'20'
        movwf   count3
; Get display address from local variable pic_ad
        movf    pic_ad,w ; First display RAM address to W
        movwf   FSR               ; W to FSR
getchar
        movf    INDF,w  ; get character from display RAM
                        ; location pointed to by file select
                        ; register
        call    send8             ; 4-bit interface routine
; Test for 16 characters displayed
        decfsz  count3,f          ; Decrement counter
        goto    nextchar          ; Skipped if done
        return
nextchar:
        incf    FSR,f             ; Bump pointer
        goto    getchar


;=========================
;   send 2 nibbles in
;     4-bit mode
;=========================
; Procedure to send two 4-bit values to Port-B lines
; 7, 6, 5, and 4. High-order nibble is sent first
; ON ENTRY:
;        w register holds 8-bit value to send
send8:
        movwf   store1            ; Save original value
        call    merge4            ; Merge with Port-B
; Now w has merged byte
        movwf   PORTD             ; w to Port-D
        call    pulseE            ; Send data to LCD
; High nibble is sent
        movf    store1,w ; Recover byte into w
        swapf   store1,w ; Swap nibbles in w
        call    merge4
        movwf   PORTD
        call    pulseE            ; Send data to LCD
        call    delay_125
        return
;=========================
;      merge bits
;=========================
```

```
; Routine to merge the 4 high-order bits of the
; value to send with the contents of Port-B
; so as to preserve the 4 low-bits in Port-B
; Logic:
;       AND value with 1111 0000 mask
;       AND Port-B with 0000 1111 mask
;       Now low nibble in value and high nibble in
;       Port-B are all 0 bits:
;            value = vvvv 0000
;           Port-B = 0000 bbbb
;       OR value and Port-B resulting in:
;                   vvvv bbbb
; ON ENTRY:
;       w contain value bits
; ON EXIT:
;       w contains merged bits
merge4:
          andlw    b'11110000'        ; ANDing with 0 clears the
                                      ; bit. ANDing with 1 preserves
                                      ; the original value
          movwf    store2             ; Save result in variable
          movf     PORTD,w            ; Port-D to w register
          andlw    b'00001111'        ; Clear high nibble in Port-b
                                      ; and preserve low nibble
          iorwf    store2,w ; OR two operands in w
          return
;===========================
;   Set address register
;       to LCD line 1
;===========================
; ON ENTRY:
;        Address of LCD line 1 in constant LCD_1
line1:
          bcf      PORTE,E_line       ; E line low
          bcf      PORTE,RS_line      ; RS line low, setup for
                                      ; control
          call     delay_5            ; busy?
; Set to second display line
          movlw    LCD_1              ; Address and command bit
          call     send8              ; 4-bit routine
; Set RS line for data
          bsf      PORTE,RS_line      ; Setup for data
          call     delay_5            ; Busy?
          return
;===============================
;   first text string procedure
;===============================
storeMS1:
```

```
; Procedure to store in PIC RAM buffer the message
; contained in the code area labeled msg1
; ON ENTRY:
;         variable pic_ad holds address of text buffer
;         in PIC RAM
;         w register holds offset into storage area
;         msg1 is routine that returns the string characters
;         and a zero terminator
;         index is local variable that holds offset into
;         text table. This variable is also used for
;         temporary storage of offset into buffer
; ON EXIT:
;         Text message stored in buffer
;
; Store offset into text buffer (passed in the w register)
; in temporary variable
        movwf    index              ; Store w in index
; Store base address of text buffer in FSR
        movf     pic_ad,w ; first display RAM address to W
        addwf    index,w            ; Add offset to address
        movwf    FSR                ; W to FSR
; Initialize index for text string access
        movlw    0                  ; Start at 0
        movwf    index              ; Store index in variable
; w still = 0
get_msg_char:
        call     msg1               ; Get character from table
; Test for zero terminator
        andlw    0x0ff
        btfsc    STATUS,Z           ; Test zero flag
        goto     endstr1            ; End of string
; ASSERT: valid string character in w
;         store character in text buffer (by FSR)
        movwf    INDF               ; store in buffer by FSR
        incf     FSR,f              ; increment buffer pointer
; Restore table character counter from variable
        movf     index,w            ; Get value into w
        addlw    1                  ; Bump to next character
        movwf    index    ; Store table index in variable
        goto     get_msg_char       ; Continue
endstr1:
        return
; Routine for returning message stored in program area
; Message has 10 characters
msg1:
        addwf    PCL,f              ; Access table
        retlw    'P'
        retlw    'o'
```

```
          retlw    't'
          retlw    ' '
          retlw    'R'
          retlw    'e'
          retlw    's'
          retlw    'i'
          retlw    's'
          retlw    't'
          retlw    ':'
          retlw    0
;========================
;     blank buffer
;========================
; Procedure to store 20 blank characters in PIC RAM
; buffer starting at address stored in the variable
; pic_ad
blank20:
          movlw    D'20'              ; Setup counter
          movwf    count1
          movf     pic_ad,w           ; First PIC RAM address
          movwf    FSR                ; Indexed addressing
          movlw    0x20               ; ASCII space character
storeit:
          movwf    INDF     ; Store blank character in PIC RAM
                            ; buffer using FSR register
          decfsz   count1,f ; Done?
          goto     incfsr   ; no
          return            ; yes
incfsr:
          incf     FSR,f    ; Bump FSR to next buffer space
          goto     storeit


;==============================
;   binary to ASCII decimal
;         conversion
;==============================
; ON ENTRY:
;         w register has binary value in range 0 to 255
; ON EXIT:
;         output variables asc100, asc10, and asc1 have
;         three ASCII decimal digits
; Routine logic:
;   The value 100 is subtracted from the source operand
;   until the remainder is < 0 (carry cleared). The number
;   of subtractions is the decimal hundreds result. 100 is
;   then added back to the subtrahend to compensate
;   for the last subtraction. Now 10 is subtracted in the
;   same manner to determine the decimal tenths result.
```

```
;     The final remainder is the decimal units result.
; Variables:
;      inNum       storage for source operand
;      asc100      storage for hundreds position result
;      asc10       storage for tens position result
;      asc1        storage for unit position reslt
;      thisDig     Digit counter
bin2asc:
        movwf    inNum              ; Save copy of source value
        clrf     asc100             ; Clear hundreds storage
        clrf     asc10              ; Tens
        clrf     asc1               ; Units
        clrf     thisDig
sub100:
        movlw    .100
        subwf    inNum,f            ; Subtract 100
        btfsc    STATUS,C           ; Did subtract overflow?
        goto     bump100            ; No. Count subtraction
        goto     end100
bump100:
        incf     thisDig,f          ; Increment digit counter
        goto     sub100
; Store 100th digit
end100:
        movf     thisDig,w          ; Adjusted digit counter
        addlw    0x30               ; Convert to ASCII
        movwf    asc100             ; Store it
; Calculate tenth position value
        clrf     thisDig
; Adjust minuend
        movlw    .100       ; Minuend
        addwf    inNum,f  ; Add value to minuend to
                            ; Compensate for last operation
sub10:
        movlw    .10
        subwf    inNum,f            ; Subtract 10
        btfsc    STATUS,C           ; Did subtract overflow?
        goto     bump10             ; No. Count subtraction
        goto     end10
bump10:
        incf     thisDig,f          ;increment digit counter
        goto     sub10
; Store 10th digit
end10:
        movlw    .10
        addwf    inNum,f            ; Adjust for last subraction
        movf     thisDig,w          ; get digit counter contents
        addlw    0x30               ; Convert to ASCII
```

```
        movwf    asc10              ; Store it
; Calculate and store units digit
        movf     inNum,w            ; Store units value
        addlw    0x30               ; Convert to ASCII
        movwf    asc1               ; Store digit
        return


;=============================================================
;                  Analog to Digital Procedures
;=============================================================
;============================
;      init A/D module
;============================
; 1. Procedure to initialize the A/D module, as follows:
;     Configure the PIC I/O lines. Init analog lines as input.
; 2. Select ports to be used by setting the PCFGx bits in the
;     ADCON1 register. Select right- or left-justification.
; 3. Select the analog channels, select the A/D conversion
;     clock, and enable the A/D module.
; 4. Wait the acquisition time.
; 5. Initiate the conversion by setting the GO/DONE bit in the
;     ADCON0 register.
; 6. Wait for the conversion to complete.
; 7. Read and store the digital result.
InitA2D:
        Bank1               ; Select bank for TRISA register
        movlw    b'00000001'
        movwf    TRISA     ; Set Port-A, line 0, as input
; Select the format and A/D port configuration bits in
; the ADCON1 register
; Format is left-justified so that ADRESH bits are the
; most significant
;   0  x  x  x  1  1  1  0  <== value installed in ADCON1
;   7  6  5  4  3  2  1  0  <== ADCON1 bits
;   |           |__|__|__|____ RA0 is analog.
;   |                          Vref+ = Vdd
;   |                          Vref- = Vss
;   |_____ 0 = left-justified
; ADCON1 is in bank 1
        movlw    b'00001110'
        movwf    ADCON1              ; RA0 is analog. All others ;
                                     ; digital
                                     ; Vref+ = Vdd
; Select D/A options in ADCON0 register
; For a 10 MHz clock the Fosc32 option produces a conversion
; speed of 1/(10/32) = 3.2 microseconds, which is within the
; recommended range of 1.6 to 10 microseconds.
;   1  0  0  0  0  0  0  1  <== value installed in ADCON0
```

```
;    7   6   5   4   3   2   1   0   <== ADCON0 bits
;    |   |   |   |   |   |       |____ A/D function select
;    |   |   |   |   |   |           1 = A/D ON
;    |   |   |   |   |   |_____ A/D status bit
;    |   |   |__|__|_____ Analog Channel Select
;    |   |                       000 = Chanel 0 (RA0)
;    |__|_____ A/D Clock Select
;                                 10 = Fosc/32
; ADCON0 is in bank 0
        Bank0
        movlw    b'10000001'
        movwf    ADCON0              ; Channel 0, Fosc/32, A/D
                            ; enabled
; Delay for selection to complete. (Existing routine provides
; more than 20 microseconds required)
        call     delayAD            ; Local procedure
        return
;============================
;        read A/D line
;============================
; Procedure to read the value in the A/D line and convert
; to digital
ReadA2D:
; Initiate conversion
        Bank0                       ; Bank for ADCON0 register
        bsf      ADCON0,GO          ; Set the GO/DONE bit
; GO/DONE bit is cleared automatically when conversion ends
convWait:
        btfsc    ADCON0,GO          ; Test bit
        goto     convWait           ; Wait if not clear
; At this point conversion has concluded
; ADRESH register (bank 0) holds 8 MSBs of result
; ADRESL register (bank 1) holds 4 LSBs.
; In this application value is left-justified. Only the
; MSBs are read
        movf     ADRESH,W           ; Digital value to w register
        return
;======================
;    delay procedure
;======================
; For a 10 MHz clock the Fosc32 option produces a conversion
; speed of 1/(10/32) = 3.2 microseconds. At 3.2 ms per bit
; 13 bits require approximately 41 ms. The instruction time
; at 10 MHz is 10 ms. 4/10 = 0.4 ms per insctruction. To delay
; 41 ms a 10 MHz PIC must execute 11 instructions. Add one
; more for safety.
delayAD:
        movlw    .12                ; Repeat 12 machine cycles
```

```
           movwf     count1               ; Store value in counter
repeat11:
           decfsz    count1,f             ; Decrement counter
           goto      repeat11             ; Continue if not 0
           return


;=============================================================
           end               ; END OF PROGRAM
;=============================================================
```

13.6.3 RTC2LCD Program

```
; File name: RTC2LCD.asm
; Last Update: June 6, 2011
; Authors: Canton and Sanchez
; Processor: 16F84A
;
; Description:
; Program to demonstrate use of the NJU6355 Real Time Clock
; IC. Program uses LCD to display results of hours, minutes,
; and seconds, as follows:
;
; Top LCD line:   H:xx M:yy S:zz
;
; Initialization values are in #define statements that start
; with i, such as iYear, iMonth, etc.
;
; For LCD display parameters see the LCDTest2 program.
; WARNING:
; Code assumes 4 MHz clock. Delay routines must be
; edited for faster clock
;
;===========================
;         switches
;===========================
; Switches used in __config directive:
;   _CP_ON         Code protection ON/OFF
; * _CP_OFF
; * _PWRTE_ON      Power-up timer ON/OFF
;   _PWRTE_OFF
;   _WDT_ON        Watchdog timer ON/OFF
; * _WDT_OFF
;   _LP_OSC        Low power crystal oscillator
; * _XT_OSC        External parallel resonator/crystal oscillator

;   _HS_OSC        High speed crystal resonator (8 to 10 MHz)
;                  Resonator: Murate Erie CSA8.00MG = 8 MHz
;   _RC_OSC        Resistor/capacitor oscillator
```

```
;  |                    (simplest, 20% error)
;  |
;  |_____ * indicates setup values presently selected

;=========================
; setup and configuration
;=========================
        processor 16f84A
        include   <p16f84A.inc>
        __config _XT_OSC & _WDT_OFF & _PWRTE_ON & _CP_OFF

        errorlevel -302
; Suppress bank-related warning
;================================================================
;                          M A C R O S
;================================================================
; Macros to select the register banks in 16F84
Bank0   MACRO              ; Select RAM bank 0
        bcf     STATUS,RP0
        ENDM

Bank1   MACRO              ; Select RAM bank 1
        bsf     STATUS,RP0
        ENDM
;======================================================
;                 constant definitions
;   for PIC-to-LCD pin wiring and LCD line addresses
;======================================================
#define E_line 1          ; |
#define RS_line 2         ; | => from circuit wiring diagram
#define RW_line 3         ; |
; LCD line addresses (from LCD data sheet)
#define LCD_1 0x80        ; First LCD line constant
#define LCD_2 0xc0        ; Second LCD line constant
; Note: The constants that define the LCD display line
;       addresses have the high-order bit set in
;       order to facilitate the controller command

; Defines from real-time clock wiring diagram
; all lines in Port-B
#define  DAT               0   ; |
#define  CLK               1   ; | => from circuit wiring diagram
#define  CE                2   ; |
#define  IO                3   ; |
;
; Defines for RTC initialization (values are arbitrary)
#define iYear    .7
#define iMonth   .6
```

```
#define iDay       .5
#define iDoW       .4
#define iHour      .3
#define iMin       .2
#define iSec       .1
;=====================================================
;                   PIC register equates
;=====================================================
;=====================================================
;               variables in PIC RAM
;=====================================================
; Reserve 16 bytes for string buffer
        cblock   0x0c
        strData
        endc
; Reserve three bytes for ASCII digits
        cblock   0x1d
        asc100
        asc10
        asc1
        endc
; Continue with local variables
        cblock   0x20              ; Start of block
        count1            ; Counter # 1
        count2            ; Counter # 2
        count3            ; Counter # 3
        pic_ad            ; Storage for start of text area
        J                 ; Counter J
        K                 ; Counter K
        index             ; Index into text table (also used
                          ; for auxiliary storage)
        store1            ; Local temporary storage
        store2            ; Storage # 2
; Storage for BCD digits
        bcdLow            ; Low-order nibble of packed BCD
        bcdHigh           ; High-order nibble
; Variables for Real-Time Clock
        year
        month
        day
        dayOfWeek         ; Sunday to Saturday (1 to 7)
        hour
        minutes
        seconds
        temp1
        counter
; Storage for BCD conversion routine
        inNum             ; Source operand
```

```
        thisDig             ; Digit counter
        endc

;============================================================
;                               program
;============================================================
        org              0          ; start at address
        goto    main
; Space for interrupt handlers
        org              0x08

main:
        movlw   b'00000000' ; All lines to output
        Bank1
        movwf   TRISA               ; in Port-A
        movwf   TRISB               ; and Port-B
        Bank0
        movlw   b'00000000'         ; All outputs ports low
        movwf   PORTA
        movwf   PORTB
; Wait and initialize HD44780
        call    delay_5             ; Allow LCD time to initialize
                                    ; itself
        call    delay_5
        call    initLCD             ; Then do forced
initialization
        call    delay_5             ; Wait again
; Store base address of text buffer in PIC RAM
        movlw   0x0c                ; Start address for buffer
        movwf   pic_ad              ; to local variable
;=====================
;    first LCD line
;=====================
; Store 16 blanks in PIC RAM, starting at address stored
; in variable pic_ad
        call    blank16
; Call procedure to store ASCII characters for message
; in text buffer
        movlw   d'0'                ; Offset into buffer
        call    storeMS1 ; Store message text in buffer
; Initialize real time clock
        call    initRTC             ; Initialize variables
        call    setRTC              ; Start clock
        call    delay_5             ; Wait for operation to
                                    ; conclude
newTime:
; Get variables from RTC
        call    Get_Time
```

```
        call    delay_5             ; Wait
        movf    hour,w              ; Get hours
        call    Bcd2asc             ; Conversion routine
; At this point three ASCII digits are stored in local
; variables. Move digits to display area
        movf    asc1,w              ; Unit digit
        movwf   .15                 ; Store in buffer
        movf    asc10,w             ; Same with other digit
        movwf   .14
        call    delay_5
        movf    minutes,w
        call    Bcd2asc             ; Conversion routine
; At this point three ASCII digits are stored in local
; variables. Move two digits to display area
        movf    asc1,w              ; Unit digit
        movwf   .20                 ; Store in buffer
        movf    asc10,w             ; same with other digit
        movwf   .19
        call    delay_5

        movf    seconds,w
        call    Bcd2asc             ; Conversion routine
; Move digits to display area
        movf    asc1,w              ; Unit digit
        movwf   .25                 ; Store in buffer
        movf    asc10,w             ; same with other digit
        movwf   .24
        call    delay_5
; Set DDRAM address to start of first line
        call    line1
; Call procedure to display 16 characters in LCD
        call    display16
        goto    newTime


;=============================================================
;                   initialize LCD for 4-bit mode
;=============================================================
initLCD:
; Initialization for Densitron LCD module as follows:
;       4-bit interface
;   2 display lines of 16 characters each
;   cursor on
;   left-to-right increment
;   cursor shift right
;   no display shift
;======================|
;   set command mode   |
;======================|
```

```
        bcf        PORTA,E_line      ; E line low
        bcf        PORTA,RS_line     ; RS line low
        bcf        PORTA,RW_line     ; Write mode
        call       delay_125         ; delay 125 microseconds
;*********************|
;     FUNCTION SET    |
;*********************|
        movlw      0x28     ; 0 0 1 0 1 0 0 0 (FUNCTION SET)
        call       send8    ; 4-bit send routine
;
; Set 4-bit mode command must be repeated
        movlw      0x28
        call       send8
;
;*********************|
; DISPLAY AND CURSOR ON |
;*********************|
        movlw      0x0e     ; 0 0 0 0 1 1 1 0 (DISPLAY ON/OFF)
        call       send8
;*********************|
;   set entry mode     |
;*********************|
        movlw      0x06     ; 0 0 0 0 0 1 1 0 (ENTRY MODE SET)
        call       send8
;
;*********************|
; cursor/display shift |
;*********************|
        movlw      0x14     ; 0 0 0 1 0 1 0 0 (CURSOR/DISPLAY
                            ; SHIFT)
        call       send8
;*********************|
;   clear display      |
;*********************|
        movlw      0x01     ; 0 0 0 0 0 0 0 1 (CLEAR DISPLAY)
        call       send8
; Per documentation
        call       delay_5  ; Test for busy
        return
;
;=====================
;  Procedure to delay
;   42 microseconds
;=====================
delay_125
        movlw      D'42'             ; Repeat 42 machine cycles
        movwf      count1            ; Store value in counter
repeat:
```

```
             decfsz   count1,f        ; Decrement counter
             goto     repeat          ; Continue if not 0
             return                   ; End of delay
;
;========================
;  Procedure to delay
;    5 milliseconds
;========================
delay_5:
             movlw    D'41'           ; Counter = 41
             movwf    count2          ; Store in variable
delay:
             call     delay_125       ; Delay
             decfsz   count2,f        ; 40 times = 5 milliseconds
             goto     delay
             return                   ; End of delay
;========================
;     pulse E line
;========================
pulseE:
             bsf      PORTA,E_line    ; Pulse E line
             nop
             bcf      PORTA,E_line
             return


;==============================
;    long delay sub-routine
;        (for debugging)
;==============================
long_delay:
                 movlw    D'200'      ; w = 200 decimal
                 movwf    J           ; J = w
jloop:
             movwf    K               ; K = w
kloop:
             decfsz   K,f             ; K = K-1, skip next if zero
             goto     kloop
             decfsz   J,f             ; J = J-1, skip next if zero
             goto     jloop
             return
;==============================
;   LCD display procedure
;==============================
; Sends 16 characters from PIC buffer with address stored
; in variable pic_ad to LCD line previously selected
display16
             call     delay_5         ; Make sure not busy
; Set up for data
```

```
        bcf       PORTA,E_line      ; E line low
        bsf       PORTA,RS_line     ; RS line high for data
; Set up counter for 16 characters
        movlw     D'16'             ; Counter = 16
        movwf     count3
; Get display address from local variable pic_ad
        movf      pic_ad,w ; First display RAM address to W
        movwf     FSR               ; W to FSR
getchar:
        movf      INDF,w   ; get character from display RAM
                            ; location pointed to by file select
                            ; register
        call      send8    ; 4-bit interface routine
; Test for 16 characters displayed
        decfsz    count3,f          ; Decrement counter
        goto      nextchar ; Skipped if done
        return
nextchar:
        incf      FSR,f             ; Bump pointer
        goto      getchar


;========================
;   send 2 nibbles in
;      4-bit mode
;========================
; Procedure to send two 4-bit values to Port-B lines
; 7, 6, 5, and 4. High-order nibble is sent first
; ON ENTRY:
;        w register holds 8-bit value to send
send8:
        movwf     store1            ; Save original value
        call      merge4            ; Merge with Port-B
; Now w has merged byte
        movwf     PORTB             ; w to Port-B
        call      pulseE            ; Send data to LCD
; High nibble is sent
        movf      store1,w ; Recover byte into w
        swapf     store1,w ; Swap nibbles in w
        call      merge4
        movwf     PORTB
        call      pulseE            ; Send data to LCD
        call      delay_125
        return
;=================
;   merge bits
;=================
; Routine to merge the 4 high-order bits of the
; value to send with the contents of Port-B
```

```
; so as to preserve the 4 low-bits in Port-B
; Logic:
;       AND value with 1111 0000 mask
;       AND Port-B with 0000 1111 mask
;       Now low nibble in value and high nibble in
;       Port-B are all 0 bits:
;            value = vvvv 0000
;           Port-B = 0000 bbbb
;       OR value and Port-B resulting in:
;                  vvvv bbbb
; ON ENTRY:
;       w contain value bits
; ON EXIT:
;       w contains merged bits
merge4:
          andlw    b'11110000'        ; ANDing with 0 clears the
                                      ; bit. ANDing with 1 preserves
                                      ; the original value
          movwf    store2            ; Save result in variable
          movf     PORTB,w           ; Port-B to w register
          andlw    b'00001111' ; Clear high nibble in Port-B
                                      ; and preserve low nibble
          iorwf    store2,w ; OR two operands in w
          return


;========================
;     blank buffer
;========================
; Procedure to store 16 blank characters in PIC RAM
; buffer starting at address stored in the variable
; pic_ad
blank16
          movlw    D'16'              ; Set up counter
          movwf    count1
          movf     pic_ad,w          ; First PIC RAM address
          movwf    FSR               ; Indexed addressing
          movlw    0x20              ; ASCII space character
storeit
          movwf    INDF       ; Store blank character in PIC RAM
                              ; buffer using FSR register
          decfsz   count1,f          ; Done?
          goto     incfsr            ; no
          return                     ; yes
incfsr
          incf     FSR,f    ; Bump FSR to next buffer space
          goto     storeit


;========================
```

```
; Set address register
;    to LCD line 1
;=========================
; ON ENTRY:
;          Address of LCD line 1 in constant LCD_1
line1:
        bcf       PORTA,E_line       ; E line low
        bcf       PORTA,RS_line      ; RS line low, set up for
                                     ; control
        call      delay_5            ; busy?
; Set to second display line
        movlw     LCD_1              ; Address and command bit
        call      send8              ; 4-bit routine
; Set RS line for data
        bsf       PORTA,RS_line      ; Setup for data
        call      delay_5            ; Busy?
        return


;================================
;   first text string procedure
;================================
storeMS1:
; Procedure to store in PIC RAM buffer the message
; contained in the code area labeled msg1
; ON ENTRY:
;          variable pic_ad holds address of text buffer
;          in PIC RAM
;          w register hold offset into storage area
;          msg1 is routine that returns the string characters
;          an a zero terminator
;          index is local variable that holds offset into
;          text table. This variable is also used for
;          temporary storage of offset into buffer
; ON EXIT:
;          Text message stored in buffer
;
; Store offset into text buffer (passed in the w register)
; in temporary variable
        movwf     index              ; Store w in index
; Store base address of text buffer in FSR
        movf      pic_ad,w ; first display RAM address to W
        addwf     index,w            ; Add offset to address
        movwf     FSR                ; W to FSR
; Initialize index for text string access
        movlw     0                  ; Start at 0
        movwf     index              ; Store index in variable
; w still = 0
get_msg_char:
```

```
            call      msg1                 ; Get character from table
; Test for zero terminator
            andlw     0x0ff
            btfsc     STATUS,Z ; Test zero flag
            goto      endstr1              ; End of string
; ASSERT: valid string character in w
;         store character in text buffer (by FSR)
            movwf     INDF                 ; store in buffer by FSR
            incf      FSR,f                ; increment buffer pointer
; Restore table character counter from variable
            movf      index,w              ; Get value into w
            addlw     1                    ; Bump to next character
            movwf     index                ; Store table index in
variable
            goto      get_msg_char         ; Continue
endstr1:
            return

; Routine for returning message stored in program area
; Message has 10 characters
msg1:
            addwf     PCL,f                ; Access table
            retlw     'H'
            retlw     ':'
            retlw     ' '
            retlw     ' '
            retlw     ' '
            retlw     'M'
            retlw     ':'
            retlw     ' '
            retlw     ' '
            retlw     ' '
            retlw     'S'
            retlw     ':'
            retlw     0


;===============================
;     BCD to ASCII decimal
;           conversion
;===============================
; ON ENTRY:
;         w register has two packed BCD digits
; ON EXIT:
;         output variables asc10, and asc1 have
;         two ASCII decimal digits
; Routine logic:
;   The low-order nibble is isolated and the value 30H
;   added to convert to ASCII. The result is stored in
```

```
;    the variable asc1. Then the same is done to the
;    high-order nibble and the result is stored in the
;    variable asc10

Bcd2asc:
        movwf   store1            ; Save input
        andlw   b'00001111'       ; Clear high nibble
        addlw   0x30              ; Convert to ASCII
        movwf   asc1              ; Store result
        swapf   store1,w    ; Recover input and swap digits
        andlw   b'00001111'       ; Clear high nibble
        addlw   0x30              ; Convert to ASCII
        movwf   asc10             ; Store result
        return


;================================================================
;                      6355 RTC procedures
;================================================================
;============================
;         init RTC
;============================
; Procedure to initialize the real time clock chip. If chip
; is not initialized it will not operate and the values
; read will be invalid.
; Because the 6355 operates in BCD format the stored values
; must be converted to packed BCD.
; According to wiring diagram
; NJU6355 Interface for setting time:
; DAT            PORTB,0           Output
; CLK            PORTB,1           Output
; CE             PORTB,2           Output
; IO             PORTB,3           Output
setRTC:
        Bank1
        movlw   b'00000000'       ; All output bits
        movlw   TRISB
        Bank0
; Writing to the 6355 requires that the CLK bit be held
; low while the IO and CE lines are high
        bcf     PORTB,CLK         ; CLK low
        call    delay_5
        bsf     PORTB,IO          ; IO high
        call    delay_5
        bsf     PORTB,CE          ; CE high
; Data is stored in RTC as follows:
;   year                8 bits (0 to 99)
;   month               8 bits (1 to 12)
;   day                 8 bits (1 to 31)
```

```
;   dayOfWeek              4 bits (1 to 7)
;   hour                   8 bits (0 to 23)
;   minutes                8 bits (0 to 59)
;                          ======
;   Total                  44 bits
; Seconds cannot be written to RTC. RTC seconds register
; is automatically initialized to zero
            movf      year,w           ; Get item from storage
            call      bin2bcd          ; Convert to BCD
            movwf     temp1
            call      writeRTC

            movf      month,w
            call      bin2bcd
            movwf     temp1
            call      writeRTC

            movf      day,w
            call      bin2bcd
            movwf     temp1
            call      writeRTC

            movf      dayOfWeek,w      ; Day of week is 4-bits
            call      bin2bcd
            movwf     temp1
            call      write4RTC

            movf      hour,w
            call      bin2bcd
            movwf     temp1
            call      writeRTC

            movf      minutes,w
            call      bin2bcd
            movwf     temp1
            call      writeRTC
; Done
            bcf       PORTB,CLK        ; Hold CLK line low
            call      delay_5
            bcf       PORTB,CE         ; and the CE line
; to the RTC
            call      delay_5
            bcf       PORTB,IO         ; RTC in output mode
            return
;============================
;     read RTC data
;============================
; Procedure to read the current time from the RTC and store
```

```
; data (in packed BCD format) in local time registers.
; According to wiring diagram
; NJU6355 Interface for read operations:
; DAT            PORTB,0           Input
; CLK            PORTB,1           Output
; CE             PORTB,2           Output
; IO             PORTB,3           Output
Get_Time
; Clear Port-B
        movlw    b'00000000'
        movwf    PORTB
; Make data line input
        Bank1
        movlw    b'00000001'
        movwf    TRISB
        Bank0
; Reading RTC data requires that the IO line be low and the
; CE line be high. CLK line is held low
        bcf      PORTB,CLK          ; CLK low
        call     delay_125
        bcf      PORTB,IO ; IO line low
        call     delay_125
        bsf      PORTB,CE ; and CE line high
; Data is read from RTC as follows:
;  year                 8 bits (0 to 99)
;  month                8 bits (1 to 12)
;  day                  8 bits (1 to 31)
;  dayOfWeek            4 bits (1 to 7)
;  hour                 8 bits (0 to 23)
;  minutes              8 bits (0 to 59)
;  seconds              8 bits (0 to 59)
;                       ======
;   Total               52 bits
        call     readRTC
        movwf    year
        call     delay_125

        call     readRTC
        movwf    month
        call     delay_125

        call     readRTC
        movwf    day
        call     delay_125

; Day of week is a 4-bit value
        call     read4RTC
        movwf    dayOfWeek
```

```
            call      delay_125

            call      readRTC
            movwf     hour
            call      delay_125

            call      readRTC
            movwf     minutes
            call      delay_125

            call      readRTC
            movwf     seconds

            bcf       PORTB,CE ; CE line low to end output
            return

;==============================
;   read 4/8 bits from RTC
;==============================
; Procedure to read 4/8 bits stored in 6355 registers
; Value returned in w register
read4RTC
            movlw     .4               ; 4 bit read
            goto      anyBits
readRTC
            movlw     .8               ; 8 bits read
anyBits:
            movwf     counter
; Read 6355 read operation requires the IO line be set low
; and the CE line high. Data is read in the following order:
; year, month, day, day-of-week, hour, minutes, seconds
readBits:
            bsf       PORTB,CLK; Set CLK high to validate data
            bsf       STATUS,C         ; Set the carry flag (bit = 1)
; Operation:
;   If data line is high, then bit read is a 1-bit
;   otherwise bit read is a 0-bit
            btfss     PORTB,DAT        ; Is data line high?
                                ; Leave carry set (1 bit) if high
            bcf       STATUS,C ; Clear the carry bit (make bit 0)
; At this point the carry bit matches the data line
            bcf       PORTB,CLK        ; Set CLK low to end read
; The carry bit is now rotated into the temp1 register
            rrf       temp1,1
            decfsz    counter,1        ; Decrement the bit counter
            goto      readBits ; Continue if not last bit
; At this point all bits have been read (8 or 4)
            movf      temp1,0          ; Result to w
```

```
        return

;==============================
;   write 4/8 bits to RTC
;==============================
; Procedure to write 4 or 8 bits to the RTC registers
; ON ENTRY:
;     temp1 register holds value to be written
; ON EXIT:
;     nothing
write4RTC
        movlw    .4              ; Init for 4 bits
        goto     allBits
writeRTC
        movlw    .8              ; Init for 8 bits
allBits:
        movwf    counter         ; Store in bit counter
writeBits:
        bcf      PORTB,CLK       ; Clear the CLK line
        call     delay_5         ; Wait
        bsf      PORTB,DAT       ; Set the data line to RTC
        btfss    temp1,0         ; Send LSB
        bcf      PORTB,DAT       ; Clear data line
        call     delay_5         ; Wait for operation to
                                 ; complete
        bsf      PORTB,CLK       ; Bring CLK line high to
                                 ; validate
        rrf      temp1,f         ; Rotate bits in storage
        decfsz   counter,1       ; Decrement bit counter
        goto     writeBits       ; Continue if not last bit
        return

;==============================
;   init time variables
;==============================
; Procedure to initialize time variables for testing
; Constants used in ininitialization are located in
; #define statements.
initRTC:
        movlw    iYear
        movwf    year
        movlw    iMonth
        movwf    month
        movlw    iDay
        movwf    day
        movlw    iDoW
        movwf    dayOfWeek
        movlw    iHour
```

```
            movwf      hour
            movlw      iMin
            movwf      minutes
            movlw      iSec
            movwf      seconds
            return
;============================
;  binary to BCD conversion
;============================
; Convert a binary number into two packed BCD digits
; ON ENTRY:
;          w register has binary value in range 0 to 99
; ON EXIT:
;          output variables bcdLow and bcdHigh contain two
;          packed unpacked BCD digits
;          w contains two packed BCD digits
; Routine logic:
;   The value 10 is subtracted from the source operand
;   until the remainder is < 0 (carry cleared). The number
;   of subtractions is the high-order BCD digit. 10 is
;   then added back to the subtrahend to compensate
;   for the last subtraction. The final remainder is the
;   low-order BCD digit
; Variables:
;      inNum       storage for source operand
;      bcdHigh     storage for high-order nibble
;      bcdLow      storage for low-order nibble
;      thisDig     Digit counter
bin2bcd:
            movwf      inNum          ; Save copy of source value
            clrf       bcdHigh        ; Clear storage
            clrf       bcdLow
            clrf       thisDig
min10:
            movlw      .10
            subwf      inNum,f        ; Subtract 10
            btfsc      STATUS,C       ; Did subtract overflow?
            goto       sum10          ; No. Count subtraction
            goto       fin10
sum10:
            incf       thisDig,f      ;increment digit counter
            goto       min10
; Store 10th digit
fin10:
            movlw      .10
            addwf      inNum,f        ; Adjust for last subratct
            movf       thisDig,w      ; get digit counter contents
            movwf      bcdHigh        ; Store it
```

```
; Calculate and store low-order BCD digit
        movf    inNum,w           ; Store units value
        movwf   bcdLow            ; Store digit
; Combine both digits
        swapf   bcdHigh,w         ; High nibble to HOBs
        iorwf   bcdLow,w          ; OR-in low nibble
        return
;=============================================================
        end     ; END OF PROGRAM
;=============================================================
```

Chapter 14

Data EEPROM

14.1 EEPROM Programming

This chapter is about *Electrically-Erasable Programmable Read-Only Memory*. EEPROM memory is used in digital devices as nonvolatile storage, such as in flash drives, BIOS chips, and in memory storage facilities. In PIC microcontrollers EEPROM memory is used as semi-permanent data storage because EEPROM can be erased and reprogrammed electrically without removing the chip. The technology used before the development of EEPROM, called EPROM, required that the chip be removed from the circuit and placed under ultraviolet light in order to erase it. In addition, EPROM required higher-than-TTL voltages for reprogramming while EEPROM does not. In this chapter we refer to data EEPROM as implemented in the mid-range PICs and in particular in the 16F84.

14.1.1 Data EEPROM

To the PIC programmer, EEPROM data memory can refer either to on-board EEPROM memory or to EEPROM memory ICs that are furnished as separate circuit components. EEPROM elements are classified according to their electrical interfaces into serial and parallel. In this context we deal only with serial EEPROMs. The storage capacity of Serial EEPROMs range from a few bytes to 128 kilobytes. In PIC technology the typical use of serial EEPROM on-board memory and EEPROM ICs is in the storage of passwords, codes, configuration settings, and other information to be remembered after the system is turned off. For example, a PIC-based automated environment sensor can use EEPROM memory to store daily temperatures, humidity, air pressure, and other values. Later on this information can be downloaded to a PC and the EEPROM storage erased and reused for new data. In personal computers, EEPROM memory is used to store BIOS code and other system data.

Some early EEPROMs could be erased and rewritten about 100 times before failing, but more recent EEPROMs tolerate thousands of erase-write cycles. EEPROM memory is different from *Random Access Memory* (RAM) in that RAM can be rewritten millions of times. Also, RAM is generally faster to write than EEPROM and considerably cheaper per unit of storage. On the other hand, RAM is volatile (the contents are lost when power is removed.)

PICs also use EEPROM-type memory internally as flash program memory and as data memory. In this context we deal with EEPROM data memory. Serial EEPROM memory is also available as separate ICs that can be placed on the circuit board and accessed through PIC ports. For example, the Microchip 24LC04B EEPROM IC is a 4K electrically erasable PROM with a two-wire serial interface that follows the I2C convention. Programming serial EEPROM ICs is not covered in this chapter.

14.2 EEPROM Programming

The 16F84 and 16F84A contain 64 bytes of EEPROM data memory. This memory is both readable and writable during normal operation. It is not mapped in the register file space but is indirectly addressed through the Special Function Registers EECON1, EECON2, EEDATA, and EEADR. The address of EEPROM memory starts at location 0x00 and extends to the maximum contained in the PIC, in this case 0x3f. The following registers relate to EEPROM operations:

- EEDATA holds the data byte to be read or written.

- EEADR contains the EEPROM address to be accessed by the read or write operation.

- EECON1 contains the control bits for EEPROM operations.

- EECON2 protects EEPROM memory from accidental access. This is not a physical register.

Figure 14.1 is a bitmap of the EECON1 register in the 16F84.

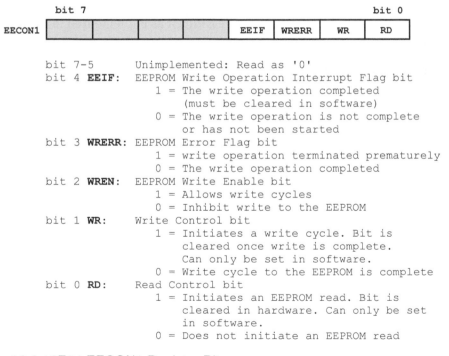

```
       bit 7                                                    bit 0
EECON1 [          |          |          |          | EEIF | WRERR |  WR  |  RD  ]

       bit 7-5      Unimplemented: Read as '0'
       bit 4 EEIF:  EEPROM Write Operation Interrupt Flag bit
                       1 = The write operation completed
                           (must be cleared in software)
                       0 = The write operation is not complete
                           or has not been started
       bit 3 WRERR: EEPROM Error Flag bit
                       1 = write operation terminated prematurely
                       0 = The write operation completed
       bit 2 WREN:  EEPROM Write Enable bit
                       1 = Allows write cycles
                       0 = Inhibit write to the EEPROM
       bit 1 WR:    Write Control bit
                       1 = Initiates a write cycle. Bit is
                           cleared once write is complete.
                           Can only be set in software.
                       0 = Write cycle to the EEPROM is complete
       bit 0 RD:    Read Control bit
                       1 = Initiates an EEPROM read. Bit is
                           cleared in hardware. Can only be set
                           in software.
                       0 = Does not initiate an EEPROM read
```

Figure 14.1 16F84 EECON1 Register Bitmap.

The CPU may continue to access EEPROM memory even if the device is code protected, but in this case the device programmer cannot access EEPROM memory.

14.2.1 Reading EEPROM Data

Reading an EEPROM data memory location in the 16F84 requires the following operations:

- Bank 0 is selected and the address of the memory to be read is stored in the EEADR register.

- Bank 1 is selected and the RD bit is set in the EECON1 register.

- Bank 0 is selected and data is read from the EEDATA register.

The following procedure returns in the w register the data stored at the specified EEPROM memory address:

```
;==============================
;        read EEPROM 16F84
;==============================
; Procedure to read EEPROM memory. Address of memory
; location to read is passed in local variable EEMemAdd
; On exit: read data in w
EERead:
        bcf   STATUS,RP0              ; Bank 0
        movf  EEMemAdd,w             ; Address to w
        movwf EEADR                   ; w to address register
        bsf   STATUS,RP0              ; Bank 1
        bsf   EECON1,RD               ; EE Read
        bcf   STATUS,RP0              ; Bank 0
        movf  EEDATA,w               ; w = EEDATA
        return
```

14.2.2 EEPROM Data Memory Write

Writing to 16F84 EEPROM data memory consists of the following operations:

- Bank 0 is selected and the address of the desired memory location is stored in the EEADR register.

- The value to be written is stored in the EEDATA register.

- Bank 1 is selected, interrupts are disabled, and the write enable bit (WREN) is set in the EECON1 register.

- The special values 0x55 and 0xaa are written consecutively to the EECON2 register.

- The WR bit is set in the EECON1 register. The EEPROM write takes place automatically after the WR bit is set.

- Interrupts are reenabled and bank 0 is selected.

The following procedure shows the processing for the EEPROM write:

```
;==============================
;        write EEPROM
;==============================
; Procedure to write asc1 byte to EEPROM memory
; Address to write passed in local variable EEMemAdd
```

```
; Data byte to write is passed in local variable EEByte
EEWrite:
; Load byte to write into EE data register
    movf EEByte,w       ; Data to w
    movwf    EEDATA          ; Write
; Set write address in EE address register
    movf     EEMemAdd,w      ; Address to w
    movwf    EEADR           ; w to address register
; Write data to EEPROM memory
    bsf      STATUS,RP0      ; Bank 1
    bcf      INTCON,GIE      ; Disable interrupts
    bsf      EECON1,WREN     ; Enable Write
    movlw    0x55            ; Code # 1
    movwf    EECON2          ; Write 0x55
    movlw    0xaa            ; Code # 2
    movwf    EECON2          ; Write 0xaa
    bsf  EECON1,WR           ; Set WR bit
; Write operation now takes place automatically
    bsf      INTCON,GIE      ; Re-enable interrupts
    bcf      STATUS,RP0      ; Bank 0
    return
```

Microchip documentation recommends that critical applications should verify the write operation by reading EEPROM memory after the write operation has taken place in order to make sure that the correct value was stored.

14.3 EEPROM Programming Application

The program EECounter, in this book's software package, is a demonstration of EEPROM memory access on the 16F84 PIC. The program keeps track of the number of times that the code has executed by storing each iteration in EEPROM data memory. The program uses the circuit shown in Figure 14-2. Note that the circuit in Figure 14-2 is compatible with the one emulated by the Virtual Board B program.

14.3.1 EECounter Program

The EECounter program, in this book's software package, demonstrates programming the onboard EEPROM data memory on the 16F84. The program contains EEPROM read and write primitives that can be reused in other applications. The program uses LCD display to output results. Code keeps track of the number of times it has been run by storing each iteration in EEPROM memory and reading back this value at every new execution cycle.

Code Details

In the EECounter program we encounter a situation that is quite common in computer programs: An application must deal with two or more different numeric formats. In the case of the EECounter program a local register is used to store and manipulate a binary value. But before this value can be displayed in the LCD, it must be converted to a string of ASCII decimal digits. The conversion routine transforms an unsigned binary value into an ASCII decimal string. The process consists of dividing the binary by 10, adding 30H to the remainder, and continuing until the original binary becomes zero. The EECounter program contains a binary-to-ASCII decimal conversion routine called bin2asc, and listed next.

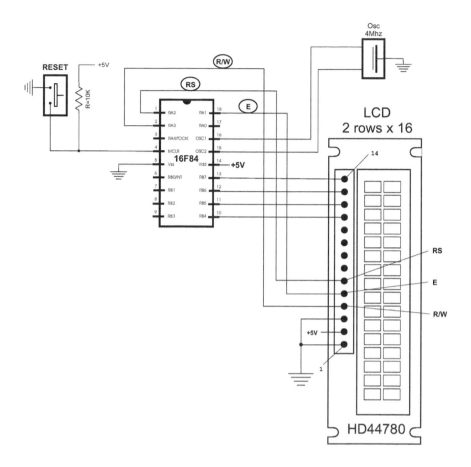

Figure 14-2 Circuit for EEPROM Demonstration Program.

```
;================================
;    binary to ASCII decimal
;         conversion
;================================
; ON ENTRY:
;        w register has binary value in range 0 to 255
; ON EXIT:
;        output variables asc100, asc10, and asc1 have
;        three ASCII decimal digits
; Routine logic:
;   The value 100 is subtracted from the source operand
;   until the remainder is < 0 (carry cleared). The number
;   of subtractions is the decimal hundreds result. 100 is
;   then added back to the subtrahend to compensate
;   for the last subtraction. Now 10 is subracted in the
;   same manner to determine the decimal tenths result.
;   The final remainder is the decimal units result.
```

```
; Variables:
;       inNum       storage for source operand
;       asc100      storage for hundreds position result
;       asc10       storage for tens position result
;       asc1        storage for unit position reslt
;       thisDig     Digit counter
bin2asc:
        movwf       inNum           ; Save copy of source value
        clrf        asc100          ; Clear hundreds storage
        clrf        asc10           ; Tens
        clrf        asc1            ; Units
        clrf        thisDig
sub100:
        movlw       .100
        subwf       inNum,f         ; Subtract 100
        btfsc       STATUS,C        ; Did subtract overflow?
        goto        bump100         ; No. Count subtraction
        goto        end100
bump100:
        incf        thisDig,f                   ;increment digit
counter
        goto        sub100
; Store 100th digit
end100:
        movf        thisDig,w       ; Adjusted digit counter
        addlw       0x30            ; Convert to ASCII
        movwf       asc100          ; Store it
; Calculate tenth position value
        clrf        thisDig
; Adjust minuend
        movlw       .100            ; Minuend
        addwf       inNum,f         ; Add value to minuend to
                                    ; compensate for last
                                    ; operation
sub10:
        movlw       .10
        subwf       inNum,f         ; Subtract 10
        btfsc       STATUS,C        ; Did subtract overflow?
        goto        bump10          ; No. Count subtraction
        goto        end10
bump10:
        incf        thisDig,f       ;increment digit counter
        goto        sub10
; Store 10th digit
end10:
        movlw       .10
        addwf       inNum,f         ; Adjust for last subratct
        movf        thisDig,w       ; get digit counter contents
```

```
        addlw    0x30              ; Convert to ASCII
        movwf    asc10             ; Store it
; Calculate and store units digit
        movf     inNum,w           ; Store units value
        addlw    0x30              ; Convert to ASCII
        movwf    asc1              ; Store digit
        return
```

Two other procedures, called EERead and EEWrite, perform the EEProm basic operations. These procedures were discussed and listed in Sections 14.4.1 and 14.2.2.

```
;============================
;    I2C read procedure
;============================
; Procedure to read one byte from 24LC04B EEPROM
; Steps:
;                   1. Send START
;                   2. Send control. Wait for ACK
;                   3. Send address. Wait for ACK
;                   4. Send RESTART + control. Wait for ACK
;                   5. Switch to receive mode. Get data.
;                   6. Send NACK
;                   7. Send STOP
;                   8. Retreive data into w register
; STEP 1:
ReadI2C
; Send RESTART. Wait for ACK
        Bank1
        bsf      SSPCON2,RSEN      ; RESTART Condition
        call     WaitI2C           ; Wait for I2C operation
; STEP 2:
; Send control byte. Wait for ACK
        movlw    LC04READ          ; Control byte
        call     Send1I2C          ; Send Byte
        call     WaitI2C  ; Wait for I2C operation
; Now check to see if I2C EEPROM is ready
        Bank1
        btfsc    SSPCON2,ACKSTAT   ; Check ACK Status bit
        goto     ReadI2C           ; ACK Poll waiting for EEPROM
                                   ; write to complete
; STEP 3:
; Send address. Wait for ACK
        Bank0
        movf     EEMemAdd,w        ; Load from address register
        call     Send1I2C          ; Send Byte
        call     WaitI2C           ; Wait for I2C operation
        Bank1
```

```
        btfsc     SSPCON2,ACKSTAT  ; Check ACK Status bit
        goto      FailI2C          ; failed, skipped if
                                   ; successful
; STEP 4:
; Send RESTART. Wait for ACK
        bsf       SSPCON2,RSEN     ; Generate RESTART Condition
        call      WaitI2C          ; Wait for I2C operation
; Send output control. Wait for ACK
        movlw     LC04WRITE        ; Load CONTROL BYTE (output)
        call      Send1I2C         ; Send Byte
        call      WaitI2C          ; Wait for I2C operation
        Bank1
        btfsc     SSPCON2,ACKSTAT  ; Check ACK Status bit
        goto      FailI2C          ; failed, skipped if
                                   ; successful
; STEP 5:
; Switch MSSP to I2C Receive mode
        bsf       SSPCON2,RCEN     ; Enable Receive Mode (I2C)
; Get the data. Wait for ACK
        call      WaitI2C          ; Wait for I2C operation
; STEP 6:
; Send NACK to acknowledge
        Bank1
        bsf       SSPCON2,ACKDT    ; ACK DATA to send is 1 (NACK)
        bsf       SSPCON2,ACKEN    ; Send ACK DATA now.
; Once ACK or NACK is sent, ACKEN is automatically cleared
; STEP 7:
; Send STOP. Wait for ACK
        bsf       SSPCON2,PEN      ; Send STOP condition
        call      WaitI2C          ; Wait for I2C operation
; STEP 8:
; Read operation has finished
        Bank0
        movf      SSPBUF,W         ; Get data from SSPBUF into W
; Procedure has finished and completed successfully.
        return


;=============================
;   I2C support procedures
;=============================
; I2C Operation failed code sequence
; Procedure hangs up. User should provide error handling.
FailI2C
        Bank1
        bsf       SSPCON2,PEN      ; Send STOP condition
        call      WaitI2C          ; Wait for I2C operation
fail:
        goto      fail
```

```
; Procedure to transmit one byte
Send1I2C
        Bank0
        movwf   SSPBUF          ; Value to send to SSPBUF
        return
; Procedure to wait for the last I2C operation to complete.
; Code polls the SSPIF flag in PIR1.
WaitI2C
        Bank0
        btfss   PIR1,SSPIF      ; Check if I2C operation done
        goto    $-1 +           ; I2C module is not ready yet
        bcf     PIR1,SSPIF      ; I2C ready, clear flag
        return
```

14.4 Demonstration Programs

The following sections contain the code listing for the programs discussed in this chapter.

14.4.1 EECounter Program

```
; File name: EECounter.asm
; Last Update: May 22, 2011
; Authors: Sanchez and Canton
; Processor: 16F84A
;
; Description:
; Program to demonstrate on chip EEPROM data memory read
; and write operation. Program uses LCD display to output
; results.
; Operation:
; The program keeps track of and displays the inNum of times
; the code has been started.
; For LCD display parameters see the LCDTest2 program.
; WARNING:
; Code assumes 4 MHz clock. Delay routines must be
; edited for faster clock
;
;============================
;         switches
;============================
; Switches used in __config directive:
;   _CP_ON         Code protection ON/OFF
; * _CP_OFF
; * _PWRTE_ON      Power-up timer ON/OFF
;   _PWRTE_OFF
;   _WDT_ON        Watchdog timer ON/OFF
; * _WDT_OFF
```

```
;    _LP_OSC          Low power crystal oscillator
; *  _XT_OSC          External parallel resonator/crystal oscillator

;    _HS_OSC          High speed crystal resonator (8 to 10 MHz)
;                     Resonator: Murate Erie CSA8.00MG = 8 MHz
;    _RC_OSC          Resistor/capacitor oscillator
; |                   (simplest, 20% error)
; |
; |_____ * indicates setup values presently selected


;=========================
; setup and configuration
;=========================
        processor 16f84A
        include   <p16f84A.inc>
        __config  _XT_OSC & _WDT_OFF & _PWRTE_ON & _CP_OFF


;========================================================
;                 constant definitions
;   for PIC-to-LCD pin wiring and LCD line addresses
;========================================================
#define E_line 1            ;|
#define RS_line 2           ;| => from wiring diagram
#define RW_line 3           ;|
; LCD line addresses (from LCD data sheet)
#define LCD_1 0x80          ; First LCD line constant
#define LCD_2 0xc0          ; Second LCD line constant
; Note: The constants that define the LCD display line
;       addresses have the high-order bit set in
;       order to faciliate the controller command
;
;========================================================
;                 variables in PIC RAM
;========================================================
; Reserve 16 bytes for string buffer
        cblock  0x0c
        strData
        endc
; Reserve three bytes for ASCII digits
        cblock  0x1d
        asc100
        asc10
        asc1
        endc
; Continue with local variables
        cblock  0x20                ; Start of block
        count1                      ; Counter # 1
        count2                      ; Counter # 2
```

```
        count3              ; Counter # 3
        pic_ad              ; Storage for start of text area
                            ; (labeled strData) in PIC RAM
        J                   ; counter J
        K                   ; counter K
        index               ; Index into text table (also used
                            ; for auxiliary storage)
        store1              ; Local temporary storage
        store2              ; Storage # 2
; EEPROM-related variables
        EEMemAdd ; EEPROM address to access
        EEByte              ; Data byte to write
; Storage for ASCII decimal conversion and digits
        inNum               ; Source operand
        thisDig             ; Digit counter
        endc


;=========================================================
;                            program
;=========================================================
        org     0           ; start at address
        goto    main
; Space for interrupt handlers
        org     0x08

main:
        movlw   b'00000000'     ; All lines to output
        tris    PORTA           ; in Port-A
        tris    PORTB           ; and Port-B
        movlw   b'00000000'     ; All output ports low
        movwf   PORTA
        movwf   PORTB
; Wait and initialize HD44780
        call    delay_5         ; Allow LCD time to initialize
                                ; itself
        call    delay_5
        call    initLCD         ; Then do forced
initialization
        call    delay_5         ; Wait again
; Store base address of text buffer in PIC RAM

        movlw   0x0c            ; Start address for buffer
        movwf   pic_ad          ; to local variable
; Initialize EEPROM data to 0x0
        clrf    EEMemAdd        ; Set address to 0
;=====================
;   first LCD line
;=====================
```

```
; Store 16 blanks in PIC RAM, starting at address stored
; in variable pic_ad
        call     blank16
; Call procedure to store ASCII characters for message
; in text buffer
        movlw    d'0'              ; Offset into buffer
        call     storeMS1
;=======================
;  Read EEPROM memory
;=======================
; EEPROM memory address to use is at 10 (0x0a). Variable
; EEMemAdd is already initialized.
; Fill data for EEPROM is 0xff. This value indicates
; the first iteration
        call     EERead            ; Local procedure. Value in w
        movwf    EEByte            ; Save result
; EEPROM data still in w
        incf     EEByte,f
        call     EEWrite
; At this point iteration inNum is stored in EEByte
; This value must be displayed on the LCD at offset 11
; of the first line. This means it must be stored at offset
; 11 in the buffer. Because the buffer starts at 0x0c the
; iteration digit must be stored at offset 0x0c+11=0x17
ShowEEData:
; Binary data in EEByte
        movf     EEByte,w ; Value to w
        call     bin2asc           ; Conversion routine
; At this point three ASCII digits are stored in local
; variables. Move digits to display area
        movf     asc1,w            ; Unit digit
        movwf    0x18              ; Store in buffer
        movf     asc10,w           ; same with other digits
        movwf    0x17
        movf     asc100,w
        movwf    0x16
; Display line
; Set DDRAM address to start of first line
showLine:
        call     line1
; Call procedure to display 16 characters in LCD
        call     display16
loopHere:
        goto     loopHere ;done


;============================================================
;                    initialize LCD for 4-bit mode
;============================================================
```

```
initLCD:
; Initialization for Densitron LCD module as follows:
;    4-bit interface
;    2 display lines of 16 characters each
;    cursor on
;    left-to-right increment
;    cursor shift right
;    no display shift
;=====================|
;    set command mode  |
;=====================|
            bcf       PORTA,E_line       ; E line low
            bcf       PORTA,RS_line      ; RS line low
            bcf       PORTA,RW_line      ; Write mode
            call      delay_125          ; delay 125 microseconds
;*********************|
;       FUNCTION SET  |
;*********************|
            movlw     0x28     ; 0 0 1 0 1 0 0 0 (FUNCTION SET)
            call      send8    ; 4-bit send routine

; Set 4-bit mode command must be repeated
            movlw     0x28
            call      send8


;*********************|
; DISPLAY AND CURSOR ON |
;*********************|
            movlw     0x0e     ; 0 0 0 0 1 1 1 0 (DISPLAY ON/OFF)
            call      send8
;*********************|
;    set entry mode   |
;*********************|
            movlw     0x06     ; 0 0 0 0 0 1 1 0 (ENTRY MODE SET)
            call      send8


;*********************|
; cursor/display shift |
;*********************|
            movlw     0x14     ; 0 0 0 1 0 1 0 0 (CURSOR/DISPLAY
                               ; SHIFT)
            call      send8
;*********************|
;    clear display    |
;*********************|
            movlw     0x01     ; 0 0 0 0 0 0 0 1 (CLEAR DISPLAY)
            call      send8
; Per documentation
```

```
        call    delay_5  ; Test for busy
        return


;=======================
;   Procedure to delay
;    42 microseconds
;=======================
delay_125
        movlw   D'42'              ; Repeat 42 machine cycles
        movwf   count1             ; Store value in counter
repeat
        decfsz  count1,f           ; Decrement counter
        goto    repeat             ; Continue if not 0
        return                     ; End of delay


;=======================
;   Procedure to delay
;    5 milliseconds
;=======================
delay_5
        movlw   D'41'              ; Counter = 41
        movwf   count2             ; Store in variable
delay
        call    delay_125          ; Delay
        decfsz  count2,f           ; 40 times = 5 milliseconds
        goto    delay
        return                     ; End of delay
;=======================
;     pulse E line
;=======================
pulseE
        bsf     PORTA,E_line       ; Pulse E line
        nop
        bcf     PORTA,E_line
        return


;============================
;    long delay sub-routine
;      (for debugging)
;============================
long_delay
        movlw   D'200'             ; w = 200 decimal
        movwf   J            ;        J = w
jloop:
        movwf   K                  ; K = w
kloop:
        decfsz  K,f                ; K = K-1, skip next if zero
        goto    kloop
```

```
          decfsz   J,f                   ; J = J-1, skip next if zero
          goto     jloop
          return
;==============================
;   LCD display procedure
;==============================
; Sends 16 characters from PIC buffer with address stored
; in variable pic_ad to LCD line previously selected
display16
          call     delay_5              ; Make sure not busy
; Set up for data
          bcf      PORTA,E_line         ; E line low
          bsf      PORTA,RS_line        ; RS line high for data
; Set up counter for 16 characters
          movlw    D'16'                ; Counter = 16
          movwf    count3
; Get display address from local variable pic_ad
          movf     pic_ad,w ; First display RAM address to W
          movwf    FSR                  ; W to FSR
getchar:
          movf     INDF,w   ; get character from display RAM
                            ; location pointed to by file select
                            ; register
          call     send8                ; 4-bit interface routine
; Test for 16 characters displayed
          decfsz   count3,f             ; Decrement counter
          goto     nextchar ; Skipped if done
          return
nextchar:
          incf     FSR,f                ; Bump pointer
          goto     getchar


;========================
;   send 2 nibbles in
;     4-bit mode
;========================
; Procedure to send two 4-bit values to Port-B lines
; 7, 6, 5, and 4. High-order nibble is sent first
; ON ENTRY:
;        w register holds 8-bit value to send
send8:
          movwf    store1               ; Save original value
          call     merge4               ; Merge with Port-B
; Now w has merged byte
          movwf    PORTB                ; w to Port-B
          call     pulseE               ; Send data to LCD
; High nibble is sent
          movf     store1,w             ; Recover byte into w
```

```
          swapf     store1,w ; Swap nibbles in w
          call      merge4
          movwf     PORTB
          call      pulseE            ; Send data to LCD
          call      delay_125
          return
;==================
;   merge bits
;==================
; Routine to merge the 4 high-order bits of the
; value to send with the contents of Port-B
; so as to preserve the 4 low-bits in Port-B
; Logic:
;      AND value with 1111 0000 mask
;      AND Port-B with 0000 1111 mask
;      Now low nibble in value and high nibble in
;      Port-B are all 0 bits:
;           value = vvvv 0000
;          Port-B = 0000 bbbb
;      OR value and Port-B resulting in:
;                 vvvv bbbb
; ON ENTRY:
;      w contain value bits
; ON EXIT:
;      w contains merged bits
merge4:
          andlw     b'11110000'       ; ANDing with 0 clears the
                                      ; bit. ANDing with 1 preserves
                                      ; the original value
          movwf     store2            ; Save result in variable
          movf      PORTB,w           ; Port-B to w register
          andlw     b'00001111'       ; Clear high nibble in Port-b
                                      ; and preserve low nibble
          iorwf     store2,w          ; OR two operands in w
          return


;========================
;     blank buffer
;========================
; Procedure to store 16 blank characters in PIC RAM
; buffer starting at address stored in the variable
; pic_ad
blank16
          movlw     D'16'             ; Set-up counter
          movwf     count1
          movf      pic_ad,w          ; First PIC RAM address
          movwf     FSR               ; Indexed addressing
          movlw     0x20              ; ASCII space character
```

```
storeit
        movwf   INDF        ; Store blank character in PIC RAM
                            ; buffer using FSR register
        decfsz  count1,f            ; Done?
        goto    incfsr              ; no
        return                      ; yes
incfsr:
        incf    FSR,f       ; Bump FSR to next buffer space
        goto    storeit

;========================
; Set address register
;    to LCD line 1
;========================
; ON ENTRY:
;        Address of LCD line 1 in constant LCD_1
line1:
        bcf     PORTA,E_line        ; E line low
        bcf     PORTA,RS_line       ; RS line low, set up for
control
        call    delay_5             ; busy?
; Set to second display line
        movlw   LCD_1               ; Address and command bit
        call    send8               ; 4-bit routine
; Set RS line for data
        bsf     PORTA,RS_line       ; Setup for data
        call    delay_5             ; Busy?
        return

;================================
;   first text string procedure
;================================
storeMS1:
; Procedure to store in PIC RAM buffer the message
; contained in the code area labeled msg1
; ON ENTRY:
;        variable pic_ad holds address of text buffer
;        in PIC RAM
;        w register holds offset into storage area
;        msg1 is routine that returns the string characters
;        and a zero terminator
;        index is local variable that holds offset into
;        text table. This variable is also used for
;        temporary storage of offset into buffer
; ON EXIT:
;        Text message stored in buffer
;
; Store offset into text buffer (passed in the w register)
```

```
; in temporary variable
        movwf    index              ; Store w in index
; Store base address of text buffer in FSR
        movf     pic_ad,w ; first display RAM address to W
        addwf    index,w            ; Add offset to address
        movwf    FSR                ; W to FSR
; Initialize index for text string access
        movlw    0                  ; Start at 0
        movwf    index              ; Store index in variable
; w still = 0
get_msg_char:
        call     msg1               ; Get character from table
; Test for zero terminator
        andlw    0x0ff
        btfsc    STATUS,Z ; Test zero flag
        goto     endstr1            ; End of string
; ASSERT: valid string character in w
;         store character in text buffer (by FSR)
        movwf    INDF               ; store in buffer by FSR
        incf     FSR,f              ; increment buffer pointer
; Restore table character counter from variable
        movf     index,w            ; Get value into w
        addlw    1                  ; Bump to next character
        movwf    index    ; Store table index in variable
        goto     get_msg_char    ; Continue
endstr1:
        return

; Routine for returning message stored in program area
; Message has 10 characters
msg1:
        addwf    PCL,f              ; Access table
        retlw    'I'
        retlw    't'
        retlw    'e'
        retlw    'r'
        retlw    '.'
        retlw    0x20
        retlw    'N'
        retlw    'o'
        retlw    '.'
        retlw    0x20
        retlw    0


;===============================
;   binary to ASCII decimal
;         conversion
;===============================
```

```
;   ON ENTRY:
;           w register has binary value in range 0 to 255
;   ON EXIT:
;           output variables asc100, asc10, and asc1 have
;           three ASCII decimal digits
;   Routine logic:
;     The value 100 is subtracted from the source operand
;     until the remainder is < 0 (carry cleared). The number
;     of subtractions is the decimal hundreds result. 100 is
;     then added back to the subtrahend to compensate
;     for the last subtraction. Now 10 is subtracted in the
;     same manner to determine the decimal tenths result.
;     The final remainder is the decimal units result.
;   Variables:
;       inNum       storage for source operand
;       asc100      storage for hundreds position result
;       asc10       storage for tens position result
;       asc1        storage for unit position result
;       thisDig     Digit counter
bin2asc:
        movwf   inNum               ; Save copy of source value
        clrf    asc100      ; Clear hundreds storage
        clrf    asc10           ; Tens
        clrf    asc1            ; Units
        clrf    thisDig
sub100:
        movlw   .100
        subwf   inNum,f             ; Subtract 100
        btfsc   STATUS,C            ; Did subtract overflow?
        goto    bump100             ; No. Count subtraction
        goto    end100
bump100:
        incf    thisDig,f           ;increment digit counter
        goto    sub100
; Store 100th digit
end100:
        movf    thisDig,w           ; Adjusted digit counter
        addlw   0x30                ; Convert to ASCII
        movwf   asc100              ; Store it
; Calculate tenth position value
        clrf    thisDig
; Adjust minuend
        movlw   .100                ; Minuend
        addwf   inNum,f             ; Add value to minuend to
                                    ; compensate for last
                                    ; operation
sub10:
        movlw   .10
```

```
        subwf    inNum,f          ; Subtract 10
        btfsc    STATUS,C         ; Did subtract overflow?
        goto     bump10           ; No. Count subtraction
        goto     end10
bump10:
        incf     thisDig,f        ;increment digit counter
        goto     sub10
; Store 10th digit
end10:
        movlw    .10
        addwf    inNum,f          ; Adjust for last subraction
        movf     thisDig,w        ; get digit counter contents
        addlw    0x30             ; Convert to ASCII
        movwf    asc10            ; Store it
; Calculate and store units digit
        movf     inNum,w    ;     Store units value
        addlw    0x30             ; Convert to ASCII
        movwf    asc1             ; Store digit
        return


;=============================================================
;                     EEPROM procedures
;=============================================================
;==============================
;       read EEPROM
;==============================
; Procedure to read EEPROM memory. Address of memory
; location to read is stored in local register EEMemAdd
; On exit: read data in w
EERead:
        bcf      STATUS,RP0       ; Bank 0
        movf     EEMemAdd,w       ; Address to w
        movwf    EEADR            ; w to address register
        bsf      STATUS,RP0       ; Bank 1
        bsf      EECON1,RD        ; EE Read
        bcf      STATUS,RP0       ; Bank 0
        movf     EEDATA,w         ; W = EEDATA
        return


;==============================
;       write EEPROM
;==============================
; Procedure to write asc1 byte to EEPROM memory
; Address to write stored in local register EEMemAdd
; Data byte to write is in local register EEByte
EEWrite:
; Load byte to write into EE data register
        movf     EEByte,w ; Data to w
```

```
        movwf    EEDATA              ; Write
; Set write address in EE address register
        movf     EEMemAdd,w          ; Address to w
        movwf    EEADR               ; w to address register
; Write data to EEPROM memory
        bsf      STATUS,RP0          ; Bank 1
        bcf      INTCON,GIE          ; Disable INTs.
        bsf      EECON1,WREN         ; Enable Write
        movlw    0x55                ; Code # 1
        movwf    EECON2              ; Write 0x55
        movlw    0xaa                ; Code # 2
        movwf    EECON2              ; Write 0xaa
        bsf      EECON1,WR           ; Set WR bit
; Write operation now takes place automatically
        bsf      INTCON,GIE          ; Re-enable interrupts
        bcf      STATUS,RP0          ; Bank 0
        return

        end
```

14.4.2 Ser2EEP Program

```
; File name: Ser2EEP.asm
; Last revision: May 22, 2011
; Authors: Canton and Sanchez
; PIC: 16F877
;
; Description:
; Receive character data through RS-232 line and store in
; EEPROM data memory. Received characters are echoed on
; the second LCD line. When <Enter> key is detected (code
; 0x0d) the text stored in EEPROM memory is retrieved and
; displayed on the LCD. On start-up, the top LCD line displays
; the prompt: 'Receiving:'. At that time a message 'Rdy-' is
; sent through the serial line so as to test the connection.
;
; Default serial line setting:
;               2400 baud
;               no parity
;               1 stop bit
;               8 character bits
;
; Program to use 4-bit PIC-to-LCD interface.
; Code assumes that LCD is driven by Hitachi HD44780
; controller and PIC 16F977. Display supports two lines,
; each one with 20 characters. The length, wiring and base
; address of each display line is stored in #define
; statements. These statements can be edited to accommodate
```

```
; a different set-up.
;
; WARNING:
; Code assumes 10 MHz clock. Delay routines must be
; edited for a different clock. Clock speed also determines
; values for baud rate setting (see spbrgVal constant).
;
;==========================
;      16F877 switches
;==========================
; Switches used in __config directive:
;   _CP_ON           Code protection ON/OFF
; * _CP_OFF
; * _PWRTE_ON        Power-up timer ON/OFF
;   _PWRTE_OFF
;   _BODEN_ON        Brown-out reset enable ON/OFF
; * _BODEN_OFF
; * _PWRTE_ON        Power-up timer enable ON/OFF
;   _PWRTE_OFF
;   _WDT_ON          Watchdog timer ON/OFF
; * _WDT_OFF
;   _LPV_ON          Low voltage IC programming enable ON/OFF
; * _LPV_OFF
;   _CPD_ON          Data EE memory code protection ON/OFF
; * _CPD_OFF
; OSCILLATOR CONFIGURATIONS:
;   _LP_OSC       Low power crystal oscillator
;   _XT_OSC       External parallel resonator/crystal oscillator

; * _HS_OSC       High speed crystal resonator
;   _RC_OSC       Resistor/capacitor oscillator
; |             (simplest, 20% error)
; |
; |_____ * indicates setup values presently selected

        processor        16f877              ; Define processor
        #include <p16f877.inc>
        __CONFIG _CP_OFF & _WDT_OFF & _BODEN_OFF & _PWRTE_ON &
_HS_OSC & _WDT_OFF & _LVP_OFF & _CPD_OFF

; __CONFIG directive is used to embed configuration data
; within the source file. The labels following the directive
; are located in the corresponding .inc file.
        errorlevel -302
; Suppress bank-related warning
;===============================================================
;                        M A C R O S
;===============================================================
```

```
; Macros to select the register banks
Bank0    MACRO                ; Select RAM bank 0
         bcf       STATUS,RP0
         bcf       STATUS,RP1
         ENDM

Bank1    MACRO                ; Select RAM bank 1
         bsf       STATUS,RP0
         bcf       STATUS,RP1
         ENDM

Bank2    MACRO                ; Select RAM bank 2
         bcf       STATUS,RP0
         bsf       STATUS,RP1
         ENDM

Bank3    MACRO                ; Select RAM bank 3
         bsf       STATUS,RP0
         bsf       STATUS,RP1
         ENDM
;=======================================================
;                   constant definitions
;   for PIC-to-LCD pin wiring and LCD line addresses
;=======================================================
#define E_line 1             ; |
#define RS_line 0            ; | => from wiring diagram
#define RW_line 2            ; |
; LCD line addresses (from LCD data sheet)
#define LCD_1 0x80           ; First LCD line constant
#define LCD_2 0xc0           ; Second LCD line constant
#define LCDlimit .20; Number of characters per line
#define  spbrgVal .64; For 2400 baud on 10 MHz clock
; Note: The constants that define the LCD display
;       line addresses have the high-order bit set
;       so as to meet the requirements of controller
;       commands.
;
;===========================================================
;                   General-Purpose Variables
;===========================================================
; Local variables
; Reserve 20 bytes for string buffer
         cblock    0x20
         strData
         endc
; Other data
         cblock    0x34               ; Start of block
         count1                       ; Counter # 1
```

```
        count2                          ; Counter # 2
        count3                          ; Counter # 3
        J                               ; Counter J
        K                               ; Counter K
        bufAdd
        index
        store1                          ; Local storage
        store2
        endc
;==============================
;       Common RAM Area
;==============================
; These GPRs can be accessed from any bank.
; 15 bytes are available, from 0x70 to 0x7f
        cblock  0x70
; For LCDscroll procedure
        LCDcount ; Counter for characters per line
        LCDline              ; Current display line (0 or 1)
; Communications variables
        newData              ; not 0 if new data received
        ascVal
        errorFlags
; EEPROM-related variables
        EEMemAdd ; EEPROM address to access
        EEByte               ; Data byte to write
        endc


;================================================================
;                         P R O G R A M
;================================================================
        org     0               ; start at address
        goto    main
; Space for interrupt handlers
        org             0x08
main:
; Wiring:
;     LCD data to Port-D, lines 0 to 7
;     E line -> Port-E, 1
;     RW line -> Port-E, 2
;     RS line -> Port-E, 0
; Set PORTE D and E for output
; First, initialize Port-B by clearing latches
        clrf    STATUS
        clrf    PORTB
; Select bank 1 to TRIS Port-D for output
        Bank1
; TRIS Port-D for output. Port-D lines 4 to 7 are wired
; to LCD data lines. Port-D lines 0 to 4 are wired to LEDs.
```

```
        movlw   B'00000000'
        movwf   TRISD     ; and Port-D
; By default Port-A lines are analog. To configure them
; as digital code must set bits 1 and 2 of the ADCON1
; register (in bank 1)
        movlw   0x06      ; binary 0000 0110  is code to
                          ; make all Port-A lines digital
        movwf   ADCON1
; Port-B, lines are wired to keypad switches, as follows:
;   7 6 5 4 3 2 1 0
;   | | | | |_|_|_|_____ switch rows (output)
;   |_|_|_|_____ switch columns (input)
; rows must be defined as output and columns as input
        movlw   b'11110000'
        movwf   TRISB
; TRIS Port-E for output
        movlw   B'00000000'
        movwf   TRISE             ; TRIS Port-E
; Enable Port-B pullups for switches in OPTION register
        movlw   b'00001000'
        movwf   OPTION_REG
; Back to bank 0
        Bank0
; Initialize serial Port for 2400 baud, 8 bits, no parity,
; 1 stop
        call    InitSerial
; Test serial transmission by sending "RDY-"
        movlw   'R'
        call    SerialSend
        movlw   'D'
        call    SerialSend
        movlw   'Y'
        call    SerialSend
        movlw   '-'
        call    SerialSend
        movlw   0x20
        call    SerialSend
; Clear all output lines
        movlw   b'00000000'
        movwf   PORTD
        movwf   PORTE
; Wait and initialize HD44780
        call    delay_5 ; Allow LCD time to initialize itself
        call    initLCD ; Then do forced initialization
        call    delay_5 ; (Wait probably not necessary)
; Clear character counter and line counter variables
        clrf    LCDcount
        clrf    LCDline
```

```
; Set display address to start of first LCD line
        call      line1
; Store address of display buffer
        movlw     0x20
        movwf     bufAdd
; Display 'Receiving: ' message prompt
        call      blank20           ; Clear buffer
        movlw     0x00              ; Offset in buffer
        call      storeMS1          ; Store message at offset
        call      display20         ; Display message
; Start address of EEPROM
        clrf      EEMemAdd
; Setup for display in second line
        call      line2
        clrf      LCDline
        incf      LCDline,f         ; Set scroll control for
                                    ; line 2
;===============================================================
;              receive serial data, store, and display
;===============================================================
receive:
; Call serial receive procedure
        call      SerialRcv
; HOB of newData register is set if new data
; received
        btfss     newData,7
        goto      scanExit
; At this point new data was received.
        movwf     EEByte            ; Save received character
; Display character on LCD
        movf      EEByte,w ; Recover character
        call      send8             ; Display in LCD
        call      LCDscroll         ; Scroll at end of line
; Store character in EEPROM at location in EEMemAdd
        call      EEWrite           ; Local procedure
        incf      EEMemAdd,f        ; Bump to next EEPROM
; Check for <Enter> key (0x0d) and execute display function
        movf      EEByte,w          ; Recover last received
        sublw     0x0d
        btfsc     STATUS,Z          ; Test if <Enter> key
        goto      isEnter           ; Go if <Enter>
; Not <Enter> key, continue processing
scanExit:
        goto      receive           ; Continue
;============================
;    display EEPROM data
;============================
; This routine receives control when the <Enter> key is
```

```
; received.
; Action:
;       1. Clear LCD
;       2. Output is set to top LCD line
;       3. Characters stored in EEPROM are displayed
;          until 0x0d code is detected
isEnter:
        call    clearLCD
; Clear character counter and line counter variables
        clrf    LCDcount
        clrf    LCDline
; Read data from EEPROM memory, starting at address 0
; and display on LCD until 0x0d terminator
        call    line1
        clrf    EEMemAdd ; Start at EEPROM 0
readOne:
        call    EERead          ; Get character
; Store character
        movwf   EEByte          ; Save character
; Test for terminator
        sublw   0x0d
        btfsc   STATUS,Z ; Test if 0x0d
        goto    atEnd           ; Go if 0x0d
; At this point character read is not 0x0d
; Display on LCD
        movf    EEByte,w ; Recover character
; Display character on LCD
        call    send8           ; Display in LCD
        call    LCDscroll       ; Scroll at end of line
        incf    EEMemAdd,f      ; Next EEPROM byte
        goto    readOne

; End of execution
atEnd:
        goto    atEnd


;=================================================================
;=================================================================
;               L O C A L   P R O C E D U R E S
;=================================================================
;=================================================================
;=========================
; init LCD for 4-bit mode
;=========================
initLCD:
; Initialization for Densitron LCD module as follows:
;   4-bit interface
;   2 display lines of 16 characters each
```

```
;    cursor on
;    left-to-right increment
;    cursor shift right
;    no display shift
;======================|
;   set command mode    |
;======================|
        bcf      PORTE,E_line      ; E line low
        bcf      PORTE,RS_line     ; RS line low
        bcf      PORTE,RW_line     ; Write mode
        call     delay_125         ; delay 125 microseconds
        movlw    0x28     ; 0 0 1 0 1 0 0 0 (FUNCTION SET)
        call     send8    ; 4-bit send routine
; Set 4-bit mode command must be repeated
        movlw    0x28
        call     send8
        movlw    0x0e     ; 0 0 0 0 1 1 1 0 (DISPLAY ON/OFF)
        call     send8
        movlw    0x06     ; 0 0 0 0 0 1 1 0 (ENTRY MODE SET)
        call     send8
        movlw    0x14     ; 0 0 0 1 0 1 0 0 (CURSOR/DISPLAY
                          ; SHIFT)
        call     send8
        movlw    0x01     ; 0 0 0 0 0 0 0 1 (CLEAR DISPLAY)
                          ;               |___ COMMAND BIT
        call     send8
        call     delay_5  ; Test for busy
        return


.;==========================
;   procedure to clear LCD
;==========================
clearLCD:
        bcf      PORTE,E_line      ; E line low
        bcf      PORTE,RS_line     ; RS line low
        bcf      PORTE,RW_line     ; Write mode
        call     delay_125         ; delay 125 microseconds
        movlw    0x01     ; 0 0 0 0 0 0 0 1 (CLEAR DISPLAY)
                          ;               |___ COMMAND BIT
        call     send8
        call     delay_5  ; Test for busy
        return


;======================
;  Procedure to delay
;   42 microseconds
;======================
delay_125:
```

```
            movlw    .105           ; Repeat 105 machine cycles
            movwf    count1         ; Store value in counter
repeat
            decfsz   count1,f       ; Decrement counter
            goto     repeat         ; Continue if not 0
            return                  ; End of delay


;========================
;   Procedure to delay
;    5 milliseconds
;========================
delay_5:
            movlw    .105           ; Counter = 105 cycles
            movwf    count2         ; Store in variable
delay:
            call     delay_125      ; Delay
            decfsz   count2,f       ; 40 times = 5 milliseconds
            goto     delay
            return                  ; End of delay
;========================
;     pulse E line
;========================
pulseE
            bsf      PORTE,E_line   ; Pulse E line
            nop
            bcf      PORTE,E_line
            return


;==============================
;   long delay sub-routine
;==============================
long_delay:
            movlw    D'200'         ; w delay count
            movwf    J              ; J = w
jloop:
            movwf    K              ; K = w
kloop:
            decfsz   K,f            ; K = K-1, skip next if zero
            goto     kloop
            decfsz   J,f            ; J = J-1, skip next if zero
            goto     jloop
            return
;========================
;   send 2 nibbles in
;      4-bit mode
;========================
; Procedure to send two 4-bit values to Port-B lines
; 7, 6, 5, and 4. High-order nibble is sent first
```

```
; ON ENTRY:
;          w register holds 8-bit value to send
send8:
          movwf     store1             ; Save original value
          call      merge4             ; Merge with Port-B
; Now w has merged byte
          movwf     PORTD              ; w to Port-D
          call      pulseE             ; Send data to LCD
; High nibble is sent
          movf      store1,w           ; Recover byte into w
          swapf     store1,w           ; Swap nibbles in w
          call      merge4
          movwf     PORTD
          call      pulseE             ; Send data to LCD
          call      delay_125
          return
;==========================
;      merge bits
;==========================
; Routine to merge the 4 high-order bits of the
; value to send with the contents of Port-B
; so as to preserve the 4 low-bits in Port-B
; Logic:
;      AND value with 1111 0000 mask
;      AND Port-B with 0000 1111 mask
;      Now low nibble in value and high nibble in
;      Port-B are all 0 bits:
;           value = vvvv 0000
;           Port-B = 0000 bbbb
;      OR value and Port-B resulting in:
;                 vvvv bbbb
; ON ENTRY:
;      w contain value bits
; ON EXIT:
;      w contains merged bits
merge4:
          andlw     b'11110000'        ; ANDing with 0 clears the
                                       ; bit. ANDing with 1 preserves
                                       ; the original value
          movwf     store2             ; Save result in variable
          movf      PORTD,w            ; Port-D to w register
          andlw     b'00001111' ; Clear high nibble in Port-b
                                 ; and preserve low nibble
          iorwf     store2,w ; OR two operands in w
          return
;==========================
;   Set address register
;      to LCD line 2
```

```
;==========================
; ON ENTRY:
;           Address of LCD line 2 in constant LCD_2
line2:
        bcf       PORTE,E_line      ; E line low
        bcf       PORTE,RS_line     ; RS line low, setup for
                                    ; control
        call      delay_5           ; Busy?
; Set to second display line
        movlw     LCD_2             ; Address with high-bit set
        call      send8
; Set RS line for data
        bsf       PORTE,RS_line     ; RS = 1 for data
        call      delay_5           ; Busy?
        return
;==========================
;    Set address register
;         to LCD line 1
;==========================
; ON ENTRY:
;           Address of LCD line 1 in constant LCD_1
line1:
        bcf       PORTE,E_line      ; E line low
        bcf       PORTE,RS_line     ; RS line low, set up for
                                    ; control
        call      delay_5           ; busy?
; Set to second display line
        movlw     LCD_1             ; Address and command bit
        call      send8             ; 4-bit routine
; Set RS line for data
        bsf       PORTE,RS_line     ; Setup for data
        call      delay_5           ; Busy?
        return


;==========================
;    scroll to LCD line 2
;==========================
; Procedure to count the number of characters displayed on
; each LCD line. If the number reaches the value in the
; constant LCDlimit, then display is scrolled to the second
; LCD line. If at the end of the second line, then LCD is
; reset to the first line.
LCDscroll:
        incf      LCDcount,f        ; Bump counter
; Test for line limit
        movf      LCDcount,w
        sublw     LCDlimit          ; Count minus limit
        btfss     STATUS,Z          ; Is count minus limit = 0
```

```
            goto      scrollExit        ; Go if not at end of line
; At this point the end of the LCD line was reached
; Test if this is also the end of the second line
            movf      LCDline,w
            sublw     0x01              ; Is it line 1?
            btfsc     STATUS,Z          ; Is LCDline minus 1 = 0?
            goto      line2End          ; Go if end of second line
; At this point it is the end of the top LCD line
            call      line2             ; Scroll to second line
            clrf      LCDcount          ; Reset counter
            incf      LCDline,f         ; Bump line counter
            goto      scrollExit
; End of second LCD line
line2End:
            call      initLCD           ; Reset
            clrf      LCDcount          ; Clear counters
            clrf      LCDline
            call      line1             ; Display to first line
scrollExit:
            return


;==============================
;    LCD display procedure
;==============================
; Sends 20 characters from PIC buffer with address stored
; in variable bufAdd to LCD line previously selected
display20:
            call      delay_5           ; Make sure not busy
; Set up for data
            bcf       PORTA,E_line      ; E line low
            bsf       PORTA,RS_line     ; RS line high for data
; Set up counter for 20 characters
            movlw     D'20'
            movwf     count3
; Get display address from local variable bufAdd
            movf      bufAdd,w          ; First display RAM address
                                        ; to W
            movwf     FSR               ; W to FSR
getchar:
            movf      INDF,w            ; get character from display
                                        ; RAM
                                        ; location pointed to by file
                                        ; select register
            call      send8             ; 4-bit interface routine
; Test for 20 characters displayed
            decfsz    count3,f          ; Decrement counter
            goto      nextchar          ; Skipped if done
            return
```

```
nextchar:
        incf    FSR,f               ; Bump pointer
        goto    getchar

;===============================
;  first text string procedure
;===============================
storeMS1:
; Procedure to store in PIC RAM buffer the message
; contained in the code area labeled msg1
; ON ENTRY:
;           variable bufAdd holds address of text buffer
;           in PIC RAM
;           w register hold offset into storage area
;           msg1 is routine that returns the string characters
;           and a zero terminator
;           index is local variable that hold offset into
;           text table. This variable is also used for
;           temporary storage of offset into buffer
; ON EXIT:
;           Text message stored in buffer
;
; Store offset into text buffer (passed in the w register)
; in temporary variable
        movwf   index               ; Store w in index
; Store base address of text buffer in FSR
        movf    bufAdd,w            ; first display RAM address
                                    ; to W
        addwf   index,w            ; Add offset to address
        movwf   FSR                ; W to FSR
; Initialize index for text string access
        movlw   0                   ; Start at 0
        movwf   index               ; Store index in variable
; w still = 0
get_msg_char:
        call    msg1                ; Get character from table
; Test for zero terminator
        andlw   0x0ff
        btfsc   STATUS,Z            ; Test zero flag
        goto    endstr1            ; End of string
; ASSERT: valid string character in w
;         store character in text buffer (by FSR)
        movwf   INDF               ; store in buffer by FSR
        incf    FSR,f              ; increment buffer pointer
; Restore table character counter from variable
        movf    index,w            ; Get value into w
        addlw   1                   ; Bump to next character
        movwf   index               ; Store table index in
```

```
                                    ; variable
        goto    get_msg_char        ; Continue
endstr1:
        return

; Routine for returning message stored in program area
; Message has 10 characters
msg1:
        addwf   PCL,f               ; Access table
        retlw   'R'
        retlw   'e'
        retlw   'c'
        retlw   'e'
        retlw   'i'
        retlw   'v'
        retlw   'i'
        retlw   'n'
        retlw   'g'
        retlw   ':'
        retlw   0


;========================
;      blank buffer
;========================
; Procedure to store 20 blank characters in PIC RAM
; buffer starting at address stored in the variable
; bufAdd
blank20:
        movlw   D'20'               ; Setup counter
        movwf   count1
        movf    bufAdd,w            ; First PIC RAM address
        movwf   FSR                 ; Indexed addressing
        movlw   0x20                ; ASCII space character
storeit:
        movwf   INDF      ; Store blank character in PIC RAM
                          ; buffer using FSR register
        decfsz  count1,f  ; Done?
        goto    incfsr    ; no
        return            ; yes
incfsr:
        incf    FSR,f     ; Bump FSR to next buffer space
        goto    storeit


;================================================================
;                  communications procedures
;================================================================
; Initiazalize serial Port for 2400 baud, 8 bits, no parity,
; 1 stop
```

```
InitSerial:
        Bank1                        ; Macro to select bank1
; Bits 6 and 7 of Port-C are multiplexed as TX/CK and RX/DT
; for USART operation. These bits must be set to input in the
; TRISC register
        movlw    b'11000000'         ; Bits for TX and RX
        iorwf    TRISC,f             ; OR into TRISc register
; The asynchronous baud rate is calculated as follows:
;                        Fosc
;              ABR = ---------
;                       S*(x+1)
; where x is value in the SPBRG register and S is 64 if the high
; baud rate select bit (BRGH) in the TXSTA control register is
; clear, and 16 if the BRGH bit is set. For setting to 2400 baud
; using a 10 MHz oscillator at a slow baud rate, the formula
; is:
;
;          10,000,000   10,000,000
;          ---------- = ---------- = 2,403.84 (0.16% error)
;          64*(64+1)       4160
;
        movlw    spbrgVal ; Value in spbrgVal = 64
        movwf    SPBRG                ; Place in baud rate generator
; Setup value: 0010 0000 = 0x20
        movlw    0x20     ; Enable transmission and high baud
                          ; rate
        movwf    TXSTA
        Bank0             ; Bank 0
; Setup value: 1001 0000 = 0x90
        movlw    0x90     ; Enable serial Port-and continuous
                          ; reception
        movwf    RCSTA
;
        clrf     errorFlags          ; Clear local error flags
                                     ; register
        return
;==============================
;       transmit data
;==============================
; Test for Transmit Register Empty and transmit data in w
SerialSend:
        Bank0                        ; Select bank 0
        btfss    PIR1,TXIF           ; check if transmitter busy
        goto     $-1                 ; wait until transmitter is
not busy
        movwf    TXREG               ; and transmit the data
        return
;==============================
```

```
;       receive data
;================================
; Procedure to test line for data received and return value
; in w. Overrun and framing errors are detected and
; remembered in the variable errorFlags, as follows:
;       7  6  5  4  3  2  1  0   <== errorFlags
;       |- not used -- |  |  |___ overrun error
;                         |_____ framing error
SerialRcv:
        clrf    newData             ; Clear new data received
register
        Bank0                       ; Select bank 0
; Bit 5 (RCIF) of the PIR1 Register is clear if the USART
; receive buffer is empty. If so, no data has been received
        btfss   PIR1,RCIF           ; Check for received data
        return                      ; Exit if no data
; At this point data has been received. First eliminate
; possible errors: overrun and framing.
; Bit 1 (OERR) of the RCSTA register detects overrun
; Bit 2 (FERR( of the RCSTA register detects framing error
        btfsc   RCSTA,OERR          ; Test for overrun error
        goto    OverErr             ; Error handler
        btfsc   RCSTA,FERR          ; Test for framing error
        goto    FrameErr ; Error handler
; At this point no error was detected
; Received data is in the USART RCREG register
        movf    RCREG,w             ; get received data
        bsf     newData,7           ; Set bit 7 to indicate new
                                    ; data
; Clear error flags
        clrf    errorFlags
        return
;=========================
;     error handlers
;=========================
OverErr:
        bsf     errorFlags,0        ; Bit 0 is overrun error
; Reset system
        bcf     RCSTA,CREN          ; Clear continuous receive bit
        bsf     RCSTA,CREN          ; Set to re-enable reception
        return
;error because FERR framing error bit is set
;can do special error handling here - this code simply clears
; and continues
FrameErr:
        bsf     errorFlags,1        ; Bit 1 is framing error
        movf    RCREG,W             ; Read and throw away bad data
        return
```

```
;===============================================================
;                  local EEPROM data procedures
;===============================================================
; GPRs used in EEPROM-related code are placed in the common
; RAM area (from 0x70 to 0x7f). This makes the registers
; accessible from any bank.
;==============================
;     read local EEPROM
;==============================
; Procedure to read EEPROM memory
; ON ENTRY:
; Address of EEPROM memory location to read is stored in
; local register EEMemAdd
; ON EXIT:
; Read data in w
EERead:
        Bank2
        movf     EEMemAdd,W         ; EEPROM address
        movwf    EEADR              ; to read from
        Bank3
        bcf      EECON1,EEPGD       ; Point to Data memory
        bsf      EECON1,RD          ; Start read
        Bank2
        movf     EEDATA,W           ; Data to w register
        Bank0
        return


;==============================
;     write local EEPROM
;==============================
; Procedure to write data byte to EEPROM memory
; ON ENTRY:
; Address to write stored in local register EEMemAdd
; Data byte to write is in local register EEByte
EEWrite:
        Bank3
Wait2Start:
        btfsc    EECON1,WR          ; Wait for
        GOTO     Wait2Start         ; write to finish
        Bank2
        movf     EEMemAdd,w         ; Address to
        movwf    EEADR              ; SFR
        movf     EEByte,w           ; Data to
        movwf    EEDATA             ; SFR
        Bank3
        bcf      EECON1,EEPGD       ; Point to Data memory
        bsf      EECON1,WREN        ; and enable writes
; Disable interrupts. Can be done in any case
```

```
        bcf        INTCON,GIE
; Write special codes
        movlw      0x55              ; First code is 0x55
        movwf      EECON2
        movlw      0xaa              ; Second code is 0xaa
        movwf      EECON2
        bsf        EECON1,WR         ; Start write operation
        nop                          ; Time for write
        nop
; Test for end of write operation
wait2End:
        btfsc      EECON1,WR         ; Wait until WR clear
        goto       wait2End
;
; Re-enable interrupts if program uses interrupts
; If not, comment out next line
;       bsf        INTCON,GIE
;
        bcf        EECON1,WREN       ; Prevent accidental writes
        Bank0
        return


;===============================================================
        end              ; END OF PROGRAM
;===============================================================
```

14.4.3 I2CEEP Program

```
; File name: I2CEEP.asm
; Last revision: May 21, 2010
; Authors: Sanchez and Canton
; Processor: 16F877
;
; Description:
; Receive character data through RS-232 line and store in
; 24LC04B EEPROM IC, using the I2C serial protocol in the
; PIC's MSSP module. Received characters are echoed on
; the second LCD line. When <Enter> key is detected (code
; 0x0d), the text stored in EEPROM memory is retrieved and
; displayed on the LCD. On start-up the top LCD line displays
; the prompt: "Receiving:". At that time a message 'Rdy-' is
; sent through the serial line so as to test the connection.
;
; Default serial line setting:
;                2400 baud
;                no parity
;                1 stop bit
;                8 character bits
```

```
;
; Wiring:
; 24LC04B SDA line is wired to PIC RC4 (MSSP SDA)
; 24LC04B SCL line is wired to PIC RC3 (MSSP SCL)
; 24LC04B A0-A2 and WP lines are not used (GND)
;
; Program to uses 4-bit PIC-to-LCD interface.
; Code assumes that LCD is driven by Hitachi HD44780
; controller and PIC 16F977. Display supports two lines,
; each one with 20 characters. The length, wiring, and base
; address of each display line is stored in #define
; statements. These statements can be edited to accommodate
; a different set-up.
;
; WARNING:
; Code assumes 10 Mhz clock. Delay routines must be
; edited for a different clock. Clock speed also determines
; values for baud rate setting (see spbrgVal constant).
;
;===========================
;        16F877 switches
;===========================
; Switches used in __config directive:
;   _CP_ON          Code protection ON/OFF
; * _CP_OFF
; * _PWRTE_ON       Power-up timer ON/OFF
;   _PWRTE_OFF
;   _BODEN_ON       Brown-out reset enable ON/OFF
; * _BODEN_OFF
; * _PWRTE_ON       Power-up timer enable ON/OFF
;   _PWRTE_OFF
;   _WDT_ON         Watchdog timer ON/OFF
; * _WDT_OFF
;   _LPV_ON         Low voltage IC programming enable ON/OFF
; * _LPV_OFF
;   _CPD_ON         Data EE memory code protection ON/OFF
; * _CPD_OFF
; OSCILLATOR CONFIGURATIONS:
;   _LP_OSC         Low power crystal oscillator
;   _XT_OSC         External parallel resonator/crystal oscillator

; * _HS_OSC         High speed crystal resonator
;   _RC_OSC         Resistor/capacitor oscillator
; |                 (simplest, 20% error)
; |
; |_____ * indicates setup values presently selected

        processor       16f877              ; Define processor
```

```
        #include <p16f877.inc>
        __CONFIG _CP_OFF & _WDT_OFF & _BODEN_OFF & _PWRTE_ON &
_HS_OSC & _WDT_OFF & _LVP_OFF & _CPD_OFF

; __CONFIG directive is used to embed configuration data
; within the source file. The labels following the directive
; are located in the corresponding .inc file.
        errorlevel -302
; Suppress bank-related warning
;===============================================================
;                         M A C R O S
;===============================================================
; Macros to select the register banks
Bank0   MACRO               ; Select RAM bank 0
        bcf     STATUS,RP0
        bcf     STATUS,RP1
        ENDM


Bank1   MACRO               ; Select RAM bank 1
        bsf     STATUS,RP0
        bcf     STATUS,RP1
        ENDM


Bank2   MACRO               ; Select RAM bank 2
        bcf     STATUS,RP0
        bsf     STATUS,RP1
        ENDM


Bank3   MACRO               ; Select RAM bank 3
        bsf     STATUS,RP0
        bsf     STATUS,RP1
        ENDM
;=====================================================
;                 constant definitions
;   for PIC-to-LCD pin wiring and LCD line addresses
;=====================================================
#define E_line 1          ;|
#define RS_line 0         ;| => from wiring diagram
#define RW_line 2         ;|
; LCD line addresses (from LCD data sheet)
#define LCD_1 0x80        ; First LCD line constant
#define LCD_2 0xc0        ; Second LCD line constant
#define LCDlimit .20; Number of characters per line
#define  spbrgVal .64; For 2400 baud on 10 MHz clock
; Note: The constant that define the LCD display
;       line addresses have the high-order bit set
;       so as to meet the requirements of controller
;       commands.
```

```
; ============================================================
;           constants for I2C initialization
; ============================================================
;  I2C connected to 24LC04B EEPROM.
;  The MSSP module is in I2C MASTER mode.
#define LC04READ 0xa0      ; I2C value for read control byte
#define LC04WRITE 0xa1     ; I2C value for write control byte


; ============================================================
;                    General Purpose Variables
; ============================================================
;  Local variables
;  Reserve 20 bytes for string buffer
        cblock   0x20
        strData
        endc
;  Other data
        cblock   0x34      ; Start of block
        count1             ; Counter # 1
        count2             ; Counter # 2
        count3             ; Counter # 3
        J                  ; Counter J
        K                  ; Counter K
        bufAdd
        index
        store1             ; Local storage
        store2
;  For LCDscroll procedure
        LCDcount           ; Counter for characters per line
        LCDline            ; Current display line (0 or 1)
        Endc
;
;==============================
;       Common RAM area
;==============================
;  These GPRs can be accessed from any bank.
;  15 bytes are available, from 0x70 to 0x7f
        cblock   0x70
;  Communications variables
        newData            ; not 0 if new data received
        ascVal
        errorFlags
;  EEPROM-related variables
        EEMemAdd           ; EEPROM address to access
        EEByte             ; Data byte to write
        endc


; ==============================================================
```

```
;                        P R O G R A M
;=============================================================
        org             0        ; start at address
        goto    main
; Space for interrupt handlers
        org             0x08
main:
; Wiring:
;     LCD data to Port-D, lines 0 to 7
;     E line -> Port-E, 1
;     RW line -> Port-E, 2
;     RS line -> Port-E, 0
; Set PORTE D and E for output
; First, initialize Port-B by clearing latches
        clrf    STATUS
        clrf    PORTB
; Select bank 1 to TRIS Port-D for output
        Bank1
; TRIS Port-D for output. Port-D lines 4 to 7 are wired
; to LCD data lines. Port-D lines 0 to 4 are wired to LEDs.
        movlw   B'00000000'
        movwf   TRISD    ; and Port-D
; By default Port-A lines are analog. To configure them
; as digital code must set bits 1 and 2 of the ADCON1
; register (in bank 1)
        movlw   0x06     ; binary 0000 0110  is code to
                         ; make all Port-A lines digital
        movwf   ADCON1
; Port-B, lines are wired to keypad switches, as follows:
;   7 6 5 4 3 2 1 0
;   | | | | |_|_|_|_____ switch rows (output)
;   |_|_|_|_____ switch columns (input)
; rows must be defined as output and columns as input
        movlw   b'11110000'
        movwf   TRISB
; TRIS Port-E for output
        movlw   B'00000000'
        movwf   TRISE              ; TRIS Port-E
; Enable Port-B pullups for switches in OPTION register
        movlw   b'00001000'
        movwf   OPTION_REG
; Back to bank 0
        Bank0
; Initialize serial port for 2400 baud, 8 bits, no parity
; 1 stop
        call    InitSerial
; Test serial transmission by sending "RDY-"
        movlw    'R'
```

```
        call    SerialSend
        movlw   'D'
        call    SerialSend
        movlw   'Y'
        call    SerialSend
        movlw   '-'
        call    SerialSend
        movlw   0x20
        call    SerialSend
; Clear all output lines
        movlw   b'00000000'
        movwf   PORTD
        movwf   PORTE
; Wait and initialize HD44780
        call    delay_5 ; Allow LCD time to initialize itself
        call    initLCD ; Then do forced initialization
        call    delay_5
; Clear character counter and line counter variables
        clrf    LCDcount
        clrf    LCDline
; Set display address to start of first LCD line
        call    line1
; Store address of display buffer
        movlw   0x20
        movwf   bufAdd
; Display "Receiving:" message prompt
        call    blank20 ; Clear buffer
        movlw   0x00     ; Offset in buffer
        call    storeMS1 ; Store message at offset
        call    display20        ; Display message
; Start address of EEPROM
        clrf    EEMemAdd
; Setup for display in second line
        call    line2
        clrf    LCDline
        incf    LCDline,f ; Set scroll control for line 2
; Initialize I2C EEPROM operation
        call    SetupI2C ; Local procedure
;===============================================================
;           receive serial data, store, and display
;===============================================================
receive:
; Call serial receive procedure
        call    SerialRcv
; HOB of newData register is set if new data
; received
        btfss   newData,7
        goto    scanExit
```

```
; At this point new data was received.
        movwf    EEByte              ; Save received character
; Display character on LCD
        movf     EEByte,w ; Recover character
        call     send8               ; Display in LCD
        call     LCDscroll           ; Scroll at end of line
; Store character in EEPROM at location in EEMemAdd
        call     WriteI2C ; Local procedure
        incf     EEMemAdd,f          ; Bump to next EEPROM
; Check for <Enter> key (0x0d) and execute display function
        movf     EEByte,w ; Recover last received
        sublw    0x0d
        btfsc    STATUS,Z ; Test if <Enter> key
        goto     isEnter             ; Go if <Enter>
; Not <Enter> key, continue processing
scanExit:
        goto     receive             ; Continue
;=============================
;    display EEPROM data
;=============================
; This routine receives control when the <Enter> key is
; received.
; Action:
;        1. Clear LCD
;        2. Output is set to top LCD line
;        3. Characters stored in EEPROM are displayed
;           until 0x0d code is detected
isEnter:
        call     clearLCD
; Clear character counter and line counter variables
        clrf     LCDcount
        clrf     LCDline
; Read data from EEPROM memory, starting at address 0
; and display on LCD until 0x0d terminator
        call     line1
        clrf     EEMemAdd            ; Start at EEPROM 0
readOne:
        call     ReadI2C             ; Get character
; Store character
        movwf    EEByte              ; Save character
; Test for terminator
        sublw    0x0d
        btfsc    STATUS,Z            ; Test if 0x0d
        goto     atEnd               ; Go if 0x0d
; At this point character read is not 0x0d
; Display on LCD
        movf     EEByte,w            ; Recover character
; Display character on LCD
```

```
              call     send8              ; Display in LCD
              call     LCDscroll          ; Scroll at end of line
              incf     EEMemAdd,f         ; Next EEPROM byte
              goto     readOne
; End of execution
atEnd:
              goto     atEnd


;================================================================
;================================================================
;                  L O C A L     P R O C E D U R E S
;================================================================
;================================================================
;==========================
; init LCD for 4-bit mode
;==========================
initLCD:
; Initialization for Densitron LCD module as follows:
;    4-bit interface
;    2 display lines of 16 characters each
;    cursor on
;    left-to-right increment
;    cursor shift right
;    no display shift
;======================|
;    set command mode  |
;======================|
              bcf      PORTE,E_line       ; E line low
              bcf      PORTE,RS_line      ; RS line low
              bcf      PORTE,RW_line      ; Write mode
              call     delay_125          ; delay 125 microseconds
              movlw    0x28               ; 0 0 1 0 1 0 0 0 (FUNCTION
SET)
              call     send8              ; 4-bit send routine
; Set 4-bit mode command must be repeated
              movlw    0x28
              call     send8
              movlw    0x0e    ; 0 0 0 0 1 1 1 0 (DISPLAY ON/OFF)
              call     send8
              movlw    0x06    ; 0 0 0 0 0 1 1 0 (ENTRY MODE SET)
              call     send8
              movlw    0x14    ; 0 0 0 1 0 1 0 0 (CURSOR/DISPLAY
                               ; SHIFT)
              call     send8
              movlw    0x01    ; 0 0 0 0 0 0 0 1 (CLEAR DISPLAY)
              call     send8
              call     delay_5 ; Test for busy
              return
```

```
.;============================
;    procedure to clear LCD
;============================
clearLCD:
        bcf       PORTE,E_line      ; E line low
        bcf       PORTE,RS_line     ; RS line low
        bcf       PORTE,RW_line     ; Write mode
        call      delay_125         ; delay 125 microseconds
        movlw     0x01              ; 0 0 0 0 0 0 0 1
        call      send8
        call      delay_5  ; Test for busy
        return

;=======================
;   Procedure to delay
;    42 microseconds
;=======================
delay_125:
        movlw     .105              ; Repeat 105 machine cycles
        movwf     count1            ; Store value in counter
repeat:
        decfsz    count1,f          ; Decrement counter
        goto      repeat            ; Continue if not 0
        return                      ; End of delay

;=======================
;   Procedure to delay
;    5 milliseconds
;=======================
delay_5:
        movlw     .105              ; Counter = 105 cycles
        movwf     count2            ; Store in variable
delay:
        call      delay_125         ; Delay
        decfsz    count2,f          ; 40 times = 5 milliseconds
        goto      delay
        return                      ; End of delay
;=======================
;     pulse E line
;=======================
pulseE
        bsf       PORTE,E_line      ; Pulse E line
        nop
        bcf       PORTE,E_line
        return

;==============================
```

```
;   long delay sub-routine
;==============================
long_delay:
        movlw   D'200'              ; w delay count
        movwf   J                   ; J = w
jloop:
        movwf   K                   ; K = w
kloop:
        decfsz  K,f                 ; K = K-1, skip next if zero
        goto    kloop
        decfsz  J,f                 ; J = J-1, skip next if zero
        goto    jloop
        return
;========================
;   send 2 nibbles in
;      4-bit mode
;========================
; Procedure to send two 4-bit values to Port-B lines
; 7, 6, 5, and 4. High-order nibble is sent first
; ON ENTRY:
;        w register holds 8-bit value to send
send8:
        movwf   store1              ; Save original value
        call    merge4              ; Merge with Port-B
; Now w has merged byte
        movwf   PORTD               ; w to Port-D
        call    pulseE              ; Send data to LCD
; High nibble is sent
        movf    store1,w ; Recover byte into w
        swapf   store1,w ; Swap nibbles in w
        call    merge4
        movwf   PORTD
        call    pulseE              ; Send data to LCD
        call    delay_125
        return
;===========================
;       merge bits
;===========================
; Routine to merge the 4 high-order bits of the
; value to send with the contents of Port-B
; so as to preserve the 4 low-bits in Port-B
; Logic:
;       AND value with 1111 0000 mask
;       AND Port-B with 0000 1111 mask
;       Now low nibble in value and high nibble in
;       Port-B are all 0 bits:
;           value = vvvv 0000
;           Port-B = 0000 bbbb
```

```
;        OR value and Port-B resulting in:
;                     vvvv bbbb
; ON ENTRY:
;      w contain value bits
; ON EXIT:
;      w contains merged bits
merge4:
        andlw    b'11110000'        ; ANDing with 0 clears the
                                    ; bit. ANDing with 1 preserves
                                    ; the original value
        movwf    store2             ; Save result in variable
        movf     PORTD,w            ; Port-B to w register
        andlw    b'00001111'        ; Clear high nibble in Port-b
                                    ; and preserve low nibble
        iorwf    store2,w           ; OR two operands in w
        return
;===========================
;   Set address register
;       to LCD line 2
;===========================
; ON ENTRY:
;        Address of LCD line 2 in constant LCD_2
line2:
        bcf      PORTE,E_line       ; E line low
        bcf      PORTE,RS_line      ; RS line low, setup for
                                    ; control
        call     delay_5            ; Busy?
; Set to second display line
        movlw    LCD_2              ; Address with high-bit set
        call     send8
; Set RS line for data
        bsf      PORTE,RS_line      ; RS = 1 for data
        call     delay_5            ; Busy?
        return
;===========================
;   Set address register
;       to LCD line 1
;===========================
; ON ENTRY:
;        Address of LCD line 1 in constant LCD_1
line1:
        bcf      PORTE,E_line       ; E line low
        bcf      PORTE,RS_line      ; RS line low, set up for
                                    ; control
        call     delay_5            ; busy?
; Set to second display line
        movlw    LCD_1              ; Address and command bit
        call     send8              ; 4-bit routine
```

```
; Set RS line for data
        bsf       PORTE,RS_line     ; Setup for data
        call      delay_5           ; Busy?
        return

;===========================
;    scroll to LCD line 2
;===========================
; Procedure to count the number of characters displayed on
; each LCD line. If the number reaches the value in the
; constant LCDlimit, then display is scrolled to the second
; LCD line. If at the end of the second line, then LCD is
; reset to the first line.
LCDscroll:
        incf      LCDcount,f        ; Bump counter
; Test for line limit
        movf      LCDcount,w
        sublw     LCDlimit          ; Count minus limit
        btfss     STATUS,Z          ; Is count minus limit = 0
        goto      scrollExit        ; Go if not at end of line
; At this point the end of the LCD line was reached
; Test if this is also the end of the second line
        movf      LCDline,w
        sublw     0x01              ; Is it line 1?
        btfsc     STATUS,Z          ; Is LCDline minus 1 = 0?
        goto      line2End          ; Go if end of second line
; At this point it is the end of the top LCD line
        call      line2             ; Scroll to second line
        clrf      LCDcount          ; Reset counter
        incf      LCDline,f         ; Bump line counter
        goto      scrollExit
; End of second LCD line
line2End:
        call      initLCD           ; Reset
        clrf      LCDcount          ; Clear counters
        clrf      LCDline
        call      line1             ; Display to first line
scrollExit:
        return


;=============================
;    LCD display procedure
;=============================
; Sends 20 characters from PIC buffer with address stored
; in variable bufAdd to LCD line previously selected
display20:
        call      delay_5           ; Make sure not busy
; Set up for data
```

```
        bcf      PORTA,E_line    ; E line low
        bsf      PORTA,RS_line   ; RS line high for data
; Set up counter for 20 characters
        movlw    D'20'
        movwf    count3
; Get display address from local variable bufAdd
        movf     bufAdd,w ; First display RAM address to W
        movwf    FSR      ; W to FSR
getchar:
        movf     INDF,w   ; get character from display RAM
                          ; location pointed to by file select
                          ; register
        call     send8    ; 4-bit interface routine
; Test for 20 characters displayed
        decfsz   count3,f ; Decrement counter
        goto     nextchar ; Skipped if done
        return
nextchar:
        incf     FSR,f    ; Bump pointer
        goto     getchar


;===============================
;  first text string procedure
;===============================
storeMS1:
; Procedure to store in PIC RAM buffer the message
; contained in the code area labeled msg1
; ON ENTRY:
;        variable bufAdd holds address of text buffer
;        in PIC RAM
;        w register holds offset into storage area
;        msg1 is routine that returns the string characters
;        and a zero terminator
;        index is local variable that holds offset into
;        text table. This variable is also used for
;        temporary storage of offset into buffer
; ON EXIT:
;        Text message stored in buffer
;
; Store offset into text buffer (passed in the w register)
; in temporary variable
        movwf    index              ; Store w in index
; Store base address of text buffer in FSR
        movf     bufAdd,w ; first display RAM address to W
        addwf    index,w            ; Add offset to address
        movwf    FSR                ; W to FSR
; Initialize index for text string access
        movlw    0                  ; Start at 0
```

```
                movwf    index              ; Store index in variable
; w still = 0
get_msg_char:
                call     msg1               ; Get character from table
; Test for zero terminator
                andlw    0x0ff
                btfsc    STATUS,Z ; Test zero flag
                goto     endstr1            ; End of string
; ASSERT: valid string character in w
;         store character in text buffer (by FSR)
                movwf    INDF               ; store in buffer by FSR
                incf     FSR,f              ; increment buffer pointer
; Restore table character counter from variable
                movf     index,w            ; Get value into w
                addlw    1                  ; Bump to next character
                movwf    index              ; Store table index in
                                            ; variable
                goto     get_msg_char       ; Continue
endstr1:
                return

; Routine for returning message stored in program area
; Message has 10 characters
msg1:
                addwf    PCL,f              ; Access table
                retlw    'R'
                retlw    'e'
                retlw    'c'
                retlw    'e'
                retlw    'i'
                retlw    'v'
                retlw    'i'
                retlw    'n'
                retlw    'g'
                retlw    ':'
                retlw    0


;========================
;     blank buffer
;========================
; Procedure to store 20 blank characters in PIC RAM
; buffer starting at address stored in the variable
; bufAdd
blank20:
                movlw    D'20'              ; Setup counter
                movwf    count1
                movf     bufAdd,w ; First PIC RAM address
```

```
        movwf    FSR       ; Indexed addressing
        movlw    0x20      ; ASCII space character
storeit
        movwf    INDF      ; Store blank character in PIC RAM
                           ; buffer using FSR register
        decfsz   count1,f  ; Done?
        goto     incfsr    ; no
        return             ; yes
incfsr:
        incf     FSR,f     ; Bump FSR to next buffer space
        goto     storeit
```

```
;================================================================
;                    communications procedures
;================================================================
; Initialize serial port for 2400 baud, 8 bits, no parity,
; 1 stop
InitSerial:
        Bank1                  ; Macro to select bank1
; Bits 6 and 7 of Port-C are multiplexed as TX/CK and RX/DT
; for USART operation. These bits must be set to input in the
; TRISC register
        movlw    b'11000000'       ; Bits for TX and RX
        iorwf    TRISC,f           ; OR into TRISc register
; The asynchronous baud rate is calculated as follows:
;                      Fosc
;            ABR =   -------
;                     S*(x+1)
; where x is value in the SPBRG register and S is 64 if the high
; baud rate select bit (BRGH) in the TXSTA control register is
; clear, and 16 if the BRGH bit is set. For setting to 2400 baud
; using a 10 MHz oscillator at a slow baud rate, the formula
; is:
;
;          10,000,000    10,000,000
;          ---------- = ----------- = 2,403.84 (0.16% error)
;          64*(64+1)       4160
;
        movlw    spbrgVal ; Value in spbrgVal = 64
        movwf    SPBRG                 ; Place in baud rate generator
; Setup value: 0010 0000 = 0x20
        movlw    0x20                  ; Enable transmission and high
                                       ; baud rate
        movwf    TXSTA
        Bank0                          ; Bank 0
; Setup value: 1001 0000 = 0x90
        movlw    0x90                  ; Enable serial port and
                                       ; continuous reception
```

```
        movwf    RCSTA
;

        clrf     errorFlags ; Clear local error flags register
        return
;==============================
;       transmit data
;==============================
; Test for Transmit Register Empty and transmit data in w
SerialSend:
        Bank0              ; Select bank 0
        btfss    PIR1,TXIF        ; check if transmitter busy
        goto     $-1       ; wait until transmitter is not busy
        movwf    TXREG     ; and transmit the data
        return


;==============================
;       receive data
;==============================
; Procedure to test line for data received and return value
; in w. Overrun and framing errors are detected and
; remembered in the variable errorFlags, as follows:
;       7  6  5  4  3  2  1  0   <== errorFlags
;       |- not used -- |  |  |___ overrun error
;                         |_____ framing error
SerialRcv:
        clrf     newData ; Clear new data received register
        Bank0              ; Select bank 0
; Bit 5 (RCIF) of the PIR1 Register is clear if the USART
; receive buffer is empty. If so, no data has been received
        btfss    PIR1,RCIF          ; Check for received data
        return                      ; Exit if no data
; At this point data has been received. First eliminate
; possible errors: overrun and framing.
; Bit 1 (OERR) of the RCSTA register detects overrun
; Bit 2 (FERR( of the RCSTA register detects framing error
        btfsc    RCSTA,OERR         ; Test for overrun error
        goto     OverErr            ; Error handler
        btfsc    RCSTA,FERR         ; Test for framing error
        goto     FrameErr           ; Error handler
; At this point no error was detected
; Received data is in the USART RCREG register
        movf     RCREG,w            ; get received data
        bsf      newData,7          ; Set bit 7 to indicate new
data
; Clear error flags
        clrf     errorFlags
        return
;==========================
```

```
;       error handlers
;===========================
OverErr:
        bsf         errorFlags,0        ; Bit 0 is overrun error
; Reset system
        bcf         RCSTA,CREN          ; Clear continuous receive bit
        bsf         RCSTA,CREN          ; Set to re-enable reception
        return
;error because FERR framing error bit is set
;can do special error handling here - this code simply clears
; and continues
FrameErr:
        bsf         errorFlags,1        ; Bit 1 is framing error
        movf        RCREG,W             ; Read and throw away bad data
        return
;=============================================================
;                   I2C EEPROM data procedures
;=============================================================
; GPRs used in EEPROM-related code are placed in the common
; RAM area (from 0x70 to 0x7f). This makes the registers
; accessible from any bank.
;=============================
;     LIST OF PROCEDURES
;=============================
; SetupI2C   --  Initialize MSSP module for I2C mode
;                in hardware master mode
;                Configure I2C lines
;                Set slew rate for 100 Kbps
;                Set baud rate for 10 MHz
; WriteI2C   --  Write byte to I2C EEPROM device
;                Data is stored in EEByte variable
;                Address is stored in EEMemAdd
; ReadI2C    --  Read byte from I2C EEPROM device
;                Address stored in EEMemAdd
;                Read data returned in w register
;=============================
;     I2C setup procedure
;=============================
SetupI2C:
        Bank1
        movlw       b'00011000'
        iorwf       TRISC,f             ; OR into TRISC
; Setup MSSP module for Master Mode operation
        Bank0
        movlw       B'00101000'; Enables MSSP and uses appropriate
;  0  0  1  0  1  0  0  0   Value to install
;  7  6  5  4  3  2  1  0   <== SSPCON bits in this operation
;  |  |  |  |  |__|__|__|___ Serial port select bits
```

```
;   |   |   |   |                    1000 = I2C master mode
;   |   |   |   |                    Clock = Fosc/(4*(SSPAD+1))
;   |   |   |   |_____ UNUSED IN MASTER MODE
;   |   |   |_____ SSP Enable
;   |   |                            1 = SDA and SCL pins as serial
;   |   |_____ Receive overflow indicator
;   |                                0 = no overflow
;   |_____ Write collision detect
;                                    0 = no collision detected
          movwf    SSPCON   ; This is loaded into SSPCON
; Input levels and slew rate as standard I2C
          Bank1
          movlw    B'10000000'
;   1   0   0   0   0   0   0   0  Value to install
;   7   6   5   4   3   2   1   0  <== SSPSTAT bits in this operation
;   |   |   |   |   |   |   |   |   |___ Buffer full status bit READ ONLY
;   |   |   |   |   |   |   |   |_____ UNUSED in present application
;   |   |   |   |   |   |   |_____ Read/write information READ ONLY
;   |   |   |   |   |   |_____ UNUSED IN MASTER MODE
;   |   |   |   |   |_____ STOP bit READ ONLY
;   |   |   |   |_____ Data address READ ONLY
;   |   |   |_____ SMP bus select
;   |   |                                0 = use normal I2C specs
;   |_____ Slew rate control
;                                     0 = disabled
          movwf    SSPSTAT
; Set-up Baud Rate
; Baud Rate = Fosc/(4*(SSPADD+1))
;     Fosc = 10 MHz
;     Baud Rate = 24 for 100 Kbps
          movlw    .24                ; Value to use
          movwf    SSPADD   ; Store in SSPADD
          Bank0
          return


;=============================
;        I2C write procedure
;=============================
; Write one byte to I2C EEPROM 24LC04B
; Steps:
;                  1. Send START
;                  2. Send control. Wait for ACK
;                  3. Send address. Wait for ACK
;                  4. Send data. Wait for ACK
;                  5. Send STOP
; STEP 1:
WriteI2C:
          Bank1
```

```
        bsf      SSPCON2,SEN        ; Produce START Condition
        call     WaitI2C ; Wait for I2C to complete
; STEP 2:
; Send control byte. Wair for ACK
        movlw    LC04READ           ; Control byte
        call     Send1I2C          .; Send Byte
        call     WaitI2C ; Wait for I2C to complete
        btfsc    SSPCON2,ACKSTAT ; Check ACK bit to see if
                                    ; I2C failed, skip if not
        goto     FailI2C
; STEP 3:
; Send address. Wait for ACK
        Bank0
        movf     EEMemAdd,w         ; Load Address Byte
        call     Send1I2C           ; Send Byte
        call     WaitI2C ; Wait for I2C operation to complete
        Bank1
        btfsc    SSPCON2,ACKSTAT ; Check ACK Status bit to see
                                    ; If I2C failed, skip if not
        goto     FailI2C
; STEP 4:
; Send data. Wait for ACK
        Bank0
        movf     EEByte,w           ; Load Data Byte
        call     Send1I2C           ; Send Byte
        call     WaitI2C ; Wait for I2C operation to complete
        Bank1
        btfsc    SSPCON2,ACKSTAT ; Check ACK Status bit to see
                                    ; if I2C failed, skip if not
        goto     FailI2C
; STEP 5:
; Send STOP. Wait for ACK
        bsf      SSPCON2,PEN        ; Send STOP condition
        call     WaitI2C ; Wait for I2C operation to complete
; WRITE operation has completed successfully.
        Bank0
        return


;============================
;    I2C read procedure
;============================
; Procedure to read one byte from 24LC04B EEPROM
; Steps:
;       1. Send START
;       2. Send control. Wait for ACK
;       3. Send address. Wait for ACK
;       4. Send RESTART + control. Wait for ACK
;       5. Switch to receive mode. Get data.
```

```
;               6. Send NACK
;               7. Send STOP
;               8. Retrieve data into w register
; STEP 1:
ReadI2C
; Send RESTART. Wait for ACK
        Bank1
        bsf     SSPCON2,RSEN ; RESTART Condition
        call    WaitI2C ; Wait for I2C operation
; STEP 2:
; Send control byte. Wait for ACK
        movlw   LC04READ        ; Control byte
        call    Send1I2C        ; Send Byte
        call    WaitI2C ; Wait for I2C operation
; Now check to see if I2C EEPROM is ready
        Bank1
        btfsc   SSPCON2,ACKSTAT ; Check ACK Status bit
        goto    ReadI2C ; ACK Poll waiting for EEPROM
                                ; write to complete
; STEP 3:
; Send address. Wait for ACK
        Bank0
        movf    EEMemAdd,w      ; Load from address register
        call    Send1I2C        ; Send Byte
        call    WaitI2C         ; Wait for I2C operation
        Bank1
        btfsc   SSPCON2,ACKSTAT  ; Check ACK Status bit
        goto    FailI2C ; failed, skipped if successful
; STEP 4:
; Send RESTART. Wait for ACK
        bsf     SSPCON2,RSEN    ; Generate RESTART Condition
        call    WaitI2C         ; Wait for I2C operation
; Send output control. Wait for ACK
        movlw   LC04WRITE       ; Load CONTROL BYTE (output)
        call    Send1I2C        ; Send Byte
        call    WaitI2C         ; Wait for I2C operation
        Bank1
        btfsc   SSPCON2,ACKSTAT ; Check ACK Status bit
        goto    FailI2C ; failed, skipped if successful
; STEP 5:
; Switch MSSP to I2C Receive mode
        bsf     SSPCON2,RCEN    ; Enable Receive Mode (I2C)
; Get the data. Wait for ACK
        call    WaitI2C         ; Wait for I2C operation
; STEP 6:
; Send NACK to acknowledge
        Bank1
        bsf     SSPCON2,ACKDT   ; ACK DATA to send is 1 (NACK)
```

```
        bsf        SSPCON2,ACKEN     ; Send ACK DATA now.
; Once ACK or NACK is sent, ACKEN is automatically cleared
; STEP 7:
; Send STOP. Wait for ACK
        bsf        SSPCON2,PEN       ; Send STOP condition
        call       WaitI2C           ; Wait for I2C operation
; STEP 8:
; Read operation has finished
        Bank0
        movf       SSPBUF,W          ; Get data from SSPBUF into W
; Procedure has finished and completed successfully.
        return


;============================
;   I2C support procedures
;============================
; I2C Operation failed code sequence
; Procedure hangs up. User should provide error handling.
FailI2C
        Bank1
        bsf        SSPCON2,PEN       ; Send STOP condition
        call       WaitI2C           ; Wait for I2C operation
fail:
        goto       fail

; Procedure to transmit one byte
Send1I2C
        Bank0
        movwf      SSPBUF            ; Value to send to SSPBUF
        return

; Procedure to wait for the last I2C operation to complete.
; Code polls the SSPIF flag in PIR1.
WaitI2C
        Bank0
        btfss      PIR1,SSPIF        ; Check if I2C operation done
        goto       $-1               ; I2C module is not ready yet
        bcf        PIR1,SSPIF        ; I2C ready, clear flag
        return


;==============================================================
```

Chapter 15

Stepper Motors

15.1 Description and Operation

A simple DC motor rotates when a voltage is applied to its terminals; therefore its control is quite simple. Stepper motors, on the other hand, convert electrical pulses into discrete units of rotational movement, which can be controlled independently and without a feedback mechanism. Typically, the *shaft* or *rotor* is toothed, while the *stator* contains several windings that are energized in a specific order. The electromagnetic attraction of the windings force the alignment of the toothed rotor thus producing rotation. Figure 15-1 is a diagram of a stepper motor with eight windings and six teeth on the rotor.

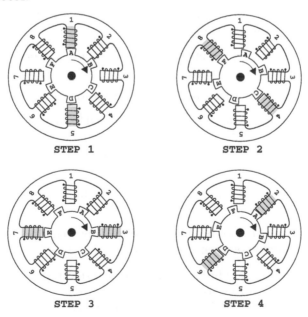

STEP 1 STEP 2

STEP 3 STEP 4

Figure 15-1 Cross-Section of a Typical Stepper Motor.

The motor of Figure 15-1 belongs to a group called *variable-reluctance motors*. Permanent magnet and hybrid stepper motors are also common. We have chosen the VR type for the illustration due to its simplicity.

Step 1 in Figure 15-1 shows that windings labeled 1 and 5 are initially energized, forcing the alignment of rotor teeth A and D. In Step 2, stator windings 2 and 6 are energized, forcing rotor teeth C and F into alignment and rotating the rotor clockwise. Steps 3 and 4 complete the sequence. At the end of Step 4, the rotor would have turned 45 degrees, which means that the rotor moves 5.625 degrees per pulse of the stator winding pairs and that, in this case, one complete revolution of the rotor requires thirty-two individual pulses.

Rotation of a stepper motor requires that the electrical pulses applied to the stator windings follow a definite sequence. In the motor of Figure 15-1, the sequence for clockwise rotation consists of applying current to the windings labeled 1-5, 4-8, 3-7, and 2-6, in that order. If the pulses were applied in some other order, the rotation of the motor would be different or would completely fail. By the same token, by applying the pulses in some other order the motor can be made to rotate in a counterclockwise direction. The speed of rotation of a stepper motors is determined by the frequency of the pulses and the amount of rotation by the number of pulses. For example, the motor in Figure 15-1 could be made to rotate 180 degrees (one-half revolution) by applying sixteen pulses only. These controls make stepper motors powerful devices for use in embedded systems and robotics.

The following are the fundamental characteristics of stepper motors in general:

- The motor has full torque at stand-still condition if the windings are energized.

- The rotation angle of the motor is determined by its design and by the number and sequence of the applied pulses.

- Positioning error of a stepper motor is in the range of 3 to 5 percent. This error does not accumulate from step to step.

- Stepper motors have long lives because they have no contact brushes to wear out.

- The position of a stepper motor can be determined from the number of input pulses applied. This feature, called *open-loop operation*, means that motor control is simple and straightforward because no feedback signals or optical senders are required.

- Stepper motors have good response at start time and can be rapidly stopped or reversed.

- Because stepper motors do not have brushes, they do not produce electrical arcs, which are undesirable or even dangerous.

Stepper motors are a good choice in applications that require control over motor speed, angle of rotation, direction of rotation, position, or synchronization. They find frequent use in robotics, office equipment such as printers and fax machines, in floppy and hard disk drives, in medical equipment, in computer control of machine tools (CNC), and in automobiles.

15.1.1 Stepper Motor Types

Although there are several possible classifications of stepper motors, the hardware design falls into one of three main types:

Variable Reluctance

This type, which has been around the longest time, consists of a soft core, which is often toothed, and a stator containing several electromagnets. The term *variable reluctance* refers to the principle that maximum attraction occurs between poles with minimum gaps. In Figure 15-1, which depicts a variable reluctance motor, you can see that in each consecutive step the stator windings closest to the rotor poles, in a clockwise direction, are energized in sequence.

Permanent Magnet

These are often referred to as *tin can motors*. This type, which is the cheapest to build and most common one, consists of a permanent magnet rotor without teeth. Figure 15-2 shows is a schematic drawing of a *permanent magnet type stepper motor*.

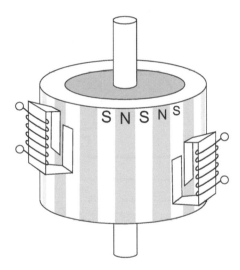

Figure 15-2 Schematic of a Permanent Magnet Stepper Motor.

In the motor of Figure 15-2, the magnetic fields in the rotor are in line with the motor shaft and the rotor does not have teeth. The rotor magnetic fields alternate north and south. As the windings in the stator are energized, the poles in the rotor are attracted to the opposite poles in the stator and the motor rotates.

Hybrid

Hybrid stepper motors use a combination of the variable reluctance and permanent magnet schemes. The rotor in the hybrid motor is both toothed and magnetized. The teeth provide a path for the magnetic flux, which increases the holding power and the magnetic characteristics of the motor. Because of their more complex construction, hybrid motors are considerably more expensive than permanent magnet types. At the same time, they have better resolution, and higher torque and speed.

15.1.2 Unipolar Stepper Motors

In electronics, a *pole* (normally labeled North or South) is a region where magnetic flux density is concentrated. Both the rotor and the stator of stepper motors have poles. Similar poles repel each other (NN or SS) and opposite poles attract (NS or SN). Thus, magnetic attraction of unlike poles will make the rotor move until its north poles are located opposite the stator south poles, and vice versa.

Although stepper motors can be cataloged in several ways, from a programming and control viewpoint the most useful classification is by the number of poles. *Unipolar stepper motors* have two windings per phase, one for the direction of each magnetic field. The magnetic poles can be reversed by selecting which winding is energized. By not having to change the direction of the magnetic field, the circuit of a unipolar stepper motor is very simple. Figure 15-3 depicts a unipolar stepper motor.

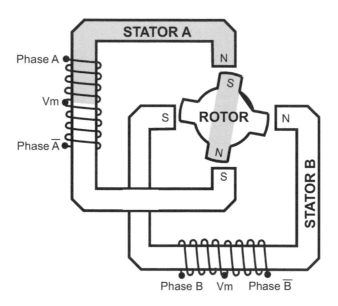

Figure 15-3 Unipolar Stepper Motor Schematics.

Note that the unipolar motor in Figure 15-3 has two windings per phase, one for each direction of the magnetic field. The center tap (labeled Vm) is connected to a common line. This arrangement allows changing the magnetic pole of the winding without reversing the direction of the current flow. The two-phase common line is sometimes joint internally, in which case the motor will have five wires. If the common lines are separate for each winding, then the motor will have six wires. In the illustration, the top half of phase A is shown activated (light gray color). The lower half (labeled phase bar A) is energized separately. Similarly, the center tap creates two separate phases in stator B, labeled phase B and phase bar B. The presence of two phases in each winding explains why unipolar motors are sometimes referred to as *four-phase motors*, while in reality they have only two phases.

15.1.3 Determining Unipolar and Bipolar Wiring

It is possible to determine the wiring of a unipolar or bipolar stepper motor with either four, five, or six leads. The only instrument needed for this test is an *ohmmeter*.

Four-Wire Motor

A four-wire motor is always a *bipolar motor*. The two windings can be easily identified with an ohmmeter: two wires that show *finite resistance* (closed circuit) are the ends of a winding.

Six-Wire Unipolar Motor

STEP 1: Find the wires belonging to each of the two windings by measuring resistance between wires. There will be two groups of three wires that show finite resistance. Label these groups of three wires as group A and group B.

STEP 2: Two wires in group A will show higher resistance than any other combination. These are the end leads of the windings and should be labeled A and B. The third wire in the group is the common tap and should be labeled VmA. The common lead will show one-half of the resistance when tested against wires A or B.

STEP 3: Similarly, two wires in group B will show higher resistance than any other combination. These are the end leads of the windings and should be labeled C and D. The third wire in the group is the common tap and should be labeled VmB.

STEP 4: Wires A and B should show infinite resistance when tested against wires C, D, or VmB. By the same token, wires C and D should show *infinite resistance* (open circuit) when tested against wires A, B, or VmA.

Five-Wire Unipolar

STEP 1: Find a single wire that shows finite resistance with all the other four wires in the motor. This is the common lead and should be labeled Vm.

STEP 2: Find two other wires, excluding Vm, that show finite resistance with each other and label them A and B, respectively.

STEP 3: Similarly, the two remaining wires will show finite resistance when tested with each other. They should be labeled C and D.

STEP 4: Wires C and D should show infinite resistance (open circuit) when tested against wires A or B. Wires A, B, C, and D should show finite resistance (closed circuit) when tested against Vm.

Figure 15-4 shows the wiring diagram for six- and five-wire unipolar motors.

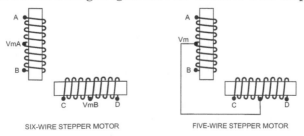

SIX-WIRE STEPPER MOTOR   FIVE-WIRE STEPPER MOTOR

Figure 15-4 Six- and Five-Wire Stepper Motors.

Some unipolar stepper motors are designed to have two individual windings for each stator phase. These motors, sometimes called *bifilar motors*, have two common leads in each winding and the motor has a total of eight wires. The advantage of this design is that these motors can be operated as unipolar or bipolar devices.

15.1.4 Bipolar Stepper Motors

Bipolar motors have a single winding per phase. The motor operates by reversing the current flow in the winding, thus reversing the magnetic poles. Because there is a single winding for each phase, there is no center tap and the motor has four wires. Figure 15-5 shows the schematic of a bipolar stepper motor.

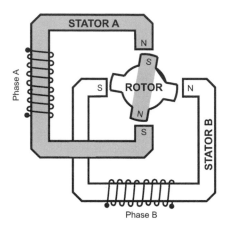

Figure 15-5 Schematic of a Bipolar Stepper Motor.

The entire winding of a bipolar motor is activated in each cycle, while only half the winding is activated in a unipolar motor. This results in bipolar motors producing more torque than unipolar motors of the same size. On the other hand, because the current flow in the winding must be reversed, bipolar motors require more complex control circuitry. The most common circuit used in reversing the polarity of bipolar motors is known as an *H bridge*. One H bridge is required for each winding in a bipolar motor. The H bridge is discussed later in this chapter.

15.2 Stepper Motor Controls

Stepper motors can be controlled by means of general electronic circuitry, by ICs specially designed for this purpose, or by signals from microprocessors or microcontrollers. Very often, a stepper motor circuit combines all three types of components. The complexity of the control circuitry of a stepper motor depends on the type of motor and on the degree of control required by the application. Unipolar motors, although less efficient, are easier to control than bipolar ones. Also, an application that is limited to turning on and off a stepper motor, at a fixed speed and rotation direction, would be simpler to design and build than one that must vary the speed, change the direction, and select among several modes of operation. Figure 15-6 is a photograph of a breadboard circuit for controlling a stepper motor.

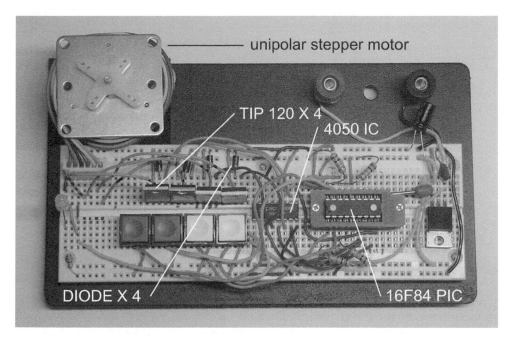

Figure 15-6 Breadboard Circuit for a Unipolar Motor.

The circuit in Figure 15-6 uses a 16F84A PIC microcontroller to handle the signals to the unipolar stepper motor. The 4050 IC is a noninverting hex buffer that serves to drive the higher current loads of the TIP 120 Darlington transistors. The four diodes are an additional safety to protect the circuit from backflows. The two white pushbuttons, when held down, activate slow rotation in the forward and reverse directions, respectively. The dark gray pushbuttons do the same in fast rotation. The circuit and software are developed and explained later in this chapter.

15.2.1 Stepping Modes

The sequence in which the windings or winding sections are activated in a unipolar or bipolar stepper motor is called the *stepping mode*. Three stepping modes are most common, although a fourth mode, called *microstepping*, is occasionally used. The general characteristics are as follows.

Wave Drive Mode

In this mode a single winding is energized at a time. The wave mode is easy to implement in the control hardware but provides significantly less than the rated torque of the motor. If the four windings of unipolar motors are labeled A, B, C, and D, then in the wave drive mode the windings are energized in the sequence: A → B → C → D. Figure 15-7 shows a unipolar motor in the first cycle of a wave mode sequence.

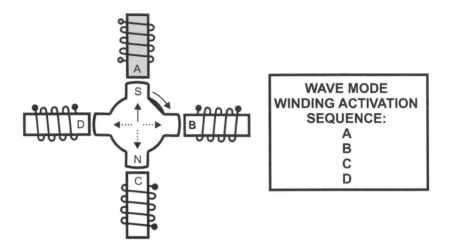

Figure 15-7 Unipolar Motor in Wave Mode Activation Sequence.

The straight arrows in Figure 15-7 show the various positions of the rotor as the windings are activated in sequence. The solid arrow represents the original position of the rotor and the dashed arrows the subsequent positions as windings B, C, and D are turned on. The activation sequence in Figure 15-7 results in a clockwise rotation. If the windings were activated in the order A → D → C → B then the motor would rotate counterclockwise. Note that in the wave mode, only 25 percent of the motor's total windings are energized at any time.

Full Step Mode

In the *full step mode* two windings are activated during each sequence step. Because the rotor is equally attracted to the two active windings, it takes an intermediate position. If the four windings of a unipolar motor are labeled A, B, C, and D, then in the full step drive mode the windings are energized in the sequence: AB → BC → CD → DA. Figure 15-8 shows a unipolar motor in the first cycle of a full step mode sequence.

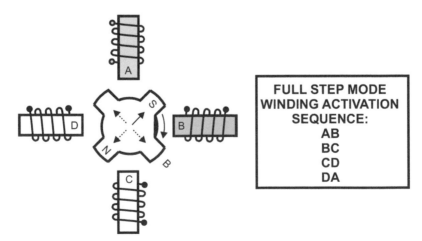

Figure 15-8 Unipolar Motor in Full Step Mode Activation Sequence.

As in Figure 15-7, the straight arrows show the various positions of the rotor as the windings are activated in sequence. The activation sequence in Figure 15-8 results in a clockwise rotation. If the windings were activated in the order AD → DC → CB → BA then the motor would rotate counterclockwise. In full step mode, 50 percent of the motor's total windings are energized at a time, resulting in higher efficiency than in the wave mode.

Half Step Mode

The half step mode combines the winding excitation sequence of the wave mode and the full step mode in order to double the number of steps in each full revolution of the rotor. This means that single winding and double winding excitation are generated alternatively, as shown in Figure 15-9.

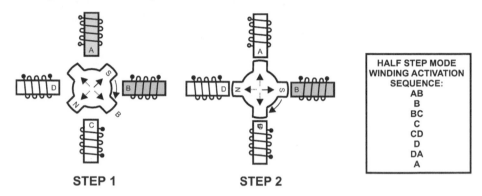

STEP 1   **STEP 2**

Figure 15-9 Unipolar Motor in Half-Step Mode Activation Sequence.

Here again the straight arrows show the various positions of the rotor as the windings are activated in sequence. In the half-step mode, the motor rotates one-half the angle during each activation cycle of the windings. The activation sequence in Figure 15-9 results in a clockwise rotation. By reversing the order of the cycles, the motor is made to rotate counterclockwise.

Microstepping

Operating a stepper motor in any of the three modes discussed in this section results in a jerkiness that varies according to the number of steps. This jerkiness is a direct result of the abrupt changes in current magnitude or direction, which apply torque to the rotor abruptly, as shown by the square wave current changes in Figure 15-10.

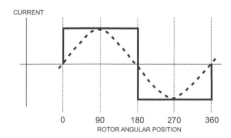

Figure 15-10 Current Transitions in Stepping.

With the waveform shown in Figure 15-10, as the voltage rises abruptly, so does the attraction of the stator windings on the motor poles. As the voltage declines abruptly at the end of each step, the attraction is instantly turned off. This results in the less-than-smooth movement sometimes associated with stepper motors. The square wave in Figure 15-10 approximately represents the level of attraction of the windings on the rotor at various positions in the step cycle.

Microstepping reduces or eliminates this effect by varying the current in the windings so as to produce a uniform attraction on the rotor throughout the step cycle. This can be achieved by modifying the current applied to the windings so as to follow the sine waveform shown by the dotted line in the illustration. In this case, the attraction on the rotor is uniform throughout the step cycle and the motor rotates smoothly. Because microstepping requires control over the current levels, it results in more complex circuitry, more sophisticated hardware, and more complex programming. Microstepping is covered in detail in Chapter 19.

Chapter 16

Stepper Motor Circuit Components

16.1 Circuit Elements

Over the years, stepper motor hardware and software have proliferated and evolved. Today there are scores, if not hundreds, of dedicated circuits and components that are intended for stepper motor applications. In addition, circuit designers have found ways of utilizing elements that were not originally designed for stepper circuits but that have resulted useful or economical. In this section we cover the most used and useful circuit elements for stepper motor driving and control. In this selection we keep in mind the book's intermediate scope as well as its focus on PIC microcontrollers. The result is that we have excluded some interesting circuits because of their complexity or limited use or because they are based on controllers or not covered in the text.

In general, microcontroller-based stepper motor circuits perform three functions:

1. **The control function**. The component or components in this specialization area of the circuit provide the signals for operating one or more stepper motors. The *control phase* is a command function and the circuit elements are usually conventional input and output components such as switches, sensors, potentiometers, keypads, and feedback devices such as LEDs and LCDs.

2. **The translator function.** Components of this function can be a microcontroller, a specialized IC, or a series of individual hardware components. The translator reads the input commands from the control stage of the circuit and generates the necessary unipolar or bipolar control signals for activating the driver function. The actions to be performed include, but are not limited to, forward or reverse rotation, speed control, stop and resume commands, and stepping mode sequence. In this book we consider only microcontrollers in the translator function. Dedicated ICs that include both translator and driver functions are classified as *drivers*.

3. **The driver function**. Microcontrollers can carry currents up to 20 mA. Because most motors require more power, the driver stage contains the hardware that supplies this power. In addition, some circuits must manipulate current polarity and intensity that are produced by the driver. In summary, the driver receives the signals generated by the translator and converts them into pulses of the adequate

voltage, current, and polarity to drive the stepper motor. The driver function can also be called the power stage of the circuit or the power driver function.

These three stages are not neatly delimited. Often, driver IC performs some functions that are typically associated with the translator stage. For example, the L297/298 IC pair, presented in Chapter 18, is a two-chip driver that includes functions typically associated with the translator. In such cases we have classified circuit components according to their principal function or to some didactical convenience.

16.1.1 Input, Output, and Feedback

We can imagine a circuit that provides no control functions. For example, a device that turns on a motor when power is applied and turns it off when power is cut. But such a simplistic operation would be unusual. In a typical circuit there will be components whose function is to turn on and off the various modes and execute the different commands and controls. Others will provide information regarding possible controls or the state of modes or devices. Finally, a third group of components can be used to furnish feedback information.

All conventional, common, and specialty circuit elements can provide input. These include toggle and pushbutton switches, jumpers, potentiometers of various designs, sensors, keypads, keyboards, or a computer connected to the circuit. Output and feedback are also provided by off-the-shelf devices such as LEDs, Seven-Segment LEDs, bar graph displays, liquid crystal displays (LCDs), or even full screens. In most circuits these devices will be connected to the input and output lines of the translator, most often a microcontroller, which will monitor and drive these elements. The applications and programming of most input devices used in motor circuits were discussed in Chapter 9.

16.2 Translator

On the Web and in the literature, there are many circuits that both control and drive stepper motors. Typically these translator/drivers are not programmable devices; they either originated at a time when microcontrollers were rare and expensive or were intended for circuits that do not require programmable logic. Today the abundance of programmable ICs and their low prices make it difficult to justify a stepper motor circuit of any complexity that does not include a microcontroller or microprocessor. In any case, because this book is also about programming, we focus on systems with a *programmable translator stage*, specifically using the Microchip microcontrollers (PIC). We classify ICs that include both translator and driver functions as drivers and reserve the term "translator" for programmable devices.

16.2.1 PIC Microcontroller as a Translator

Which PIC we select as a motor controller translator depends mostly on the circuit characteristics and the application. For example, in circuits that use microstepping (covered in detail later in the book), it is reasonable to select a PIC that provides *pulse width modulation* (PWM) functions. Although microstepping can be achieved in devices without PWM, it is the availability of PWM hardware that makes it easiest to vary the current sent to the driver or power stage. Or in a circuit that relies on analog input, such as a potentiometer for controlling motor speed, a microcontroller with an analog-to-digital module will be quite convenient.

At this time, the simplest and least expensive PIC that can be practically used for microstepping applications is the 16F684. This is a 14-pin, 8-bit device that contains an internal oscillator thus making external clocking unnecessary. The pulse width modulation is provided by an enhanced capture, compare, PMW module. The PWM element is 10 bits wide with one, two, or four output channels with a maximum frequency of 20 KHz. The IC also provides analog-to-digital conversions (A/D module), which is convenient for driving analog-based motor speed control devices. The most important limitation of the 16F684 is that there are only twelve I/O channels available; six channels are mapped to port A and six to port C. A complex application with multiple input and output requirements may run out of port lines in a 16F684. Several other mid-range PICs provide one or more PWM modules, in addition to more I/O lines and other specialized functions. For example, the 16F87x line has twenty eight or forty pins, three or five ports, and two PWM modules. In this line, the forty-pin 16F877 has been a popular favorite over the years.

Many other basic and mid-range PICs are suitable for use in the simpler stepper motor circuits. Controlling a stepper motor directly with a PIC microcontroller, or using a PIC to send commands to a dedicated motor controller IC, are quite popular and common techniques. In these cases the only requirement in the PIC is the availability of sufficient ports for the input and output lines needed by the circuit. Although the 16F84 has been deprecated by Microchip, it is , by far, the most popular PIC used as a stepper motor controller. A large percentage of the circuits available on the Web and in the popular literature uses the 16F84 or the 16F84A. However, circuits that require more complex operations, such as driving several input and output devices, communicating with an LCD, operating a keypad, or driving several motors, usually require more powerful PICs. In these cases, the 16F877 is often a suitable alternative.

16.3 Translator/Drivers

There are many dedicated ICs, usually called *stepper motor controllers*, that provide translator as well as motor driving functions. These devices are sometimes referred to as "translators" in the specialized literature. The advantage of using these ICs is that they take care of generating the required signals in the appropriate sequence, thus offloading this task from the microcontroller. It is easier for the PIC application to set a single line low or high to select between full-step or half-step modes, than to manipulate several lines to produce the required sequence of steps. In this section we list and describe some of the more popular and useful stepper motor controllers. Many others are available that further simplify specific circuits or that provide functions usually furnished during the driver stage. The listed controllers are a mere sampling.

16.3.1 UCN 5804

This translator/driver is one of the first of its kind and still quite popular. It is furnished as a sixteen-pin IC intended as a driver for unipolar stepper motors. Its maximum current rating (which is one of its greatest limitations) is 1.25 A per phase at 35V. The IC will drive four output lines with continuous output current. The logic includes enabling output, direction control, half-step and full-step modes, and a step input line that is typically driven by a clock IC or a microcontroller. Thermal protection disables the IC when chip temperature is excessive. Internal fly-back diodes provide protection against transients, although external diodes are often included as a safeguard. One of the circuits in Chapter 17 is based on the UCN 5804. Figure 16-1 is a schematic of the UCN 5804.

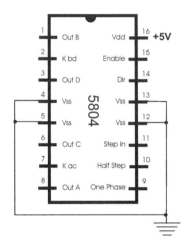

Figure 16-1 The UCN 5804 Stepper Motor Controller Pinout.

Although the 5804 is no longer in production, it is still readily available in the United States. The current limitation of 1.25 A has led to other designs in which the driver is brought outside the chip. The Septronics StepGenie is a popular substitute for the 5804. In the StepGenie, the driver consists of four externals HEXFETs that allow a current rating of 15 A.

16.3.2 L297

Another factor in the demise of the 5804 is the fact that the IC can only drive a unipolar motor. A more flexible circuit design is to use a chip, such as the L297, which is also compatible with bipolar drivers. Because unipolar motors can be driven in bipolar mode (with some possible gain in power), this approach is often preferable.

Typically, the L297 receives command signals from a microcontroller and furnishes all the necessary drive signals to the power element. In driving bipolar motors, it is usually combined with the L298 or the L293E, which are dual, full-bridge drivers. A quad Darlington array or HEXFETs can also be used in the power stage. The input signals into the L297 are step (labeled CLOCK), direction, half-step or full-step, enable, and a reset line. The ENABLE line is active high and can be used to turn on and off the chip's output. There is also an output line labeled HOME that signals that the IC is in its initial state.

A SYNC line is an output for the chip's *chopper oscillator* (choppers are covered in Chapter 19) and can also be used to input an external clock signal, although the L297 has its own oscillator. The motor output lines are labeled A, B, C, and D. Several other pin functions are also present, including two inhibit lines, labeled INH1 and INH2, which are active low controls for the drivers. INH1 inhibits the A and B phases, and INH2 the C and D phases.

The minimal circuit configuration for the L297 is shown in Figure 16-2.

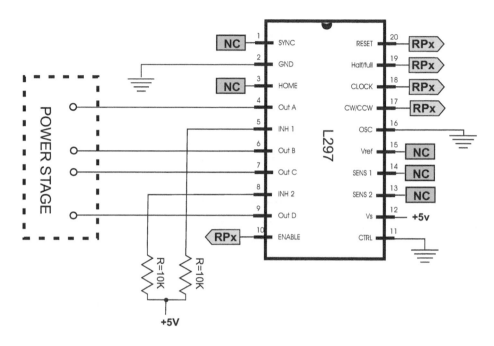

Figure 16-2 Minimal Circuit Wiring for the L297 IC.

In Figure 16-2 the lines labeled RPx are possible connections to a microcontroller port. The lines labeled NC are not connected in this minimal circuit configuration and are left floating. In other circuits, shown later in this chapter, the unconnected lines of Figure 16-2 are wired to pins in the driver stage. Two lines inhibit control of the two driver stages. These are labeled INH1 and INH2 in the diagram. Because these lines are active low, they must be held high in the minimal configuration. Other unused lines are either left floating or brought to ground, as shown in the illustration.

Although the L297 is often employed in simple drivers, it is actually most useful when the circuit takes advantage of all the capabilities of the IC. In one application the L297 can be configured in a "chopper" scheme, consisting of a closed-loop feedback system. This action is based on the fact that bipolar motors require a high current at the start of the step, but at some point in the cycle this current can be reduced to a minimum until the next step begins. The L297 accomplishes the chopping function by comparing the current through the motor coil with a reference voltage on its pin 15 (labeled Vref in Figure 16-2). When the coil current exceeds the reference value it is "chopped off" for the remainder of the power step. A trimmer potentiometer is included in the circuit so as to adjust the reference voltage to provide maximum torque with minimal waste of power. Later in this chapter we develop a complete circuit that uses the L297 in this manner

16.3.3 EDE1204

This is a bipolar controller somewhat different from the L297. The EDE1204 IC is furnished in an 18-pin package that provides both external controls and self-clocking. Figure 16-3 is a functional diagram of the 1204.

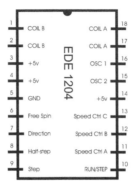

Figure 16-3 EDE 1204 Stepper Motor Controller Pinout.

The full-step run mode is activated by setting pin 10 low. In this mode, the direction pin (pin 7) makes the motor rotate clockwise or counterclockwise; the half-step control can be used to double the resolution, and speed control pins A, B, and C (pins 11, 12 and 13) allow selecting eight possible speeds. This appears to result in sixteen possible speeds, eight in full-step mode and eight in half-step mode. In reality, three speeds in full-step mode are duplicated in the half-step mode, so the total number of different speeds is actually thirteen. When the run pin is set high, then the 1204 goes into the step mode, and the motor speed is determined with the clock signal received on the step pin (pin 9). In the step mode the direction and half-step pins continue to be active. The free spin pin (pin 6), which is active low, serves to deactivate both motor coils. This results in a suspension of the breaking effect that is characteristic of stepper motors. Another interesting feature is that the IDE 1204 can change the stepping rate while the motor is running.

The self-clocking feature of the 1204, which results in speed control without an external clock signal, allows for simplification of some simple stepper motor circuits. For example, it is possible to provide speed selection and rotation direction by directly reading input devices, thus eliminating the microcontroller element from the circuit. Circuits in which the microcontroller is overburdened by other tasks can also benefit from this simplification.

16.3.4 SLA7060 and SLA7024

The SLA7060M, from Allegro Microsystems, provides both control and drive functions for two-phase, unipolar stepper motors. Their principal feature is that they include the translator and the driver in a single IC. The SLA7060 can use pulse width modulation (PWM) to control the output current, thus supporting microstepping. Microstepping circuits are covered in Chapter 19. The SLA7024M, SLA7026M, and SMA7029M, also from Allegro, are also two-phase, unipolar stepper motor controller/driver ICs, which include NMOS FETs that support high-current and high-voltage output. The differences in the three ICs are current ratings and package styles.

16.4 Power Driver

The *driver phase* of a stepper motor circuit provides power to the motor windings at the necessary current. The specific components used in this stage depend on the motor characteristics as well as the circuit design. The driver phase components are determined by the motor type: unipolar or bipolar.

16.4.1 Unipolar Drivers

Unipolar circuits require simple drivers consisting of transistors that serve as current amplifiers, and possibly diodes to prevent damaging backflows. The most common transistors used in unipolar drivers are Darlington transistors. Internally the Darlington consists of two bipolar NPN transistors connected back-to-back so the current output of the first one is further amplified by the second one. This double-stage design is called a *Darlington pair*. Darlington pairs come in an array packaged in an integrated circuit, such as the ULN2803, or as individual transistors such as the TIP120.

PIC Microcontroller as a Driver

We have mentioned that a PIC microcontroller can conveniently serve as a translator by furnishing the required sequence of power signals for a unipolar stepper motor. But using a PIC microcontroller to drive a stepper motor presents difficulties. The main concern is that the current-carrying capacity of a PIC port pin is 20 mA. This means that only a very, very small motor could be driven directly.

Most real-world circuits require additional amperage for the driver stage. The most common solution is to use several transistors. But here again, the current-carrying capacity of the base pin of some transistors exceeds the 20-mA limitation of the PIC port. For example, the TIP120 Darlington transistor, frequently used in stepper motor circuits, can handle up to 120 mA base current. Although with small loads it is possible to drive a TIP120 transistor directly from a PIC port, a more common design is to use a *current gain device*. In this application the CMOS 4050 hex non-inverting buffer IC can be placed between the PIC and the TIP120 because the 4050 furnishes sufficient current gain (called *fanout*) to safely drive the base pin of a TIP120 transistor, even at maximum loads. One of the circuits developed later in this chapter uses a PIC16F84 translator and a 4050 IC to drive four TIP120 transistors.

ULN2803A

This is one of five members of the ULN280x family; the 2803 is compatible with 5V TTL/CMOS inputs and therefore can be connected directly to a PIC microcontroller port. The IC packages eight Darlington pairs in a DIP 18 IC that also includes integral suppression diodes. All transistors share a common emitter, thus saving electronic hardware. The current rating of the 2803 is of 500 mA at a maximum of 50V. For driving a unipolar stepper motor only four of the eight diodes are used. Figure 16-4 shows the pinout of the ULN2303.

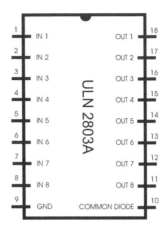

Figure16-4 ULN 2803 Pinout.

In a typical system, the common diode line of the 2803 (pin 10) is wired to the motor's power line (which we have labeled Vm). Inputs from the translator stage are wired to the IN x lines and the OUT x lines are connected to the motor windings. The mapping of the IN lines, to the motor windings is determined from the motor's step sequence.

TIP 120

The TIP120, already mentioned, is an NPN Darlington transistor in a TO-220 package. It is rated at 5 A and 60V. The TIP120 is described as a general-purpose amplifier at low switching speeds. Figure 16- 5 shows the TIP120 transistor and its electronic symbol.

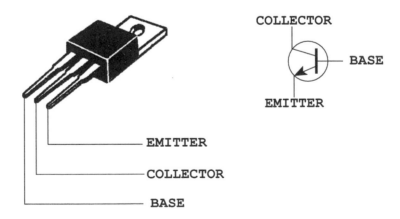

Figure 16-5 TIP120 Amplifier Transistor.

The TIP120 contains an internal diode but some circuits include a fast, external one on the emitter pin as additional protection. In a typical circuit, the base pin provides input from the controller, the collector serves as output to the motor winding, and the emitter line is set to ground.

16.4.2 Bipolar Drivers

Bipolar motors have a single winding per phase, and rotation is achieved by reversing the current flow in each winding. The most common device for reversing the current flow in a DC circuit is called an *H bridge*.

In operation, the H bridge can be visualized as four switches placed along the vertical arms of an H-shaped component, as shown in Figure 16-6.

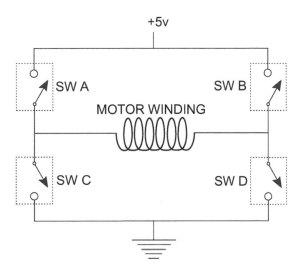

Figure 16-6 H Bridge Circuit Visualization.

In an actual circuit, the four switches, labeled SW A, SW B, SW C, and SW D in Figure 16-6, take the place of four transistors. In H bridge terminology, switches SW A and SW C are said to be on the left side of the bridge, while SW B and SW D are on the right side. Switches SW A and SW B are on the high side and SW C and SW D on the low side of the bridge. When SW A and SW D are closed, the current flows through the motor winding in one direction. When SW A and SW D are open and SW B and SW C are closed, the current flows in the opposite direction. When all four switches are open, there is no flow through the circuit. Also note that if SW A and SW C are open (or SW B and SW D), the input voltage source would be in short circuit and damage to the components is likely. Because a bipolar stepper motors has two motor windings, two H bridges are required to drive it.

H bridges can be built using individual components, usually MOSFET transistors, and are also furnished in dedicated ICs. The L298, discussed later in this section, is one such H bridge IC.

16.4.3 Transistorized H Bridge

When individual transistors are used it is common to select two P-channel MOSFETs for the high side of the bridge, and two N-channel MOSFETs for the low side. Figure 16-7 shows a popular H bridge design using P- and N-channel MOSFETs.

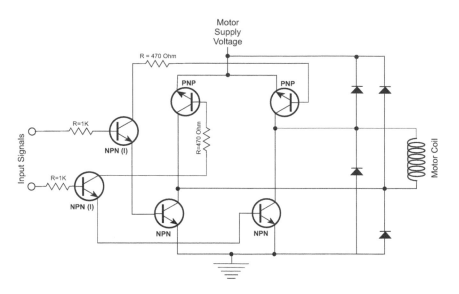

Figure 16-7 NPN-PNP Transistor H Bridge.

The two transistors labeled NPN(I) provide an interface with the IC that supplies the input signals. These transistors can be the 2N2222 or any small-signal equivalent. Many NPN-PNP transistor pairs are suitable for the H bridge itself. The matched pairs with identical characteristics are usually called *complementary transistors*. The NPN ZTX690B and the PNP ZTX790B in a TO-92 packages, are a matched pairs that deliver up to 2 Amps per coil. The TIP 3055 (NPN) and the TIP2955 (PNP) are also suitable and provide a 15 Amp current gain. Integrated circuits, such as the BC847BPN, are transistor pairs that include the NPN and PNP components. Note that the H bridge in Figure 16-7 supplies a single motor coil. To drive a bipolar motor, two such H-bridges are required. A more efficient, albeit more complicated, circuit uses N-channel MOSFETs for both the high and the low side.

Snubber Diodes

When inductors are energized and de-energized the magnetic field in the coil builds up (in the energizing phase) and collapses (in the de-energizing phase). During the collapsing phase the changes in the magnetic field in the coil cause a current to be induced and flow in the opposite direction. This flow reversal may damage the switching transistors and other logic components in the circuit. The problem can be prevented by the installation of diodes in such a way so that they will not conduct during normal operation. But when the falling magnetic field causes the current flow to reverse, then diodes conduct the current to ground and away from the transistors and other sensitive components. These are usually caller "snubber" or "clamping" diodes.

The snubber diodes are shown in the H bridge in Figure 16-7. In selecting these diodes, their current carrying capacity and switching speed must be considered. We address this problem in greater detail in forthcoming sections related to motor driver circuits. The four snubber diodes in Figure 16-7. can be either the 1N4007 or the 1N5408.

16.4.4 H Bridge ICs

An alternative to the transistorized H bridge described previously is the use of an integrated circuit driver that furnishes the H bridge function, usually in addition to other driver-level controls. The circuit designer often looks at these additional functionalities in order to determine if an IC driver is appropriate, as the H-bridge component by itself is quite compatible with its transistor-based counterpart, as in Figure 16-7.

A consideration sometimes mentioned in favor of the integrated circuit versions of the H bridge, versus the transistorized option, is the possibility of damage due to improper bridge switch activation with the transistorized variation. This possibility is precluded with the IC component because the damaging connections that can possibly damage circuit components are not allowed by the chip's logic.

Driver ICs are sometimes designed to complement the functionalities of a specific stepper motor controller. This is the case of the L297 (mentioned earlier in this chapter) and the L298 driver IC. The L297/298 pair provides a powerful set of features as shown by some circuits presented later in this chapter.

There are hundreds of stepper motor driver ICs available on the market. A single company (Allegro Microsystems) produces over twenty different bipolar stepper motor drivers. A listing of their various stepper motor control products by specifications and applications is available at:

http://www.allegromicro.com/en/Products/Categories/ICs/motor.asp

L293D

The L293D from ST Microelectronics, SGS Thomson, and other vendors, is a two H-bridge, four-channel driver that includes the appropriate snubber diodes. The SN754410 from Texas Instruments is reputedly an improved substitute for the L293. The L293 is furnished in a sixteen pin DIP. It can drive 1.2 A current per channel; it is typically used to drive NEMA teen-size stepper motors. The L293 can be used to run two DC motors bi-directionally (not covered in this book) or to control the two windings of a bipolar stepper motor. Figure 16-8 is a pin diagram of the L293D.

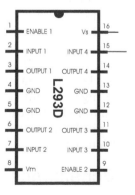

Figure 16-8 L293D Bipolar Motor Driver Pinout.

One advantage of the L293 is its low price (currently under $2.00); another is its simplicity. There are four input lines that are typically wired to the translator phase. In order to turn the motor on and off, two enable lines are either held high by wiring to the positive power supply or are controlled by the translator. The four output lines are wired to the motor coils. Four ground lines are on the center of the IC, which facilitates a PCB large ground that serves as a heat sink. The IC pin 16 is an input for the logic voltage (typically 5V) and pin 8 is an input for the motor power. A circuit using the L293D driven by a 16F84 PIC (circuit SMB-L293D-1) is presented in Chapter 18. A second circuit (SMB-L297-293D-1) in which an L297 is used in minimal configuration to drive an L293D is also described.

L298

The L298 is a popular high-current dual bridge driver often used in bipolar stepper motor circuits. The supply current can go up to 46V and 4 A, which allows the L298, with appropriate heat sinks, to drive NEMA 23 and larger stepper motors. The IC is offered in Multiwatt15 (vertical and horizontal) and PowerSO20 Packages. Figure 16-9 shows the pinout and mechanical data for the L298 driver IC in the vertical configuration.

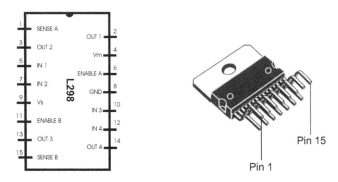

Figure 16-9 Pinout and Mechanical Data for L298 in Mutiwatt15 Package.

Circuits that use the L298 as a bipolar driver range widely in complexity. In its simplest circuit version, the L298 receives coil inputs through its IN 1 to IN 4 lines and sends output to the motor through the OUT 1 to OUT 4 lines (see Figure 16-9). The controller in these circuits can be a microcontroller or another IC. In these simple configurations the ENABLE A and ENABLE B lines are held high while the SENSE A and SENSE B lines are wired to ground. In all implementations the Vs line is wired to the circuit's logic voltage supply and the Vm line to the motor power supply. Because the L298 does not include snubber diodes, these must be provided separately. Circuit SMB-298-1, presented later in Chapter 18, shows a simple circuit in which the L298 is used mostly as a high-current capacity double-H bridge. Other more complex L298 circuits are combined with the L297, or other controllers, to provide chopping, pulse width modulation, and microstepping. These more advanced circuits are discussed in Chapter 19.

16.5 Modules in Circuit Schematics

Previously we classified stepper motor circuit functions into three types: control, translator, and driver. We also noted that many commercial devices include functions from more than one group. In this context we find many circuits in which an IC performs both translator and driver functions, or those in which a microcontroller is programmed to drive a stepper motor, sometimes with no other support than a digital buffer. For practical purposes rather than to divide circuits into control, translator, and driver stages, it is preferable to modularize as follows:

- The control and translator stage include input and output devices and digital logic. In this stage the input data is processed and output is sent to the driver stage or to other circuit devices.

- The driver stage includes the power driver function as well as motor-specific control functions.

Two advantages of this modularization are reusability and the possibility of shuffling stages to suit a particular design. The control and translator stage can be used with several compatible drivers by developing ad-hoc software. For example, a circuit module that contains several toggle and pushbutton switches and a 16F84 microcontroller can be paired with either a unipolar or bipolar driver. In either case, the program running on the microcontroller makes the circuit suitable for the particular motor type.

16.5.1 Example 16F84 Translator Modules

For many stepper motor control applications, the microprocessor provides *state code* for each of four motor coil lines. When a motor controller IC is present in the circuit, the microprocessor provides a step pulse and one or more motor controls, such as forward or reverse direction, mode selection, or enable and disable controls. A useful circuit can be designed around the 16F84 PIC and several switches, such as the one in Figure 16-10.

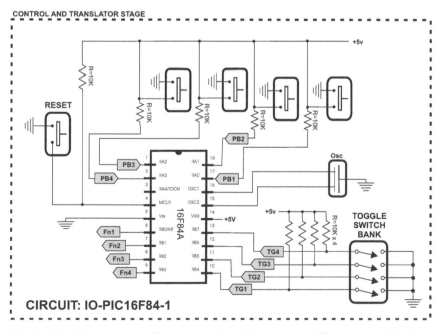

Figure 16-10 Pushbutton and Toggle-switch Control and Translator Module.

In circuit IO-PIC16F84-1 (Figure 16-10), input can come from any combination of four pushbutton switches, wired to port A lines RA0 to RA3, or from any of the four toggle switches in the bank wired to port B lines RB4 to RB7. Four output lines are available to communicate with devices or to provide controls for the driver stage. These four lines are wired to port B, lines B0 to RB3. The circuit is intended as a general-purpose experimenter.

The circuit designer can easily modify it to suit a specific purpose by eliminating unneeded components or replacing others. In the circuit in Figure 16-11 we have replaced the pushbutton switches with a second bank of toggle switches.

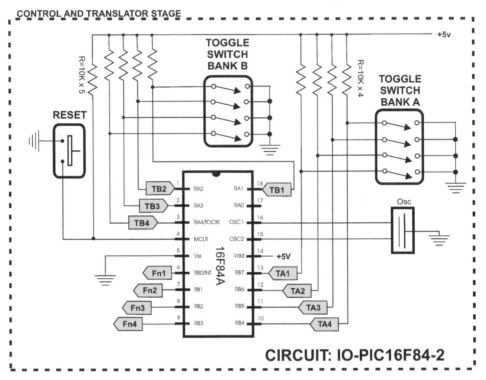

Figure 16-11 Toggle-Switch Based Control and Translator Module.

In most of the circuit schematics for motor controls presented in the following chapters, we have separated the control and translator stage from the driver stage. However, we have abstained from using generic stages that do not exactly fit the circuit at hand. The use of generic stages, such as the ones in Figure 16-10 and Figure 16-11, would have resulted in circuits that contain unused components and therefore are inefficient and confusing. The developer wishing to experiment with motor controls can build the generic control and translator stage circuits in this present section, or the ones contained in other circuit schematics, and use them to control the compatible driver stages. The number and type of outputs from a control and translator stage will clearly define if it is compatible with a particular driver. Circuit designations in the schematics include both the control and translator stage, and the driver stage.

Chapter 17

Unipolar Motor Circuits and Programs

17.1 Stepper Motor Control Circuits

The circuits and programs presented in this chapter relate to unipolar motors. Chapter 18 is devoted to basic circuits and programs for driving bipolar motors. The reader should keep in mind that unipolar motors can be used with bipolar circuits by ignoring the center taps, which can be left floating. The same is not true of bipolar motors, which require variations in the current polarity and, therefore, are not compatible with unipolar circuits.

In general, bipolar motors and circuits provide better performance and efficiency than the unipolar ones. On the other hand, unipolar circuits are easier to design, code, and fabricate. For this reason, if the requirements of the application allow, a unipolar circuit may sometimes be preferable, despite its lower efficiency and performance.

17.1.1 Stepper Motor Circuit Schematic Conventions

The naming convention that we have adopted for stepper motor circuits always starts with the letters SM, followed by U for unipolar applications and B for bipolar. The remainder of the circuit designation includes the names of the one or more ICs in the circuit. For example: SMU-5804-1 is a unipolar stepper motor controller that uses the 5804 driver IC. The last digit is the circuit version, notwithstanding that often there is only one version of the circuit. Each circuit is furnished with one or more programs that exercise the circuit's basic operations. The names of the source files for the programs also follow these conventions. For example, the program named SMU_5804.asm is the source files for driving a unipolar stepper motor circuit based on the 5804 IC. The code listing and the circuit diagrams are cross-referenced.

In order to keep the circuit diagrams as useful as possible we have sometimes structured or simplified the schematics. For example, in most circuits we have separated the schematics into stages that are relatively independent. In these cases the circuit diagrams contain dashed-line boxes that indicate the individual stages. Figure 17-1 shows a circuit that is separated into two stages.

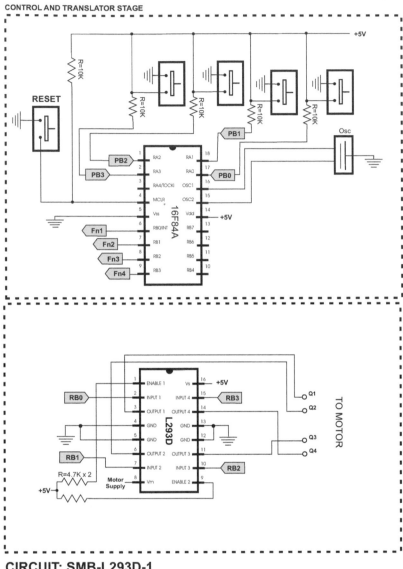

CIRCUIT: SMB-L293D-1
CODE: SMB_L293D.asm

Figure 17-1 Example of Circuit Schematic.

The circuit in Figure 17-1 is that of a bipolar stepper motor controller described in Chapter 18. Each stage in a circuit diagram is identified with a variation of the circuit name. In the case of the circuit in Figure 17-1, the translator and control stage is labeled, at the bottom of the illustration, SMB-L297-298-1 and the driver stage is labeled SMB-L297-298-2. The separation of a circuit into stages is complemented by using directed flags that reference the corresponding element in each stage. The flags are labeled FnX, where X is a sequential digit indicating a particular function. For example, in the diagram of Figure 17-1, the flag labeled Fn1 in the translator and control stage is shown to access the L297 ENABLE function in the driver stage. Also note that not all outputs from the schematics of the translator and control stage are used in the driver stage of a particular circuit.

17.2 Motor Speed Control

Stepper motor speed is determined by two factors: the *code sequence* sent to the motor coils and the *pulse rate*. These two factors are related: At the same pulse rate a stepper motor in wave drive mode runs twice as fast as one in the half step mode. Later in this book we discuss microstepping techniques that reduce the motor speed even further for the same pulse rate.

In any case, within a specific drive mode (wave, full, half, or microstep), the motor speed is determined by the rate at which the pulses are sent to the motor coils. When a microcontroller is used in the driver stage, as is the case with the circuits and programs in this book, then software will determine the pulse rate. In code the actual pulse rate can be obtained using simple delay counters or by Interrupt-Driven routines. Timer-related issues were discussed in Chapter 11. At present we are concerned with hardware devices that furnish a way of inputting motor speed information into the circuit.

The first issue in implementing a motor speed control mechanism is the device or devices that provides speed selection information to the circuit. In other words, the circuit typically includes input devices that allow selecting the motor speed. These devices are either digital or analog in nature. A toggle switch used to select between a low and a high motor speed can be considered a digital input device because the switch can be wired so that it reports either a high or low state. A potentiometer that allows controlling motor speed within a certain range can be considered an analog input device because the varying resistance read from the potentiometer must be converted to a digital value; this value is then used by the microcontroller to select a motor pulse rate.

17.2.1 Speed Control from Digital Input

The simplest (but not always the most convenient) mechanism for controlling motor speed is an input device that directly provides digital data or can be easily converted to digital form. In the sample circuit SMU-PIC16F84-1 (Figure 17-3), slow and fast motor rates are determined by the state pushbutton switches. The switches are wired so that the pushbuttons reports a high or low value to a microcontroller port. Because the state of the switch can be interpreted as a binary zero or one, the information provided by the pushbutton can be considered digital input. In this same context, sample circuit SMU-5804-1 (Figure 17-4) uses a bank of four toggle switches to provide data that allows selecting sixteen different motor speeds. The program reads the state of the four switches and the binary value is then used to select the corresponding delay in any one of sixteen possible rates.

The use of a bank of switches to select motor speed has its advantages and its drawbacks. With this scheme the number of individual switches in the bank determines the number of possible speeds. One toggle switch allows selecting two speeds, two switches four speeds, three switches eight speeds, and so on. One possible advantage of toggle-switch-based controls is that the motor speed is selected on the board and cannot be changed accidentally. It is a suitable option for circuits in which the motor speed is not changed often and can be scaled in discrete steps. By the same token, the toggle switch bank approach is not suitable when the motor speed must be easily changed or must have a continuous range. Another drawback of switch controls is that each switch requires using a microcontroller port. Sample circuits SMU-PIC16F84-1 and SMU-5804-1 use digital inputs to control motor speed. In one case (SMU-PIC16F84-1), pushbutton switches select slow or fast motor speeds, in the other case (SMU-5804-1), a battery of four toggle switches (see Figure 14-4) is used to select one of sixteen possible motor speeds.

17.2.2 Analog Input Speed Control

An analog input device provides an alternative way of controlling motor speed. The *potentiometer* (also called a *pot*) performs as a variable resistor and acts as a voltage divider. Another analog device for varying circuit resistance is the *rheostat*. The rheostat is a two-terminal variable resistor that allows handling much larger currents than a potentiometer.

Potentiometers are often used to adjust an analog signal that is then fed into a higher-wattage device (such as a TRIAC) or converted to a digital scale. This second option is the one most often found in microcontroller-driven stepper motor circuits, which also explains why rheostats are not commonly found in these circuits. The typical circuit reads the resistance offered by the current setting of the potentiometer, converts it to a digital value, and uses this value to determine the pulse rate sent to the motor coils. The one new element in a pot-based circuit is the conversion of the analog value read from the potentiometer into a scale of digital values that can be used by the microcontroller.

In this application the circuit designer can select one of two options: use an analog-to-digital conversion device, such as the ADC0831, or select a microcontroller that contains an analog-to-digital converter. These internal analog-to-digital devices are referred to in the literature as an A/D or ADC module. Selecting a microcontroller with an ADC module is often more economical and easier to implement than using a separate analog-to-digital IC.

Many PIC microcontrollers contain an ADC module. The most suitable one in each case depends on the circuit requirements. Unfortunately, the popular 16F84, used in many circuits in this book, does not contain an ADC module. On the other hand, the 16F87x family, which includes the forty-pin 16F877 used in many of this book's circuits, contains an ADC module. An interesting option for compact stepper motor circuits is the PIC16F684, which also includes an ADC component. The 16F686 is a small, inexpensive device with the same instruction set as the 16F84 and very similar architecture and memory structure. The chip provides twelve I/O ports: six mapped to port A and six to Port C. An internal oscillator is programmable from 31 KHz to 8 MHz, the default being 4 MHz. This saves having to use an external timing device. The 16F684 has two timer modules and a comparator. The ADC converter has a resolution of ten binary digits. The circuit diagram of the 16F684 is shown in Figure 17-2.

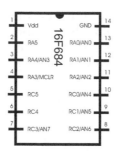

Figure 17-2 Diagram of the Fourteen-Pin Versions of the PIC 16F684.

Later in this chapter we develop the circuit named SMU-PIC16F684-1, which contains a 16F684 PIC. Two sample programs (SMU_PIC16F684.asm and SMU_PIC16F684_INT.asm) use the PIC's ADC to convert the resistance value read from the potentiometer to digital, which is then used to control motor speed.

17.3 Unipolar Motor Control Circuits

Unipolar motor control circuits require fewer components and are easier to design and manufacture than their bipolar counterparts. In the simplest possible implementation, a microcontroller (such as a PIC 16F84 or 16F684) can be used as a controller, usually combined with some simple devices that augment the current capacity to match the needs of the motor. Other more complex circuits use dedicated translators and drivers to relieve the microcontroller of some functions or to provide additional functionalities.

17.3.1 Matching Circuit to Motor Power

Stepper motors come in many sizes and power ranges. Sometimes the motor's datasheet expresses the motor's power requirements as its maximum current rating per coil. When this value is known, it can be directly compared to the ratings of the circuit components. For example, a motor rated at 0.8 A can be safely used with a 5804 motor driver IC that is rated at 1.25 A.

If the datasheet is not available or if the motor's current rating is not listed, then we can determine the motor's power requirements by measuring the coil resistance with an ohmmeter or a multimeter. In the case of a unipolar motor, make sure that the measured value is between the ends of the coil and not from the coil center taps. If in doubt use the larger resistance value.

For example, suppose you have a 14V stepper motor and that the measured resistance of the windings is 5 Ohm. Ohm's law allows us to calculate:

$$E = \frac{I}{R}$$
$$E = \frac{14}{5}$$
$$E = 2.8 A$$

In the case of a 5804 driver, this value considerably exceeds the 1.25 A at which it is rated. Two possible solutions are available when the motor current exceeds the capacity of the circuit component: the first one is to select components with a higher power rating. The second possible solution is to add resistance to the motor coils. In this case Ohm's law can be used to determine the total coil resistance necessary to meet the limits of the hardware. In the previous example we can calculate

$$R = \frac{E}{I}$$
$$R = \frac{14}{1.25}$$
$$R = 11.2\,\Omega$$

Because the measured coil resistance is 5 Ohm we would need external resistors of 6 or 7 Ohms so that the total coil resistance would be approximately 11.2 Ohms. Also note that the wattage of the resistor must also be calculated. In this case the external resistors should be rated for 17 W. Resistors of 20 W are commonly available.

In the unipolar circuits that follow, we have included a resistor labeled Rx connected to the center tap of the motor coils. This resistor will only be necessary if the motor rating exceeds the circuit capacity as determined at its output.

17.3.2 16F84 Unipolar Circuit

The first circuit is a simple unipolar motor driver that uses the 16F84 PIC as a translator with a 4050 hex buffer IC and four TIP 120 NPN Darlington transistors as amplifiers. Although the TIP 120 contains internal snubber diodes, externals ones are provided in the circuit as additional protection. Figure 17-3 shows the circuit schematics.

The circuit SMU-PIC16F84-1/2 in Figure 17-3 receives input from four pushbutton switches wired to the microcontroller port A lines 0 to 3. Port B lines 0 to 3 furnish outputs to the four coils of the unipolar motor via the 4050 hex buffer and the base pins of the TIP 120 transistors. Output from the four TIP 120 transistors are wired to the respective ends of the motor coils. The center taps of the coils are connected to the motor power source. The resistors labeled Rx are only necessary if the motor exceeds the capacity of the TIP 120 transistors, which is 6 A. The calculation of the Rx resistors was discussed earlier in Section 17.3.1.

Sample Program SMU_PIC16F84.asm

The program SMU_PIC16F84.asm, in this book's software package, is a driver that exercises the circuit. The code assumes a unipolar motor in one-half step mode and uses a lookup table for the corresponding coil sequence codes. When held down, the pushbuttons wired to ports A3 and A2 activate slow rotation in the forward and reverse direction. Pushbuttons wired to ports A0 and A1 activate fast forward and reverse rotation. The pointer to the lookup table is either incremented or decremented according to the selected direction. This ensures that changes in the direction of rotation are executed smoothly. Fast and slow execution are determined by the value of a local variable read by the delay routine.

CONTROL AND TRANSLATOR STAGE

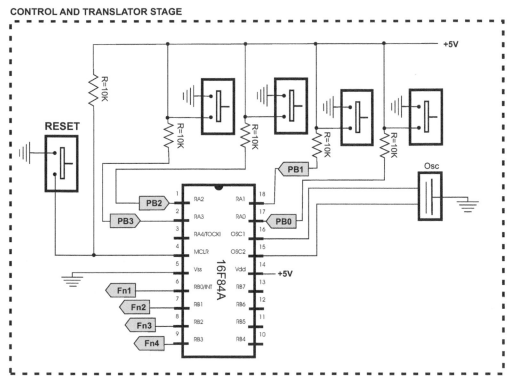

DRIVER STAGE

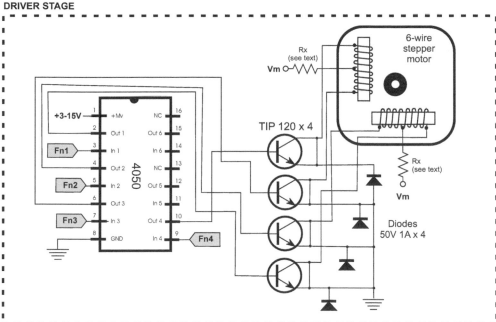

CIRCUIT: SMU-PIC16F84-2 CODE: SMU_PIC16F84.asm

Figure 17-3 Unipolar Motor Driver with the 16F84 PIC.

The code table matches the requirement of the 42M048A19 unipolar motor manufactured by Airpak. The motor is rated for 5VDC at 9.1 Ohms per coil. The degrees per step is 7.5. The table is coded as follows:

```
CodeTable:
      addwf     PCL,f           ; Add w to program counter
      retlw     B'00001010'     ; cycle = 0
      retlw     B'00001000'     ;         1
      retlw     B'00001001'     ;         2
      retlw     B'00000001'     ;         3
      retlw     B'00000101'     ;         4
      retlw     B'00000100'     ;         5
      retlw     B'00000110'     ;         6
      retlw     B'00000010'     ;         7
      retlw     B'00000000'     ; Table terminator
```

Processing consists of reading the state of the four pushbutton switches wired to port A lines and setting the values of the variables that control direction and motor speed. The switches are debounced in software to avoid spurious values. The driver then goes to a routine that obtains the corresponding table code and writes it to port B lines 0 to 3. The procedure named OneTick obtains the corresponding code from the table in forward or reverse rotation and writes the code to port B. In forward direction, the table pointer is incremented and it is decremented in reverse rotation. The procedure then calls a Delay procedure that waits doing nothing. The duration of the wait, thus the speed of the motor, is determined by the value stored in a local variable. Code is as follows:

```
;===========================
;    forward or reverse
;    single cycle procedure
;===========================
OneTick:
; Test for change of direction
      btfss     direction,0     ; Test forward bit
      goto      Forward
; At this point direction rotation is reverse
      goto      Reverse
; Foward direction routine
Forward:
      movf      this_cycle,w  ; Get current cycle
      call      CodeTable     ; Get code from table
      movwf     PORTB         ; Store code in port
      call      Delay
; Bump cycle counter
      incf      this_cycle,f  ; Add one
; Test for cycle number 7 (last one in sequence)
      btfsc     this_cycle,3  ; 1000 = 8
      goto      Recycle       ; Reset if at end of cycle
      return
Recycle:
      clrf      this_cycle
      return
; Reverse direction routine
Reverse:
      movf      this_cycle,w  ; Get current cycle
      call      CodeTable     ; Get code from table
      movwf     PORTB         ; Store code in port
      call      Delay
```

```
; Bump cycle counter
      decf       this_cycle,f  ; Subtract one
; Test for cycle number 0xff (overflow from cycle # 0)
      btfsc      this_cycle,4  ; Bit 4 set indicates overflow
      goto       Recycle2      ; Reset if at end of cycle
      return
Recycle2:
      movlw      .7            ; First reverse cycle
      movwf      this_cycle    ; To cycle counter
      return
;================================
;         delay sub-routine
;================================
Delay:
      movf       delay,w       ; Load delay value
      movwf      j             ; j = w
Jloop:
      movwf      k             ; k = w
Kloop:
      decfsz     k,f           ; k = k-1, skip next if zero
      goto       Kloop
      decfsz     j,f           ; j = j-1, skip next if zero
      goto       Jloop
      return
```

The complete listing for the SMU_PIC16F84 program is found in this book's software resource.

17.3.3 5804 Unipolar Circuit

Easy interfacing with small-size stepper motors can be achieved with circuits that employ the 5804 IC. As mentioned in Chapter 16, the 5804 allows enabling output, and controlling direction and half-step or full-step modes. A step input line, which is typically driven by a clock IC or a microcontroller, can be programmed to set the motor speed. Several speed control mechanisms are examined in the following sections. The 5804 provides *thermal protection* so as to disable the chip when the safe operating temperature is exceeded. Internal *fly-back diodes* are also part of the IC, although external diodes are sometimes included as a safeguard.

One limitation often mentioned in relation to circuits based on the 5804 is the IC's limitation of 1.25A per phase. However, most teen-size stepper motors of 5V and greater fall within this range. For example, the Airpax 9123 stepper motor, which is size NEMA 16, is rated at 5 V and 9.1 Ohms per coil. This results in a current value of 0.55 A, well below the 1.2 A rating of the circuit. Even some larger-than-teen stepper motors are compatible with the 5804 circuit. For example, a NEMA size 22 unipolar stepper motor (2.2 inches wide) rated at 12V, draws only 0.6 A per coil.

The circuit in Figure 17-4 can receive input from two sets of toggle switches. One set of three switches is wired to 16F84 port A lines 2 to 4. These three switches are used to determine the 5804 motor controls. Another set of four toggle switches is wired to the 16F84 port B lines 4 to 7 and is used to determine the motor speed. Two port A lines (RA0 and RA1) are unused in the circuit. The developer can use these lines for additional controls.

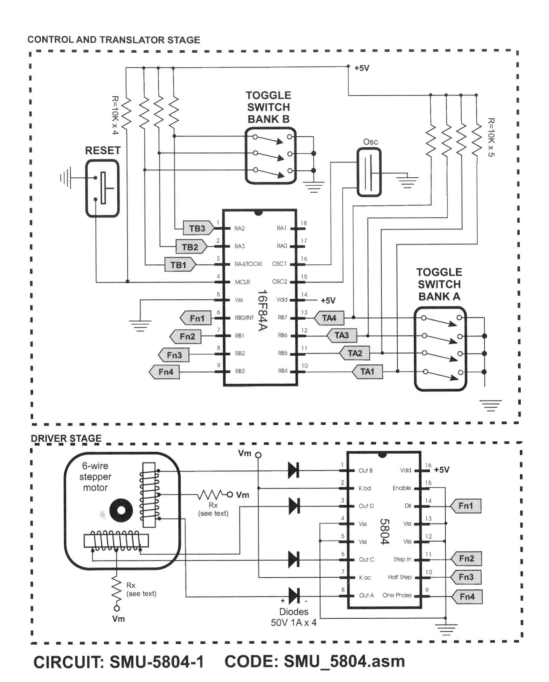

CIRCUIT: SMU-5804-1   CODE: SMU_5804.asm

Figure 17-4 5804-Based Unipolar Motor Driver Circuit.

In circuit SMU-5804-1/2 (Figure 17-4), there are four control lines from the 16F84 to the 5804. The line labeled Step In provides the strobe pulse to the 5804. The line labeled Dir furnishes direction control, while the lines labeled Half Step and One Phase allow selecting four drive sequences, as follows:

```
Half Step      One Phase      Drive mode
0              0              = two phase
0              1              = wave drive
1              0              = half step
1              1              = step inhibit
```

Note that the diodes in the circuit can be omitted and that the resistors labeled Rx are necessary only if the motor exceeds the 1.2 A.

Sample Program SMU_5804.asm

The program SMU_5804.asm, in this book's software package, is a driver that exercises the SMU-5804-1 circuit (Figure 17-4). The code reads the three toggle switches wired to port A lines 2, 3, and 4 and selects the speed, direction, and drive mode, as follows:

```
;   PORTA line    SW      ACTION
;       4         1       Fast/slow delay rate
;                         delay = 100 / delay = 50
;       3         2       Forward/Reverse
;                         Set PORTB to: xxxx xxx0
;                                   or: xxxx xxx1
;       2         3       Half step / single step
;                         Set PORTB to: xxxx x10x
;                                   or: xxxx x01x
```

The circuit contains pushbuttons wired to ports A0 and A1, which are not used by the sample program.

The program first reads port B lines 4 to 7, trissed for input and wired to the four toggle switch bank. The value in the range 0 to 15 read from the four toggle switches is used to obtain a delay code from a local lookup table to determine slow or fast speed. Code is as follows:

```
;==============================
; Read speed control toggles
;==============================
; Read four speed control toggle switches on port
; B lines 4 to 7. The swapf instruction allows swapping
; nibbles as they are read from the port
      swapf      PORTB,w             ; Read port B and swap nibbles
      andlw      B'00001111'         ; Mask out 4 high bits
      call       SpeedTable          ; Use value in w to obtain delay
      movwf      delay               ; Store code in local variable

 The delay code table is coded as follows:
;==============================
;   Delay table for speed
;        control
;==============================
; Values in table range from 255 to 30 (15 units apart)
; Table has 16 entries corresponding to the values
; represented by the setting of the 4 speed control
; toggle switches
SpeedTable:
      addwf      PCL,f       ; Add w to program counter
      retlw      .255        ; Slowest speed 0000 = 0
      retlw      .240        ;                    0001 = 1
      retlw      .225        ;                    0010 = 2
```

```
    retlw      .210       ;                        0011 = 3
    retlw      .195       ;                        0100 = 4
    retlw      .180       ;                        0101 = 5
    retlw      .165       ;                        0110 = 6
    retlw      .150       ;                        0111 = 7
    retlw      .135       ;                        1000 = 8
    retlw      .120       ;                        1001 = 9
    retlw      .105       ;                        1010 = 10
    retlw      .90        ;                        1011 = 11
    retlw      .75        ;                        1100 = 12
    retlw      .60        ;                        1101 = 13
    retlw      .45        ;                        1110 = 14
    retlw      .30        ;        Fastest speed 1111 = 15
    retlw      .0         ; Terminator
```

The developer can edit the table to suit the particular requirements of a motor or circuit.

After the delay rate has been set, code reads the three port A lines wired that will determine the settings of the 5804 direction, one-half step, and one phase controls. Code is as follows:

```
;=============================
;   Read 5804 control lines
;=============================
; PORTA bit 4 (wired to toggle # 1) controls forward
; or reverse direction by means of PORTB bit 1 (wired
; to 5804 direction line on pin 14).
    btfsc      PORTA,4             ; Direction control switch
    goto       ReverseRot          ; Engage reverse rotation
; At this point rotation is forward
    bcf        PORTB,1             ; Clear bit `
    goto       HalfStepMode
ReverseRot:
    bsf        PORTB,1             ; Set reverse
HalfStepMode:
; PORTA bit 3 (wired to toggle # 2) selects half-step
; mode by means of PORTB bit 2 (wired to 5804 half-step
; line on pin 10).
    btfsc      PORTA,3             ; Half-step mode control
    goto       NoHalfStep          ; Disable half step
; At this point half step is selected
    bsf        PORTB,2             ; Set bit `
    goto       OnePhaseMode
NoHalfStep:
    bcf        PORTB,2             ; Turn off half-step control
OnePhaseMode:
; PORTA bit 2 (wired to toggle # 3) selects one phase
; mode by means of PORTB bit 3 (wired to 5804 single
; phase line on pin 9).
    btfsc      PORTA,2             ; One phase mode control
    goto       NoOnePhase          ; Disable single step
; At this point single phase is selected
    bsf        PORTB,3             ; Set single phase `
    goto       PulseAndDelay
NoOnePhase:
    bcf        PORTB,3             ; Turn off one phase
    goto       PulseAndDelay
```

Note that the code for reading pushbutton switches (in the sample program SMU_16F84.asm) includes debouncing. However, we have found that debouncing switch action can usually be omitted when reading toggle switches in motor-driving circuits. For this reason switch debouncing operations are not used in the program SMU_5804.asm.

Generating the Motor Pulses

Note the difference between the sample program SMU_16F84 and SMU_5804 regarding the generation of the motor-driving pulses. The first program (SMU_16F84) reads the drive sequence codes from a local table and writes these codes to the port B lines that are wired to the motor coils. In this case the speed of the motor is determined by the rate at which these coil codes are updated, which, in turn, depends on the delay routine. On the other hand, the sample program SMU_5804 relies on the 5804 IC to produce the corresponding codes. The actual sequence generated by the 5804 depends on whether the half-step or single-step mode has been previously selected. This means that on the 5804 circuit and program, the motor coils are driven by the 5804 IC and not by the microcontroller. The developer should not assume that the code sequence generated by the controller IC is always the same one expected by the motor. The motor's datasheet usually lists the coil driving sequence for each supported mode. This sequence should match the one generated by the controller chip. If this is not the case, then the motor coil taps must be relabeled so that they are in accordance with the sequence generated by the 5804.

When the motor is driven by a dedicated controller, such as the 5804, a microcontroller is typically required to produce the pulses or strobes that are impressed on the controller's step line. In circuits without a microcontroller (not covered in this book), a clock or timer IC can be used to generate the strobe pulses. In the circuit SMU-5804-1, the microcontroller's port B line 1 is wired to the 5804 line labeled Step In.

The pulses sent to a specific controller can be positive-going or negative-going. In the *positive-going pulse*, the line is held low and then strobed high. In a *negative-going pulse*, the line is held high and strobed low. In either case there is a minimum time during which the line state (high or low) must be maintained. Ideally the device datasheet will provide sufficient information regarding its pulse requirements, but this is not always the case. For instance, the 5804 datasheet does not explicitly state that the pulse must be positive-going, although it is indirectly suggested by the statement in the datasheet that that states that the *step input line* must be low when changing state or direction. In addition, there is no specific information in the 5804 datasheet regarding the duration of the *high strobe*. The sample program SMU_5804 contains a sub-routine to pulse the 5804 Step In line. We have coded ten *no-operation codes* (**nop**) to produce a delay of as many machine cycles. Code is as follows:

```
;=================================
;   routine to pulse the 5804
;=================================
Pulse:
     bsf        PORTB,1        ; Bring step line high
     nop
```

```
        nop
        nop
        nop
        nop
        nop
        nop
        nop
        nop
        nop
        bcf       PORTB,1        ; Step line low
        return
```

Clearly the delay could have been accomplished more compactly with a timed or counter-based loop. Note that although this routine works well with the 5804, it is not adequate for a device that requires a negative-going pulse or different timing. Later in this chapter we will find that the L298 driver requires a negative-going pulse that must be held low for one machine cycle. A different pulse routine is developed for L298 circuits.

Interrupt-Driven Motor Pulsing

The timing of motor pulses using a delay routine is simple to code and in some cases serves its purpose. However, there are more accurate and efficient ways of generating a timed pulse even with the simplest microcontrollers. Timing circuits and code were covered in Chapter 11 so we discuss only the implementation of interrupt-based timers in the context of motor controls. The methods described in this section can be implemented in any PIC microcontroller covered in the book, although sometimes minor modifications to the code will be required to accommodate the various hardware.

Interrupt-Driven timers have several advantages over polled routines: One is that the timing operation is independent of application code. Because the timing takes place as a background operation, changes in the application itself do not affect the accuracy of the timer. Another advantage is that the application can continue to do other work in the foreground without concern for the accuracy of the timing routine.

The Timer0 module, which is available in all mid-range PICs, is particularly suited for implementing an Interrupt-Driven motor pulsing routine. In this application Timer0 has the following useful features:

- The timer register is readable and writeable by software
- Can be driven by an internal or external clock
- Edge of timing pulse can be selected on the high-to-low or low-to-high transition
- 8-bit prescaler is available
- Can be Interrupt driven

In a simplest implementation, the program sets up the Timer0 interrupt to take place on register overflow, selects a suitable prescaler, and chooses the internal clock source. The sample program SMU_5804_INT.asm, in this book's software package, proceeds as follows:

```
; Clear the Watchdog timer and reset prescaler
```

```
          clrf TMR0
          clrwdt
; Set up the OPTION register
      movlw     b'11010000'
;     7  6  5  4  3  2  1  0 <= OPTION bits
;     |  |  |  |  |  |  |__|__|_____ PS2-PS0 (prescaler bits)
;     |  |  |  |  |  |           Values for Timer0
;     |  |  |  |  |  |           *000 = 1:2    001 = 1:4
;     |  |  |  |  |  |           010 = 1:8    011 = 1:16
;     |  |  |  |  |  |           100 = 1:32   101 = 1:64
;     |  |  |  |  |  |           110 = 1:128 111 = 1:256
;     |  |  |  |  |  |_____ PSA (prescaler assign)
;     |  |  |  |  |           1 = to WDT
;     |  |  |  |  |           *0 = to Timer0
;     |  |  |  |  |_____ TOSE (Timer0 edge select)
;     |  |  |  |           0 = increment on low-to-high
;     |  |  |  |           *1 = increment on high-to-low
;     |  |  |_____ TOCS (TMR0 clock source)
;     |  |  |           *0 = internal clock
;     |  |  |           1 = RA4/TOCKI bit source
;     |  |_____ INTEDG (Edge select)
;     |  |           *0 = falling edge
;     |_____ RBPU (Pullup enable)
;     |           0 = enabled
;     |           *1 = disabled
; Note that OPTION register is in bank 1 and its register
; name is OPTION_REG
      bsf       STATUS,RP0        ; RP0 is bank select bit
      movwf     OPTION_REG        ; Copy w to OPTION
      bcf       STATUS,RP0        ; Bank 0
;==============================
;      setup interrupts
;==============================
; Clear external interrupt flag (INTF = bit 1)
      bcf       INTCON,INTF       ; Clear flag
; Enable global interrupts (GIE = bit 7)
; Enable RB0 interrupt (inte = bit 4)
      bsf       INTCON,GIE        ; Enable global int (bit 7)
      bsf       INTCON,T0IE       ; Enable TMR0 overflow interrupt
```

Once the Timer0 interrupt is set up and initialized, the interrupt service routine receives control every time the timer register overflows. The service routine uses similar processing for pulsing the motor as the in-line delay routines covered previously. One change is that two counters are now required: one that is set from the input provided by switches or a potentiometer, and a second, running counter that is decremented during each iteration of the service routine. In the sample program SMU_5804_INT.asm, the ISR is coded as follows:

```
;========================================================
;              Interrupt Service Routine
;========================================================
; Service routine receives control when timer register
; TMR0 overflows, that is, when 256 timer beats have
; elapsed
IntServ:
; First test if source is a Timer0 interrupt
      btfss     INTCON,T0IF   ; T0IF is Timer0 interrupt
      goto      notTOIF       ; Go if not RB0 origin
; If so clear the timer interrupt flag so that count continues
```

```
        bcf   INTCON,T0IF   ; Clear interrupt flag
; Save context
        movwf      old_w                ; Save w register
        swapf      STATUS,w             ; STATUS to w
        movwf      old_STATUS           ; Save STATUS
;==========================
;    interrupt action
;==========================
; Decrement the iteration counter. Exit if not zero
        decfsz     iteration,f
        goto       exitISR              ; Continue if counter not zero
; At this point the delay count has expired so the programmed
; time has elapsed.
; First reset the iteration counter
        movf       delay,w              ; Read delay
        movwf      iteration            ; Reset iteration counter
;=================================
;      Pulse motor and delay
;=================================
; Make sure 5804 Step In line is low before strobe
; cycle
        bcf        PORTB,0              ; Set low
        call       Pulse                ; Local pulse motor routine
;==========================
;         exit ISR
;==========================
exitISR:
; Restore context
        swapf      old_STATUS,w         ; Saved STATUS to w
        movfw      STATUS               ; To STATUS register
        swapf      old_w,f              ; Swap file register in itself
        swapf      old_w,w              ; re-swap back to w
; Reset,interrupt
notTOIF:
        retfie
;=================================
;   routine to pulse the 5804
;=================================
; Note that 5804 requires a positive-going pulse
; therefore the line is held high during the strobe
; cycle
Pulse:
        bsf        PORTB,0       ; Bring step line high
        nop
        nop
        nop
        nop
        nop
        nop
        nop
        nop
        nop
        nop
        bcf        PORTB,0       ; Step line low
        return
```

In the Interrupt-Driven sample program named SMU_5804_INT.asm, motor speed is determined by the delay value read from the lookup table and by the value selected for the prescaler. Note that the prescaled delay can be eliminated by assigning the prescaler to the Watchdog timer. When the prescaler is assigned to Timer0, the lowest reduction is in the rate 1:2. However, when the prescaler is assigned to the Watchdog timer no delay of the Timer0 interrupt takes place. In the case of the sample program, we could eliminate the prescaled delay by setting bit 3 of the OPTION register. This would in fact generate the interrupt at double the speed.

17.3.4 16F686 PIC Circuit

By reading the state of each switch in a bank of four toggle switche,s we can obtain sixteen discrete values that can be used to set the motor speed in as many steps (circuit SMU-5804-1, Figure 17-4). However, it is sometimes necessary to devise a circuit that allows controlling motor speed with more precision or more conveniently than is provided by the discrete steps of one or more digital input devices, for example, when the circuit has to provide a finer degree of motor speed control or a more user-friendly device than toggle switches.

Speed control potentiometers are often suitable for this purpose. If the motor speed control is to be available during normal use, then a conventional knob-operated potentiometer can be selected. If the motor speed is to be set during installation or initialization, then a *trimmer pot* on the board may be more suitable. In either case, the analog resistance reading provided by the potentiometer or trimmer must be converted to digital so that it can be manipulated by code. The PIC 16F684 IC contains an ADC module that can be used for this purpose. Figure 17-5 is a circuit with potentiometer speed control and a 16F684 PIC.

One advantage of potentiometer or trimmer control over toggle switches is that the pot or trimmer requires a single input line. On the other hand, a circuit that contains a potentiometer must include a microcontroller with ADC or a separate analog-to-digital conversion IC.

17.3.5 16F686 Programming

Transitioning from the 16F84 PIC to the 16F684 is straightforward but not without some complications. Although both PICs belong to the same mid-range family and use the same instruction set, there are architectural differences. In the first place, the 16F684 is a fourteen-pin device in PDIP, SOIC, and TSSOP configurations, while the 16F84 has eighteen pins. This means that the latter PIC cannot be replaced with the former one without making circuit changes. The following are the most notable differences between these PICs:

- The 16F684 has twelve I/O ports, six assigned to port C and six to port A. The 16F84 has thirteen ports: five assigned to port A and eight to port B. There is no port B in the 16F684.

CONTROL AND TRANSLATOR STAGE

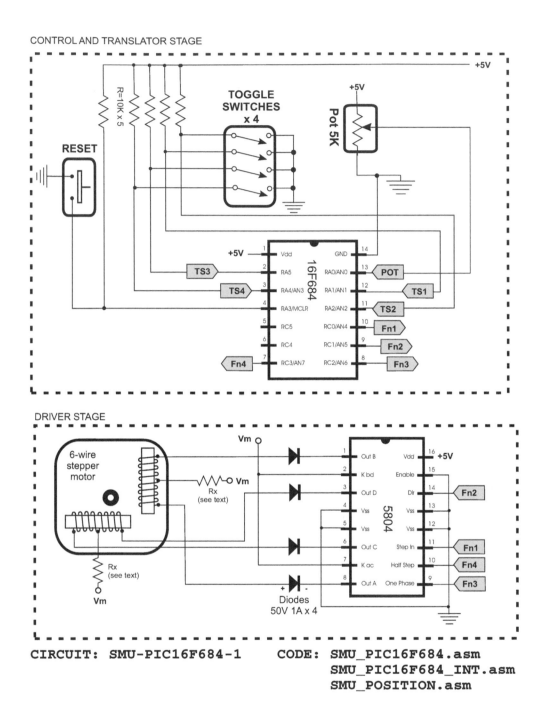

CIRCUIT: **SMU-PIC16F684-1**   CODE: **SMU_PIC16F684.asm**
                                       SMU_PIC16F684_INT.asm
                                       SMU_POSITION.asm

Figure 17-5 Unipolar Motor Driver Circuit with 16F684 PIC.

- The general-purpose registers (GPR) are mapped to addresses 0x0c to 0x4f in the 16F84 (68 registers). In the 16F684 there are ninety-six GPRs mapped to addresses 0x20 to 0x7f and thirty-two additional registers mapped to 0xa0 to 0xbf in bank 1.

- The 16F84 requires an external clock source or oscillator while the 16F684 has an internal precision oscillator. The frequency of the internal oscillator in the 16F684 can be selected by software between 8 MHz and 125 KHz. There is also a 31-KHz internal oscillator.

- The 16F84 requires that the master clear line (mapped to pin 5 and labeled MCLR) be held high for IC operation. In the 16F684 the master clear pin (mapped to pin 4) is multiplexed with port A line 3. During initialization of the 684, software can select between the master clear and the general-purpose I/O function for this line.

- The 16F684 includes several peripherals that are not available in the 16F84. These include an Analog Comparator, and A/D Converter, and an Enhanced Capture, Compare, and PWM module. Other devices, such as the timer, have additional features and functionalities in the 16F684.

In addition to the features compared in the preceding list, the 16F684 datasheet claims many other refinements and advantages over the 16F84. In the context of stepper motor software the A/D converter and the PWM module in the 16F684 are particularly useful. Porting code from the 16F84 to the 16F684 consists mostly of taking into account the differences between the two ICs. In addition, code that uses particular features of the 16F684 must initialize and operate these functions.

Sample Program SMU_PIC16F684.asm

This sample program is designed for the SMU-PIC16F684-1 circuit. The program drives a unipolar stepper motor in half-step mode. The 16F684 PIC controls a UCN5804 driver, which furnishes the motor control. Three toggle switches on the circuit are wired to ports RA1 to RA3. These are used to turn on and off the direction, half-step and single-phase modes on the 5804. A 5K potentiometer on port RA0 provides input to the chip's ADC. The resulting resistance value is used to select the delay rate, which, in turn, sets the motor speed in a delay routine.

Program variables are defined starting at address 0x20, as follows:

```
; Declare variables at 2 memory locations
j           equ      0x20
k           equ      0x21
delay       equ      0x22 ; Delay count
```

In the sample program we use the **banksel** directive to select between the two memory banks. Alternatively we could have used bank selection macros or manipulated the RP0 bit directly in code, as discussed earlier in the book.

Initialization starts by "trissing" the ports as required by the hardware devices. In this case, port C is set to output and port A to input. Port A line 0 is selected for analog input because this line is connected to the potentiometer in the circuit. Code is as follows:

```
Main:
; Initialize all line in port C for output
    banksel    PORTC
    clrf       PORTC
    banksel    TRISC                  ; Prepare to tris
    clrf       TRISC                  ; All lines to 0 (output)
```

```
; Set PORT A line 0 as analog
      movlw     B'00000001'          ; Line 0 is analog
      banksel   ANSEL
      movwf     ANSEL                ; Select as analog
; Turn off comparator
      movlw     B'00000111'          ; Setting bits 0, 1, and 2
                                     ; sets comparators off and
                                     ; Port C IN pins to digital
      banksel   CMCON0
      movwf     CMCON0
; Tris port A for input
      banksel   TRISA
      movlw     B'00111111'          ; All port A lines for input
      clrf      TRISA
      movwf     TRISA
```

The second initialization step consists of configuring the oscillator. In this case we have set the oscillator to 4 MHz and set the conversion clock use by the analog-to-digital converter to the value recommended in the 16F684 datasheet. Code is as follows:

```
; Configure oscillator
      movlw     B'01100001'
;                     |||   |_____ internal clock
;                     |||_____ 4 MHz
      banksel   OSCCON
      movwf     OSCCON
; Set conversion clock to Tad = 8 * Tosc (ADCS = 001)
      movlw     B'00010000'
;                     |||_____ ADCS bits
      banksel   ADCON1
      movwf     ADCON1
```

Finally, the A/D converter module is initialized and turned on, as follows:

```
; Select analog line (AN0) as the analog input
; channel. Select Vcc as the voltage reference.
; Left justify result
      movlw     B'00000001'; Bitmap:
;                             0 0 0 0 0 0 0 1
;                             | |   | | |   x___ 1 = turn on ADC
;                             | |   x_x_x_____ 000 = select AN0
;                             | x_____ 0 = voltage ref Vcc
;                             x_____ 1 = left justify
      banksel   ADCON0
      movwf     ADCON0
```

The A/D converter module in the 16F684 provides a 10-bit result that is stored in two dedicated registers labeled ADRESH and ADRESL. The high-order bit of the ADCON0 register, labeled the ADFM bit, is used to select between left- or right-justification of the 10-bit result. In either case, there are six bits that are not significant and are set to zero by the conversion module. The structures of the result registers are shown in Figure 17-6.

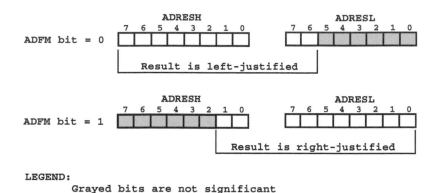

Figure 17-6 Conversion Options in the 16F684 A/D Module.

Once the analog-tdo-igital module has been initialized and an analog line is defined and wired to the potentiometer, we can perform the read-and-convert operation. The first step consists of setting bit 1 of the ADCON0 register (labeled the GO bit) to start the conversion. Code then tests the state of this bit to determine when the conversion has completed. This is signaled by the A/D module clearing the GO bit. Processing is as follows:

```
; 5K pot is wired to port A line 0 which has been set
    banksel    ADCON0              ; Prepare to sample
    bsf        ADCON0,GO           ; Start operation
Wait4ADC:
    btfsc      ADCON0,GO           ; Wait for completion
    goto       Wait4ADC            ; Loop back if not done
```

Using a potentiometer to regulate motor speed is usually based on using the resistance value read from the pot to determine the delay between motor pulses. Thus, the higher the resistance, the larger the delay and the slower the motor speed. By the same token, the lower the resistance read from the pot, the faster the motor turns. In most cases the resistance value needs to be scaled so that the resulting delay between pulses matches the minimum and maximum motor speeds desired.

In the case of the 16F684 analog-to-digital module, the 10-bit result stored in the ADRESH and ADRESL registers has a numeric range from 0 to 1023. In most cases we would not need this many motor steps. Program code can manipulate the conversion results to adjust for the desired motor speed range. For example, by selecting left-justification and reading only the ADRESH register, the two low-order bits are eliminated. This will produce a result in the range 0 to 255 instead of 0 to 1023. Often the conversion needs to be scaled further before it can be used as a *delay value*. This requires performing *binary arithmetic* on the results of the analog-to-digital conversion. *Shifting bits* provides an easy way to multiply or divide by two. Adding or subtracting a constant serves to transpose the scale to a higher or lower range.

The sample program SMU_PIC16F684.asm reads the high-order nibble of a left-justified result and halves this value by shifting the bits to the right one position. Then a value of 40 is added to limit the fastest motor speed. Code is as follows:

```
        banksel   STATUS
        bcf       STATUS,C       ; Clear carry to shift
        banksel   ADRESH
        rrf       ADRESH,w       ; Rotate in w
                                 ; Range is now 0 to 128
        addlw     .40            ; Add constant
                                 ; Range is now 40 to 168
        banksel   delay              ; Delay variable
        movwf     delay
```

This code ensures that the value stored in the variable named *delay* is in the range 40 to 168. This range and scale worked well for the particular motor used in testing this circuit; however, this code will need to be modified to suit the characteristics of a different motor or the speed range requirements of the application.

Sample Program SMU_PIC16F684_INT.asm

The interrupt system of the 16F684 is downward compatible with that of the 16F84. For this reason, the Interrupt-Driven motor controls developed in the sample program SMU_5804_INT (which uses the 16F84) can be easily ported to an application that uses the 16F684.

The program SMU_PIC16F684_INT.asm, in this book's software package, is an Interrupt-Driven version of the program SMU_PIC16F684.asm described in the previous section. The only modification required in the interrupt service routine is replacing the designation for the port wired to the 16F84 step line by the one used by the 16F684.

17.3.6 Stepper Motor Position Control

All the programs and examples considered to this point assume that the stepper motor rotates continuously in one direction or the other, that is, that it performs the conventional function of a classical motor. The control functions discussed so far include speed, direction, and step sequence mode. The motor speed is controlled by the frequency with which the step commands are sent to the hardware. However, because each step code sent to a stepper motor turns the rotor by a fixed angle, it is possible to control the rotor position by counting the number of steps sent to the drive. For example, if a given motor in full step mode turns by 2 degrees for each pulse received, then the software can make the rotor turn by 20 degrees by sending ten consecutive pulses in the selected direction.

Although *servo motors* (not covered in this book) are designed to provide position control and do so effectively, stepper motors also have this capability. The use of conventional stepper motors in position control functions is convenient in situations in which the device needs to be located at a certain position by the operator. This can be accomplished by "jogging" the motor, typically by operating one or more pushbuttons. Another example of position control is to provide a *"slewing" rate or mode* so as to re-position a device, for example, to correct the location of a stepper motor-controlled device by adding steps in a given direction.

Sample Program SMU_POSITION.asm

Position control operations are mostly accomplished in software. The circuit typically includes an input source (analog or digital) that provides information regarding the direction, amount, or speed of the movement required. The sample program SMU_PIC16F84.asm, described earlier in this chapter, can be considered a position control application. In this case, the circuit SMU-PIC16F84-1 contains four pushbuttons and the program reads the pushbuttons to jog the motor in the forward or reverse direction at a slow or fast slew rate.

The sample program SMU_POSITION.asm, in this book's software package, uses the circuit SMU-PIC16F84-1 (Figure 17-5) to illustrate position control. In this case the 5K potentiometer in the circuit is read by the PIC's ADC module. The resulting resistance value is used to turn the motor a specific number of steps in either direction. Program logic is as follows:

1. During initialization, the current position of the potentiometer arm is stored as the *local reference point* (LRP).

2. The potentiometer is read by code and designated as new reference point (NRP).

3. NRP and LRP are compared. If they are the same, no action takes place. Execution continues at Step 2.

4. If the NRP is greater than the LRP, then the difference is the number of steps the motor is turned in the clockwise direction.

5. If the NRP is smaller than LRP then the difference is the number of steps the motor is turned in the counterclockwise direction.

6. LRP is now set to NRP.

7. Processing continues at Step 2.

In the program SMU_POSITION.asm, the values of NRP and LRP are stored in local variables called *nrp* and *lrp*, respectively. The 16F684 PIC is initialized so that PORTC is output and PORTA is input. PORTA line 0 is designated as analog input and wired to the potentiometer. The ADC hardware is set to left-justify the result because the program only used the eight high-order bits of the resistance reading. Once the local variables are initialized, code proceeds as follows:

```
;=================================
;      Read pot and pulse motor
;=================================
ReadPotAndMove:
    call      Delay           ; Wait to update
    call      ReadPot
    banksel   nrp
    movwf     nrp             ; Store result
;
; Compare new value with old value by subtraction.
; Subtraction performs w (new reference position)
; minus local reference position (old value).
; If Z flag set nrp == lrp
; If carry flag set lrp < nrp
;
    subwf     lrp,w           ; Subtract w (nrp) from lrp
```

```
        banksel   STATUS
        btfsc     STATUS,Z  ; Test zero flag
        goto      NoChange      ; Go if Z flag is set
; At this point nrp not equal to lrp
; Test the carry flag to determine if lrp < nrp
        btfss     STATUS,C       ; Is C flag set?
        goto      MoveLeft       ; Go if new bigger than old
        goto      MoveRight
;
;=============================
;          move left
;=============================
; Move left by difference between new and old readings
; First obtain absolute value of difference. At this point
; lrp > nrp
MoveLeft:
        banksel   lrp
        movf      lrp,w          ; Get lrp
        subwf     nrp,w          ; old (in w) minus new
        banksel   ticks
        movwf     ticks          ; Store difference
; Set controller to reverse rotation
        banksel   PORTC
        bsf       PORTC,1        ; Set reverse
TickLeft:
        banksel   PORTC
        bcf       PORTC,0        ; Set low
        call      Pulse
        call      Delay
        banksel   ticks          ; Counter
        decfsz    ticks,f
        goto      TickLeft       ; Loop
        goto      Update
;
;=============================
;          move right
;=============================
; Move to the right by number of ticks in w
MoveRight:
        banksel   ticks
        movwf     ticks          ; Store difference
; Set forward rotation
        banksel   PORTC
        bcf       PORTC,1        ; Clear bit
TickRight:
        banksel   PORTC
        bcf       PORTC,0        ; Set low
        call      Pulse
        call      Delay
        banksel   ticks          ; Counter
        decfsz    ticks,f
        goto      TickRight ; Loop
;
;=============================
;    Update position control
;=============================
Update:
        banksel   nrp
        movf      nrp,w          ; New value to w
        movwf     lrp
```

```
NoChange:
     goto ReadPotAndMove
```

The local procedures ReadPot, Delay, and Pulse are not listed and can be found in the sample code.

17.4 Demonstration Programs

The programs listed in the following section demonstrate the programming discussed in this chapter.

17.4.1 SMB_297_293D.asm

```
;=============================================================
; File: SMB_297_293D.asm
; Date: November 6, 2010
; Update: November 13, 2010
; Authors: Sanchez and Canton
; Processor: 16F84A
; Reference circuit: SMB-L297-293D-1
;
; Program Description:
; Program to drive a bipolar stepper motor using a 16F84 PIC
; as a controller, wired to an L297 and a 293D driver. The
; program reads the toggle switch wired to port RA4 to select
; between clockwise or counterclockwise rotation. The toggle
; switch wired to line RA3 allows selecting between half- or
; full-step modes. Slow and fast speeds are determined by the
; setting of the toggle switch on port line RA3. The pushbutton
; switches on circuit SMB-L297-293D-1 are not used by the
program.
; Circuit wiring is as follows:
;
; SLOW/FAST RA2->|--------|
; FULL/HALF RA3->|        |
; CW/CCW    RA4->|        |<------- OSC
;          RESET->|        |<------- OSC
;            GND |        | +5v
; ENABLE    RB0<-|        |
; CW / CCW  RB1<-|        |
;     STEP  RB2<-|        |
; HALF/FULL RB3<-|        |
;                |--------|
;
;========================
; setup and configuration
;========================
         processor 16f84A
         include   <p16f84A.inc>
         __config  _XT_OSC & _WDT_OFF & _PWRTE_ON & _CP_OFF
```

```
        errorlevel      -302
;================================================================
;                      constant definition
;================================================================
;
;================================================================
;                      variables in PIC RAM
;================================================================
; Declare variables at 2 memory locations
j       equ     0x0c
k       equ     0x0d
bounce  equ     0x0e    ; For debounce routine
delay   equ     0x0f    ; Delay count


;================================================================
;              m a i n    p r o g r a m
;================================================================
        org             0          ; start at address 0
        goto    Main
;=============================
; space for interrupt handler
;=============================
        org             8
;=============================
;       main program
;=============================
Main:
; Clear bit 7 in OPTION_REG to enable port B pullups
        bcf             OPTION_REG,7
; Note: although enabling the weak pullups on port B
;       should allow the circuit to operate without the
;       conventional 10K pullup resistors, we have
;       found that the weak pullups are unreliable on
;       the 16F84. Therefore we advise that the board
;       include the 10K pullups.
; Tris PORT A for input
        movlw   B'00011111'        ; w = 00011111 binary
        tris    PORTA              ; Set up port A for input
; Tris lines 0 to 3 in port B for output and
; lines 4 to 7 (toggle switch # 2) for input
        movlw   B'11110000'        ; RB7-RB4 for input, rest
                                   ; for output
        tris    PORTB              ; Set up port B
        clrf    PORTB              ; Clear all lines
; Set default state for L297 control lines, wired as
; follows:
;    PORT B BITS
;    7 6 5 4 3 2 1 0                        Default
```

```
;      |   |   |   |   |   |   |   |____ ENABLE        1 (active)
;      |   |   |   |   |   |   |_____ CW/CCW         0 (CW)
;      |   |   |   |   |   |_____ CLOCK           1
;      |   |   |   |   |_____ mode             1 (half step)
;      |__|__|__|_____ NOT USED

        movlw     B'00001101'       ; Half step, cw direction
        movwf     PORTB
; Initialize default value for delay variable
        movlw     .40
        movwf     delay             ; To delay variable
;================================================================
;                  main control monitoring routine
;================================================================
ControlRtn:
; Call procedure to set L297 lines according to state of three
; toggle switches wired to port lines A2 to A4
        call      ReadAToggles
        call      Delay             ; Delay now
        call      Pulse             ; Local pulse motor routine
        goto      ControlRtn


;===========================================================
;       Auxiliary procedure to read toggle switches
;===========================================================
ReadAToggles:
; Read PORTA toggle switches
;       PORT A      TOGGLE
;       line        SW      ACTION
;        4          1       0 = CW rotation 1 = CCW
;        3          2       0 = full step 1 = half step
;        2          3       1 = fast 0 = slow
;=============================
;   set CW or CCW rotation
;=============================
; Bit 4 on port A is wired to toggle # 1
; Bit 1 on port B is wired to pin 17 of the L297, which
; controls CW and CCW rotation.
        btfsc     PORTA,4
        goto      goCCW                 ; Set rotation bit to CCW
; Bit clear. Set CW
        bcf       PORTB,1               ; Clear port B line 1
        goto      SetStep
goCCW:
        bsf       PORTB,1
;=============================
;   set half or full step
;=============================
```

```
SetStep:
; Toggle switch # 2, on port A, line 3, selects between
; half step and full step modes. Port B bit 3 is wired
; to L297 line 19, which selects half or full step modes
        btfsc   PORTA,3
        goto    HalfStep            ; Bit is set. Set half step
mode
; Bit clear. Set full step mode
        bcf     PORTB,3             ; Clear port B line 2
        goto    SetSpeed
HalfStep:
        bsf     PORTB,3
;=============================
;  set fast or slow speed
;=============================
SetSpeed:
; Toggle switch # 3, on port A, line 2, selects between
; fast and slow motor speed. If the toggle is on the delay
; value it is set to 40, if it is off it is set to 80
        btfsc   PORTA,2
        goto    FastSpeed           ; Bit is set. Set fast delay
; Bit clear. Set slow speed
        movlw   .80
        movwf   delay               ; To delay variable
        goto    Enable
FastSpeed:
        movlw   .40
        movwf   delay               ; To delay variable
;=============================
;  enable motor control line
;=============================
; The L297 is always in the ENABLE state
Enable:
        bsf     PORTB,0             ; Set bit to enable
        return


;================================
;  routine to pulse the motor
;================================
; L297 CLOCK line is wired to RB2
Pulse:
        bsf     PORTB,2             ; Bring step line high
        nop
        nop
        nop
        nop
        nop
        nop
```

```
        nop
        nop
        nop
        nop
        bcf             PORTB,2         ; Step line low
        return

;===============================
;       delay sub-routines
;===============================
Delay:
        movf    delay,w                 ; Delay to w register
        movwf   j                       ; j = w
Jloop:
        movwf   k                       ; k = w
Kloop:
        decfsz  k,f                     ; k = k-1, skip next if zero
        goto    Kloop
        decfsz  j,f                     ; j = j-1, skip next if zero
        goto    Jloop
        return
        end                     ;END OF PROGRAM
```

17.4.2 SMU_PIC16F84.asm Program

```
;================================================================
; File: SMU_PIC16F84.asm
; Date: October 3, 2010
; Update: November 27, 2010
; Authors: Sanchez and Canton
; Processor: 16F84A
; Reference circuit: SMU-PIC16F84-1
;
; Program Description:
; Program to drive a unipolar stepper motor in one-half step
; mode with a 16F84 wired to 4050 hex buffer IC and TIP 120
; transistors.
; Program uses a lookup table for the sequence codes.
;
; INPUT:
; Pushbuttons wired to ports A3 and A2 activate slow rotation
; in forward and reverse direction when held down. Pushbuttons
; wired to ports A0 and A1 activate forward and reverse fast
; rotation. Code uses an access code table with a pointer, that
; is either incremented or decremented according to the
; selected direction. This ensures that reverses in rotation
; are executed smoothly. Fast and slow executions are determined
; by the value of a local variable read by the delay routine.
```

```
;============================
;    demo board circuit
;============================
; Port A lines 0 to 3 trised for input:
;   0 ---> Pushbutton switch A, active low
;   1 ---> Pushbutton switch B, active low
;   2 ---> Pushbutton switch C, active low
;   3 ---> Pushbutton switch D, active low
; Port B lines 0 to 3 trissed for output and wired
; as follows:
; RB0 ---> Q1 motor winding
; RB1 ---> Q4 motor winding
; RB2 ---> Q2 motor winding
; RB3 ---> Q3 motor winding


;=========================
; setup and configuration
;=========================
        processor 16f84A
        include   <p16f84A.inc>
        __config  _XT_OSC & _WDT_OFF & _PWRTE_ON & _CP_OFF
        errorlevel      -302
;==============================================================
;                    variables in PIC RAM
;==============================================================
; Declare variables at 2 memory locations
j                equ        0x0c
k                equ        0x0d
this_cycle       equ        0x0e      ; Current cycle counter
direction        equ        0x0f      ; Direction control
                                      ; 0 = forward
                                      ; 1 = reverse
bounce           equ        0x10      ; Debounce counter
delay            equ        0x11      ; Delay counter
;==============================================================
;              m a i n   p r o g r a m
;==============================================================
        org     0             ; start at address 0
        goto    Main
;=============================
; space for interrupt handler
;=============================
        org     0x08
;==================================
;    look-up table for motor
;==================================
; Series 42M048C unipolar stepper motors require the
; following one-half step drive sequence for clockwise
```

```
; rotation:
;                   |   w  i  n  d  i  n  g  s   |
;           STEP   Q1      Q2      Q3      Q4      BIN
;            1     ON      OFF     ON      OFF     1010
;            2     ON      OFF     OFF     OFF     1000
;            3     ON      OFF     OFF     ON      1001
;            4     OFF     OFF     OFF     ON      0001
;            5     OFF     ON      OFF     ON      0101
;            6     OFF     ON      OFF     OFF     0100
;            7     OFF     ON      ON      OFF     0110
;            8     OFF     OFF     ON      OFF     0010
; Table is placed low in memory to avoid page overlaps.
;
CodeTable:
        addwf   PCL,f               ; Add w to program counter
        retlw   B'00001010'         ; cycle = 0
        retlw   B'00001000'         ;          1
        retlw   B'00001001'         ;                      2
        retlw   B'00000001'         ;                      3
        retlw   B'00000101'         ;                      4
        retlw   B'00000100'         ;                      5
        retlw   B'00000110'         ;                      6
        retlw   B'00000010'         ;                      7
        retlw   B'00000000'         ; Table terminator


;==============================
;         main program
;==============================
Main:
; Tris PORT A for input
        movlw   B'00011111'         ; w = 00011111 binary
        tris    PORTA               ; Set up port A for input
; Initialize all lines in port B for output
        movlw   B'00000000'         ; w = 00000000 binary
        tris    PORTB               ; Set up port B for output
        clrf    PORTB               ; Clear all lines
; Initialize control variables
        clrf    this_cycle
        clrf    direction           ; Assume forward direction
        movlw   .100                ; Initial delay for fast
        movwf   delay               ; To variable
;==============================
;   read and debounce PORTA
;       lines 0 to 3
;==============================
; Switches are active low, so changes in direction take place
; if port line is clear. Software debouncing is used to make
; sure that spurious switch reads are ignored.
```

```
TestPortA:
; First read and debounce PORTA-3
        movlw    .10              ; Counter
        movwf    bounce
Wait43:
        btfsc    PORTA,3
        goto     No3Action        ; Go if bit is set
        decfsz   bounce,f         ; Decrement counter
        goto     Wait43
; At this point switch held zero for 10 tests
        clrf     direction        ; 0 is forward on direction
switch
        movlw    .100             ; Slow delay rate
        movwf    delay            ; To delay counter
        call     OneTick          ; Tick motor
        goto     TestPortA
;==============================
;   read and debounce PORTA-2
;==============================
No3Action:
        movlw    .10              ; Counter
        movwf    bounce
Wait42:
        btfsc    PORTA,2
        goto     No2Action        ; Go if bit is set
        decfsz   bounce,f         ; Decrement counter
        goto     Wait42
; At this point switch held zero for 10 tests
        movlw    .1               ; 1 is reverse on direction
                                  ; switch
        movwf    direction        ; Code to switch
        movlw    .100             ; Slow delay rate
        movwf    delay            ; To delay counter
        call     OneTick          ; Tick motor
        goto     TestPortA
No2Action:
;================================
;   read and debounce PORTA-1
;================================
; Debounce PORTA-1
TestPortA1:
        movlw    .10              ; Counter
        movwf    bounce
Wait41:
        btfsc    PORTA,1
        goto     No1Action        ; Go if bit is set
        decfsz   bounce,f         ; Decrement counter
        goto     Wait41
```

```
; At this point switch held zero for 10 tests
        clrf      direction          ; 0 is forward on direction
switch
        movlw     .40                ; Fast delay rate
        movwf     delay              ; To delay counter
        call      OneTick            ; Tick motor
        goto      TestPortA
;================================
;   read and debounce PORTA-1
;================================
No1Action:
; Debounce PORTA-0
        movlw     .10                ; Counter
        movwf     bounce
Wait40:
        btfsc     PORTA,0
        goto      No0Action          ; Go if bit is set
        decfsz    bounce,f ; Decrement counter
        goto      Wait40
; At this point switch held zero for 10 tests
        movlw     .1                 ; 1 is reverse on direction
                                     ; switch
        movwf     direction          ; Code to switch
        movlw     .40                ; Fast delay rate
        movwf     delay              ; To delay counter
        call      OneTick            ; Tick motor
        goto      TestPortA
No0Action:
        goto      TestPortA
;==========================
;    forward or reverse
;   single cycle procedure
;==========================
OneTick:
; Test for change of direction
        btfss     direction,0        ; Test forward bit
        goto      Forward
; At this point direction rotation is reverse
        goto      Reverse
; Forward direction routine
Forward:
        movf      this_cycle,w       ; Get current cycle
        call      CodeTable          ; Get code from table
        movwf     PORTB              ; Store code in port
        call      Delay
; Bump cycle counter
        incf      this_cycle,f       ; Add one
; Test for cycle number 7 (last one in sequence)
```

```
        btfsc   this_cycle,3 ; 1000 = 8
        goto    Recycle           ; Reset if at end of cycle
        return
Recycle:
        clrf    this_cycle
        return
; Reverse direction routine
Reverse:
        movf    this_cycle,w      ; Get current cycle
        call    CodeTable         ; Get code from table
        movwf   PORTB             ; Store code in port
        call    Delay
; Bump cycle counter
        decf    this_cycle,f ; Subtract one
; Test for cycle number 0xff (overflow from cycle # 0)
        btfsc   this_cycle,4 ; Bit 4 set indicates overflow
        goto    Recycle2 ; Reset if at end of cycle
        return
Recycle2:
        movlw   .7                ; First reverse cycle
        movwf   this_cycle        ; To cycle counter
        return


;===============================
;       delay sub-routine
;===============================
Delay:
        movf    delay,w ; Load delay value
        movwf   j                 ; j = w
Jloop:
        movwf   k                 ; k = w
Kloop:
        decfsz  k,f               ; k = k-1, skip next if zero
        goto    Kloop
        decfsz  j,f               ; j = j-1, skip next if zero
        goto    Jloop
        return

        end                     ;END OF PROGRAM
```

17.4.3 SMU_5804.asm

```
;=============================================================
; File: SMU_5804.asm
; Date: October 3, 2010
; Update: December 4, 2010
; Authors: Sanchez and Canton
; Processor: 16F84A
; Reference circuit: SMU-5804-1
```

```
;
;  Program Description:
;  Program to drive a unipolar stepper motor mode with a 16F84
;  PIC controlling a UCN5804 driver.
;  Three toggle switches wired to ports A2 to A4 turn on and
;  off the direction, half step, and single phase modes on the
;  5804.
;  Four toggle switches wired to ports B4 to B7 provide input
;  in the range 0 to 15. This value is used to select the delay
;  rate that determines motor speed.
;===========================
;    demo board circuit
;===========================
;  Port A lines 0 to 4 trissed for input:
;   0 -----> NOT WIRED
;   1 -----> NOT WIRED
;   2 -----> Toggle switch 3
;   3 -----> Toggle switch 2
;   4 -----> Toggle switch 1
;  Port B lines 0 to 3 trissed for output and wired
;  as follows:
;    PIC          5804
;    RB0 -----> pin 11 - Step input
;    RB1 -----> pin 14 - Direction
;    RB2 -----> pin 10 - Half step
;    RB3 -----> pin 9  - One phase
;  Port B lines 4 to 7 are trissed for input
;    RB4 --------------|
;    RB5 --------------|---- motor speed control
;    RB6 --------------|        toggle switches
;    RB7 --------------|
;  Settings for phase control:
;   RB1 (half step)  RB3 (one phase)
;           0                0            = two phase
;           0                1            = wave drive
;           1                0            = half step
;           1                1            = step inhibit
;  Toggle switch controls
;   PORTA line    SW       ACTION
;        4        TS3      Direction (forward/reverse)
;                          Set PORTB to: xxxx xx0x
;                                    or: xxxx xx1x
;        3        TS2      One-half step
;                          Set PORTB to: xxxx x0xx
;                                    or: xxxx x1xx
;        2        TS1      One phase
;                          Set PORTB to: xxxx 0xxx
;                                    or: xxxx 1xxx
;
```

```
;==========================
; setup and configuration
;==========================
        processor 16f84A
        include   <p16f84A.inc>
        __config  _XT_OSC & _WDT_OFF & _PWRTE_ON & _CP_OFF
        errorlevel        -302
;=============================================================
;                       variables in PIC RAM
;=============================================================
; Declare variables at 2 memory locations
j         equ       0x0c
k         equ       0x0d
delay     equ       0x11      ; Delay count
;=============================================================
;                   m a i n   p r o g r a m
;=============================================================
        org     0             ; start at address 0
        goto    Main
;=============================
; space for interrupt handler
;=============================
        org     8
;=============================
;   Delay table for speed
;          control
;=============================
; Values in table range from 250 to 25 (15 units apart)
; Table has 16 entries corresponding to the values
; represented by the setting of the 4 speed control
; toggle switches
SpeedTable:
        addwf   PCL,f             ; Add w to program counter
        retlw   .255              ; Slowest speed  0000 = 0
        retlw   .240              ;                0001 = 1
        retlw   .225              ;                0010 = 2
        retlw   .210              ;                0011 = 3
        retlw   .195              ;                0100 = 4
        retlw   .180              ;                0101 = 5
        retlw   .165              ;                0110 = 6
        retlw   .150              ;                0111 = 7
        retlw   .135              ;                1000 = 8
        retlw   .120              ;                1001 = 9
        retlw   .105              ;                1010 = 10
        retlw   .90               ;                1011 = 11
        retlw   .75               ;                1100 = 12
        retlw   .60               ;                1101 = 13
        retlw   .45               ;                1110 = 14
```

```
            retlw    .30                 ; Fastest         1111 = 15
            retlw    .0                         ; Terminator


;===============================
;        main program
;===============================
Main:
; Tris PORT A for input
            movlw    B'00011111'        ; w = 00011111 binary
            tris     PORTA              ; Set up port A for input
; Tris port B low nibble for output and high nibble
; for input
            movlw    B'11110000'        ; w = 11110000 binary
            tris     PORTB              ; Port B lines 0 to 3 are
                                        ; output and lines 4 to 7
                                        ; are input
            clrf     PORTB              ; Clear all lines
; Set default state for 5801 control lines, wired as
; follows:
;     PORT B BITS
;     7  6  5  4  3  2  1  0                    Default
;     |  |  |  |  |  |  |  |____ Step input       0
;     |  |  |  |  |  |  |_____ Direction        0
;     |  |  |  |  |  |_____ One half step    1
;     |  |  |  |  |_____ One phase         0
;     |__|__|__|_____ speed control input
; Note that 5804 Enable line is wired to ground
            movlw    B'00000100'        ; Half step, cw direction
            movwf    PORTB
; Initialize default value for delay variable
            movlw    .100
            movwf    delay              ; To delay variable
;===============================
;      Pulse motor and delay
;===============================
PulseAndDelay:
; Make sure 5804 Step In line is low before strobe
; cycle
            bcf      PORTB,0            ; Set low
            call     Pulse              ; Local pulse motor routine
            call     Delay
;===============================
; Read speed control toggles
;===============================
; Read four speed control toggle switches on port
; B lines 4 to 7. The swapf instruction allows swapping
; nibbles as they are read from the port
            swapf    PORTB,w            ; Read port B and swap nibbles
```

```
        andlw    B'00001111'        ; Mask out 4 high bits
        call     SpeedTable         ; Use value in w to obtain
                                    ; delay
        movwf    delay              ; Store code in local variable
;=============================
;   Read 5804 control lines
;=============================
; PORTA bit 4 (wired to toggle # 1) controls forward
; or reverse direction by means of PORTB bit 1 (wired
; to 5804 direction line on pin 14).
        btfsc    PORTA,4            ; Direction control switch
        goto     ReverseRot         ; Engage reverse rotation
; At this point rotation is forward
        bcf      PORTB,1            ; Clear bit `
        goto     HalfStepMode
ReverseRot:
        bsf      PORTB,1            ; Set reverse
HalfStepMode:
; PORTA bit 3 (wired to toggle # 2) selects half-step
; mode by means of PORTB bit 2 (wired to 5804 half-step
; line on pin 10).
        btfsc    PORTA,3            ; Half step mode control
        goto     NoHalfStep         ; Disable half step
; At this point half step is selected
        bsf      PORTB,2            ; Set bit `
        goto     OnePhaseMode
NoHalfStep:
        bcf      PORTB,2            ; Turn off half step control
OnePhaseMode:
; PORTA bit 2 (wired to toggle # 3) selects one phase
; mode by means of PORTB bit 3 (wired to 5804 single
; phase line on pin 9).
        btfsc    PORTA,2            ; One phase mode control
        goto     NoOnePhase         ; Disable single step
; At this point single phase is selected
        bsf      PORTB,3            ; Set single phase `
        goto     PulseAndDelay
NoOnePhase:
        bcf      PORTB,3            ; Turn off one phase
        goto     PulseAndDelay
;================================
;   routine to pulse the 5804
;================================
; Note that 5804 requires a positive-going pulse
; therefore the line is held high during the strobe
; cycle
Pulse:
        bsf      PORTB,0            ; Bring step line high
```

```
        nop
        nop
        nop
        nop
        nop
        nop
        nop
        nop
        nop
        nop
        bcf       PORTB,0           ; Step line low
        return


;===============================
;       delay sub-routine
;===============================
Delay:
        movf      delay,w           ; Delay to w register
        movwf     j                 ; j = w
Jloop:
        movwf     k                 ; k = w
Kloop:
        decfsz    k,f               ; k = k-1, skip next if zero
        goto      Kloop
        decfsz    j,f               ; j = j-1, skip next if zero
        goto      Jloop
        return

        end                       ;END OF PROGRAM
```

17.4.4 SMU_5804_INT.asm

```
;================================================================
; File: SMU_5804_INT.asm
; Date: December 10, 2010
; Update:
; Authors: Sanchez and Canton
; Processor: 16F84A
; Reference circuit: SMU-5804-1
;
; Program Description:
; This is a version SMU_5804.asm that uses the Timer0
; interrupt to send pulses to the stepper motor.
; Three toggle switches wired to ports A2 to A4 turn on and
; off the direction, half step, and single phase modes on the
; 5804.
; Four toggle switches wired to ports B4 to B7 provide input
; in the range 0 to 15. This value is used to select the delay
```

```
; rate from a lookup table in RAM. This delay rate is used in
; the interrupt handler to determine when a pulse is sent to
; the stepper motor.
;===========================
;    demo board circuit
;===========================
; Port A lines 0 to 4 trissed for input:
;   0 -----> NOT WIRED
;   1 -----> NOT WIRED
;   2 -----> Toggle switch 3
;   3 -----> Toggle switch 2
;   4 -----> Toggle switch 1
; Port B lines 0 to 3 trissed for output and wired
; as follows:
;   PIC           5804
;   RB0 -----> pin 11 - Step input
;   RB1 -----> pin 14 - Direction
;   RB2 -----> pin 10 - Half step
;   RB3 -----> pin 9  - One phase
; Port B lines 4 to 7 are trissed for input
;   RB4 --------------|
;   RB5 --------------|---- motor speed control
;   RB6 --------------|     toggle switches
;   RB7 --------------|
; Settings for phase control:
;   RB1 (half step)  RB3 (one phase)
;        0                0          = two phase
;        0                1          = wave drive
;        1                0          = half step
;        1                1          = step inhibit
; Toggle switch controls:
;   PORTA line     SW       ACTION
;      4           TS3      Direction (forward/reverse)
;                           Set PORTB to: xxxx xx0x
;                                     or: xxxx xx1x
;      3           TS2      One-half step
;                           Set PORTB to: xxxx x0xx
;                                     or: xxxx x1xx
;      2           TS1      One phase
;                           Set PORTB to: xxxx 0xxx
;                                     or: xxxx 1xxx
;========================
; setup and configuration
;========================
        processor 16f84A
        include   <p16f84A.inc>
        __config  _XT_OSC & _WDT_OFF & _PWRTE_ON & _CP_OFF
        errorlevel        -302
```

```
;=================================================================
;                       variables in PIC RAM
;=================================================================
; Declare variables at 2 memory locations
j               equ     0x0c
k               equ     0x0d
delay           equ     0x11    ; Delay count
old_w           equ     0x12    ; Context saving
old_STATUS      equ     0x13    ; Idem
iteration       equ     0x14    ; Iteration counter used by
                                ; interrupt handler
;=================================================================
;                   m a i n   p r o g r a m
;=================================================================
        org     0               ; start at address 0
        goto    Main
;==============================
; space for interrupt handler
;==============================
        org     0x04
        goto    IntServ
;==============================
;   Delay table for speed
;         control
;==============================
; Values in table range from 250 to 25 (15 units apart)
; Table has 16 entries corresponding to the values
; represented by the setting of the 4 speed control
; toggle switches
SpeedTable:
        addwf   PCL,f   ; Add w to program counter
        retlw   .255    ; Slowest speed 0000 = 0
        retlw   .240    ;                0001 = 1
        retlw   .225    ;                0010 = 2
        retlw   .210    ;                0011 = 3
        retlw   .195    ;                0100 = 4
        retlw   .180    ;                0101 = 5
        retlw   .165    ;                0110 = 6
        retlw   .150    ;                0111 = 7
        retlw   .135    ;                1000 = 8
        retlw   .120    ;                1001 = 9
        retlw   .105    ;                1010 = 10
        retlw   .90     ;                1011 = 11
        retlw   .75     ;                1100 = 12
        retlw   .60     ;                1101 = 13
        retlw   .45     ;                1110 = 14
        retlw   .30     ; Fastest speed 1111 = 15
        retlw   .0      ; Terminator
```

```
;===============================
;         main program
;===============================
Main:
; Tris PORT A for input
        movlw     B'00011111'        ; w = 00011111 binary
        tris      PORTA              ; Set up port A for input
; Tris port B low nibble for output and high nibble
; for input
        movlw     B'11110000'        ; w = 11110000 binary
        tris      PORTB              ; Port B lines 0 to 3 are
                                     ; output and lines 4 to 7
                                     ; are input
        clrf      PORTB              ; Clear all lines
; Clear the watchdog timer and reset prescaler
        clrf      TMR0
        clrwdt
; Set up the OPTION register
        movlw     b'11010000'
;    7  6  5  4  3  2  1  0 <= OPTION bits
;    |  |  |  |  |  |  |__|__|_____ PS2-PS0 (prescaler bits)
;    |  |  |  |  |  |              Values for Timer0
;    |  |  |  |  |  |              *000 = 1:2    001 = 1:4
;    |  |  |  |  |  |              010 = 1:8    011 = 1:16
;    |  |  |  |  |  |              100 = 1:32   101 = 1:64
;    |  |  |  |  |  |              110 = 1:128 111 = 1:256
;    |  |  |  |  |  |_____ PSA (prescaler assign)
;    |  |  |  |  |                1 = to WDT
;    |  |  |  |  |                *0 = to Timer0
;    |  |  |  |  |_____ TOSE (Timer0 edge select)
;    |  |  |  |                   0 = increment on low-to-high
;    |  |  |  |                   *1 = increment on high-to-low
;    |  |  |  |_____ TOCS (TMR0 clock source)
;    |  |  |                      *0 = internal clock
;    |  |  |                      1 = RA4/TOCKI bit source
;    |  |  |_____ INTEDG (Edge select)
;    |  |                         *0 = falling edge
;    |  |_____ RBPU (Pullup enable)
;    |                            0 = enabled
;    |                            *1 = disabled
; Note that OPTION register is in bank 1 and its register
; name is OPTION_REG
        bsf       STATUS,RP0         ; RP0 is bank select bit
        movwf     OPTION_REG         ; Copy w to OPTION
        bcf       STATUS,RP0         ; Bank 0
;=============================
;      set up interrupts
```

```
;=============================
; Clear external interrupt flag (INTF = bit 1)
        bcf       INTCON,INTF      ; Clear flag
; Enable global interrupts (GIE = bit 7)
; Enable RB0 interrupt (inte = bit 4)
        bsf       INTCON,GIE       ; Enable global int (bit 7)
        bsf       INTCON,T0IE      ; Enable TMR0 overflow
                                   ; interrupt
; Set default state for 5801 control lines, wired as
; follows:
;     PORT B BITS
;     7  6  5  4  3  2  1  0                  Default
;     |  |  |  |  |  |  |  |____ Step input       0
;     |  |  |  |  |  |  |_____ Direction       0
;     |  |  |  |  |  |_____ One half step    1
;     |  |  |  |  |_____ One phase         0
;     |__|__|__|_____ speed control input
; Note that 5804 Enable line is wired to ground
        movlw     B'00000100'                ; Half step, cw
direction
        movwf     PORTB
; Initialize default value for delay variable
        movlw     .100
        movwf     delay            ; To delay variable
        movwf     iteration        ; To iteration counter
;=============================
; Read speed control toggles
;=============================
; Read four speed control toggle switches on port
; B lines 4 to 7. The swapf instruction allows swapping
; nibbles as they are read from the port
ReadControls:
        swapf     PORTB,w          ; Read port B and swap nibbles
        andlw     B'00001111'      ; Mask out 4 high bits
        call      SpeedTable       ; Use value in w to obtain
delay
        movwf     delay            ; Store code in local variable
;=============================
;   Read 5804 control lines
;=============================
; PORTA bit 4 (wired to toggle # 1) controls forward
; or reverse direction by means of PORTB bit 1 (wired
; to 5804 direction line on pin 14).
        btfsc     PORTA,4          ; Direction control switch
        goto      ReverseRot       ; Engage reverse rotation
; At this point rotation is forward
        bcf       PORTB,1          ; Clear bit `
        goto      HalfStepMode
```

```
ReverseRot:
        bsf       PORTB,1            ; Set reverse
HalfStepMode:
; PORTA bit 3 (wired to toggle # 2) selects half-step
; mode by means of PORTB bit 2 (wired to 5804 half-step
; line on pin 10).
        btfsc     PORTA,3            ; Half step mode control
        goto      NoHalfStep         ; Disable half step
; At this point half step is selected
        bsf       PORTB,2            ; Set bit `
        goto      OnePhaseMode
NoHalfStep:
        bcf       PORTB,2            ; Turn off half step control
OnePhaseMode:
; PORTA bit 2 (wired to toggle # 3) selects one phase
; mode by means of PORTB bit 3 (wired to 5804 single
; phase line on pin 9)
        btfsc     PORTA,2            ; One phase mode control
        goto      NoOnePhase         ; Disable single step
; At this point single phase is selected
        bsf       PORTB,3            ; Set single phase `
        goto      ReadControls
NoOnePhase:
        bcf       PORTB,3            ; Turn off one phase
        goto      ReadControls


;===========================================================
;                 Interrupt Service Routine
;===========================================================
; Service routine receives control when the timer
; register TMR0 overflows, that is, when 256 timer beats
; have elapsed
IntServ:
; First test if source is a Timer0 interrupt
        btfss     INTCON,T0IF        ; T0IF is Timer0 interrupt
        goto      notT0IF            ; Go if not RB0 origin
; If so clear the timer interrupt flag so that count continues
        bcf       INTCON,T0IF        ; Clear interrupt flag
; Save context
        movwf     old_w              ; Save w register
        swapf     STATUS,w           ; STATUS to w
        movwf     old_STATUS         ; Save STATUS
;==========================
;    interrupt action
;==========================
; Decrement the iteration counter. Exit if not zero
        decfsz    iteration,f
        goto      exitISR            ; Continue if counter not zero
```

```
; At this point the delay count has expired so the programmed
; time has elapsed.
; First reset the iteration counter
        movf    delay,w             ; Read delay
        movwf   iteration           ; Reset iteration counter
;=================================
;     Pulse motor and delay
;=================================
; Make sure 5804 Step In line is low before strobe
; cycle
        bcf     PORTB,0             ; Set low
        call    Pulse               ; Local pulse motor routine
;=========================
;        exit ISR
;=========================
exitISR:
; Restore context
        swapf   old_STATUS,w        ; Saved STATUS to w
        movfw   STATUS              ; To STATUS register
        swapf   old_w,f             ; Swap file register in itself
        swapf   old_w,w             ; re-swap back to w
; Reset,interrupt
notTOIF:
        retfie
;=================================
;   routine to pulse the 5804
;=================================
; Note that 5804 requires a positive-going pulse
; therefore the line is held high during the strobe
; cycle
Pulse:
        bsf     PORTB,0             ; Bring step line high
        nop
        nop
        nop
        nop
        nop
        nop
        nop
        nop
        nop
        nop
        bcf     PORTB,0             ; Step line low
        return

                end                 ;END OF PROGRAM
```

17.4.5 SMU_PIC16F684.asm

```
; ================================================================
; File: SMU_PIC16F684.asm
; Date: December 9, 2010
; Update: December 30, 2010
; Authors: Sanchez and Canton
; Processor: 16F684A
; Reference circuit: SMU-PIC16F684-1
;
; Program Description:
; Program to drive a unipolar stepper motor in half step
; mode, with a 16F684 PIC controlling a UCN5804 driver.
; Three toggle switches wired to ports RA1 to RA3 turn on and
; off the direction, half step, and single phase modes on the
; 5804.
; A 5K potentiometer on port RA0 provides input to the chip's
; ADC. The resulting resistance value is used to select the
; delay rate which, in turn, determines the motor speed.
;
;============================
;    demo board circuit
;============================
; Port A lines 0 to 4 trised for input:
;   0 ---> 5K potentiometer  => Analog input
;   1 ---> Toggle switch 3   |
;   2 ---> Toggle switch 2   |=> Digital input
;   3 ---> MCLR function     |
;   4 ---> Toggle switch 4   |
;   5 ---> NOT WIRED
; Port C lines 0 to 3 trissed for output and wired
; as follows:
;    PIC          5804
;    RC0 ---> pin 11 - Step input
;    RC1 ---> pin 14 - Direction
;    RC2 ---> pin 9  - One phase
;    RC3 ---> pin 10 - Half Step
; Settings for phase control:
;    RC3 (half step)  RC2 (one phase)
;           0                0          = two phase
;           0                1          = wave drive
;           1                0          = half step
;           1                1          = step inhibit
; Toggle switch controls
;    PORTC line     SW      ACTION
;       1           TS3     Direction (forward/reverse)
;                           Set PORTC to: xxxx xx0x
;                                     or: xxxx xx1x
```

```
;       2       TS2     One phase
;                       Set PORTC to: xxxx x0xx
;                                 or: xxxx x1xx
;       3       TS1     One half step
;                       Set PORTC to: xxxx 0xxx
;                                 or: xxxx 1xxx
;=========================
; setup and configuration
;=========================
        processor 16f684
        include    <p16f684.inc>
        __CONFIG _MCLRE_OFF & _CP_OFF & _CPD_OFF & _BOD_OFF &
_WDT_OFF & _PWRTE_ON & _INTOSCIO & _FCMEN_OFF & _IESO_OFF

        errorlevel      -302    ; No register in bank error
                                ; messages
        errorlevel      -312    ; No page or bank messages
;============================================================
;                   variables in PIC RAM
;============================================================
; Declare variables at 2 memory locations
j       equ     0x20
k       equ     0x21
delay   equ     0x22    ; Delay count
;============================================================
;               m a i n   p r o g r a m
;============================================================
        org             0       ; start at address 0
        goto    Main
;============================
; space for interrupt handler
;============================
;============================
;       main program
;============================
        org     0x08
Main:
; Initialize all lines in port C for output
        banksel PORTC
        clrf    PORTC
        banksel TRISC           ; Prepare to tris
        clrf    TRISC           ; All lines to 0 (output)
; Set PORT A line 0 as analog
        movlw   B'00000001'                     ; Line 0 is
analog
        banksel ANSEL
        movwf   ANSEL           ; Select as analog
; Turn off comparator
```

```
        movlw    B'00000111'        ; Setting bits 0, 1, and 2
                                     ; sets
comparators off and
                                     ; Port C IN
pins to digital
        banksel  CMCON0
        movwf    CMCON0
; Tris port A for input
        banksel  TRISA
        movlw    B'00111111'        ; All port A lines for input
        clrf     TRISA
        movwf    TRISA
; Configure oscillator
        movlw    B'01100001'
;                    |||   |____ internal clock
;                    |||_____ 4 MHz
        banksel  OSCCON
        movwf    OSCCON
; Set conversion clock to Tad = 8 * Tosc (ADCS = 001)
        movlw    B'00010000'
;                    |||_____ ADCS bits
        banksel  ADCON1
        movwf    ADCON1
;==============================
;       set ADC hardware
;==============================
; Select analog line (AN0) as the analog input
; channel. Select Vcc as the voltage reference.
; Left justify result
        movlw    B'00000001'; Bitmap:
;                     0 0 0 0 0 0 0 1
;                     | |___| | | |___x___ 1 = turn on ADC
;                     | |___x_x_x_____ 000 = select AN0
;                     | x_____ 0 = voltage ref Vcc
;                     x_____ 0 = left justify
        banksel  ADCON0
        movwf    ADCON0
; Set default state for 5804 control lines, wired as
; follows:
;    PORT C BITS
;    5 4 3 2 1 0                        Default
;    | | | | | |____ Step input         0
;    | | | | |_____ Direction          1
;    | | | |_____ One phase          0
;    | | |_____ Half step          1
;    | |_____ MCLR
;    |_____ NOT USED
; Note that 5804 Enable line is wired to ground
```

```
            banksel PORTC
            movlw   B'00001010'        ; Half step, ccw direction
            movwf   PORTC
; Initialize default value for delay variable
            movlw   .100
            banksel delay
            movwf   delay              ; To delay variable
;=================================
;      Pulse motor and delay
;=================================
PulseAndDelay:
; Make sure 5804 Step In line is low before strobe
; cycle
            banksel PORTC
            bcf     PORTC,0            ; Set low
            call    Pulse              ; Local pulse motor routine
            call    Delay
;=============================
;   Read speed control pot
;=============================
; 5K pot is wired to port A line 0, which has been set
            banksel ADCON0             ; Prepare to sample
            bsf     ADCON0,GO          ; Start operation
Wait4ADC:
            btfsc   ADCON0,GO          ; Wait for completion
            goto    Wait4ADC           ; Loop back if not done
;=================================
;      store and display result
;=================================
; At this point digital values of ADC operation are
; found in the PICs ADRESH and ADRESL registers.
; Program uses the 8 high bits in ADRESH, which are
; divided by 2 by shifting right one bit. A constant of
; 40 is then added to scale and limit the fastest speed.
            banksel STATUS
            bcf     STATUS,C           ; Clear carry to shift
            banksel ADRESH
            rrf     ADRESH,w           ; Rotate in w
                                       ; Range is now 0 to 128
            addlw   .40                ; Add constant
                                       ; Range is now 40 to 168
            banksel delay              ; Delay variable
            movwf   delay
;=============================
;   Read 5804 control lines
;=============================
; PORTA bit 4 (wired to toggle # 1) controls forward
; or reverse direction by means of PORTC bit 1 (wired
```

```
; to 5804 direction line on pin 14).
        banksel PORTA
        btfsc   PORTA,4          ; Direction control switch
        goto    ReverseRot       ; Engage reverse rotation
; At this point rotation is forward
        bcf     PORTC,1          ; Clear bit `
        goto    HalfStepMode
ReverseRot:
        bsf     PORTC,1          ; Set reverse
HalfStepMode:
; PORTA bit 2 (wired to toggle # 2) selects half-step
; mode by means of PORTC bit 3 (wired to 5804 half-step
; line on pin 10).
        btfsc   PORTA,2 ; Half step mode control
        goto    NoHalfStep       ; Disable half step
; At this point half step is selected
        bsf     PORTC,3          ; Set bit
        goto    OnePhaseMode
NoHalfStep:
        bcf     PORTC,3          ; Turn off half step control
OnePhaseMode:
; PORTA bit 1 (wired to toggle # 3) selects one phase
; mode by means of PORTC bit 2 (wired to 5804 single
; phase line on pin 9).
        btfsc   PORTA,1          ; One phase mode control
        goto    NoOnePhase       ; Disable single step
; At this point single phase is selected
        bsf     PORTC,2          ; Set single phase
        goto    PulseAndDelay
NoOnePhase:
        bcf     PORTC,2          ; Turn off one phase
        goto    PulseAndDelay
;================================
;   routine to pulse the 5804
;================================
; Note that 5804 requires a positive-going pulse
; therefore the line is held high during the strobe
; cycle
Pulse:
        banksel PORTC
        bsf     PORTC,0          ; Bring step line high
        nop
        nop
        nop
        nop
        nop
        nop
        nop
```

```
            nop
            nop
            nop
            bcf       PORTC,0              ; Step line low
            return
```

```
;================================
;          delay sub-routines
;================================
Delay:
            banksel delay
            movf      delay,w             ; Delay to w register
            movwf     j                   ; j = w
Jloop:
            movwf     k                   ; k = w
Kloop:
            decfsz    k,f                 ; k = k-1, skip next if zero
            goto      Kloop
            decfsz    j,f                 ; j = j-1, skip next if zero
            goto      Jloop
            return

            end                   ;END OF PROGRAM
```

17.4.6 SMU_PIC16F684_INT.asm

```
;================================================================
; File: SMU_PIC16F684_INT.asm
; Date: December 9, 2010
; Update: December 29, 2010
; Author: Julio Sanchez
; Processor: 16F684A
; Reference circuit: SMU-PIC16F684-1
;
; Program Description:
; This is a version SMU_PIC16F684.asm that uses the Timer0
; interrupt to send pulses to the stepper motor.
; Three toggle switches wired to ports RA1 to RA3 turn on and
; off the direction, half step, and single phase modes on the
; 5804.
; A 5K potentiometer on port RA0 provides input to the chip's
; ADC. The resulting resistance value is used to select the
; delay rate, which, in turn, sets the motor speed.
;
;============================
;    demo board circuit
;============================
; Port A lines 0 to 4 trissed for input:
;   0 -----> 5K potentiometer  -- Analog input
```

```
;   1 -----> Toggle switch 3  |
;   2 -----> Toggle switch 2  |-- digital input
;   3 -----> MCLR function     |
;   4 -----> Toggle switch 4  |
;   5 -----> NOT WIRED
; Port C lines 0 to 3 trissed for output and wired
; as follows:
;   PIC        5804
;   RC0 -----> pin 11 - Step input
;   RC1 -----> pin 14 - Direction
;   RC2 -----> pin 9  - One phase
;   RC3 -----> pin 10 - Half Step
; Settings for phase control:
;   RC3 (half step)  RC2 (one phase)
;           0                0           = two phase
;           0                1           = wave drive
;           1                0           = half step
;           1                1           = step inhibit
; Toggle switch controls
;   PORTC line      SW       ACTION
;       1           TS3      Direction (forward/reverse)
;                            Set PORTC to: xxxx xx0x
;                                      or: xxxx xx1x
;       2           TS2      One phase
;                            Set PORTC to: xxxx x0xx
;                                      or: xxxx x1xx
;       3           TS1      One half step
;                            Set PORTC to: xxxx 0xxx
;                                      or: xxxx 1xxx
;=========================
; setup and configuration
;=========================
      processor 16f684
      include      <p16f684.inc>
      __CONFIG   _MCLRE_OFF & _CP_OFF & _CPD_OFF & _BOD_OFF &
_WDT_OFF & _PWRTE_ON & _INTOSCIO & _FCMEN_OFF & _IESO_OFF

      errorlevel -302  ; No register in bank error messages
      errorlevel  -312 ; No page or bank messages


;=============================================================
;                      variables in PIC RAM
;=============================================================
; Declare variables at 2 memory locations
j                 equu         0x20
k                 equ          0x21
delay      equ          0x22  ; Delay count
old_w      equ          0x23  ; Context saving
```

```
old_STATUS          equ             0x24  ; Idem
iteration           equ             0x25  ; Iteration counter used by
                                          ; interrupt handler
;=============================================================
;                   m a i n   p r o g r a m
;=============================================================
        org         0         ; start at address 0
        goto        Main
;=============================
; space for interrupt handler
;=============================
        org         0x04
        goto        IntServ
;=============================
;       main program
;=============================
Main:
; Initialize all lines in port C for output
        banksel     PORTC
        clrf        PORTC
        banksel     TRISC         ; Prepare to tris
        clrf        TRISC         ; All lines to 0 (output)
; Set PORT A line 0 as analog
        movlw B'00000001'         ; Line 0 is analog
        banksel     ANSEL
        movwf ANSEL               ; Select as analog
; Turn off comparator
        movlw B'00000111' ; Setting bits 0, 1, and 2
                                        ; sets comparators off and
                                        ; Port C IN pins to digital
        banksel     CMCON0
        movwf CMCON0
; Tris port A for input
        banksel     TRISA
        movlw B'00111111' ; All port A lines for input
        clrf        TRISA
        movwf TRISA
; Configure oscillator
        movlw B'01100001'
;                       |||   |____ internal clock
;                       ||_____ 4 MHz
        banksel     OSCCON
        movwf OSCCON
; Set conversion clock to Tad = 8 * Tosc (ADCS = 001)
        movlw B'00010000'
;                       |||_____ ADCS bits
        banksel     ADCON1
        movwf ADCON1
```

```
;==============================
;   setup for interrupt
;==============================
; Clear external interrupt flag (INTF = bit 1)
      banksel    INTCON
      bcf            INTCON,INTF ; Clear flag
; Enable global interrupts (GIE = bit 7)
; Enable RB0 interrupt (inte = bit 4)
      bsf            INTCON,GIE  ; Enable global int (bit 7)
      bsf            INTCON,T0IE ; Enable TMR0 overflow interrupt
;==============================
; Set up the OPTION register
;==============================
      movlw b'11010000'
;    7  6  5  4  3  2  1  0 <= OPTION bits
;    |  |  |  |  |  |  |__|__|_____ PS2-PS0 (prescaler bits)
;    |  |  |  |  |  |              Values for Timer0
;    |  |  |  |  |  |              *000 = 1:2    001 = 1:4
;    |  |  |  |  |  |              010 = 1:8    011 = 1:16
;    |  |  |  |  |  |              100 = 1:32  101 = 1:64
;    |  |  |  |  |  |              110 = 1:128 111 = 1:256
;    |  |  |  |  |_____ PSA (prescaler assign)
;    |  |  |  |                     1 = to WDT
;    |  |  |  |                     *0 = to Timer0
;    |  |  |  |_____ TOSE (Timer0 edge select)
;    |  |  |                        0 = increment on low-to-high
;    |  |  |                        *1 = increment on high-to-low
;    |  |  |_____ TOCS (TMR0 clock source)
;    |  |                        *0 = internal clock
;    |  |                        1 = RA4/TOCKI bit source
;    |  |_____ INTEDG (Edge select)
;    |                         *0 = falling edge
;    |_____ RBPU (Pullup enable)
;                          0 = enabled
;                          *1 = disabled
      banksel    OPTION_REG
      movwf OPTION_REG ; Copy w to OPTION
;==============================
;      set ADC hardware
;==============================
; Select analog line (AN0) as the analog input
; channel. Select Vcc as the voltage reference.
; Left justify result
      movlw B'00000001'; Bitmap:
;                        0 0 0 0 0 0 0 1
;                        | |___| | |___x___ 1 = turn on ADC
;                        | |___x_x_x_____ 000 = select AN0
;                        | x_____ 0 = voltage ref Vcc
```

```
;                               x_____ 0 = left justify
        banksel     ADCON0
        movwf ADCON0

; Set default state for 5804 control lines, wired as
; follows:
;     PORT C BITS
;     5   4   3   2   1   0                       Default
;     |   |   |   |   |   |____ Step input        0
;     |   |   |   |   |_____ Direction         1
;     |   |   |   |_____ One phase         0
;     |   |   |_____ Half step         1
;     |   |_____ MCLR
;     |_____ NOT USED
; Note that 5804 Enable line is wired to ground
        banksel     PORTC
        movlw B'00001010'        ; Half step, ccw direction
        movwf PORTC
; Initialize default value for delay variable
        movlw .100
        banksel     delay
        movwf delay         ; To delay variable
        movwf iteration          ; To iteration counter
;=========================================================
;                 command monitoring
;=========================================================
;============================
;   Read speed control pot
;============================
; 5K pot is wired to port A line 0, which has been set
; for analog input.
ReadInputs:
        banksel     ADCON0              ; Prepare to sample
        bsf         ADCON0,GO           ; Start operation
Wait4ADC:
        btfsc ADCON0,GO          ; Wait for completion
        goto  Wait4ADC                  ; Loop back if not done
;============================
;   store value in variable
;============================
; At this point digital values of ADC operation are
; found in the PICs ADRESH and ADRESL registers.
; Program uses the 8 high bits in ADRESH, which are
; divided by 2 by shifting right one bit. A constant of
; 40 is then added to scale and limit the fastest speed.
        banksel     STATUS
        bcf         STATUS,C            ; Clear carry to shift
        banksel     ADRESH
```

```
        rrf             ADRESH,w            ; Rotate in w
                                            ; Range is now 0 to 128
        addlw .40                   ; Add constant
                                            ; Range is now 40 to 168
        banksel     delay           ; Delay variable
        movwf delay
;==============================
;   Read 5804 control lines
;==============================
; PORTA bit 4 (wired to toggle # 1) controls forward
; or reverse direction by means of PORTC bit 1 (wired
; to 5804 direction line on pin 14).
        banksel     PORTA
        btfsc PORTA,4               ; Direction control switch
        goto        ReverseRot ; Engage reverse rotation
; At this point rotation is forward
        bcf         PORTC,1             ; Clear bit
        goto        HalfStepMode
ReverseRot:
        bsf         PORTC,1             ; Set reverse
HalfStepMode:
; PORTA bit 2 (wired to toggle # 2) selects half-step
; mode by means of PORTC bit 3 (wired to 5804 half-step
; line on pin 10).
        btfsc PORTA,2               ; Half step mode control
        goto        NoHalfStep ; Disable half step
; At this point half step is selected
        bsf         PORTC,3             ; Set bit
        goto  OnePhaseMode
NoHalfStep:
        bcf         PORTC,3             ; Turn off half step control
OnePhaseMode:
; PORTA bit 1 (wired to toggle # 3) selects one phase
; mode by means of PORTC bit 2 (wired to 5804 single
; phase line on pin 9).
        btfsc PORTA,1               ; One phase mode control
        goto        NoOnePhase ; Disable single step
; At this point single phase is selected
        bsf         PORTC,2             ; Set single phase
        goto  ReadInputs
NoOnePhase:
        bcf         PORTC,2             ; Turn off one phase
        goto        ReadInputs
;===============================
;   routine to pulse the 5804
;===============================
; Note that 5804 requires a positive-going pulse
; therefore the line is held high during the strobe
```

```
; cycle
Pulse:
     banksel    PORTC
     bsf        PORTC,0          ; Bring step line high
     nop
     nop
     nop
     nop
     nop
     nop
     nop
     nop
     nop
     nop
     bcf        PORTC,0          ; Step line low
     return


;============================================================
;              Interrupt Service Routine
;============================================================
; Service routine receives control when the timer
; register TMR0 overflows, that is, when 256 timer beats
; have elapsed
IntServ:
; First test if source is a Timer0 interrupt
     banksel    INTCON
     btfss INTCON,T0IF ; T0IF is Timer0 interrupt
     goto         notT0IF      ; Go if not RB0 origin
; If so clear the timer interrupt flag so that count continues
     bcf          INTCON,T0IF ; Clear interrupt flag
; Save context
     banksel    old_w
     movwf old_w        ; Save w register
     swapf STATUS,w    ; STATUS to w
     movwf old_STATUS  ; Save STATUS
;=========================
;   interrupt action
;=========================
; Decrement the iteration counter. Exit if not zero
     banksel    iteration
     decfsz     iteration,f
     goto         exitISR          ; Continue if counter not zero
; At this point the delay count has expired so the programmed
; time has elapsed.
; First reset the iteration counter
     movf         delay,w          ; Read delay
     movwf iteration   ; Reset iteration counter
;=================================
```

```
;      Pulse motor and delay
;=================================
; Make sure 5804 Step In line is low before strobe
; cycle
      banksel    PORTC
      bcf        PORTC,0            ; Set low
      call       Pulse       ; Local pulse motor routine
;=========================
;        exit ISR
;=========================
exitISR:
; Restore context
      banksel    old_STATUS
      swapf old_STATUS,w  ; Saved STATUS to w
      movfw STATUS       ; To STATUS register
      swapf old_w,f      ; Swap file register in itself
      swapf old_w,w      ; re-swap back to w
; Reset,interrupt
notTOIF:
      retfie
      end             ;END OF PROGRAM
```

17.4.7 SMU_POSITION.asm

```
;============================================================
; File: SMU_POSITION.asm
; Date: January 16, 2011
; Update: January 18, 2011
; Authors: Sanchez and Canton
; Processor: 16F684A
; Reference circuit: SMU-PIC16F684-1
;
; Program Description:
; Program to drive a unipolar stepper motor in half step
; mode, with a 16F684 PIC controlling a UCN5804 driver.
; A 5K potentiometer on port RA0 provides input to the chip's
; ADC. The resulting resistance value is used to turn the
; motor a specific number of steps. Program operates as
; follows:
; 1. The initial position of the pot arm is stored as the
;    local reference point (LRP).
; 2. The pot value is read by code and designated as new
;    reference point (NRP).
; 3. NRP and LRP are compared.
;    If they are the same, no action takes place.
; 3. If the NRP is greater than the LRP, then the difference
;    is the amount of steps the motor is turned to the
;    right (clockwise).
```

```
; 5. If the NRP is smaller than LRP, then the difference is
;     the number of steps the motor is turned to the left
;     (counterclockwise).
; 6. LRP is now set to NRP.
;     Processing continues at STEP 2.
;
;===========================
;    demo board circuit
;===========================
; Port A lines 0 to 4 trissed for input:
;   0 ----------> 5K potentiometer  -- Analog input
;   1 to 4 -----> NOT WIRED
; Port C lines 0 to 3 trissed for output and wired
; as follows:
;   PIC        5804
;   RC0 -----> pin 11 - Step input
;   RC1 -----> pin 14 - Direction
;
; Settings for phase control:
;   RC3 (half step)  RC2 (one phase)
;          0                0          = two phase
;          0                1          = wave drive
;          1                0          = half step
;          1                1          = step inhibit


;========================
; setup and configuration
;========================
      processor 16f684
      include        <p16f684.inc>
      __CONFIG   _MCLRE_OFF & _CP_OFF & _CPD_OFF & _BOD_OFF &
_WDT_OFF & _PWRTE_ON & _INTOSCIO & _FCMEN_OFF & _IESO_OFF

      errorlevel -302  ; No register in bank error messages
      errorlevel  -312  ; No page or bank messages


;=============================================================
;                     variables in PIC RAM
;=============================================================
; Declare variables at 2 memory locations
j         equ        0x20
k         equ        0x21
delay     equ        0x22  ; Delay count
lrp       equ        0x23
nrp       equ        0x24
ticks     equ        0x25
;=============================================================
;                m a i n   p r o g r a m
```

```
;=================================================================
        org         0          ; start at address 0
        goto        Main
;=============================
; space for interrupt handler
;=============================
;=============================
;      main program
;=============================
        org         0x08
Main:
; Initialize all lines in port C for output
        banksel     PORTC
        clrf        PORTC
        banksel     TRISC              ; Prepare to tris
        clrf        TRISC              ; All lines to 0 (output)
; Set PORT A line 0 as analog
        movlw       B'00000001' ; Line 0 is analog
        banksel     ANSEL
        movwf       ANSEL              ; Select as analog
; Turn off comparator
        movlw B'00000111'          ; Setting bits 0, 1, and 2
                                   ; sets comparators off and
                                   ; Port C IN pins to digital
        banksel     CMCON0
        movwf       CMCON0
; Tris port A for input
        banksel     TRISA
        movlw       B'00111111' ; All port A lines for input
        clrf        TRISA
        movwf       TRISA
; Configure oscillator
        movlw       B'01100001'
;                     |||   |____ internal clock
;                     |||_____ 4 MHz
        banksel     OSCCON
        movwf       OSCCON
; Set conversion clock to Tad = 8 * Tosc (ADCS = 001)
        movlw       B'00010000'
;                     |||_____ ADCS bits
        banksel     ADCON1
        movwf       ADCON1
;=============================
;     set ADC hardware
;=============================
; Select analog line (AN0) as the analog input
; channel. Select Vcc as the voltage reference.
; Left justify result
```

```
        movlw               B'00000001'; Bitmap:
;                           0 0 0 0 0 0 0 1
;                                 | |___| | |___x___ 1 = turn on ADC
;                           | |___x_x_x_____ 000 = select AN0
;                           | x_____ 0 = voltage ref Vcc
;                           x_____ 0 = left-justify
        banksel     ADCON0
        movwf       ADCON0
; Set default state for 5804 control lines, wired as
; follows:
;    PORT C BITS
;    5  4  3  2  1  0                      Default
;    |  |  |  |  |  |____ Step input       0
;    |  |  |  |  |_____ Direction        1
;    |  |  |  |_____ One phase        0
;    |  |  |_____ Half step        1
;    |  |_____ MCLR
;    |_____ NOT USED
; Note that 5804 Enable line is wired to ground
        banksel     PORTC
        movlw       B'00001010' ; Half step, ccw direction
        movwf       PORTC
; Initialize default value for delay variable
        movlw       .40
        banksel     delay
        movwf       delay                 ; To delay variable
; Read initial pot position and store
        call        ReadPot        ; Value is returned in w
        banksel     nrp
        movwf       nrp                   ; new reference position
        movwf       lrp            ; local reference position


;================================
;      Read pot and pulse motor
;================================
ReadPotAndMove:
        call        Delay                 ; Wait to update
        call        ReadPot
        banksel     nrp
        movwf       nrp                   ; Store result
; Compare new value with old value by subtraction.
; Subtraction performs w (new reference position)
; minus local reference position (old value).
; If Z flag set nrp == lrp
; If carry flag set lrp < nrp
        subwf       lrp,w          ; Subtract w (nrp) from lrp
        banksel     STATUS
        btfsc       STATUS,Z       ; Test zero flag
```

```
        goto        NoChange            ; Go if Z flag is set
; At this point nrp not equal to lrp
; Test the carry flag to determine if lrp < nrp
        btfss       STATUS,C            ; Is C flag set?
        goto        MoveLeft            ; Go if new bigger than old
        goto        MoveRight
;============================
;          move left
;============================
; Move left by difference between new and old readings
; First obtain absolute value of difference. At this point
; lrp > nrp
MoveLeft:
        banksel     lrp
        movf        lrp,w               ; Get lrp
        subwf       nrp,w               ; old (in w) minus new
        banksel     ticks
        movwf       ticks               ; Store difference
; Set controller to reverse rotation
        banksel     PORTC
        bsf         PORTC,1             ; Set reverse
TickLeft:
        banksel     PORTC
        bcf         PORTC,0             ; Set low
        call        Pulse
        call        Delay
        banksel     ticks               ; Counter
        decfsz      ticks,f
        goto        TickLeft            ; Loop
        goto        Update
;===============================
;          move right
;===============================
; Move to the right by number of ticks in w
MoveRight:
        banksel     ticks
        movwf       ticks               ; Store difference
; Set forward rotation
        banksel     PORTC
        bcf         PORTC,1             ; Clear bit
TickRight:
        banksel     PORTC
        bcf         PORTC,0             ; Set low
        call        Pulse
        call        Delay
        banksel     ticks               ; Counter
        decfsz      ticks,f
        goto  T     ickRight                    ; Loop
```

```
;===============================
;    Update position control
;===============================
Update:
      banksel     nrp
      movf        nrp,w                    ; New value to w
      movwf       lrp
NoChange:
      goto        ReadPotAndMove


;================================
;   routine to pulse the 5804
;================================
; Note that 5804 requires a positive-going pulse
; therefore the line is held high during the strobe
; cycle
Pulse:
      banksel     PORTC
      bsf         PORTC,0          ; Bring step line high
      nop
      nop
      nop
      nop
      nop
      nop
      nop
      nop
      nop
      nop
      bcf         PORTC,0          ; Step line low
      return


;================================
;        delay sub-routines
;================================
Delay:
      banksel     delay
      movf        delay,w       ; Delay to w register
      movwf       j             ; j = w
Jloop:
      movwf       k             ; k = w
Kloop:
      decfsz      k,f           ; k = k-1, skip next if zero
      goto        Kloop
      decfsz      j,f           ; j = j-1, skip next if zero
      goto        Jloop
      return
```

```
;=============================================================
;     Local procedure to read speed potentiometr on board
;=============================================================
ReadPot:
; 5K pot is wired to port A line 0, which has been set for
; digital input
      banksel    ADCON0               ; Prepare to sample
      bsf        ADCON0,GO            ; Start operation
Wait4ADC:
      btfsc      ADCON0,GO            ; Wait for completion
      goto       Wait4ADC            ; Loop back if not done
;==================================
;     store result
;==================================
; At this point digital values of ADC operation are
; found in the PICs ADRESH and ADRESL registers.
; Program uses the 8 high bits in ADRESH.
      banksel    ADRESH
      movf       ADRESH,w                  ; Value to w
      return

      end          ;END OF PROGRAM
```

Chapter 18

Constant-Voltage Bipolar Motor Controls

18.1 Unipolar versus Bipolar

In Chapter 15 we discussed the hardware characteristics of bipolar and unipolar motors. From the circuit designer's viewpoint, the fundamental difference between these motor types relates to the fact that bipolar motors contain a single winding per phase. Because there is a single winding, the current flow must be reversed in order for the rotor to turn. Another consequence of the bipolar motor structure is that the entire winding is activated during each cycle, while only half the winding is active in the unipolar design. Consequently, a bipolar motor generates more torque than a unipolar motor of the same size.

The bipolar design also has some drawbacks. One of them is that, in order to reverse current flow in the windings, bipolar motors require more complex control circuitry. The H bridge (discussed in Chapter 16) is a common circuit component that is used for this purpose. There is little difference in cost between unipolar and bipolar motors of the same weight. However, this difference is more significant if the increased torque and efficiency of the bipolar design is taken into account. A final consideration in choosing between unipolar and bipolar motors is that a unipolar motor can be used as a bipolar motor by ignoring the center taps in the coils. In summary, the differences can be listed as follows:

- Bipolar motors require more complex drive circuitry than unipolar.
- Bipolar motors produce higher torque than unipolar motors of the same weight.
- There is not much difference in cost between unipolar or bipolar motors of the same weight.
- Standard unipolar motors can be used in bipolar circuits.

18.1.1 Bipolar Drive Circuits

Bipolar motor circuits require changing current polarity during stepping operations and that this is usually accomplished with an H bridge of one type or another. We must now add that bipolar stepper motor performance is dependent on the drive circuit to a high degree. Recall that higher speeds are achieved by reversing the stator poles more quickly, which also results in a gain in torque. However, increasing the reversal rate of the poles also increases inductance in the wiring. Here again, the higher inductance can be overcome by increasing voltage, but this requires limiting the current at these higher voltages.

These considerations have led to two different types of bipolar stepper motor circuits: one in which the voltage is held constant and another one in which the current is held constant. The constant-voltage type is sometimes known as the *L/R design*. Here L represents the winding inductance and R the resistance. Constant-current circuit types are referred to as *chopper circuits*. In chopper drives, the current in each winding is held approximately constant by varying the voltage. Initially, a high voltage is applied to each winding, which causes the current to rise quickly. When the current reaches a specified limit, the voltage is turned off, or chopped. The current now drops until it reaches another low limit, in which case the voltage is turned back on. Chopper drives require additional components in order to measure the winding current and to control switching, but the payoff is that stepper motors can be driven at higher speeds and torques than with constant-voltage circuits.

The control and translator stages in many circuits described in Chapter 17, in the context of unipolar motors, are compatible with the bipolar drivers described in this chapter. For example, the control and translator stage of circuit SMU-PIC16F84-1 can be coupled with several driver stages of bipolar circuits shown in this chapter. Also consider that in this chapter we only discuss bipolar circuits of the constant-voltage type. Chopper circuits are presented in Chapter 19, together with microstepping.

18.2 Simple, L293 Bipolar Circuit

In Chapter 16 we presented the L293 IC, which is a two H-bridge, four-channel driver that includes *snubber diodes*. The L293 can drive 1.2 A per channel; it is suitable for driving NEMA teen-size stepper motors. The L293, and equivalent ICs, are little more than a double H-bridge and they require a separate translator to provide the motor coil sequence via four input lines. Two enable lines are held high by wiring to the positive power supply. These lines can also be wired to the translator and used to turn the motor on and off. The four output lines from the L293 are wired to the motor coils. Four ground lines are located on the center of the IC. This design facilitates a large ground on the PCB, which serves as a heat sink

Figure 18-1 is a simple bipolar circuit that uses a microcontroller translator and an L293D driver IC. The SMB-L293D-1 circuit is the bipolar version of the circuit SMU-PIC16F84-1 presented in Chapter 17 (see Figure 17-3).

The circuit in Figure 18-1 uses a *microcontroller-based translator*. Input is provided by means of four pushbutton switches, and motor speed, direction, and sequence mode are determined in software. Alternatively, the circuit can be easily modified to read other input devices, such as toggle switches or potentiometers. These inputs can then be used to control motor operation.

CONTROL AND TRANSLATOR STAGE

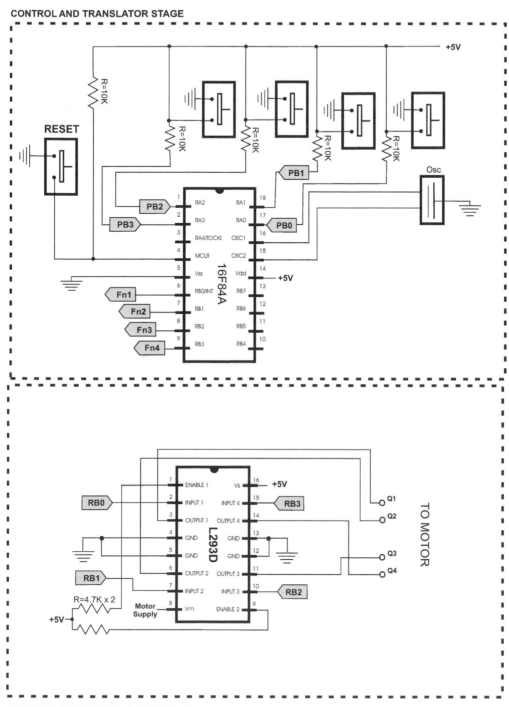

CIRCUIT: SMB-L293D-1
CODE: SMB_L293D.asm

Figure 18-1 L293D-Based Bipolar Motor Driver.

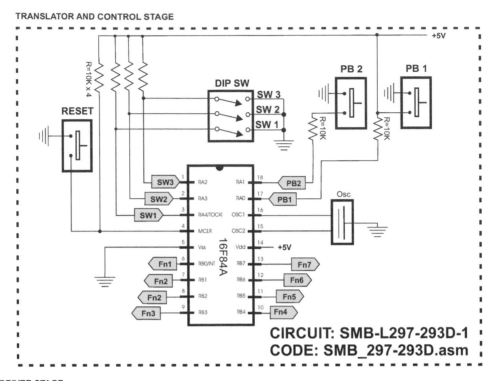

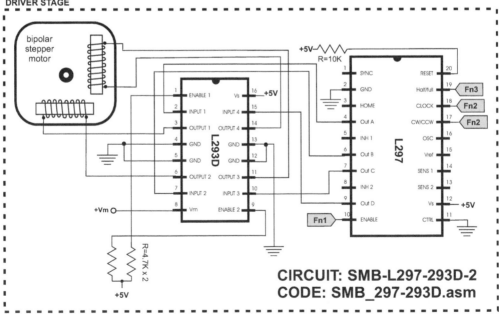

Figure 18-2 Bipolar Motor Driver with L297 and L293 ICs.

18.2.1 L297- and L293-Based Circuit

For circuits in which the microprocessor is overloaded, or to implement additional or finer motor controls, it is possible to insert an L297 motor driver IC between the translator and the H-bridge. The L297 hardware is described in Chapter 16. Although this chip is sometimes employed in simple circuits, such as the one presently described, it is most useful when the circuit takes advantage of all the chip's capabilities. For example, the L297 can be configured in a "chopper" scheme, described previously. In Chapter 19 we develop a circuit that uses the L297 in this manner. The circuit designer should evaluate using the L297, or similar ICs, in constant-voltage circuits where most of the IC's functionalities are unused. Figure 18-2 shows a simple circuit that uses the L293 and L297 ICs.

Note that many of the L297 pins are left floating in circuit SMB-L297-293D-2. These pins correspond to features and functionalities of the L297 that are not used in the circuit.

18.2.2 Minimal L297- and L298-Based Circuit

The major limitation of the L293 IC is its limitation of 1.2 A peak output current per channel. The L298, on the other hand, allows up to 3.5 A per channel, which triples the current-carrying capacity of the L293. A simple L297/L298 circuit is shown in Figure 18-3.

Circuit SMB-L297-298-1 shown in Figure 18-3 uses few of the features of both the L297 and the L298. The principal characteristic of this circuit is its higher capacity as the L298 allows an operating supply voltage of up to 46V and 4 A. The L298 IC includes capabilities not used in this circuit, such as sensing the output voltage. This feature can be used in input voltage chopping, which is implemented in circuits developed in Chapter 19. Because chopping is not supported by circuit SMB-L297-298-1, the SENSE A and SENSE B lines of the L298 are wired to ground. The L298 is also used in driving relays, solenoids, and DC motors.

Additional toggle switches in the control and translator stage could be used to control other modes or stages. In circuit SMB-l297-298-1, one toggle switch controls direction, another one selects between full- and half-step modes, a third one is wired to the L297 ENABLE line and allows turning the motor on and off. An additional toggle switch could be added to the circuit to provide selecting between high and low speeds. Because there are several unused ports in the translator, it is easy to add more input sources in order to control other features of the L297. Also note that the ENABLE, RESET, and both INH lines of the L297 are active low. In the circuit these lines are disabled either by holding them high or by leaving them floating. Similarly unused lines in the L298 are either grounded or held high with 4.7K pull-ups. Most of these lines are put to good use in the more powerful circuits developed in Chapter 19.

18.3 Demonstration Programs

The programs listed in the following section demonstrate the programming discussed in this chapter.

CONTROL AND TRANSLATOR STAGE

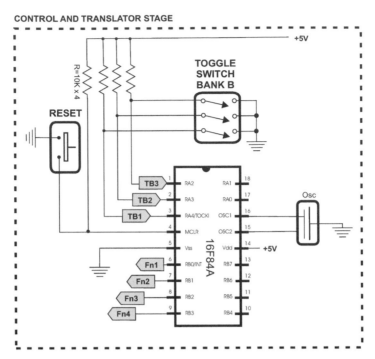

DRIVER STAGE

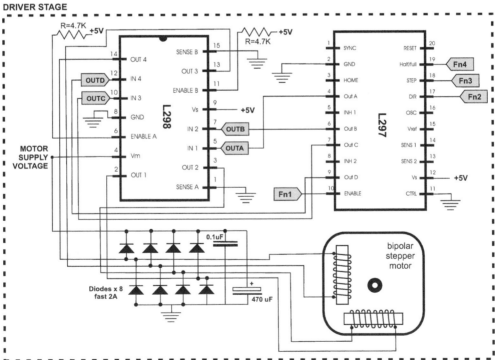

CIRCUIT: SMB-L297-298-1   CODE: SMB_297_298.asm

Figure 18-3 Bipolar Motor Circuit with L297 and L298 Drivers.

18.3.1 SMB_L293D.asm

The program SMB_L293D.asm, listed below and in this book's software resource, drives a bipolar stepper motor in a one-half step mode. The program uses a lookup table for the sequence codes. Two pushbutton switches, wired to ports A3 and A2, activate slow rotation in the forward or reverse direction. Two other pushbuttons, wired to ports A0 and A1, control slow and fast rotation. Code uses an access code table with a pointer, that is either incremented or decremented according to the selected direction. Fast and slow execution are determined by a local variable.

```
;=================================================================
; File: SMB_L293D.asm
; Date: November 12, 2010
; Update: January 3, 2011
; Authors: Sanchez and Canton
; Processor: 16F84A
; Other ICs: L293D bipolar stepper motor controller
; Reference circuit: SMB-L293D-1
;
; Program Description:
; Program to drive a bipolar stepper motor in a one-half step
; mode with a 16F84 PIC and an L293D motor driver.
; Program uses a lookup table for the sequence codes.
; Pushbuttons wired to ports A3 and A2 activate slow rotation
; in forward and reverse direction. Pushbuttons wired to ports
; A0 and A1 perform likewise for fast rotation.
; Code uses a access code table with a pointer, that is either
; incremented or decremented according to the selected
; direction.
; This ensures that reverses in rotation are executed smoothly.
; Fast and slow execution are determined by the value of a local
; variable read by the delay routine.
;=============================
;     demo board circuit
;=============================
; Port A lines 0 to 3 trissed for input:
;   0 -----> Push button switch A, active low
;   1 -----> Push button switch B, active low
;   2 -----> Push button switch C, active low
;   3 -----> Push button switch D, active low
; Port B lines 0 to 3 trissed for output and wired
; as follows:
; RB0 -----> A1 motor winding
; RB1 -----> A2 motor winding
; RB2 -----> B1 motor winding
; RB3 -----> B2 motor winding

;=========================
; setup and configuration
```

```
;========================
      processor 16f84A
      include       <p16f84A.inc>
      __config  _XT_OSC & _WDT_OFF & _PWRTE_ON & _CP_OFF
      errorlevel -302
;=============================================================
;                      variables in PIC RAM
;=============================================================
; Declare variables at 2 memory locations
j           equ         0x0c
k           equ         0x0d
this_cycle equ          0x0e  ; Current cycle counter
direction   equ         0x0f  ; Direction control
                              ; 0 = forward
                              ; 1 = reverse
bounce      equ         0x10     ; Debounce counter
delay       equ         0x11  ; Delay counter
;=============================================================
;                   m a i n    p r o g r a m
;=============================================================
      org         0         ; start at address 0
      goto        Main
;=============================
; space for interrupt handler
;=============================
      org         0x08
;===================================
;    lookup table for motor
;===================================
; Bipolar stepper motors require the following one-half
; step drive sequence for clockwise rotation:
;              |  w i n d i n g s  |
;      STEP  A1    A2    B1    B2      BIN
;       1    OFF   ON    OFF   ON      0101
;       2    OFF   ON    OFF   OFF     0100
;       3    OFF   ON    ON    OFF     0110
;       4    OFF   OFF   ON    OFF     0010
;       5    ON    OFF   ON    OFF     1010
;       6    ON    OFF   OFF   OFF     1000
;       7    ON    OFF   OFF   ON      1001
;       8    OFF   OFF   OFF   ON      0001
; Table is placed low in memory to avoid page overlaps.
;
CodeTable
      addwf PCL,f        ; Add w to program counter
      retlw B'00000101'        ; cycle = 0
      retlw B'00000100'        ;            1
      retlw B'00000110'        ;            2
```

```
        retlw B'00000010'        ;            3
        retlw B'00001010'        ;            4
        retlw B'00001000'        ;            5
        retlw B'00001001'        ;            6
        retlw B'00000001'        ;            7
        retlw B'00000000'        ; Table terminator


;===============================
;       main program
;===============================
Main:
; Tris PORT A for input
        movlw       B'00011111'       ; w = 00011111 binary
        tris        PORTA             ; Set up port A for input
; Initialize all line in port B for output
        movlw       B'00000000'       ; w = 00000000 binary
        tris        PORTB             ; Set up port B for output
        clrf        PORTB             ; Clear all lines
; Initialize control variables
        clrf        this_cycle
        clrf        direction         ; Assume forward direction
        movlw       .100              ; Initial delay for fast
        movwf       delay             ; To variable
;===============================
;   read and debounce PORTA
;       lines 0 to 3
;===============================
; Switches are active low, so changes in direction take place
; if port line is clear. Software debouncing is used to make
; sure that spurious switch reads are ignored.
TestPortA:
; First read and debounce PORTA-3
        movlw       .10               ; Counter
        movwf       bounce
Wait43:
        btfsc       PORTA,3
        goto        No3Action         ; Go if bit is set
        decfsz      bounce,f          ; Decrement counter
        goto        Wait43
; At this point switch held zero for 10 tests
        clrf        direction         ; 0 is foward on
                                      ; direction switch
        movlw .     100               ; Slow delay rate
        movwf       delay             ; To delay counter
        goto        OneTick           ; Tick motor
;===============================
;   read and debounce PORTA-2
;===============================
```

```
No3Action:
      movlw         .10                  ; Counter
      movwf         bounce
Wait42:
      btfsc         PORTA,2
      goto          No2Action            ; Go if bit is set
      decfsz        bounce,f             ; Decrement counter
      goto  Wait42
; At this point switch held zero for 10 tests
      movlw         .1                   ; 1 is reverse on
                                         ; direction switch
      movwf         direction            ; Code to switch
      movlw         .100                 ; Slow delay rate
      movwf         delay                ; To delay counter
      goto          OneTick              ; Tick motor
No2Action:
;=================================
;   read and debounce PORTA-1
;=================================
; Debounce PORTA-1
TestPortA1:
      movlw         .10                  ; Counter
      movwf         bounce
Wait41:
      btfsc         PORTA,1
      goto          No1Action            ; Go if bit is set
      decfsz        bounce,f             ; Decrement counter
      goto          Wait41
; At this point switch held zero for 10 tests
      clrf          direction            ; 0 is forward on
                                         ; direction switch
      movlw         .40                  ; Fast delay rate
      movwf         delay                ; To delay counter
      goto          OneTick              ; Tick motor
;=================================
;   read and debounce PORTA-1
;=================================
No1Action:
; Debounce PORTA-0
      movlw         .10                  ; Counter
      movwf         bounce
Wait40:
      btfsc         PORTA,0
      goto          No0Action            ; Go if bit is set
      decfsz        bounce,f             ; Decrement counter
      goto          Wait40
; At this point switch held zero for 10 tests
      movlw         .1                   ; 1 is reverse on
```

```
                                        ; direction switch
        movwf       direction           ; Code to switch
        movlw       .40                 ; Fast delay rate
        movwf       delay               ; To delay counter
        goto        OneTick             ; Tick motor
No0Action:
        goto        TestPortA
;==========================
;    forward or reverse
;    single cycle routine
;==========================
OneTick:
; Test for change of direction
        btfss       direction,0         ; Test forward bit
        goto        Forward
; At this point direction rotation is reverse
        goto        Reverse
; Forward direction routine
Forward:
        movf        this_cycle,w        ; Get current cycle
        call        CodeTable           ; Get code from table
        movwf       PORTB               ; Store code in port
        call        Delay
; Bump cycle counter
        incf        this_cycle,f        ; Add one
; Test for cycle number 7 (last one in sequence)
        btfsc       this_cycle,3        ; 1000 = 8
        goto        Recycle             ; Reset if at end of cycle
        goto        TestPortA
Recycle:
        clrf        this_cycle
        goto        TestPortA
; Reverse direction routine
Reverse:
        movf        this_cycle,w        ; Get current cycle
        call        CodeTable           ; Get code from table
        movwf       PORTB               ; Store code in port
        call        Delay
; Bump cycle counter
        decf        this_cycle,f        ; Subtract one
; Test for cycle number 0xff (overflow from cycle # 0)
        btfsc       this_cycle,4        ; Bit 4 set indicates overflow
        goto        Recycle2            ; Reset if at end of cycle
        goto        TestPortA
Recycle2:
        movlw       .7                  ; First reverse cycle
        movwf       this_cycle          ; To cycle counter
        goto        TestPortA
```

```
;===================================
;        delay sub-routine
;===================================
Delay:
     movf         delay,w           ; Load delay value
     movwf        j                 ; j = w
Jloop:
     movwf        k                 ; k = w
Kloop:
     decfsz       k,f               ; k = k-1, skip next if zero
     goto         Kloop
     decfsz       j,f               ; j = j-1, skip next if zero
     goto  Jloop
     return

     end          ;END OF PROGRAM
```

18.3.2 SMB_297_293D.asm

The sample program SMB_297_293D.asm, listed below and in this book's software resource, drives a bipolar stepper motor using a 16F84 PIC as a translator, wired to an L297 and a 293D driver. The program reads the toggle switch wired to port RA4 to select between clockwise or counterclockwise rotation. A toggle switch wired to line RA3 allows selecting between half- or full-step modes. Slow and fast speeds are determined by the setting the toggle switch on port line RA3. The pushbutton switches on circuit SMB-L297-293D-1 are not used by the program.

```
;=============================================================
; File: SMB_297_293D.asm
; Date: November 6, 2010
; Update: November 13, 2010
; Authors: Sanchez and Canton
; Processor: 16F84A
; Reference circuit: SMB-L297-293D-1
;
; Program Description:
; Program to drive a bipolar stepper motor using a 16F84 PIC
; as a controller, wired to an L297 and a 293D driver. The
; program reads the toggle switch wired to port RA4 to select
; between clockwise or counterclockwise rotation. The toggle
; switch wired to line RA3 allows selecting between half- or
; full-step modes. Slow and fast speeds are determined by the
; setting of the toggle switch on port line RA3. The pushbutton
; switches on circuit SMB-L297-293D-1 are not used by the
; program.
; Circuit wiring is as follows:
;
```

```
;  SLOW/FAST RA2<-|--------|
;  FULL/HALF RA3<-|        |
;  CW/CCW    RA4- |        |<------- OSC
;         RESET->|        |<------- OSC
;             GND |        | +5V
;  ENABLE    RB0<-|        |
;  CW / CCW  RB1<-|        |
;      STEP  RB2<-|        |
;  HALF/FULL RB3<-|        |
;                 |--------|
;
;========================
; setup and configuration
;========================
      processor 16f84A
      include      <p16f84A.inc>
      __config  _XT_OSC & _WDT_OFF & _PWRTE_ON & _CP_OFF
      errorlevel -302
;=============================================================
;                    constant definition
;=============================================================


;=============================================================
;                   variables in PIC RAM
;=============================================================
; Declare variables at 2 memory locations
j           equu         0x0c
k           equ          0x0d
bounce      equ          0x0e  ; For debounce routine
delay       equ          0x0f  ; Delay count


;=============================================================
;              m a i n   p r o g r a m
;=============================================================
      org       0       ; start at address 0
      goto      Main
;=============================
; space for interrupt handler
;=============================
      org       8
;=============================
;      main program
;=============================
Main:
; Clear bit 7 in OPTION_REG to enable port B pullups
      bcf          OPTION_REG,7
; Note: although enabling the weak pullups on port B
;       should allow the circuit to operate without the
```

```
;           conventional 10K pullup resistors, we have
;           found that the weak pullups are unreliable on
;           the 16F84. Therefore we advise that the board
;           include the 10K pullups.
; Tris PORT A for input
      movlw        B'00011111' ; w = 00011111 binary
      tris         PORTA             ; Set up port A for input
; Tris lines 0 to 3 in port B for output and
; lines 4 to 7 (toggle switch # 2) for input
      movlw        B'11110000' ; RB7-RB4 for input, rest for
                               ; output
      tris         PORTB             ; Set up port B
      clrf         PORTB             ; Clear all lines
; Set default state for L297 control lines, wired as
; follows:
;     PORT B BITS
;     7  6  5  4  3  2  1  0                    Default
;     |  |  |  |  |  |  |  |____ ENABLE         1 (active)
;     |  |  |  |  |  |  |_____ CW/CCW         0 (CW)
;     |  |  |  |  |  |_____ CLOCK          1
;     |  |  |  |  |_____ mode           1 (half step)
;     |__|__|__|_____ NOT USED

      movlw        B'00001101'       ; Half step, cw direction
      movwf        PORTB
; Initialize default value for delay variable
      movlw        .40
      movwf        delay              ; To delay variable
;===============================================================
;                   main control monitoring routine
;===============================================================
ControlRtn:
; Call procedure to set L297 lines according to state of three
; toggle switches wired to port lines A2 to A4
      call         ReadAToggles
      call         Delay              ; Delay now
      call         Pulse              ; Local pulse motor routine
      goto         ControlRtn


;=========================================================
;       Auxiliary procedure to read toggle switches
;=========================================================
ReadAToggles:
; Read PORTA toggle switches
;     PORT A    TOGGLE
;     line      SW    ACTION
;     4         1     0 = CW rotation 1 = CCW
;     3         2     0 = full step 1 = half step
```

```
;         2        3      1 = fast 0 = slow
;=============================
;   set CW or CCW rotation
;=============================
; Bit 4 on port A is wired to toggle # 1
; Bit 1 on port B is wired to pin 17 of the L297 which
; controls CW and CCW rotation.
      btfsc         PORTA,4
      goto          goCCW              ; Set rotation bit to CCW
; Bit clear. Set CW
      bcf           PORTB,1            ; Clear port B line 1
      goto          SetStep
goCCW:
      bsf           PORTB,1
;=============================
;   set half or full step
;=============================
SetStep:
; Toggle switch # 2, on port A, line 3, selects between
; half step and full step modes. Port B bit 3 is wired
; to L297 line 19, which selects half or full step modes
      btfsc         PORTA,3
      goto          HalfStep           ; Bit is set. Set half
                                       ; step mode
; Bit clear. Set full step mode
      bcf           PORTB,3            ; Clear port B line 2
      goto          SetSpeed
HalfStep:
      bsf           PORTB,3
;=============================
;   set fast or slow speed
;=============================
SetSpeed:
; Toggle switch # 3, on port A, line 2, selects between
; fast or slow motor speed. If the toggle is on, the delay
; value is set to 40, if it is off, it is set to 80
      btfsc         PORTA,2
      goto          FastSpeed          ; Bit is set. Set fast delay
; Bit clear. Set slow speed
      movlw         .80
      movwf         delay              ; To delay variable
      goto          Enable
FastSpeed:
      movlw         .40
      movwf         delay              ; To delay variable
;=============================
;   enable motor control line
;=============================
```

```
; The L297 is always in the ENABLE state
Enable:
      bsf           PORTB,0                      ; Set bit to enable
      return


;=================================
;   routine to pulse the motor
;=================================
; L297 CLOCK line is wired to RB2
Pulse:
      bsf           PORTB,2          ; Bring step line high
      nop
      nop
      nop
      nop
      nop
      nop
      nop
      nop
      nop
      nop
      bcf           PORTB,2          ; Step line low
      return


;=================================
;        delay sub-routines
;=================================
Delay:
      movf          delay,w      ; Delay to w register
      movwf         j            ; j = w
Jloop:
      movwf         k            ; k = w
Kloop:
      decfsz        k,f          ; k = k-1, skip next if zero
      goto          Kloop
      decfsz        j,f          ; j = j-1, skip next if zero
      goto          Jloop
      return


      end           ;END OF PROGRAM
```

18.3.3 SMB_297_298.asm

The program SMB_297_298.asm, listed below and in this book's software package, is a simple, noninterrupt-driven control program for circuit SMB-L297-298-1 shown in Figure 18-3. The program uses a delay routine to time the motor pulses. The three toggle switches are read by code and the corresponding L297 control lines are set or reset accordingly. The program could be improved by making it interrupt driven.

```
;=================================================================
; File: SMB_297_298.asm
; Date: November 6, 2010
; Update: January 22, 2011
; Authors: Canton and Sanchez
; Processor: 16F84A
; Reference circuit: SMB-L297-298-1
;                    SMB-l297-298-2 (chopper)
;
; Program Description:
; Program to drive a bipolar stepper motor mode with a 16F84
; PIC controlling an L297 wired to an L298 driver.
; Three toggle switches wired to port RA1, RA2, and RA3
; control direction, full or half step modes, and set or clear
; the L297 ENABLE line. Program supports chopper hardware.
;
;===========================
;    demo board circuit
;===========================
; All port A lines trissed for input:
;  0 <----- NOT USED
;  1 <----- Toggle switch 1 - CW or CCW direction
;  2 <----- Toggle switch 2 - Half/full step
;  3 <----- Toggle switch 3 - ENABLE line on/off
;  4 <----- Toggle switch 4 - CONTROL line on/off
; Port B lines
;   PIC          L297
;   RB0 -----> pin 10 - ENABLE
;   RB1 -----> pin 17 - Direction
;   RB2 -----> pin 18 - Step input
;   RB3 -----> pin 19 - Full / half step
;   RB4 <----- pin 11 - CTRL line
;   RB5 <-----  |
;   RB6 <-----  |--- NOT USED
;   RB7 <-----  |
;     PORT A    TOGGLE
;      line      SW      ACTION
;       1         1      0 = CW rotation 1 = CCW
;       2         2      0 = full step 1 = half step
;       3         3      1 = ENABLED 0 = DISABLED
;       4         4      0 = CTRL OFF  1 = CTRL ON

;===========================
; setup and configuration
;===========================
      processor 16f84A
      include        <p16f84A.inc>
      __config  _XT_OSC & _WDT_OFF & _PWRTE_ON & _CP_OFF
```

```
       errorlevel  -302

;===============================================================
;                         variables in PIC RAM
;===============================================================
; Declare variables at 2 memory locations
j           equ           0x0c
k           equ           0x0d
delay       equ           0x0e  ; Delay count


;===============================================================
;                m a i n    p r o g r a m
;===============================================================
       org         0          ; start at address 0
       goto        Main
;=============================
; space for interrupt handler
;=============================
       org         8
;=============================
;       main program
;=============================
Main:
; Tris PORT A for input
       movlw       B'00011111' ; w = 00011111 binary
       tris        PORTA                ; Set up port A for input
; Tris port B for output
       movlw       B'00000000' ; All lines for output
       tris        PORTB                ; Set up port B
       clrf        PORTB                ; Clear all lines
; Set default state for L297 control lines
       movlw       B'00001101'
                      | | | | |_____ 1 = ENABLE
                      | | | |_____ 0 = CW
                      | | |_____ 1 = clock pulse
                      | |_____ 1 = Half step
                      |_____ 0 = Ctrl line off
       movwf       PORTB
; Initialize default value for delay variable
       movlw       .100
       movwf       delay                ; To delay variable
;===============================================================
;                 main control monitoring routine
;===============================================================
ControlRtn:
; Call procedure to set L297 lines according to state of
; toggle switches wired to port A
;       call        ReadAToggles
```

```
        call        Delay               ; Delay now
        call        Pulse               ; Local pulse motor routine
        goto        ControlRtn
;===========================================================
;       Auxiliary procedure to read toggle switches
;===========================================================
ReadAToggles:
; Read PORTA toggle switches
;       PORT A      TOGGLE
;       line        SW      ACTION
;        1          1       0 = CW rotation 1 = CCW
;        2          2       0 = full step 1 = half step
;        3          3       1 = Set or clear ENABLE line
;==============================
;   set CW or CCW rotation
;==============================
; Bit 1 on port A is wired to toggle switch # 1
; Bit 1 on port B is wired to pin 17 of the L297, which
; controls CW and CCW rotation.
        btfsc       PORTA,1
        goto        goCCW               ; Set rotation bit to CCW
; Bit clear. Set CW
        bcf         PORTB,1             ; Clear port B line 1
        goto        SetStep
goCCW:
        bsf         PORTB,1
;==============================
;   set half or full step
;==============================
SetStep:
; Toggle switch # 2, on port A, line 2, selects between
; half step and full step modes. Port B bit 3 is wired
; to L297 line 19, which selects half or full step modes
        btfsc       PORTA,2
        goto        HalfStep            ; Bit is set. Set half step
                                        ; mode
; Bit clear. Set full step mode
        bcf         PORTB,3             ; Clear port B line 2
        goto        TestEnable
HalfStep:
        bsf         PORTB,3
;==============================
;   enable motor control line
;==============================
; Toggle switch # 3, on port A, line 3, controls the
; L297 ENABLE line, on pin No. 10
TestEnable:
        btfsc       PORTA,3
```

```
        goto        Enable          ; Bit is set. Set ENABLE line
; Bit clear. Disable
        bcf         PORTB,0         ; Clear the ENABLE line
        return
Enable:
        bsf         PORTB,0         ; Set bit to enable
        return
;================================
;   routine to pulse the motor
;================================
; L297 CLOCK line is wired to RB2.
; Pulse is negative going
Pulse:
        bcf         PORTB,2         ; Bring step line high
        nop
        nop
        nop
        nop
        bsf         PORTB,2         ; Step line low
        return


;================================
;       delay sub-routines
;================================
Delay:
;       bsf         PORTB,2              ; Step line low
        movf        delay,w         ; Delay to w register
        movwf       j               ; j = w
Jloop:
        movwf       k               ; k = w
Kloop:
        decfsz      k,f             ; k = k-1, skip next if zero
        goto        Kloop
        decfsz      j,f             ; j = j-1, skip next if zero
        goto        Jloop
        return


        end
```

Chapter 19

Advanced Motor Controls

19.1 Choppers and Microstepping

In Chapter 18 we discussed circuits and programs for bipolar motors that operate with a constant voltage through the coils. In this chapter we discuss controls in which the current through the coils is held approximately constant, instead of the voltage. One type of constant-current design is referred to as a "chopper" circuit. In the chopper the current in each winding is held approximately constant by varying the voltage. In operation, a high voltage is initially applied to each winding, causing the current to rise quickly. When the current reaches a predetermined limit, the voltage is "chopped" off and the current drops. The cycle concludes when a pre-determined low current limit is reached and the voltage is turned back on. Chopper drives require additional components in order to measure the current in the windings and to perform the necessary switching operations. The advantage is that motors can be driven at higher speeds and torques than in constant-voltage circuits.

In Chapter 15 we discussed the various stepping modes used in driving stepper motors. These modes define the order in which the various coils are activated. The most common ones are the wave-, full-, and half-step modes. In that context, we also mentioned an additional stepping mode usually called microstepping, in which the current in the windings is progressively varied in order produce a more uniform attraction of the rotor and reduce the jerkiness that results from the conventional stepping modes. Because microstepping requires control over the current levels applied to the windings, it can be implemented as a variation of the chopper techniques previously mentioned. Circuits and programs that implement microstepping are also covered in this chapter.

19.2 Chopper Circuit Fundamentals

Although the general concept of a chopper circuit is not very well defined, it can be viewed as a switching operation used to control one signal with a second one. Power supplies, amplifiers, switched capacitor filters, and variable-frequency drives are common uses of chopper circuits. In the context of motor controls, a *chopper circuit* is one in which a controlled switch is used to limit the current flowing through a coil.

The operation requires some way of sensing the current so that the chopper switch can be turned on and off as the predetermined current limits are reached.

Based on this scheme, a circuit can be designed to control the current in order to limit power dissipation and vary the torque in a motor winding. One way of limiting the current is by controlling the supply voltage, which usually requires adding external resistance. The alternative method of current control is to vary, for a short time period, the supply voltage to values several times higher than the nominal voltage of the motor. Figure 19-1 shows the simplified schematics of chopper-based current regulation.

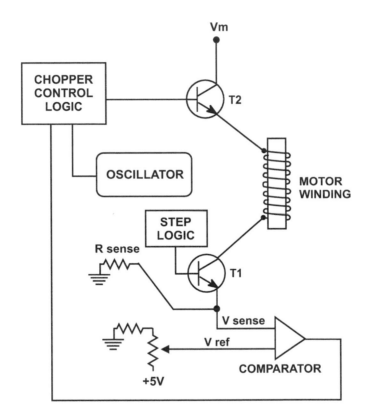

Figure 19-1 Schematics of Chopper-Based Regulation.

Current is sensed by measuring the voltage drop across a known resistor, labeled R sense in Figure 19-1. Transistor T1, which is controlled by the system's step logic, turns the flow on to produce the motor step. While the step-on state is active, the following operations take place:

- Transistor T2 turns on the current flow to the motor winding.

- At each pulse of the chopper's clock oscillator, the resistor labeled R sense provides a reference for measuring the current flow in the winding. This results from the fact that as the current increases, so does the voltage drop across R sense.

- This voltage is fed to the comparator where it is matched to an analog reference voltage determined by the setting of a trimmer pot. The trimmer pot is initially set to match the electrical characteristics of the motor in the circuit.

- If the value of V sense is the same or exceeds the reference voltage (V ref), then the chopper control logic turns off transistor T2, which, in turn, shuts off current to the winding.

- As the voltage drop across R sense decreases, the comparator output changes states and the control logic turns on transistor T2.

- The cycle continues at each pulse of the oscillator and while the step-on signal is high.

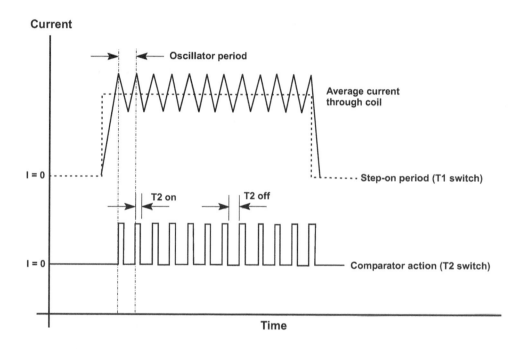

Figure 19-2 shows the various waveforms in a chopper circuit.

Figure 19-2 Chopper Circuit Waveforms.

In Figure 19-2 the waveform at the top of the illustration shows the sawtooth form of the current that results from the action of the comparator, as determined by the reference voltage. The activity of the T2 switch (see Figure 19-1) is shown by the waveform at the bottom of the figure. The dash-dot vertical lines show the oscillator period, which is typically 20 KHz in stepper motor circuits. The dashed line represents the period during which switch T1 is on, which corresponds to the steps generated by the control logic.

Current control methods such as chopper circuits provide several advantages in stepper motor applications:

- Overheating is reduced.

- Torque is increased.

- Top speed is increased.

- Efficiency is higher.

- Motors of different characteristics can be used with the same circuit.

The circuit in the following section allows implementing a bipolar stepper motor control chopper circuit. Note that chopper circuits can also be used with unipolar motors, although bipolar ones are by far more common. The Allegro SLA7024M is a chopper control IC for unipolar motors.

19.3 L297/298 Chopper Circuit

In Chapter 18 we presented a simple circuit based on the L297/298 pair. This circuit, labeled SMB-L297-298-1 and shown in Figure 18-3, leaves several unused lines in either IC. In fact, in the driver stage of circuit SMB-L297-298-1, the only connection between the 297 and the 298 chips are the four 297 output lines. The oscillator, voltage reference, sense, and inhibit lines of the 297 are left floating, and the matching lines on the 298 are either held high or wired to ground. Figure 19-3 shows the driver stage of a circuit that can be used to implement chopping.

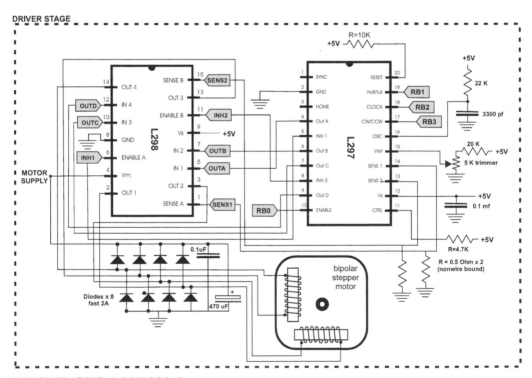

CIRCUIT: SMB-L297-298-2

Figure 19-3 Bipolar Motor Control Circuit with Chopping.

The control and translator stage of circuit SMB-L297-298-1 (in Figure 18-3) can be used with the driver stage of the circuit in Figure 19-3. The following lines of the L297 relate to the chopper function:

CONTROL: Defines the action of the chopper. When held high, the chopper acts on the phase lines A, B, C, and D. When low, the chopper acts on the INH1 and INH2 lines. In the present circuit, the CONTROL line is held high.

A-B-C-D: Motor phase lines connecting to the power stage.

ENABLE: When low, disables the L297 functions and brings down the A-B-C-D lines and the INH1 and INH2. The ENABLE line is pulled low during system initialization.

OSC: Connected to an RC network to produce a clock signal that determines the chopping rate.

INH1: Active low control for stages A and B. With a bipolar H bridge, this function ensures the fast decay of the load current when a winding is de-energized. Operates in half-step and wave drive modes.

INH2: Same as INH1 for control stages C and D.

SENS1 and SENS2: Senses the winding current for stages A and B (SENS1) and for stages C and D (SENS2). Value from the SENS1 and SENS2 lines are fed to the internal comparator to determine the maximum load current during chopping.

RESET: Active low on this line resets the L297 to its home state.

SYNC: Provides an output of the oscillator pulse useful in circuits with multiple L297 chips. Left floating in circuit SMB-L297-L298-2.

HOME: Open collector output indicating that the L297 is in its initial state. This port can be used in counting the number of steps made by the motor. It is left floating in circuit SMB-L297-L298-2 because the step count can be kept internally by the microcontroller.

In building the circuit note that the sense resistors must not be wire bound and that the diodes must be fast and have a 2-amps capacity.

19.3.1 Setting the Reference Voltage

The maximum winding current must be chosen for the particular motor wired to the board. This value determines the reference voltage that is set by means of the trimmer on the Vref line. The general formula is

$$Vref = I \bullet R$$

where I is the rated current of the motor in use and R is value of the sense resistors, in this case 0.51 Ohms. For example, with a motor rated for 0.5 A, the formula is evaluated as follows:

$$Vref = 0.5 \bullet 0.51 = 0.255$$

In order to set the circuit to the correct voltage reference, a voltmeter is connected to the trimmer pot and +5 volts applied to the board. This should be done with the L297 removed from the board. Alternatively, the board may contain a jumper between the trimmer's moveable arm and the L297 Vref pin. In this case, the adjustment can be made without having to remove the L297 by opening the jumper connection and holding low the ENABLE line. In either case the trimmer pot is adjusted so that the voltage on the Vref pin is as determined by the previous formula.

19.4 Chopper-Based Demo Board

A demonstration board for testing and adjusting chopper-based circuits is a valuable convenience for the developer, especially if a multimeter and an oscilloscope are available. The board is particularly useful in developing the software that will operate the circuit, for checking motors of unknown condition or parameters, and for experimenting before committing to design and hardware decisions. Figure 19-4 shows the schematics for a chopper-based demo board intended for bipolar motors that do not exceed the capacity of the L297/298 drivers.

The driver stage of circuit DEMO-SMB-01, in Figure 19-4, is almost identical to the driver stage of circuit SMB-L297-298-2 in Figure 19-3. The only difference is the jumper labeled J1 on the L297 Vref pin, which allows setting the reference voltage without removing the L297 from its socket. The control and translator stage is based on the 16F84A PIC and contains an eight-lever dip switch wired to ports A and B. The LED on port RB7 can be used in testing that the microcontroller program is executing by means of a short program that flashes the LED. It can also be used to report the state of a program function or control. The use of each of the eight DIP switches is left to the developer.

Materials list, building instructions, and PCB images for the DEMO-SMB-01 board can be found in online files. Note that the PCB version of the board includes a second jumper labeled J2. This jumper allows powering the motor from the same source used for the board or from a separate power supply. The PCB version also contains a conventional 78L05 logic power source and a standard wall adapter inlet. Ideally, the L297 should be mounted on an 20-pin IC socket and the microcontroller on a ZIF adapter.

19.4.1 Motor Circuit Power Requirements

Stepper motor circuits often have different power requirements for the various components. For example, the circuit SMB-L297-298-2 in Figure 19-3 requires DC power for the PIC 16F84A microcontroller, for the L297, the L298, and the stepper motor. Three of these components can operate at a nominal 5 volts: the 16F84, the L297 and the 298. Stepper motors, on the other hand, have a wide range of voltage requirements, ranging from 5 to 40 volts.

Because the classic 7805 regulator operates at 7.5 to 20 volts, it may be possible to power all components from a single source. This would be the case if the stepper motor is in the range of 8 to 20 volts and if the source available has sufficient amperage to cover the needs of both the three ICs and the motor. For example, if the motor requires 12 volts at 1 amp, a 12-volt 2-amp source could be tapped to feed both the motor and the 7805 regulator. In this case, the output of the 7805 provides 5 volts for the circuit ICs while the motor receives the full power from the source. However, if the motor requirements exceed the 20-volt limit of the 7805, then, circuit requires two power sources, or the voltage of the power source must be reduced to the 20-volt limit of the 7805.

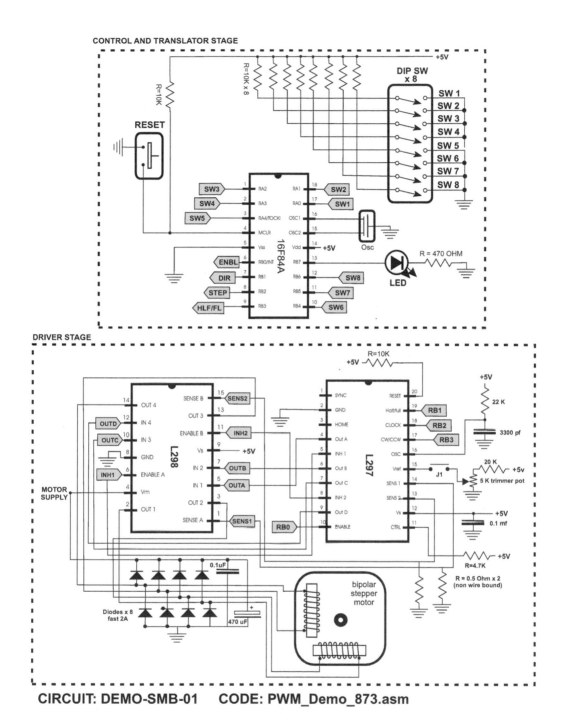

CIRCUIT: DEMO-SMB-01   CODE: PWM_Demo_873.asm

Figure 19-4 Chopper-Based Demo Board Schematics.

Demo board DEMO-SMB-01 contains a jumper that allows selecting power to the motor directly from the source used by the board, or from a separate supply. If jumper J2 (not shown in the circuit schematics) is closed, then the motor power comes from the same source used for the board. If J2 is open, then motor power must be fed through separate taps on the bottom-left corner of the demo board.

In determining the power requirements of a stepper motor, keep in mind that two different ratings are sometimes listed in the datasheet: the maximum voltage and the operating voltage. In theory, the maximum voltage should never be exceeded, at the risk of destroying the internal insulation in the windings. But chopper and microstepping circuits often use five times or more than the rated voltage for short time periods in order to increase performance and maximize torque. The practical rule is that as long as the maximum amperage of the motor is not exceeded, current limiting techniques allow fluctuations in voltage that produce no harm.

19.4.2 Chopper Demo Program

The sample program PIC_Chopper.asm, listed later in this chapter and in this book's software resource, allows testing the chopper hardware in the DEMO-SMB-01 circuit shown in Figure 19-4. The program drives the motor in the forward or reverse direction and in full- or half-step modes, but its execution proves that the chopper circuit is performing its expected functions.

19.5 Microstepping

Abrupt changes in magnitude or direction or the current driving a stepper motor result in jerkiness and uneven rotation, which is especially objectionable at lower motor speeds. This jolting action relates to stepping modes, being less in half- than in full- or wave-step modes. In some applications the abrupt changes in rotor speed can have deleterious effects. In fact, the development of microstepping is related to the abrupt motion of a turntable used in medical tests, which resulted in spilled samples. Larry Durkos, an engineer for the company that manufactured the equipment, came up with a computer-controlled solution to the problem based on the technique now called "microstepping." Figure 19-5 shows rotor action in several stepping modes.

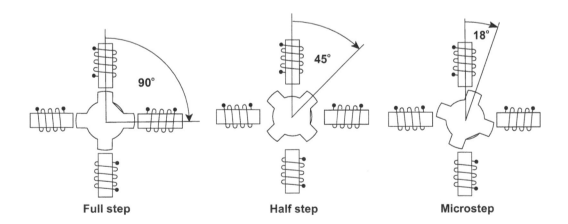

Figure 19-5 Stepper Motor Rotation in Various Stepping Modes.

In the left-hand stepper motor of Figure 19-5, the rotor moves through an angle of 90 degrees for each pulse in the full-step mode. In the stepper motor at the center of the figure the rotor moves 45 degrees in the half-step mode. The rotor at the right side of Figure 19-5 rotates through an angle of 18 degrees in the microstep mode. By reducing the angle of rotation for each motor step, the movement of the rotor is made smoother and less abrupt. In this example, one full rotation requires four pulses in the full-step mode, eight pulses in the half-step mode, and twenty pulses in the microstep mode. Note that the motor depicted in the illustration does not correspond to a real device because step angles are usually much smaller in actual motors.

Another factor that affects stepper motor rotation relates to the conventional current waveforms. Figure 19-6 shows the current transitions in a conventional full-step mode with maximum torque.

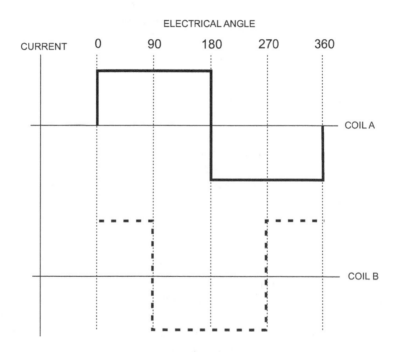

Figure 19-6 Current Transitions in High-Torque Stepping Mode.

The waveform in Figure 19-6 shows a voltage that rises abruptly at the start of each step; consequently, so does the attraction of the stator windings on the motor poles. As the voltage declines abruptly at the end of each step, the attraction is instantly turned off. This results in the less-than-smooth movement associated with stepper motors in conventional pulse modes. The square wave approximately represents the attraction of the windings on the rotor at various positions in the step cycle.

The ideal waveform for driving a stepper motor so as to produce uniform attraction on the windings is a sine wave, as the ones shown in Figure 19-7.

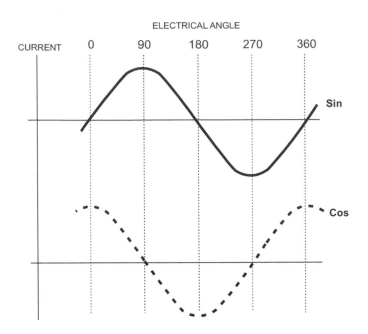

Figure 19-7 Sine/Cosine Wave-Shaped Coil Current.

Two sine waves out of phase by 90 degrees (sine and cosine of the electrical angle) ensure that the attraction of the windings on the rotor will be uniform and that the motor will move smoothly and quietly. This is due to the fact that as the current in one coil increases, the current decreases in the other coil; therefore the torque is correctly proportioned between the two coils at any angle. Trigonometrically, the power in each coil (torque) is the square root of the sum of the squares of the current in each coil.

Microstepping reduces or eliminates this effect by varying the current in the windings so as to produce a uniform attraction on the rotor throughout the step cycle. This can be achieved by modifying the current applied to the windings so as to follow the sine waveform shown by the dotted line in the illustration. In this case the attraction on the rotor is uniform throughout the step cycle and the motor rotates smoothly. Because microstepping requires control over the current levels it results in more complex circuitry, more sophisticated hardware, and more complicated programming.

Microstepping improves stepper motor performance by offering the following advantages:

- The motor rotates more smoothly at slow speeds.

- The resolution of the rotor positioning is increased due to a smaller step angle.

- Torque is higher at both low and high step rates.

Microstepping is usually implemented by controlling the current in both motor coils. Some microcontroller-based circuits control the current using pulse width modulation methods. These circuits and controls are described in the sections that follow.

19.5.1 Microstepping Fundamentals

In the field of stepper motor controls, the *step rate* is defined as the speed at which the coils are turned on or off. Step rate is measured in steps per unit of time, usually in steps per second or per microsecond. The maximum speed of a motor is achieved when the step rate is highest. This maximum speed is determined by the *inductance* in the motor windings because a device with higher winding inductance takes longer to reach the rated coil current. If the time between steps is less than the time required for current build-up, then the motor misses a step or "slips."

Because the step rate is a measure of the speed at which the motor winding current is turned on and off, we can see that it doubles in the half-step mode compared to the full-step mode. By the same token, when the rotor is moved in microsteps, the step rate is increased by a factor determined by the number of microsteps in a full step. In the case depicted in Figure 19-5, the step rate in the microstepping example is increased by a factor of 20 compared to the full-step rate because there are twenty microsteps in one full step in this case.

Microstepping Theory

It is easy to see how reducing the size of each step improves the smoothness of stepper motor rotation. However, given a two-pole motor, it is not possible to further reduce the number of steps beyond one-half the stepping rate by manipulating the pulse sequence. We define one full electrical cycle as consisting of four full steps. This means that a full electrical cycle consists of 360 degrees of electrical angle, which is not the same as 360 degrees of mechanical rotation. At any step angle one full electrical cycle of a stepper motor consists of four full steps. Hence, one full step corresponds to 90 degrees of electrical angle.

In microstepping, the motor's electrical angle is divided into smaller, equal units by varying the current in the stator windings. Although the resulting curve does not perfectly match a sine wave, it can be approximated by choosing small-enough steps, as shown in Figure 19-8.

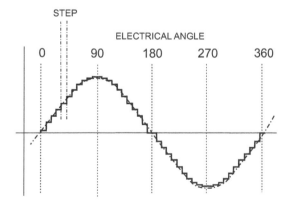

Figure 19-8 Sine Wave Approximation by Discrete Current Steps.

In a stepper motor, energizing a winding produces a flux in the air gap that is proportional to the current in the winding. If variation is introduced in more than one winding, then the flux in the air gap is the vector sum of the current in the various windings. The rotation takes place in the direction of the resultant vector. In the full step mode, the maximum rated current is applied to the windings, resulting in a flux in the air gap of 90 degrees electrical with each pulse. In the half step mode, the flux rotation is of 45 degrees electrical.

In microstepping, the current in the winding is applied at a fraction of the rated value. This means that the flux changes by a fraction of 90 degrees electrical in the resulting direction. Usually a full step is divided into 4, 8, 16, or 32 changes in current intensity, which are the microsteps. Although it is possible to divide a full step into more than thirty-two microsteps, it is generally considered that a greater number of steps provides no further improvement in rotor motion. In a *two-winding motor*, the following formula can be used to calculate the required current in the winding to achieve a required rotation in the flux:

$$Ia = Irate \bullet \sin\theta$$

$$Ib = Irate \bullet \cos\theta$$

where Ia and Ib are the current in stator windings A and B, respectively, Irate is the rated current of the motor, and θ represents the microstep angle. The vector sum of the individual winding currents is expressed as follows:

$$S = \sqrt{((Irate \bullet \sin\theta)^2 + (Irate \bullet \cos\theta)^2)}$$

where S is the resultant stator current. Figure 19-9 shows the ideal stator current variations in the two windings of a stepper motor.

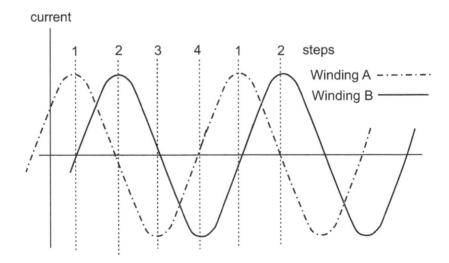

Figure 19-9 Stator Current During Microstepping.

As the current varies in each winding (labeled A and B in Figure 19-9), there is a corresponding variation in the rotating flux in the air gap. For each step, the actual value of the flux in the air gap corresponds to the value for Irate in the previous equations. It is this effect that ensures a constant rotating torque in microstepping. However, in implementing microstepping it is more effective to keep the current in each winding constant during more than half the step, while the current in the other winding is varied as function of the sine of the angle of rotation, as shown in Figure 19-10.

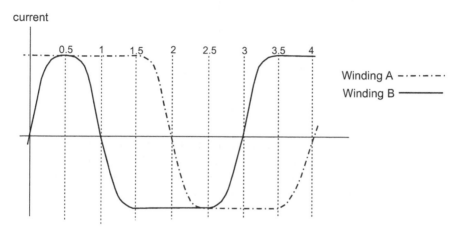

Figure 19-10 One-Winding-Held-Constant Phase-Current Diagram.

Holding the current constant during one-half the step cycle results in the following relation:

$$S = \sqrt{((Irate^2 + (Irate \bullet \sin\theta)^{2)}}$$

where S is the resultant current, Irate is the rated current of the motor, and θ represents the microstep angle. This scheme is usually called *high-torque microstepping*.

Pulse Width Modulation (PWM)

From the previous discussion we can see that in order to implement microstepping, we must find a way of varying the current through the coils during the step cycle. Although there are several ways to accomplish this, the one most suitable for micrcontroller-based circuits is to manipulate the duty cycle using a technique called pulse width modulation or PWM. The method is particularly convenient when the microcontroller contains a PWM module. Several mid-range PICs, including the 16F684, the 16F873, the 16F887, and most high-end and DSP devices, include a *Capture/Compare/ PWM* (CCP) module that provides the required functionality. An enhanced version of the CCP module named ECCP can be found in the current generation of PICs.

The basic idea of pulse width modulation is to control the average value of a current fed to a load by rapidly turning on and off the power source. In this context, a duty cycle is defined as the percentage of the PWM control period during which the signal is held high. For example, a 10V signal that is held on 50 percent of the time (50 percent duty cycle) results in an average voltage of 5V. Figure 19-11 shows the fundamental elements in the concept of pulse width modulation.

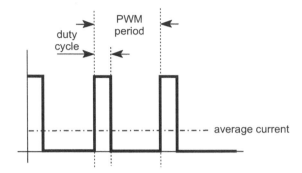

Figure 19-11 Pulse Width Modulation Schematics.

In Figure 19-11 the duty cycle is 25 percent of the PWM period, therefore the resulting average current is 25 percent of its maximum value. In any case, the resulting current is in the same ratio to the maximum current as the duty cycle is to the PWM period. For example, if the rated current is 10 A and the duty cycle is 10 percent, then the resulting current will be 1 A.

Note that PWM is effective in devices that respond to the current average and not to its instantaneous value. These include light bulbs and DC motors. Commercial dimmers used in controlling household lights use a form of pulse width modulation.

19.6 Programming PWM

The CCP and ECCP modules available in many mid-range, high-end, and DSP PICs provide the hardware for easy implementation of pulse width modulation as required for stepper motor control. Many devices contain several CCP or ECCP modules. In addition to PWM, the CCP and ECCP modules provide capture and compare operations, as indicated by the letter C in the device's name.

Three PICs covered in this book contain one or another version of the CCP module: the 16F684, 16F877, and 16F873. When these modules are not available, it is possible to implement PWM in software as described in the literature available from Microchip. In this section we discuss *pulse width modulation* using the hardware modules mentioned previously. We do not describe the refinements provided by the enhanced version of the module.

19.6.1 CCP Module

The mid-range PICs that provide PWM hardware contain up to two CCP or ECCP modules. When more than one CCP module is furnished, all of them share a single clock source; therefore, the functionality of these modules is limited because they all depend on a single clock cycle. The CCPxCON registers control the operation of the CCP devices. Here x can have the values 1 or 2. If a device contains more than one module, then all CCP-related registers use the digits 1 or 2 to identify each module. For example, the CCPxL registers are designated either as CCP1L or CCP2L. We will use the x designation in the present descriptions. The following are the CCP basic registers:

CCPxCON: One or two CCP control registers. This register provides mode selection and access to the two least-significant bits of the duty cycle.

CCPRxL: One or two CCPR registers that store the low byte value in capture or compare mode and the duty cycle in PWM mode.

CCPRxH: One or two CCP registers that store the high byte value in capture or compare mode. It is used by the hardware for latching the duty cycle in PWM mode. In PWM mode, CCPRxH is read-only.

Figure 19-12 is a bitmap of the CCPxCON register.

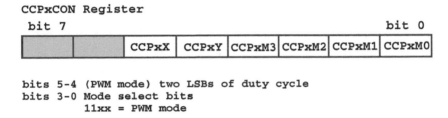

```
CCPxCON Register
 bit 7                                                      bit 0
┌──────┬──────┬──────┬──────┬──────┬──────┬──────┬──────┐
│      │      │CCPxX │CCPxY │CCPxM3│CCPxM2│CCPxM1│CCPxM0│
└──────┴──────┴──────┴──────┴──────┴──────┴──────┴──────┘

 bits 5-4 (PWM mode) two LSBs of duty cycle
 bits 3-0 Mode select bits
          11xx = PWM mode
```

Figure 19-12 Bitmap of the CCPxCON Register for PWM Operation.

In PWM mode, two port pins are multiplexed for pulse width modulation. The CCPx pins provide the PWM output in a 10-bit resolution, although applications sometimes ignore the two low-order bits and thus reduce the resolution to 8 bits. In any case, the port pins mapped to the PWM function (PORT C in the 16F87x PICs) must have their *tris* registers set for output. PWM uses Timer 2 in free running mode for a clock signal. The PR2 register is used to set the PWM period, and the CCPRxL register is used to store the high-order 8 bits of the duty cycle while the two low-order bits are in CCPxCON register bits 5 and 4 (see Figure 19-12). The duration of the PWM period and duty cycle are determined by the system's clock frequency and the prescaler setting. Figure 19-13 is a bitmap of the PR2 and CCPRxL registers.

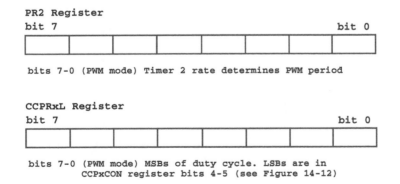

```
PR2 Register
 bit 7                                         bit 0
┌──────┬──────┬──────┬──────┬──────┬──────┬──────┬──────┐
│      │      │      │      │      │      │      │      │
└──────┴──────┴──────┴──────┴──────┴──────┴──────┴──────┘

 bits 7-0 (PWM mode) Timer 2 rate determines PWM period


CCPRxL Register
 bit 7                                         bit 0
┌──────┬──────┬──────┬──────┬──────┬──────┬──────┬──────┐
│      │      │      │      │      │      │      │      │
└──────┴──────┴──────┴──────┴──────┴──────┴──────┴──────┘

 bits 7-0 (PWM mode) MSBs of duty cycle. LSBs are in
          CCPxCON register bits 4-5 (see Figure 14-12)
```

Figure 19-13 Bitmap of the PR2 and CCPRxL Registers for PWM Operation.

The PWM period and duty cycle are determined by the following elements:

- The system clock rate and the Timer 2 prescaler which together with the value in the PR2 register determine the Timer 2 rate. This is the PWM period.

- The stetting in bits 5-4 of the CCPxCON register. These are the two low-order bits of the duty cycle.

- The setting of the CCPRxL register which determines the eight high-order bits of the duty cycle.

For example, assume a PIC microcontroller running at 4 MHz and a Timer 2 prescaler of 1:1. In this case the PWM period is determined by the value in the PR2 register. The actual formula for the PWM period is as follows:

$$PWM \ period = [(PR2) + 1] \bullet 4 \bullet Tosc \bullet (TMR2 \ preset \ value)$$

The duty cycle is determined by the bits 5-4 of the CCPxCON (low-order bits) and the value in the CCPRxL register (high-order bits). The actual formula is as follows:

$$PWM \ duty \ cycle = (CCPRxL : CCPxCON < 5 : 4 >)$$

Thus, if we store 99 decimal in PR2, clear bits 5-4 in the CCPxCON register, and store the value 25 decimal in the CCPRxL register, then the duty cycle will be 25 percent of the PWM period.

In operation, the CCP module in PWM mode proceeds as follows during the PWM cycle:

- The CCPx pin mapped to the PWM function is brought high.

- The duty cycle is loaded into the CCPRxH register.

- While the count in CCPRxH remains greater than the Timer 2 count, the output on the CCPx pin is held high.

- When the count in the CCPRxH matches the one in Timer 2, the CCPx output is brought low.

- A new cycle starts when the value in PR2 matches the Timer 2 count.

A change in the duty cycle will not take effect until the start of the next cycle.

19.6.2 PWM Circuit and Software

Because programming PWM using the CCP hardware is not without complications, it is a good idea to develop a simple board that tests the software routines without the controls and refinements of the final motor control circuit. One simple test circuit can be based on connecting an LED to the PWM line and some simple means of varying the duty cycle. As the duty cycle is increases, so does the brightness of the LED, and vice versa. Figure 19-14 shows such a circuit.

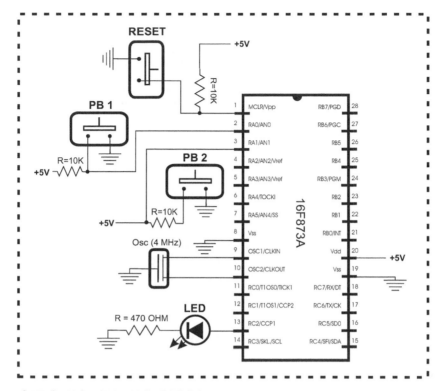

CIRCUIT: PWM-DEMO-1
CODE: PWM_Demo_873.asm

Figure 19-14 Circuit for Testing Pulse Width Modulation.

In the circuit in Figure 19-14 two pushbutton switches are wired to ports RA0 and RA1, respectively. The pushbutton on port RA0 (labeled PB 1) is used to increase the duty cycle and the one on port RA1 to decrease it. The PWM output is produced on the CCP1 pin to which an LED is wired.

The program PWM_Demo_873.asm, in this book's software resource, uses the CCP module of the 16F873 PIC to produce pulse width modulation. The program stores the duty cycle in a local variable and reads the action on the pushbutton switches to increase or decrease its value. Initialization is as follows:

```
;=============================
;      main program
;=============================
main:
    banksel   ADCON1                ; Select bank
    movlw     b'00000111'     ; Turn off A/D, port A
    movwf     ADCON1
    banksel   PORTA
    movlw     b'00000111'     ; Input lines on PORTA
    tris      PORTA
    movlw     B'00000000'     ; w = 00000000 binary
    tris      PORTC                 ; Set up port C for output
```

```
        bcf         PORTC,2             ; CCP1 pin low
; Set up registers for PWM
        banksel     INTCON              ; Select bank
        bcf         INTCON,GIE          ; Disable global interrupts
        bcf         INTCON,PEIE         ; And peripheral interrupts
        banksel     PIE1
        bcf         PIE1,TMR2IE         ; Disable timer 2 interrupts
        bcf         PIE1,CCP1IE         ; Disable CCP1 interrupts
        banksel     CCP1CON
        clrf        CCP1CON             ; Turn off CCP1 module
        banksel     PR2
        movlw       .255                ; value to load
        movwf       PR2                 ; into PWM period register
        banksel     CCP1CON
        bcf         CCP1CON,5           ; Duty cycle LSB
        bcf         CCP1CON,4           ; Duty cycle MSB
; Set initial duty cycle
        banksel     dutyCycle
        movlw       .128                ; Set initial duty cycle to 50%
        movwf       dutyCycle           ; Store in local variable
        banksel     CCPR1L
        movwf       CCPR1L              ; CCPr1l ;bits 9-2 of duty cycle
; Set prescaler to 1:1, no postscaler
        banksel     T2CON
        movlw       b'00000000'
        movwf       T2CON               ; Prescaled set and TMR2 off
        banksel     TMR2
        clrf        TMR2                ; Clear timer 2
        banksel     CCP1CON
        movlw       b'00001100'         ; CCP1 in PWM mode and turn on
        movwf       CCP1CON
        banksel     T2CON
        bsf         T2CON,2             ; Timer 2 on
```

Once the CCP hardware has been initialized to the PWM mode, the program proceeds to turn on the LED to the default duty cycle value, which is 50 percent of maximum. Code then monitors action on the two pushbutton switches with a delay loop. If action on pushbutton # 1 is detected, the duty cycle is tested for maximum and if not, it is incremented by one unit. Similarly, pushbutton # 2 decrements the duty cycle. Code is as follows:

```
;=================================================================
;                    input monitoring routine
;=================================================================
LEDon:
; Turn on line 2 in port C. All others remain off
        banksel     PORTC
        movlw       B'00000100'         ; LED ON
        movwf       PORTC
        call        PBAction            ; Test pushbutton action
        call        delay               ; Local delay routine
        goto        LEDon


;===============================
;    monitor pushbuttons
;    and update duty cycle
;===============================
PBAction:
        banksel     PORTA
```

```
        btfss       PORTA,2             ; Decrease button pressed?
        goto        DutyDown            ; Yes, decrement duty cycle
        btfss       PORTA,0             ; Increase button pressed?
        goto        DutyUp              ; Yes, increment duty cycle
        return                          ; No action on switches
DutyUp:
        banksel     dutyCycle
        movf        dutyCycle,w         ; Duty cycle to W
        sublw       .255                ; Test for maximum
        banksel     STATUS
        btfsc       STATUS,Z            ; Return if at maximum
        return
; Duty cycle not at maximum value
        banksel     dutyCycle
        incf        dutyCycle,f         ; Increment duty cycle
        movf        dutyCycle,w         ; Duty cycle to W
        banksel     CCPR1L
        movwf       CCPR1L              ; To control register
        return                          ; Done
DutyDown:
        banksel     dutyCycle
        movf        dutyCycle,w         ; Duty cycle to W
        sublw       0x00                ; Test for minimum
        banksel     STATUS
        btfsc       STATUS,Z            ; Return if at minimum
        return
; Duty cycle not a minimum value
        banksel     dutyCycle
        decf        dutyCycle,f         ; Decrement duty cycle
        movf        dutyCycle,w         ; New duty cycle to W
        banksel     CCPR1L
        movwf       CCPR1L              ; To control register
        return                          ; Done
;================================
;        delay sub-routine
;================================
delay:
        banksel   j
        movlw     .200          ; w = 200 decimal
        movwf     j             ; j = w
jloop:
        movwf     k             ; k = w
kloop:
        decfsz    k,f           ; k = k-1, skip next if zero
        goto      kloop
        decfsz    j,f           ; j = j-1, skip next if zero
        goto      jloop
        return
```

19.6.3 Microstepping by PWM

Earlier in this chapter we discussed how microstepping produces a uniform attraction on the rotor throughout the step cycle by modifying the current applied to the windings. The ideal modification results in a sine-shaped waveform. The practical result is that the attraction of the coils on the rotor is uniform throughout the step cycle and the motor rotates smoothly. In PIC-based circuits, microstepping can be implemented by controlling the current in both motor coils. A convenient way of accomplishing this is by pulse width modulation.

In Section 19.5 we saw how microstepping can be achieved by varying the current in both coils of a stepper motor in a sine wave pattern. Figure 19-7 shows how the sine wave can be offset by 90 degress to generate the sine/cosine pattern that produces a uniform attraction on the rotor. In practice, microstepping can be more effectively implemented by keeping the current in one winding constant while the current in the other one follows the sine wave shape. This is shown in Figure 19-10, earlier in this chapter. This scheme is called high-torque microstepping.

Figure 19-15 shows an example of microstepping using the high-torque method. The example matches the code in the sample program PWM_Micstep_1.asm, discussed later in this chapter and found in this book's software resource.

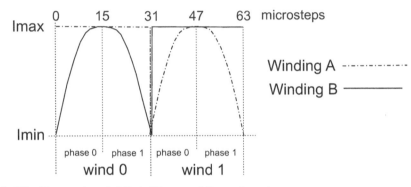

Figure 19-15 Example of High-Torque Microstepping.

In Figure 19-15, while the current in winding B cycles from 0 to Imax (phase 0) and back to 0 (phase 1), the current in winding A is held high. In wind 1 the current in the motor windings are reversed, that is, the current cycles from 0 to max and back to 0 in motor winding A while motor winding B is held high. In the example in Figure 19-15, one complete iteration consists of 32 steps (0 to 31). The first 32 (0 to 31) steps are said to be in wind 0, and the second 32 (32 to 63) are in wind 1. Both wind 0 and wind 1 have two phases (labeled phase 0 and phase 1 in the illustration).

The example assumes a duty cycle table that encodes sixteen current values. During phase 0, the duty cycle table is read top-to-bottom, that is, the values go from low to high. During phase 1 the duty cycles are read bottom-to-top, that is, the values go from high to low. Table 19-1 shows the values in windings A and B for phase 1 of wind 1.

19.6.4 Microstepping Sample Program

The sample program PWM_Micstep_1.asm, listed later in this chapter and in this book's software resource, uses the PWM-DEMO-1 circuit in Figure 19-14 to simulate microstepping using PWM. The program manipulates the current on two LEDs wired to PWM lines CCP1 and CCP2. One LED demonstrates the current in the first winding of a stepper motor, and the other LED the current on the second winding.

Table 19.1

Values for Phase 1, Wind 1, in Microstepping Example in Figure 19-15

I<===== microStep (phase 0)				
I	WINDING		DUTY CYCLE	
I	A	B	CCP2	CCP1
0	+1	+sin 5.6	100%	9.8%
1	+1	+sin 11.25	100	20
2	+1	+sin 16.8	100	29
3	+1	+sin 22.5	100	38
4	+1	+sin 28	100	47
5	+1	+sin 33.75	100	56
6	+1	+sin 39	100	63
7	+1	+sin 45	100	71
8	+1	+sin 50.6	100	77
9	+1	+sin 56.25	100	83
·10	+1	+sin 61.8	100	88
11	+1	+sin 67.5	100	93
12	+1	+sin 73.1	100	95.6
13	+1	+sin 78.75	100	98
14	+1	+sin 84.35	100	99.5
15	+1	+sin 90	100	100

When running the program, while one of the LEDs is at maximum brightness the other one is cycled through the low-to-high-to-low current values, with the corresponding changes in brightness. The table with the sixteen duty cycle changes is coded as follows:

```
;*******************************
;  Duty cycles table
;*******************************
dcTable16:
     addwf     PCL,f     ; PCL is program counter latch
     retlw     .25       ; 0 - 9.8% of 255
     retlw     .51       ; 1 - 20%
     retlw     .74       ; 2 - 29%
     retlw     .97       ; 3 - 38%
     retlw     .120      ; 4 - 47%
     retlw     .143      ; 5 - 56%
     retlw     .161      ; 6 - 63%
     retlw     .181      ; 7 - 71%
     retlw     .196      ; 8 - 77%
     retlw     .212      ; 9 - 83%
     retlw     .224      ; 10 - 88%
     retlw     .237      ; 11 - 93%
     retlw     .244      ; 12 - 95.6%
     retlw     .250      ; 13 - 98%
     retlw     .254      ; 14 - 99.5%
     retlw     .255      ; 15 - 100%
     retlw     .0        ; end marker
```

The values returned by the table were obtained by scaling the range of a byte-size variable (256 bits) into sixteen equally spaced units that follow a sine curve.

Assuming suitable hardware, the program PWM_Micstep_1.asm can be used to drive a bipolar stepper motor in microstep mode. In this case, lines CCP1 and CCP2 will be connected to the corresponding motor windings. In controlling a stepper motor, variations in the delay loop will change the motor speed.

19.7 Microstepping ICs

Previously in this chapter we saw that high-performance stepper motor controls require circuits and software that ensure current control (chopper action) as well as microstepping. Each one of these methods is not without problems and complications, but combining them in a single circuit, driven by a single program, using nonspecialized components, can be a daunting task. The lack of online examples of circuits and programs that combine chopping and microstepping seems to confirm this thought.

One approach is to use PWM techniques to vary the reference voltage (Vref in Section 19.3.1) so that it will follow the sine curve required in microstepping. However, because the reference voltage is dependent on the particular motor used in the circuit, the actual duty cycles table would have to be developed to suit the specific hardware. This means that motor, circuit, and software driver would have to be matched.

Fortunately, manufacturers of electronic components have developed integrated circuits that simplify circuit design and programming. Allegro's A3955 and its announced update the A4975 are two such chips. Because the A4975 is not available at this time, we will discuss the A3955, with the caveat that the A4975 is currently under development.

19.7.1 Allegro 3955 IC

The A3955 is supplied in sixteen-pin DIP and a sixteen-pin SOIC packages. The chip is designed to drive a single winding of a bipolar stepping motor. Two 3955 ICs will be required in the typical motor driver circuit. The chip's output is rated for 1.5 A and a 50V operating voltage of continuous load. The IC is rated for continuous output currents to ±1.5 A and operating voltage to 50V. Stepping control is available in four selectable resolutions: full (2 steps), half (4 steps), quarter (8 steps), and eighth (16 steps). The 3955 contains *internal pulse width modulation hardware* that allows current control. There is also an internal, three-bit, nonlinear digital-to-analog converter that allows selecting full-, half-, quarter-, or eighth-step modes.

Internal circuitry in the 3955 determines whether the PWM current-control circuitry operates in a *slow recirculating current-decay mode*, a *fast regenerative current-decay mode*, or in a *mixed mode*. In the mixed mode, the off-time is divided into a period of fast current decay and one of slow current decay. The powerful motor control provided by the 3955 results from a combination of a user-selectable current-sensing resistor, control over the reference voltage, a digitally selected output current ratio, and a selectable slow, fast, or mixed current-decay mode. Figure 19-16 shows the pinout of the Allegro 3955.

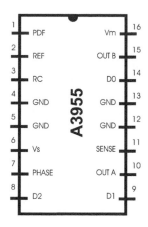

Figure 19-16 Allegro 3955 Microstepping Motor Controller Pinout.

The pin labeled PDF (*percent fast decay*) in Figure 19-16 is an analog input that allows selecting one of the available current decay modes; the current allows selecting the percentage of fast decay used by the chip, as follows:

```
Vdpf >= 3.5 V ============> slow current decay mode

Vpdf <= 0.8 V ============> fast current decay mode

Vpdf >= 1.1 V <= 3.1 V ===> mixed current decay mode
```

In a circuit with one or more 3955 ICs, the current decay mode is selected by means of two external resistors, as shown in Figure 19-17.

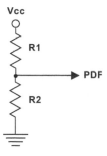

Figure 19-17 Current Decay Mode Selector Circuit.

The following formula can be applied for finding the values of R1 and R2 in Figure 19-17.

$$PDF = 100 \ln \frac{0.6(R1 + R2)}{R2}$$

The equation assumes that Vcc = 5.0V.

Alternatively, the PDF line can be connected to any other analog output, which can be a microcontroller port line. Controlling the voltage on this output allows selecting any of the three decay modes. One scheme, suggested by Allegro, is to select the slow current decay mode when the load current is increasing. This decay selection limits the switching losses in the driver, the iron losses in the motor, and improves the maximum rate at which the load current can increase. When the load current is decreasing, the mixed current-decay mode is then selected. This provides a way of regulating the load current to the desired level and prevents tailing of the current profile caused by the motor's back-EMF voltage. The sample program PIC873_3955_1.asm in this book's software resource uses this method of decay mode selection.

Three bits in the 3955, labeled D0, D1, and D2 in the chip's pinout (Figure 19-16), are DAC bits used to control the output current to the motor winding. Table 19.2 shows the various currents according to the DAC data.

Table 19.2

3955 DAC Bits

D2	D1	D0	CURRENT RATIO (IN %)
H	H	H	100
H	H	L	92.4
H	L	H	83.1
H	L	L	70.7
L	H	H	55.5
L	H	L	38.2
L	L	H	19.5
L	L	L	All Outputs Disabled

In the circuit developed later in this chapter, code reads the current from a lookup table and sets the 3955 DAC bits accordingly.

The 3955 input pin labeled PHASE (see Figure 19-16) controls the direction of current flow. These bits must be manipulated during microstepping in order to go from one phase of the sine curve to the next one.

An external current-sense resistor, the value input on the reference voltage line (labeled REF in Figure 19-16), and the setting of the three DAC bits, determine the peak motor current. When the three DAC bits are set , the maximum current limiting value is determined by the current-sense resistor and the voltage on the 3955 REF line.

19.7.2 3955-Based Circuit

The circuit described in this section contains two Allegro 3955 microstepping ICs that are driven by a PIC 16F873. The only input control device is a four-stage dip switch that selects forward and reverse rotation and one of the four stepping modes. Figure 19-18 shows the SMB-PIC873-3955-1 circuit.

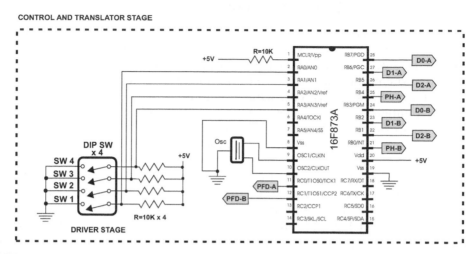

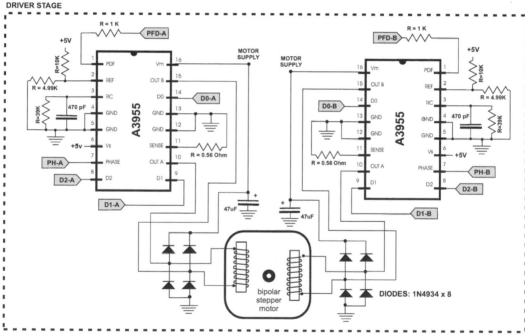

CIRCUIT: SMB-PIC873-3955-1
CODE: PIC873_3955_1.asm

Figure 19-18 Allegro 3955-Based Microstepping Circuit.

Note in the circuit in Figure 19-18 that microcontroller PORT B is wired to eight control lines in the 3955. The four high-order bits of port B go to the three DAC bits and the PHASE bit of winding A in the 3955. The four low-order bits connect to the same control lines in winding B. Bit number 0 in the PIC port C is wired to the *percent fast decay* (PDF) line of the 3955 controlling winding A, and bit 1 of port C connects to the PDF control line of the second 3955.

19.7.3 3955 Motor Driver Program

The program named PIC873_3955_1.asm, listed later in this chapter and in this book's software resource, drives a bipolar stepper motor using microstepping and automatic control of the current recirculation path. The program is interrupt driven and uses the hardware in the circuit named SMB-PIC873-3955-1 shown in Figure 19-18. Code assumes that four lines in port A are wired to a four-stage dip switch bank. One switch stage (bit number 0) selects forward or reverse motor rotation. The remaining three stages in the dip switch, which are wired to PORT B bits 1 to 3, allow selecting one of four stepping rates or modes supported by the 3955.

The program's operation starts by initializing ports A, B, and C and setting up a timer-activated interrupt linked to the processor's Timer0 line. The main program loop reads the state of the four dip switches and sets up the corresponding local variables. Motor driving takes place within the interrupt handler when the preset delay counter has expired. Code then tests if rotation is forward or reverse and directs control to the corresponding routine.

The program uses a thirty-two-step table for selecting control codes and another one for the PFD values during each cycle. Each control code table entry has eight significant bits that determine the setting of the three port bits mapped to the DAC registers and the phase bit for each 3955 IC, as in the following code fragment:

```
; Phase controls (PH-A and PH-B) and DAC codes (DA-x and DB-x) are
; mapped to PORT B lines, as follows:
;             7  6  5  4  3  2  1  0 <====== PORT B BITS
;             |  |  |  |  |  |  |  |___ PHASE B
;             |  |  |  |  |  |  |_____ DB-2
;             |  |  |  |  |  |_____ DB-1
;             |  |  |  |  |_____ DB-0
;             |  |  |  |_____ PHASE A
;             |  |  |_____ DA-2
;             |  |_____ DA-1
;             |_____ DA-0
```

The actual bit values are shown in Table 19-3.

Refer to Table 19.2 to determine the current applied to each winding by the DAC values in Table 19.3.

The PIC873_3955_1.asm program uses a variable (named *skip*) to determine how many steps are skipped during each iteration. The skip value is added to the table pointers so that a skip value of 1 moves the pointer to the next entry, a skip value of 2 skips one table entry, and so forth. By selecting values of 1, 2, 4, and 8 for the *skip* variable code can use the same thirty-two entry table to execute in each of the four supported modes. A second table, also with thirty-two entries, contains the PFD control codes. The *skip* variable is also added to the pointer when reading this table. The values from the PFD table are read into port C, which has port lines 0 and 1 wired to the corresponding controls in the 3955.

Table 19.3

Control Data for 32 Microsteps

WINDING A					WINDING B					STEP
D0-A	D1-A	D2-A	PH-A	PFDA	D0-B	D1-B	D2-B	PH-B	PFDB	
0	0	1	1	0	0	0	1	1	1	0
1	1	0	1	0	1	0	1	1	1	1
0	1	0	1	0	0	1	1	1	1	2
1	0	0	1	0	1	1	1	1	1	3
0	0	0	?	0	1	1	1	1	1	4
1	0	0	0	0	1	1	1	1	1	5
0	1	0	0	0	0	1	1	1	1	6
1	1	0	0	0	1	0	1	1	1	7
0	0	1	0	1	0	0	1	1	0	8
1	0	1	0	1	1	1	0	1	0	9
0	1	1	0	1	0	1	0	1	0	10
1	1	1	0	1	1	0	0	1	0	11
1	1	1	0	1	0	0	0	?	0	12
1	1	1	0	1	1	0	0	0	0	13
0	1	1	0	1	0	1	0	0	0	14
1	0	1	0	1	1	1	0	0	0	15
0	0	1	0	0	0	0	1	0	1	16
1	1	0	0	0	1	0	1	0	1	17
0	1	0	0	0	0	1	1	0	1	18
1	0	0	0	0	1	1	1	0	1	19
0	0	0	?	0	1	1	1	0	1	20
1	0	0	1	0	1	1	1	0	1	21
0	1	0	1	0	0	1	1	0	1	22
1	1	0	1	0	1	0	1	0	1	23
0	0	1	1	1	0	0	1	0	0	24
1	0	1	1	1	1	1	0	0	0	25
0	1	1	1	1	0	1	0	0	0	26
1	1	1	1	1	1	0	0	0	0	27
1	1	1	1	1	?	0	0	0	0	28
1	1	1	1	1	1	0	0	1	0	29
0	1	1	1	1	0	1	0	1	0	30
1	0	1	1	1	1	1	0	1	0	31

Note: Entries labeled ? are not significant because they correspond
to cycle steps with no current in the motor windings.

A practical application of the PIC873_3955_1.asm program and the SMB-PIC873-3955-1 circuit will probably require some form of motor speed control. One possible variation could be a second dip switch block to allow selecting among several possible speeds or a continuous control that could be potentiometer based. In either case several port lines are still free in the circuit. Because the program is interrupt-driven, speed control can easily be achieved by varying the value stored in the variable named *iteration*, which in the default version of the program is read from the constant named DELAY. Additional speed control may require modifying the Timer0 prescaler or the timer parameters.

19.8 Demonstration Programs

The programs listed in the following section demonstrate the the programming discussed in this chapter.

19.8.1 PWM_DEMO_873.asm

```
;=================================================================
; File: PWM_Demo_873.asm
; Date: February 26, 2011
; Authors: Canton and Sanchez
; Processor: 16F873A
;
; Program Description:
; Demonstration program for Pulse Width Modulation (PWM) on
; the 16F873 PIC. Program uses PWM output on the CCP1 pin to
; control the brightness of an LED wired to this same line.
; Two pushbutton switches on ports RA0 and RA1 control the
; duty cycle:
;       pushbutton on RA0 --> increases duty cycle
;       pushbutton on RA1 --> decreases duty cycle.
; Oscillator: Murata Erie 4 MHz
;
; Reference Circuit: PWM-DEMO
;
;                           16F873
;                    +------------------+
; +5v--res0--------| 1 !MCLR     RB7 28|---
;           PB 0 -->| 2 RA0       RB6 27|---
;           PB 1 -->| 3 RA1       RB5 26|---
;                ---| 4 RA2       RB4 25|---
;                ---| 5 RA3       RB3 24|---
;                ---| 6 RA4       RB2 23|---
;                ---| 7 RA5       RB1 22|---
;          GRND ---| 8 Vss        RB0 21|---
;           OSC ---| 9 OSC1           20|--- +5v
;           OSC ---|10 OSC2           19|--- GRND
;                ---|11 RC0        RD7 18|---
;                ---|12 RC1/CCP2  RD6 17|---
;     LED (PWM) <--|13 RC2/CCP1  RD5 16|---
;                ---|14 RC3        RD4 15|---
;                    +------------------+
; Legend:
; res0 = 10K resistor          OSC = 4 MHz oscillator
; GRND = ground

;===========================
; configuration switches
;===========================
```

```
; Switches used in __config directive:
;   _CP_ON          Code protection ON/OFF
; * _CP_OFF
; * _PWRTE_ON       Power-up timer ON/OFF
;   _PWRTE_OFF
;   _WDT_ON         Watchdog timer ON/OFF
; * _WDT_OFF
;   _LP_OSC         Low power crystal oscillator
;   _XT_OSC         External parallel resonator oscillator
; * _HS_OSC         High speed crystal resonator (8 to 10 MHz)
;                   Resonator: Murate Erie CSA8.00MG = 8 MHz
;   _RC_OSC         Resistor/capacitor oscillator (simplest,
; |                 20% error)
; |
; |_____ * indicates setup values
;
;==========================
; setup and configuration
;==========================
      processor 16f873A
      include      <p16f873A.inc>
      __config _HS_OSC & _WDT_OFF & _PWRTE_ON & _CP_OFF
      errorlevel -302
;=============================================================
;                   variables in PIC RAM
;=============================================================
j           equ          0x20
k           equ          0x21
dutyCycle   equ          0x22
;=============================================================
;                m a i n   p r o g r a m
;=============================================================
      org        0        ; start at address 0
      goto       main
;=============================
; space for interrupt handler
;=============================
      org        0x08
;=============================
;      main program
;=============================
main:
      banksel    ADCON1              ; Select bank
      movlw      b'00000111'         ; Turn off A/D, port A
      movwf      ADCON1
      banksel    PORTA
      movlw      b'00000111'         ; Input lines on PORTA
      tris       PORTA
```

```
        movlw       B'00000000' ; w = 00000000 binary
        tris        PORTC               ; Set up port C for output
        bcf         PORTC,2             ; CCP1 pin low
; Set up registers for PWM
        banksel     INTCON              ; Select bank
        bcf         INTCON,GIE          ; Disable global interrupts
        bcf         INTCON,PEIE         ; And peripheral interrupts
        banksel     PIE1
        bcf         PIE1,TMR2IE ; Disable timer 2 interrupts
        bcf         PIE1,CCP1IE ; Disable ccp1 interrupts
        banksel     CCP1CON
        clrf        CCP1CON             ; Turn off CCP1 module
        banksel     PR2
        movlw       .255                ; value to load
        movwf       PR2                 ; into PWM period register
        banksel     CCP1CON
        bcf         CCP1CON,5           ; Duty cycle LSB
        bcf         CCP1CON,4           ; Duty cycle MSB
; Set initial duty cycle
        banksel     dutyCycle
        movlw       .128                ; Set initial duty cycle to 50%
        movwf       dutyCycle           ; Store in local variable
        banksel     CCPR1L
        movwf       CCPR1L              ;ccpr1l
                                        ;bits 9-2 of duty cycle
; Set prescaler to 1:1, no postscaler
        banksel     T2CON
        movlw       b'00000000'
        movwf       T2CON           ; Prescaled set and TMR2 off
        banksel     TMR2
        clrf        TMR2                ; Clear timer 2
        banksel     CCP1CON
        movlw       b'00001100'         ; CCP1 in PWM mode and turn on
        movwf       CCP1CON
        banksel     T2CON
        bsf         T2CON,2             ; Timer 2 on
;================================================================
;                   input monitoring routine
;================================================================
LEDon:
; Turn on line 2 in port C. All others remain off
        banksel     PORTC
        movlw       B'00000100' ; LED ON
        movwf       PORTC
        call        PBAction            ; Test pushbutton action
        call        delay               ; Local delay routine
        goto    LEDon
```

```
;===============================
;     monitor pushbuttons
;     and update duty cycle
;===============================
PBAction:
      banksel     PORTA
      btfss       PORTA,2           ; Decrease button pressed?
      goto        DutyDown          ; Yes, decrement duty cycle
      btfss       PORTA,0           ; Increase button pressed?
      goto        DutyUp            ; Yes, increment duty cycle
      return                        ; No action on switches
DutyUp:
      banksel     dutyCycle
      movf        dutyCycle,w; Duty cycle to W
      sublw       .255              ; Test for maximum
      banksel     STATUS
      btfsc       STATUS,Z          ; Return if at maximum
      return
; Duty cycle not at maximum value
      banksel     dutyCycle
      incf        dutyCycle,f; Increment duty cycle
      movf        dutyCycle,w; Duty cycle to W
      banksel     CCPR1L
      movwf       CCPR1L            ; To control register
      return                        ; Done
DutyDown:
      banksel     dutyCycle
      movf        dutyCycle,w; Duty cycle to W
      sublw       0x00              ; Test for minimum
      banksel     STATUS
      btfsc       STATUS,Z          ; Return if at minimum
      return
; Duty cycle not a minimum value
      banksel     dutyCycle
      decf        dutyCycle,f; Decrement duty cycle
      movf        dutyCycle,w; New duty cycle to W
      banksel     CCPR1L
      movwf       CCPR1L            ; To control register
      return                        ; Done
;===============================
;        delay sub-routine
;===============================
delay:
          banksel     j
          movlw       .200          ; w = 200 decimal
          movwf       j             ; j = w
jloop:
          movwf       k             ; k = w
```

```
kloop:
            decfsz      k,f         ; k = k-1, skip next if zero
            goto        kloop
            decfsz      j,f         ; j = j-1, skip next if zero
            goto        jloop
            return

            end         ;END OF PROGRAM
```

19.8.2 PIC_Chopper.asm

```
;=================================================================
; File: PIC_Chopper.asm
; Date: September 15, 2011
; Update:
; Authors: Sanchez and Canton
; Processor: 16F84A
; Reference circuit: DEMO-SMB-01
;
; Program Description:
; Program to drive a bipolar stepper motor mode with a 16F84A
; PIC controlling an L297 wired to an L298 driver.
; Three toggle switches wired to port RA1, RA2, and RA3
; control direction, full or half step modes, and set or clear
; the L297 ENABLE line. Program supports chopper hardware.
;
;============================
;    demo board circuit
;============================
; All port A lines trissed for input:
;   0 <----- NOT USED
;   1 <----- Toggle switch 1 - CW or CCW direction
;   2 <----- Toggle switch 2 - Half/full step
;   3 <----- Toggle switch 3 - ENABLE line on/off
; Port B lines
;   PIC          L297
;   RB0 -----> pin 10 - ENABLE
;   RB1 -----> pin 17 - Direction
;   RB2 -----> pin 18 - Step input
;   RB3 -----> pin 19 - Full / half step
;   RB4 <-----  |
;   RB5 <-----  |
;   RB6 <-----  |--- NOT USED
;   RB7 <-----  |
;     PORT A      TOGGLE
;      line       SW      ACTION
;       1          1      0 = CW rotation 1 = CCW
```

```
;        2        2        0 = full step 1 = half step
;        3        3        1 = ENABLED 0 = DISABLED


;==========================
; setup and configuration
;==========================
      processor 16f84A
      include        <p16f84A.inc>
      __config  _XT_OSC & _WDT_OFF & _PWRTE_ON & _CP_OFF
      errorlevel -302


;=======================================================
;              variables in PIC RAM
;=======================================================
; Declare variables at 2 memory locations
j             equu         0x0c
k             equ          0x0d
delay         equ          0x0e ; Delay counter


;============================================================
;                m a i n   p r o g r a m
;============================================================
      org          0        ; start at address 0
      goto         Main
;============================
; space for interrupt handler
;============================
      org          0x04
;============================
;      main program
;============================
Main:
; Tris PORT A for input
      banksel      PORTA
      movlw        B'00011111'; w = 00011111 binary
      tris         PORTA               ; Set up port A for input
; Tris port B for output
      movlw        B'00000000'; All lines for output
      tris         PORTB               ; Set up port B
      clrf         PORTB               ; Clear all lines
; Set defalt state for L297 control lines
      movlw        B'00001101'
;                      ||||_____ 1 = ENABLE
;                      |||_____ 0 = CW
;                      ||_____ 1 = clock pulse
;                      |_____ 1 = Half step
;
      movwf        PORTB
```

```
; Initialize default value for delay variable
      banksel     delay
      movlw       .120
      movwf       delay         ; To delay variable
;================================================================
;                   main control monitoring routine
;================================================================
ControlRtn:
; Call procedure to set L297 lines according to state of
; toggle switches wired to port A
      call        ReadAToggles
      call        Delay              ; Delay now
      call        Pulse              ; Local pulse motor routine
      goto        ControlRtn
;==========================================================
;       Auxiliary procedure to read toggle switches
;==========================================================
ReadAToggles:
; Read PORTA toggle switches
;      PORT A     TOGGLE
;       line      SW      ACTION
;        1        1       0 = CW rotation 1 = CCW
;        2        2       0 = full step 1 = half step
;        3        3       1 = Set or clear ENABLE line
;==============================
;   set CW or CCW rotation
;==============================
; Bit 1 on port A is wired to toggle switch # 1
; Bit 1 on port B is wired to pin 17 of the L297 which
; controls CW and CCW rotation.
      banksel     PORTA
      btfsc       PORTA,1
      goto        goCCW                ; Set rotation bit to CCW
; Bit clear. Set CW
      bcf         PORTB,1              ; Clear port B line 1
      goto        SetStep
goCCW:
      bsf         PORTB,1
;==============================
;   set half or full step
;==============================
SetStep:
; Toggle switch # 2, on port A, line 2, selects between
; half step and full step modes. Port B bit 3 is wired
; to L297 line 19, which selects half or full step modes
      btfsc PORTA,2
      goto        HalfStep    ; Bit is set. Set half step mode
; Bit clear. Set full step mode
```

```
        bcf         PORTB,3             ; Clear port B line 2
        goto        TestEnable
HalfStep:
        bsf         PORTB,3
;=============================
;   enable motor control line
;=============================
; Toggle switch # 3, on port A, line 3, controls the
; L297 ENABLE line, on pin No. 10
TestEnable:
        btfsc       PORTA,3
        goto        Enable              ; Bit is set. Set ENABLE line
; Bit clear. Disable
        bcf         PORTB,0             ; Clear the ENABLE line
        return
Enable:
        bsf         PORTB,0             ; Set bit to enable
        return
;================================
;   routine to pulse the motor
;================================
; L297 CLOCK line is wired to RB2.
; Pulse is negative going
Pulse:
        banksel     PORTB
        bcf         PORTB,2             ; Bring step line high
        nop
        nop
        nop
        nop
        bsf         PORTB,2             ; Step line low
        return


;================================
;        delay sub-routines
;================================
Delay:
        banksel     PORTB
        bsf         PORTB,2             ; Step line low
        banksel     delay
        movf        delay,w             ; Delay to w register
        movwf       j                   ; j = w
Jloop:
        movwf       k                   ; k = w
Kloop:
        decfsz      k,f                 ; k = k-1, skip next if zero
        goto        Kloop
        decfsz      j,f             ; j = j-1, skip next if zero
```

```
        goto        Jloop
        return


        end
```

19.8.3 PWM_Micstep.asm

```
;================================================================
; File: PWM_Micstep.asm
; Date: September 19, 2011
; Authors: Sanchez and Canton
; Processor: 16F873A
; Test circuit: PWM-DEMO-1
;
;                           16F873
;                   +------------------+
; +5v--res0--------| 1 !MCLR    RB7 28|---
;              ---| 2 RA0      RB6 27|---
;              ---| 3 RA1      RB5 26|---
;              ---| 4 RA2      RB4 25|---
;              ---| 5 RA3      RB3 24|---
;              ---| 6 RA4      RB2 23|---
;              ---| 7 RA5      RB1 22|---
;         GR ---| 8 Vss      RB0 21|---
;        OSC ---| 9 OSC1         20|--- +5V
;        OSC ---|10 OSC2         19|--- GR
;              ---|11 RC0      RD7 18|---
;  GR-res1-LED ---|12 RC1/CCP2 RD6 17|---
;  GR-res1-LED ---|13 RC2/CCP1 RD5 16|---
;              ---|14 RC3      RD4 15|---
;                   +------------------+
; Legend:
; res0 = 10K resistor        OSC = 4 MHz oscillator
; res1 = 470 Ohm             GR = ground
;
; Oscillator: Murata Erie 4 MHz
;
; Program Description:
; Test program for microstepping using PWM on the 16F873 PIC.
; This test program uses a time-delay loop to cycle through
; 16 steps of the duty cycle of a bipolar stepper motor.
; Duty cycle is increased until it reaches the maximum, then
; decreased until the minimum.
; The LEDs wired to lines CCP1 and CCP2 demonstrate the
; variations in current that result from the PWM action.
;
; Current flow in high torque microstepping
```

```
; Chart shows 16 microsteps per step.
; Legend: - = Winding A (CCP1)
;         o = Winding B (CCP2)
;
;         0......15......31......47......63 <= microsteps
;  Imax   ----------------ooooooooooooooooo--------------
;                    o o                --                  o
;                  o   o               -  -               o  o
;                 o     o             -    -             o    o
;                o       o           -      -           o      o
;               o         o         -        -         o        o
;              o           o       -          -       o          o
;             o             o     -            -     o            o
;  I = 0 o                   o                   - o                o
;        |                   |                   |
;        | --  wind 0 -- | --    wind 1 --|
;        | ph 0 | ph 1 | ph 0  |  ph 1 |
;
; Diagram:
; One complete iteration consists of 64 steps (0 to 63). The
; first 32 (0 to 31) steps are in wind 0 and the second 32
; (31 to 63) are in wind 1. Both wind 0 and wind 1 have two
; phases (ph 0 and ph 1).
;
; In operation:
; While the current in winding B cycles from to 0 to Imax and
; back to 0 (wind 0), the current in winding A is held high.
; In wind 1 the current windings are switched, that is, the
; current cycles from 0 to Imax and back to 0 in winding A
; while winding B is held high. A table in memory holds the
; values for 16 duty cycles required for a PWM-generated sine
; curve. During phase 0, the duty cycle is read top-to-bottom
; from the duty cycle table, that is, current values go
; from low to high. During phase 1 the duty cycles are read
; bottom-to-top, that is, current values go from high to low.
; Phase changes are as follows:
;
; Startup value = 0
;    PHASE            microStep RANGE         DIRECTION
;      0                 0 to 15             increasing
;      1                15 to 0              decreasing
;
; Values in duty cycle table for 16 microsteps:
;
;   |<------ microStep (phase 0)
;   |      WINDING              DUTY CYCLE
;   |    A        B          CCP2   CCP1
;   0    +1       +sin 5.6          100    9.8
```

```
;    1       +1       +sin 11.25           100      20
;    2       +1       +sin 16.8            100      29
;    3       +1       +sin 22.5            100      38
;    4       +1       +sin 28              100      47
;    5       +1       +sin 33.75           100      56
;    6       +1       +sin 39              100      63
;    7       +1       +sin 45              100      71
;    8       +1       +sin 50.6            100      77
;    9       +1       +sin 56.25           100      83
;   10       +1       +sin 61.8            100      88
;   11       +1       +sin 67.5            100      93
;   12       +1       +sin 73.1            100      95.6
;   13       +1       +sin 78.75           100      98
;   14       +1       +sin 84.35           100      99.5
;   15       +1       +sin 90              100      100
; In wind 0 winding B is held high and winding A changes
; In wind 1 winding A is held high and winding B changes
; In either cycle:
;               Phase 0 counts up from microStep 0 to 15
;               Phase 1 counts down from microStep 15 to 0
;
;===============================================================
;               16F873 configuration options
;===============================================================
; Switches used in __config directive:
;   _CP_ON          Code protection ON/OFF
; * _CP_OFF
; * _PWRTE_ON       Power-up timer ON/OFF
;   _PWRTE_OFF
;   _BODEN_ON       Brown-out reset enable ON/OFF
; * _BODEN_OFF
; * _PWRTE_ON       Power-up timer enable ON/OFF
;   _PWRTE_OFF
;   _WDT_ON         Watchdog timer ON/OFF
; * _WDT_OFF
;   _LPV_ON         Low voltage IC programming enable ON/OFF
; * _LPV_OFF
;   _CPD_ON         Data EE memory code protection ON/OFF
; * _CPD_OFF
; OSCILLATOR CONFIGURATIONS:
;   _LP_OSC         Low power crystal oscillator
;   _XT_OSC         External parallel crystal oscillator
; * _HS_OSC         High speed crystal resonator
;   _RC_OSC         Resistor/capacitor oscillator
; |                 (simplest, 20% error)
; |
; |_____ * indicates setup values presently selected
```

```
        processor    16f873A              ; Define processor
        #include     <p16f873A.inc>
        __CONFIG  _CP_OFF  &  _WDT_OFF  &  _BODEN_OFF  &  _PWRTE_ON  &
_XT_OSC & _WDT_OFF & _LVP_OFF & _CPD_OFF

; Turn off banking error messages
        errorlevel       -302
;============================================================
;                  variables in PIC RAM
;============================================================
; Declare variables at 2 memory locations
j             equ         0x20
k             equ         0x21
microStep     equ         0x22  ; Range 0 to 15
dutyCycle     equ         0x23
phase         equ         0x24        ; 0 = increasing 1 = decreasing
wind          equ         0x25         ; 0 = winding B held high
                                       ; 1 = winding A held high
;============================================================
;                m a i n   p r o g r a m
;============================================================
      org         0       ; start at address 0
      goto        main
;==============================
; space for interrupt handler
;==============================
      org         0x04
;==============================
;      main program
;==============================
main:
      banksel     ADCON1               ; Select bank
      movlw       b'00000110'          ; Turn off A/D, port A
      movwf       ADCON1
      banksel     PORTA
      movlw       b'00000101'          ; Input lines on PORTA
      tris        PORTA
; Make CCP1 (RC2) and CCP2 (RC1) pins output
      movlw       B'00000000'; w = 00000000 binary
      tris        PORTC                ; Set up port C for output
      bcf         PORTC,2              ; CCP1 pin low
; Set up registers for PWM
      banksel     INTCON               ; Select bank
      bcf         INTCON,GIE           ; Disable global interrupts
      bcf         INTCON,PEIE          ; And peripheral interrupts
      banksel     PIE1
      bcf         PIE1,TMR2IE ; Disable timer 2 interrupts
      bcf         PIE1,CCP1IE ; Disable ccp1 interrupts
```

```
        banksel     CCP1CON
        clrf        CCP1CON            ; Turn off CCP1 module
        banksel     PR2
        banksel     CCP2CON
        clrf        CCP2CON            ; Turn off CCP2 module
        banksel     PR2
        movlw       .255               ; value to load
        movwf       PR2                ; into PWM period register
; Set initial micro step to 15 (maximum)
        banksel     microStep
        movlw       .0                 ; Set initial step number
        movwf       microStep          ; Store in local variable
        call        dcTable16          ; Duty cycle now in w
; Set duty cycle for both PWM lines
        movwf       dutyCycle
        banksel     CCPR1L
        movwf       CCPR1L             ; ccpr1l bits 9-2 of duty cycle
        banksel     CCPR2L
        movwf       CCPR2L             ; ccpr1l bits 9-2 of duty cycle
        banksel     CCP1CON
        bcf         CCP1CON,5          ; Duty cycle LSB
        bcf         CCP1CON,4          ; Duty cycle MSB
        banksel     CCP2CON
        bcf         CCP2CON,5          ; Duty cycle LSB
        bcf         CCP2CON,4          ; Duty cycle MSB
; Set prescaler to 1:1, no postscaler
        banksel     T2CON
        movlw       b'00000000'
        movwf       T2CON              ; Prescaler set and TMR2 off
        banksel     TMR2
        clrf        TMR2               ; Clear timer 2
        banksel     CCP1CON
        movlw       b'00001100'        ; CCP1 in PWM mode and turn on
        movwf       CCP1CON
        banksel     CCP2CON
        movlw       b'00001100'        ; CCP1 in PWM mode and turn on
        movwf       CCP2CON
; Set timer 2 control
        banksel     T2CON
        bsf         T2CON,2            ; Timer 2 on
; Starts up at lowest duty cycle. Set direction to increasing
        banksel     phase
        clrf        phase              ; Increasing DC mode
        clrf        wind               ; Winding A held high, winding
                                       ; goes low-high-low
;=================================================================
;                   routine to update duty cycle
;=================================================================
```

```
; In wind = 0 winding A (CCP1) is held high and winding B
;          changes
; In wind = 1 winding B (CCP2) is held high and winding A
;          changes
; In either cycle:
;          phase = 0 counts down from microStep 15 to 0
;          phase = 1 counts up from microStep 0 to 15
; wind variable toggles when phase variable changes from 1 to 0
NextCycle:
      call        delay
; Test for up or down direction of duty cycle update
; phase   = 0 to increase duty cycle
;         = 1 to decrease duty cycle
      banksel     phase
      movf        phase,w           ; Direction switch to W
      btfsc       phase,0           ; Test low-order bit
; Decrease duty cycle if bit 0 is set
      goto        DecreaseDC        ; Goes if bit set to
                                    ; decrease duty cycle routine
      goto        IncreaseDC
      goto        NextCycle         ; Not needed
;===============================
;     increase the duty cycle
;===============================
IncreaseDC:
; Increment step count if not at maximum
      banksel     microStep
      movf        microStep,w       ; Step count to W
      sublw       .15               ; Test for maximum
      banksel     STATUS
      btfsc       STATUS,Z          ; Return if at maximum
      goto        Reset2Down        ; Reverse direction
; Step count not at maximum value
;========================
; next higher microstep
;========================
      banksel     microStep
      incf        microStep,f ; Bump step count
      movf        microStep,w ; Step count to w
      call        dcTable16         ; Duty cycle from table
      movwf       dutyCycle         ; Store duty cycle
      call        NewDC
      goto        NextCycle
Reset2Down:
; Clear direction control (phase) and reset step
; count. Variable wind is toggled.
      banksel     phase
      bsf         phase,0           ; Set bit 0 to decrease
```

```
        banksel     microStep
        movlw       .15             ; Set step number
        movwf       microStep       ; Store in local variable
; Toggle variable wind
        banksel     wind
        movf        wind,w          ; Variable to w
        xorlw       b'00000001'     ; XORing with a 1 bit toggles
                                    ; the operand
        movwf       wind            ; Put back in register
        goto        NextCycle
;===============================
;   decrease the duty cycle
;===============================
DecreaseDC:
; Increment step count if not at maximum
        banksel     microStep
        movf        microStep,w     ; Step count to W
        sublw       .0              ; Test for minimum
        banksel     STATUS
        btfsc       STATUS,Z        ; Change direction at maximum
        goto        Reset2Up        ; Reverse direction
; Step count not at minimum value
;========================
; next lower microstep
;========================
        banksel     microStep
        decf        microStep,f ; Bump step count down
        movf        microStep,w ; Step count to w
        call        dcTable16       ; Duty cycle from table
        movwf       dutyCycle       ; Store duty cycle
        call        NewDC           ; Local procedure
        goto        NextCycle
Reset2Up:
; Set direction control (phase) and reset step
; count
        banksel     phase
        bcf         phase,0         ; Clear bit 0 to increase
        banksel     microStep
        movlw .0                    ; Set step number
        movwf       microStep       ; Store in local variable
        goto        NextCycle


;========================================================
;          routine to update PWM registers
;========================================================
; On entry:
;       w = new duty cycle
;     wind = update mode
```

```
;               if wind = 0 winding A is held high and B is cycled
;               if wind = 1 winding B is held high and A is cycled
NewDC:
        banksel    wind
        btfsc      wind,0                ; Test low-order bit
; Wind mode 0 if bit clear
        goto       WindMode1    ; Goes if bit set
; wind mode 0 processing
; Duty cycle (still in w) to CCP1
        banksel    CCPR1L
        movwf      CCPR1L                ; To control register
        banksel    CCP1CON
        bcf        CCP1CON,5            ; Duty cycle LSB
        bcf        CCP1CON,4            ; Duty cycle LSB
; Set duty cycle to max in CCP2 pin
        movlw .255
        banksel    CCPR2L
        movwf      CCPR2L                ; To control register
        banksel    CCP2CON
        bcf        CCP2CON,5            ; Duty cycle LSB
        bcf        CCP2CON,4            ; Duty cycle LSB
        return
WindMode1:
; wind mode 0 processing
; Duty cycle (still in w) to CCP2
        banksel    CCPR2L
        movwf      CCPR2L                ; To control register
        banksel    CCP2CON
        bcf        CCP2CON,5            ; Duty cycle LSB
        bcf        CCP2CON,4            ; Duty cycle LSB
; Set duty cycle to max in CCP1 pin
        movlw .255
        banksel    CCPR1L
        movwf      CCPR1L                ; To control register
        banksel    CCP1CON
        bcf        CCP1CON,5            ; Duty cycle LSB
        bcf        CCP1CON,4            ; Duty cycle LSB
        return
;================================
;       delay sub-routine
;================================
delay:
        banksel    j
        movlw      .200        ; w = 200 decimal
        movwf      j           ; j = w
jloop:
        movwf      k           ; k = w
kloop:
```

```
          decfsz      k,f           ; k = k-1, skip next if zero
          goto        kloop
          decfsz      j,f           ; j = j-1, skip next if zero
          goto        jloop
          return
;*******************************
;   Duty cycles table
;*******************************
dcTable16:
          addwf PCL,f ; PCL is program counter latch
          retlw       .25           ; 0 -- 9.8% of 255
          retlw       .51           ; 1 -- 20%
          retlw       .74           ; 2 -- 29%
          retlw       .97           ; 3 -- 38%
          retlw       .120          ; 4 -- 47%
          retlw       .143          ; 5 -- 56%
          retlw       .161          ; 6 -- 63%
          retlw       .181          ; 7 -- 71%
          retlw       .196          ; 8 -- 77%
          retlw       .212          ; 9 -- 83%
          retlw       .224          ; 10 - 88%
          retlw       .237          ; 11 - 93%
          retlw       .244          ; 12 - 95.6%
          retlw       .250          ; 13 - 98%
          retlw       .254          ; 14 - 99.5%
          retlw       .255          ; 15 - 100%
          retlw       .0            ; end marker

          end               ;END OF PROGRAM
```

19.8.4 PIC873_3955.asm

```
;=============================================================
; File: PIC873_3955.asm
; Date: April 19, 2011
; Update: September 21, 2011
; AuthorS: Sanchez and Canton
; Processor: 16F873A
; Reference circuit: SMB-PIC873-3955-1
;
; Program Description:
; 16F873-based bipolar motor control program using two Allegro
; A3955 drivers to implement microstepping and automatic control
; of the current recirculation path.
; Program uses two 32 entry lookup tables. One for the phase
; input flow and A3955 DAC control codes and a second table for
; the percent fast decay (PFD) codes.
; PORT B and PORT C lines 0 to 7 are trissed for output and wired
```

```
; as shown in the diagrams below.
; Phase controls (PH-A and PH-B) and DAC codes (DA-x and DB-x)
; are mapped to PORT B lines, as follows:
;
;             7   6   5   4   3   2   1   0 <====== PORT B BITS
;             |   |   |   |   |   |   |   |___ PHASE B
;             |   |   |   |   |   |   |_____ DB-2
;             |   |   |   |   |   |_____ DB-1
;             |   |   |   |   |_____ DB-0
;             |   |   |   |_____ PHASE A
;             |   |   |_____ DA-2
;             |   |_____ DA-1
;             |_____ DA-0
;
; Percent fast decay codes (PFDA and PFDB) are mapped to PORT C
; as follows:
;             7   6   5   4   3   2   1   0 <====== PORT C BITS
;             |   |   |   |   |   |   |   |___ PFD A
;             |   |   |   |   |   |   |_____ PFD B
;             |__|__|__|__|__|_____ NOT USED
;
; PORT A is trissed for input and lines 0 to 3 are wired to a
; toggle switch. Toggle switch number 1 (TS0) determines forward
; and reverse rotation. Toggle switches 2, 3, and 4 determine
; mode, as follows:
;
;         1   2   3   4
;       |--------------| ON
;       |  |   |   |   |  |
;       |  v   v   v   v  |
;       |--------------| OFF
;         |   |   |   |_____ ? ? 1 = 2 microsteps (skip = 4)
;         |   |   |_____ ? 1 ? = 4 microsteps (skip = 2)
;         |   |_____ 1 ? ? = 8 microsteps (skip = 1)
;         |                          0 0 0 = full step (skip = 8)
;         |_____ 1 = forward rotation
;                                    0 = reverse rotation
;
; THIS VERSION IS INTERRUPT DRIVEN.
;
; Microprocessor wiring diagram:
;                             16F873
;                     +------------------+
;     +5v--res0-------| 1 !MCLR   RB7 28|--- D0-A
;  TS0 (active) ----->| 2 RA0     RB6 27|--> D1-A
;  TS1 (not active)-->| 3 RA1     RB5 26|--> D2-A
;  TS2 (not active)-->| 4 RA2     RB4 25|--> PHASE A
;  TS3 (not active)-->| 5 RA3     RB3 24|--> D0-B
```

```
;                      ---| 6 RA4        RB2 23|--> D1-B
;                      ---| 7 RA5        RB1 22|--> D2-B
;            GRND ---| 8 Vss        RB0 21|--> PHASE B
;            OSC  ---| 9 OSC1           20|--- +5v
;            OSC  ---|10 OSC2           19|--- GRND
;            PFB-A <--|11 RC0        RD7 18|---
;            PFB-B <--|12 RC1/CCP2 RD6 17|---
;                      ---|13 RC2/CCP1 RD5 16|---
;                      ---|14 RC3       RD4 15|---
;                          +-----------------+
; Legend:
; res0 = 10K resistor          OSC = 4 MHz oscillator
; GRND = ground
;
;===============================================================
;               16F873 configuration options
;===============================================================
; Switches used in __config directive:
;   _CP_ON          Code protection ON/OFF
; * _CP_OFF
; * _PWRTE_ON      Power-up timer ON/OFF
;   _PWRTE_OFF
;   _BODEN_ON      Brown-out reset enable ON/OFF
; * _BODEN_OFF
; * _PWRTE_ON      Power-up timer enable ON/OFF
;   _PWRTE_OFF
;   _WDT_ON        Watchdog timer ON/OFF
; * _WDT_OFF
;   _LPV_ON        Low voltage IC programming enable ON/OFF
; * _LPV_OFF
;   _CPD_ON        Data EE memory code protection ON/OFF
; * _CPD_OFF
; OSCILLATOR CONFIGURATIONS:
;   _LP_OSC        Low power crystal oscillator
;   _XT_OSC        External parallel crystal oscillator
; * _HS_OSC        High speed crystal resonator
;   _RC_OSC        Resistor/capacitor oscillator
; |               (simplest, 20% error)
; |
; |_____ * indicates setup values presently selected

     processor   16f873A          ; Define processor
     #include    <p16f873A.inc>
     __CONFIG _CP_OFF & _WDT_OFF & _BODEN_OFF & _PWRTE_ON &
_XT_OSC & _WDT_OFF & _LVP_OFF & _CPD_OFF

; Turn off banking error messages
     errorlevel       -302
```

```
;
;================================================================
;                     c o n s t a n t s
;================================================================
DELAY        set          .5              ; Delay counter

;================================================================
;                     variables in PIC RAM
;================================================================
; Declare local variables starting at 0x20
w_temp             equ    0x20            ; For context saving
status_temp equ    0x21
pclath_temp equ    0x22
j                  equ    0x23            ; Delay routine counters
k                  equ    0x24
this_cycle         equ    0x25            ; Current cycle counter
iteration          equ    0x26            ; Iteration counter used by
                                          ; interrupt handler
bounce             equ    0x27            ; Debounce counter
direction          equ    0x28            ; Direction control
skip               equ    0x29            ; Number of steps to skip in
                                          ; microstepping modes
mode               equ    0x2a            ; Storage for current mode
w_temp1            equ    0xa0            ; Must also be defined in bank
1
;
;================================================================
;                     m a i n   p r o g r a m
;================================================================
Start:
      org          0                      ; start at address
      goto         Main
;=============================
; space for interrupt handler
;=============================
      org          0x04
      goto         IntServ
;=============================
;      main program
;=============================
Main:
; Tris PORT A for input
      banksel      PORTA
      clrf         PORTA
; Configure port A pins as input
      banksel      ADCON1
      movlw        0x06                   ; All pins as digital inputs
      movwf        ADCON1
```

```
        banksel    TRISA
        movlw      0xcf                ; Data pattern for tris
                                       ; registers
        movwf      TRISA
; Tris ports B and C for output
        banksel    PORTB
        movlw      B'00000000' ;All lines to output
        tris       PORTC
        tris       PORTB
        clrf       PORTB               ; Clear all lines
        clrf       PORTC
; Initialize control variables
        banksel    this_cycle          ; Select bank 0
        clrf       this_cycle
; Initialize the iteration counter
        movlw      DELAY               ; Read delay
        movwf      iteration           ; Set iteration counter
;===============================
;   setup for interrupt
;===============================
; Clear external interrupt flag (INTF = bit 1)
        banksel    INTCON
        bcf        INTCON,INTF ; Clear flag
; Enable global interrupts (GIE = bit 7)
; Enable RB0 interrupt (inte = bit 4)
        bsf        INTCON,GIE  ; Enable global int (bit 7)
        bsf        INTCON,T0IE ; Enable TMR0 overflow interrupt
;===============================
; Set up the OPTION register
;===============================
        movlw      b'01010000'
;   7  6  5  4  3  2  1  0 <= OPTION bits
;   |  |  |  |  |  |  |__|__|_____ PS2-PS0 (prescaler bits)
;   |  |  |  |  |  |              Values for Timer0
;   |  |  |  |  |  |              *000 = 1:2   001 = 1:4
;   |  |  |  |  |  |              010 = 1:8   011 = 1:16
;   |  |  |  |  |  |              100 = 1:32  101 = 1:64
;   |  |  |  |  |  |              110 = 1:128 111 = 1:256
;   |  |  |  |  |_____ PSA (prescaler assign)
;   |  |  |  |                    1 = to WDT
;   |  |  |  |                    *0 = to Timer0
;   |  |  |  |_____ TOSE (Timer0 edge select)
;   |  |  |                      0 = increment on low-to-high
;   |  |  |                      *1 = increment on high-to-low
;   |  |  |_____ TOCS (TMR0 clock source)
;   |  |                      *0 = internal clock
;   |  |                      1 = RA4/TOCKI bit source
;   |  |_____ INTEDG (Edge select)
```

```
;    |                                    *0 = falling edge
;    |_____ RBPU (Pullup enable)
;                                          *0 = enabled
;                                           1 = disabled
      banksel      OPTION_REG
      movwf        OPTION_REG         ; Copy w to OPTION
; Clear direction
      banksel      direction
      clrf         direction          ; Set direction to CW
; Set initial step skip factor
      movlw        .1                 ; For 8 microsteps
      movwf        skip
      clrf         mode               ; Clear mode
      bsf          mode,1             ; Default mode is 8
                                      ; microsteps
;=================================================================
;                   main control monitoring routine
;=================================================================
ControlRtn:
      call         ReadAToggles
      goto         ControlRtn


;===========================================================
;      Auxiliary procedure to read toggle switches
;===========================================================
ReadAToggles:
; Read PORTC toggle switches
;      PORT A      TOGGLE
;      line        SW      ACTION
;       0          1       0 = CW rotation 1 = CCW
;       1          2       1 = 2 microsteps (skip = 4)
;       2          3       1 = 4 microsteps (skip = 2)
;       3          4       1 = 8 microsteps (skip = 1)
;      1-2-3      2-3-4    0 = full step (skip = 8)
;============================
;   set CW or CCW rotation
;============================
; Bit 0 on port A is wired to toggle # 1
; controls CW and CCW rotation.
      btfsc        PORTA,0
      goto         goCCW              ; Set rotation bit to CCW
; Bit clear. Set CW
      clrf         direction
      goto         SetMode
goCCW:
      movlw        .1                 ; 1 = counterclockwise
      movwf        direction  ;
; Read PORT A bits 1, 2, and 3 to determine mode, as follows:
```

```
;              1  2  3
;              |  |  |_____    ? ? 1 = 2 microsteps (skip = 4)
;              |  |_____       ? 1 ? = 4 microsteps (skip = 2
;              |_____      1 ? ? = 8 microsteps (skip = 1)
;                                     0 0 0 = full step (skip = 8)
;
; Variable named mode stores currently selected mode, as follows:
; 7  6  5  4  3  2  1  0 <== mode variable bitmap
; |  |  |  |  |  |  |  |  |___ NOT USED
; |  |  |  |  |  |  |  |_____ 1 = 8 MS mode active
; |  |  |  |  |  |  |_____ 1 = 4 MS mode active
; |  |  |  |  |  |_____ 1 = 2 MS mode active
; |  |  |  |  |_____ 1 = full step mode active
; |__|__|_____ NOT USED
;
; Routine logic:
; Mode control toggle switches are tested low-to-high. If the
; first switch found high corresponds to the currently selected
; mode (as stored in the mode variable bits) then no action is
; taken. Otherwise, the current mode is set and the skip factor
; variable is initialized. The cycle counter variable is
; cleared and the current mode is stored in the mode variable.
SetMode:
; Bits are tested high-to-low
        btfsc      PORTA,1            ; Test 8 MS switch
        goto       Try4
; 8 microsteps mode switch is ON
; Test to see if this is the current mode
        btfsc mode ,1                 ; Is 8 MS mode active?
        goto       ExitMode           ; Exit if active
; At this point 8 MS mode is new mode
        movlw      .1                 ; Skip factor for 8 MS
        movwf      skip               ; Store in variable
        clrf       this_cycle         ; Reset counter
        clrf       mode               ; Reset current mode
        bsf        mode,1
        goto       ExitMode
Try4:
        btfsc      PORTA,2            ; Test 4 MS switch
        goto       Try2
; 4 microsteps mode switch is ON
; Test to see if this is the current mode
        btfsc mode ,2                 ; Is 4 MS mode active?
        goto       ExitMode           ; Exit if active
; At this point 4 MS mode is new mode
        movlw      .2                 ; Skip factor for 8 MS
        movwf      skip               ; Store in variable
        clrf       this_cycle         ; Reset counter
```

```
        clrf       mode                ; Reset current mode
        bsf        mode,2
        goto       ExitMode
Try2:
        btfsc      PORTA,3             ; Test 2 MS switch
        goto       FullStep        ; No mode bits are set
; 2 microsteps mode switch is ON
; Test to see if this is the current mode
        btfsc      mode,3              ; Is 2 MS mode active?
        goto       ExitMode            ; Exit if active
; At this point 2 MS mode is new mode
        movlw      .4                  ; Skip factor for 4 MS
        movwf      skip                ; Store in variable
        clrf       this_cycle          ; Reset counter
        clrf       mode                ; Reset current mode
        bsf        mode,3
        goto       ExitMode
;
FullStep:
; At this point all three mode control switches are OFF.
; Set full step mode.
; First test to see if this is the current mode
        btfsc      mode,4              ; Is full step mode active?
        goto       ExitMode            ; Exit if active
; Set controls for full step mode
        movlw      .8                  ; Skip factor for full step
        movwf      skip                ; Store in variable
        clrf       this_cycle          ; Reset counter
        clrf       mode                ; Reset current mode
        bsf        mode,4
        goto       ExitMode
ExitMode:
        return
;=============================================================
;                   interrupt service routine
;=============================================================
; Service routine receives control when the timer register
; TMR0 overflows, that is, when 256 timer beats have elapsed
IntServ:
; First test if source is a Timer0 interrupt
        banksel    INTCON
        btfss      INTCON,T0IF ; T0IF is Timer0 interrupt
        goto       notT0IF             ; Go if not RB0 origin
; If so clear the timer interrupt flag so that count continues
        bcf        INTCON,T0IF ; Clear interrupt flag
; Save context
        banksel    w_temp
        movwf      w_temp              ; Save w register
```

```
        swapf       STATUS,w            ; STATUS to w
        clrf        STATUS
        movwf       status_temp ; Save STATUS
        movf        PCLATH,w            ; Save PCLATH
        movwf       pclath_temp
        clrf        PCLATH
;=========================
;   interrupt action
;=========================
; Decrement the iteration counter. Exit if not zero
        banksel     iteration
        decfsz      iteration,f
        goto        exitISR             ; Continue if counter not zero
; At this point the delay count has expired so the programmed
; time has elapsed.
; First reset the iteration counter
        movlw       DELAY               ; Read delay
        movwf       iteration           ; Reset iteration counter
;==========================
;   determine direction
;==========================
; Read code and PDF data from table and store in ports
        movf        this_cycle,w        ; Get current cycle
        call        CodeTable           ; Get control code from table
        movwf       PORTB               ; Write code to port
        movf        this_cycle,w        ; Get current cycle
        call        PFDTable            ; Get PFD code from table
        movwf       PORTC               ; Write code to port
; Test direction switch
        btfss       direction,0 ; Test forward bit
        goto        Forward
; At this point direction rotation is reverse
        goto        Reverse
; Forward direction routine
Forward:
; Index cycle counter according to skip factor
        movf        skip,w              ; Skip factor to w
        addwf       this_cycle,f        ; Add to cycle counter
; Test for cycle number 31 (last one in sequence)
        btfsc       this_cycle,5        ; 100000 = 32
        goto        Recycle             ; Reset if at end of cycle
        goto        exitISR             ; Done
Recycle:
        clrf        this_cycle
        goto        exitISR
; Reverse direction routine
Reverse:
; Index cycle counter according to skip factor
```

```
      movf          skip,w              ; Skip factor to w
      subwf         this_cycle,f        ; Subtract from counter
; Test for cycle number 0xff (overflow from cycle # 0)
      btfsc         this_cycle,7        ; High bit set indicates over-
flow
      goto          Recycle2            ; Reset if at end of cycle
      goto          exitISR
Recycle2:
      movlw         .31                 ; First reverse cycle
      movwf         this_cycle          ; To cycle counter
      goto          exitISR
;========================
;        exit ISR
;========================
exitISR:
; Restore context
      movf          pclath_temp,w       ; Restore
      movwf         PCLATH
      swapf         status_temp,w
      movwf         STATUS
      swapf         w_temp,f            ; Swap file register in itself
      swapf         w_temp,w            ; re-swap back to w
; Reset interrupt
notTOIF:
      retfie
;
;===================================
;  Control data for 32 microsteps
;===================================
; Each full step consists of 8 microsteps, as follows:
;             WINDING A                           WINDING B         STEP
; D0-A   D1-A   D2-A   PH-A  PFDA    D0-B   D1-B   D2-B   PH-B  PFDB
;  0      0      1      1     0  |    0      0      1      1     1       0
;  1      1      0      1     0  |    1      0      1      1     1       1
;  0      1      0      1     0  |    0      1      1      1     1       2
;  1      0      0      1     0  |    1      1      1      1     1       3
;  0      0      0      ?     0  |    1      1      1      1     1       4
;  1      0      0      0     0  |    1      1      1      1     1       5
;  0      1      0      0     0  |    0      1      1      1     1       6
;  1      1      0      0     0  |    1      0      1      1     1       7
;--------------------------------|---------------------------------------
;  0      0      1      0     1  |    0      0      1      1     0       8
;  1      0      1      0     1  |    1      1      0      1     0       9
;  0      1      1      0     1  |    0      1      0      1     0      10
;  1      1      1      0     1  |    1      0      0      1     0      11
;  1      1      1      0     1  |    0      0      0      ?     0      12
;  1      1      1      0     1  |    1      0      0      0     0      13
;  0      1      1      0     1  |    0      1      0      0     0      14
```

```
;   1       0       1       0       1   |   1       1       0       0       0       15
;-----------------------------------|-----------------------------------
;   0       0       1       0       0   |   0       0       1       0       1       16
;   1       1       0       0       0   |   1       0       1       0       1       17
;   0       1       0       0       0   |   0       1       1       0       1       18
;   1       0       0       0       0   |   1       1       1       0       1       19
;   0       0       0       ?       0   |   1       1       1       0       1       20
;   1       0       0       1       0   |   1       1       1       0       1       21
;   0       1       0       1       0   |   0       1       1       0       1       22
;   1       1       0       1       0   |   1       0       1       0       1       23
;-----------------------------------|-----------------------------------
;   0       0       1       1       1   |   0       0       1       0       0       24
;   1       0       1       1       1   |   1       1       0       0       0       25
;   0       1       1       1       1   |   0       1       0       0       0       26
;   1       1       1       1       1   |   1       0       0       0       0       27
;   1       1       1       1       1   |   ?       0       0       0       0       28
;   1       1       1       1       1   |   1       0       0       1       0       29
;   0       1       1       1       1   |   0       1       0       1       0       30
;   1       0       1       1       1   |   1       1       0       1       0       31
;==================================|==================================
;
; Phase controls (PH-A and PH-B) and DAC codes (DA-x and DB-x)
; are mapped to PORT B lines, as follows:
;
;               7   6   5   4   3   2   1   0 <====== PORT B BITS
;               |   |   |   |   |   |   |   |   |___ PHASE B
;               |   |   |   |   |   |   |   |_____ DB-2
;               |   |   |   |   |   |   |_____ DB-1
;               |   |   |   |   |   |_____ DB-0
;               |   |   |   |   |_____ PHASE A
;               |   |   |   |_____ DA-2
;               |   |   |_____ DA-1
;               |   |_____ DA-0
;
; Percent fast decay codes (PFDA and PFDB) are mapped to PORT C
; as follows:
;               7   6   5   4   3   2   1   0 <====== PORT C BITS
;               |   |   |   |   |   |   |   |   |___ PFD A
;               |   |   |   |   |   |   |   |_____ PFD B
;               |__|__|__|__|__|__|_____ NOT USED
;
; Codes for PORT B lines are stored in the lookup table
; CodeTable
; Codes for PORT C lines are stored in the lookup table
; PFDTable
;
CodeTable
```

```
        addwf           PCL,f                           ; Add w to program coun-
ter
        retlw           B'00110011'         ;    cycle = 0
        retlw           B'11011011'         ;             1
        retlw           B'01010111'         ;             2
        retlw           B'10011111'         ;             3
        retlw           B'00001111'         ;             4
        retlw           B'10001111'         ;             5
        retlw           B'01000111'         ;             6
        retlw           B'11001011'         ;             7
; -----------------------------------------------------------------
        retlw           B'00100011'         ;             8
        retlw           B'10101101'         ;             9
        retlw           B'01100101'         ;            10
        retlw           B'11101001'         ;            11
        retlw           B'11100000'         ;            12
        retlw           B'11101000'         ;            13
        retlw           B'01100100'         ;            14
        retlw           B'10101100'         ;            15
;------------------------------------------------------------------
        retlw           B'00100010'         ;            16
        retlw           B'11001010'         ;            17
        retlw           B'01000110'         ;            18
        retlw           B'10001110'         ;            19
        retlw           B'00001110'         ;            20
        retlw           B'10011110'         ;            21
        retlw           B'01010110'         ;            22
        retlw           B'11011010'         ;            23
;------------------------------------------------------------------
        retlw           B'00110010'         ;            24
        retlw           B'10111100'         ;            25
        retlw           B'01110100'         ;            26
        retlw           B'11111000'         ;            27
        retlw           B'11110000'         ;            28
        retlw           B'11111001'         ;            29
        retlw           B'01110101'         ;            30
        retlw           B'10111101'         ;            31
        retlw           0x0                 ;        Safety entry

PFDTable
        addwf           PCL,f                           ; Add w to program coun-
ter
        retlw           B'00000001'         ;    cycle =  0
        retlw           B'00000001'         ;             1
        retlw           B'00000001'         ;             2
        retlw           B'00000001'         ;             3
        retlw           B'00000001'         ;             4
        retlw           B'00000001'         ;             5
```

```
        retlw       B'00000001'          ;                    6
        retlw       B'00000001'          ;                    7
;   --------------------------------------------------------
        retlw       B'00000010'          ;                    8
        retlw       B'00000010'          ;                    9
        retlw       B'00000010'          ;                   10
        retlw       B'00000010'          ;                   11
        retlw       B'00000010'          ;                   12
        retlw       B'00000010'          ;                   13
        retlw       B'00000010'          ;                   14
        retlw       B'00000010'          ;                   15
;--------------------------------------------------------
        retlw       B'00000001'          ;                   16
        retlw       B'00000001'          ;                   17
        retlw       B'00000001'          ;                   18
        retlw       B'00000001'          ;                   19
        retlw       B'00000001'          ;                   20
        retlw       B'00000001'          ;                   21
        retlw       B'00000001'          ;                   22
        retlw       B'00000001'          ;                   23
;   --------------------------------------------------------
        retlw       B'00000010'          ;                   24
        retlw       B'00000010'          ;                   25
        retlw       B'00000010'          ;                   26
        retlw       B'00000010'          ;                   27
        retlw       B'00000010'          ;                   28
        retlw       B'00000010'          ;                   29
        retlw       B'00000010'          ;                   30
        retlw       B'00000010'          ;                   31
        retlw       0x0                      ;          Safety entry

        end
```

Chapter 20

Communications

In this chapter we focus on digital communications techniques used in PIC interfacing with I/O devices, integrated circuits, and other forms of programmable logic. Communications, in general, refer to the exchange of information following rules, sometimes called a *protocol*. Digital and computer communications come in two flavors: serial and parallel. *Serial communications* take place when the data is sent one bit at a time over the communications channel. In *parallel communications* all the bits that compose a single symbol or character are sent simultaneously.

Popular lore regards serial communications as slower than parallel communications, but with modern-day technologies this is often not the case, as serial techniques often match or even excel parallel methods in speed and performance. Computer networks such as *Ethernet* and fiber-optic links are able to achieve high performance even though they use *serial bit streams*. The preference for serial over parallel communications is often more related to hardware, because parallel transmissions require more communication lines than serial transmissions.

20.1 PIC Communications Overview

Many communications standards were created with other interface and hardware requirements in mind and are not ideally suited for PIC applications. For example, RS-232-C, a serial protocol developed over 35 years ago, originated in an age of teletypewriters and modems. The voltage levels and circuit requirements of RS-232-C are not suited for PIC hardware. The more modern USB standard is more suited to PIC interfacing, but adopting a standard, RS-232-C, EIA-485, USB, or any other convention, requires adhering to special configurations in hardware and the use of ad hoc software protocols. This compliance with a standard comes at a price of added hardware components and increased software complexity.

When PIC-based circuits must interface with other systems or devices that follow these standards, then there is no alternative but to design circuits and write programs that comply with the standards. On the other hand, when the communications take place in dedicated circuits, which do not interface with devices or systems that follow standard communications protocols, then pure PIC communications techniques and hardware are often simpler and more effective. In other words, adhering to a communications protocol usually implies an additional cost in software and hardware complexity. Here are two examples: a PIC-based circuit that interfaces with a PC through the RS-232-C port would be a case where compliance with RS-232-C is required. Another case would be a PIC-based circuit that sends serial data to an onboard LCD display. In this case, the circuit and the software need not comply with any communications standards or protocols. Programmers often refer to techniques that use serial communications without the presence of specialized hardware, such as UART or USART chips, as *bit-banging*.

In the following sections, we discuss serial and parallel communications at their most essential level. In the general literature, communications concerns often focus on transmission speeds, system performance, and minimum processing time. Typically, PIC applications do not transfer large data files or communicate interactively on the Internet or in networks. In a typical PIC application, communication functions are used to upload stored data to a PC, sometimes called *data-logging*, or to receive small data sets or commands from a host machine. In this context there are no major concerns regarding super-fast transmission rates or maximum performance.

20.2 Serial Data Transmission

Serial communications take place by transmitting and receiving data in a stream of consecutive electrical pulses that represent data bits and control codes. The *Electronic Industries Association* (EIA) has sponsored the development of several standards for serial communications, such as RS-232-C, RS-422, RS-423, RS 449, EIA232E, and EIA232F, among others. In this designation the characters RS stand for the words *Recommended Standard*. The oldest, simplest to implement, and most-used serial communications standard is the *RS-232-C voltage level convention*. In the following sections we present the essential concepts of the RS-232-C standard. Most of the material also applies to the various updates of the standard. Later in the chapter we briefly discuss the EIA485 Standard.

20.2.1 Asynchronous Serial Transmission

The information in a *serial bit stream* is contained in a time-dependent waveform, that is, each bit code (data, control, or error) is transmitted for a fixed time period, known as the *baud period*. The word *baud* was chosen to honor the French scientist and inventor *Jean Maurice Emile Baudot* who studied various serial encodings in the late nineteenth century.

The serial bit streams used in data transmission follow a very simple encoding: one bit is transmitted during each baud period. A binary 1 bit is represented by a negative voltage level and a binary 0 bit by a positive voltage. The line condition during the logic 1 transmission is called a *marking state*, and the one for a logic 0 a *spacing state*. The baud rate is equal to the number of bits per second being transmitted or received. Note that the voltage levels that represent a 1 and a 0 bit in RS232 are somewhat counter-intuitive, as one would expect a logic 1 to be represented with a positive voltage, and not a negative one.

One possible approach to sending information bit-by-bit is based on the transmitter and receiver clocks being synchronized at the same frequency. That is, both receiver and transmitter operate at the same baud rate. Note that the expression "synchronized at the same frequency" implies not only that their clocks have the same speed, but that the high and the low portions of the waveform coincide.

In typical *asynchronous serial communications*, bits are transmitted as separate groups, usually seven to ten bits long. Each group is called a *character*. The name "character" relates to the fact that in alphanumeric transmissions each bit group represents one numeric or alphabetic symbol. In reality, the term "character" is also applied to control codes, error codes, and other nonalphanumeric encodings.

Each character is sent in a *frame* consisting of a *start bit*, followed by a set of *character bits*, followed (optionally) by a *parity bit*, and finalized by one or more *stop bits*. The serial line is normally held *marking*, that is, at a logic 1 state. The change from logic high to logic low, signaled by the start bit, tells the receiver that a *frame* follows. The receiver reads the number of character bits expected according to the adopted protocol until a logic high, represented by one or more stop bits, marks the end of the frame.

Figure 20-1 shows the different elements in a serial communications bit stream. The term *asynchronous* reflects the fact that the time period separating characters is variable. The transmitter holds the line to logic high (marking state) until it is *ready to send*. The start bit (*spacing state*) is used to signal the start of a new character. The start bit is also used by the receiver to synchronize with the transmitter. The logic high and low regions of the signal wave occur at the same time. This compensates for drifts and small errors in the baud rate.

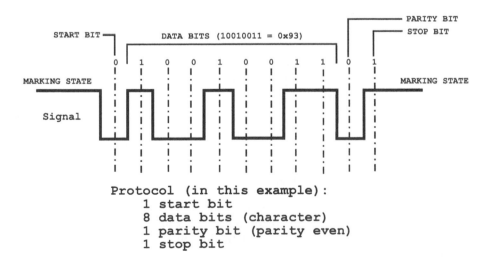

Figure 20-1 Serial Communications Bit Stream.

This form of transmitting serial data is called *asynchronous* because the receiver resynchronizes itself to the transmitter using the start bit of each frame. The lack of synchronization does not refer to the bits within each frame, which must be in fact "synchronized," but to the fact that characters need not come at a fixed time interval.

20.2.2 Synchronous Serial Transmission

An alternative approach to asynchronous serial data transmission is one in which the characters are sent in blocks with no framing bits surrounding them. In asynchronous communications, each character is framed by a start and a stop signal so that the receiver can know exactly where the character bits are located. In *synchronous communications*, the sender and receiver are synchronized with a clock or a signal that is part of the data stream.

In theory, synchronous communications implies that characters are sent out at a constant rate, in step with a clock signal. This scheme assumes that a separate line (or wire) is used for the clock signal, although, in some variations, the clock signal is contained in the transmitted characters. Alternatively, a clock line can be used to synchronize the moment in time at which the receiver reads the data line. In either case, it is this contained clock or command signal that identifies a synchronous transmission.

Most legacy PC communications systems are asynchronous, although the EIA232F standard supports both synchronous and asynchronous methods. The most common chip used in PC communications is the *UART* (*Universal Asynchronous Receiver and Transmitter*). An alternative chip called the USRT is used for synchronous communications and the USART (*Universal Synchronous/Asynchronous Receiver and Transmitter*) supports both.

Synchronous communications can be block or bit based. The *block-based modes* are also called *character-based*. In this mode, characters are grouped in blocks with each block having a starting flag, similar to the start bit used in asynchronous communications. Once the receiver and the transmitter are synchronized, the transmitter inserts two or more control characters known as *synchronous idle characters*, or SYNs. Then the block is sent and the receiver places the data in a memory storage area for later processing. *Bit-oriented methods*, on the other hand, are used for the transmission of binary data that is not tied to any particular character set.

20.2.3 PIC Serial Communications

Serial communications are often used in PIC programming, mostly due to the scarcity of available port lines. For example, an application in which a 16F84 PIC needs to read data in parallel from eight DIP switches and display the result, also in parallel, in eight LEDs, requires a total of 16 available port lines. But the 16F84 has only thirteen lines, eight in Port-B and five in Port-A; therefore, the application would not be feasible.

One possible solution is to find some way of reading the DIP switches serially; this requires three lines at most. Alternatively, the output data to the LEDs could be transmitted serially, thus reducing the total lines required from sixteen for parallel transmission, to six, or even less for serial transmission.

PIC communications can be designed both asynchronously and synchronously. Asynchronous modes are used when the same or compatible clock signals are available to both receiver and transmitter. For example, two PICs both running at the same clock rate can transmit and receive data using a single communications line, plus a common ground. PIC-to-PIC asynchronous data transmission mode is demonstrated later in this chapter with both circuit and code.

Asynchronous communications can be implemented by incorporating a dedicated IC, such as a UART or USART chip, in the circuit. PCs usually have one of these ICs, or functionally equivalent ones, in their implementation of the serial port. Some PICs include one or more serial circuits, which sometimes include a USART module. For example, the 16F877 PIC has two serial communication modules. One of them is the *Master Asynchronous Serial Port*, or MSSP. The other one is a USART. Later in this chapter we present serial communications programming examples using the USART module in the 16F877 PIC. Programs using the MSSP module are found in the chapter on EEPROM programming.

When communications take place between a PIC and a device that does not contain a clock, or whose clock runs at a different speed than the PIC's, then synchronous communications is used. For example, a circuit can be designed using a shift register IC, such as the 74HC164, that performs an 8-bit serial-in, parallel-out function. In the previous example, it is possible to reduce the number of transmission lines by connecting the eight LEDs to the output ports of the 74HC164. But the 74HC164 contains no internal clock that runs at the speed of the 16F84. Thus, communications between the PIC and the shift register IC (74HC164 in this case) require a clock or command signal transmitted through a separate line; that is, a synchronous serial transmission. In this chapter we present circuits and sample code showing synchronous communications between a PIC and one or more shift register ICs.

20.2.4 RS-232-C Standard

RS-232-C was developed jointly by the Electronic Industries Association (EIA), the Bell Telephone System, and modem and computer manufacturers. The standard has achieved such widespread acceptance that its name is often used as a synonym for the serial port. EIA232F, published in 1997, is the latest update of RS-232-C. Today, RS-232-C is gradually being replaced by USB for local communications. USB is faster, has lower voltage levels, and uses smaller connectors that are easier to wire. USB has software support in most PC operating systems. On the other hand, USB is a more complex standard, requiring more complex software. Furthermore, serial ports are used to directly control hardware devices, such as relays and lamps, because the RS-232-C control lines can be easily manipulated by software. This is not feasible with USB.

In the following sections we describe the essential terminology and communications principles of RS-232-C.

Essential Concepts

The RS-232-C convention specifies that, with respect to ground, a voltage more negative than – 3 V is interpreted as a 1 bit and a voltage more positive than +3 V as a 0 bit. Serial communications, according to RS-232-C, require that transmitter and receiver agree on a communications protocol. The following terminology refers to the RS-232-C communications protocol:

- **Baud period:** The rate of transmission measured in bits per second, also called the baud rate. In serial protocols, the transmitter and the receiver clocks must be synchronized to the same baud period.

- **Marking state:** The time period during which no data is transmitted. During the marking period, the transmitter holds the line at a steady high voltage, indicating logic 0.

- **Spacing state:** The time period during which data is transmitted. During the spacing period, the transmitter holds the line at a steady low voltage, indicating logic 1.

- **Start bit:** The transition that indicates that data transmission is about to start. The voltage low state that occurs during the start bit is called the spacing state.

- **Character bits:** The data stream composed of five, six, seven, or eight bits that encode the character transmitted. The least significant bit is the first one transmitted.

- **Parity bit:** An optional bit, transmitted following the character bits, used in checking for transmission errors. If *even parity* is chosen, the transmitter sets or clears the parity bit so as to make the sum of the character's 1 bits and the parity bit an even number. In *odd parity*, the sum of 1 bits is an odd number. If parity is not correct, the receiver sets an error flag in a special register.

- **Stop bits:** One or more logic high bits inserted in the stream following the character bits or the parity bit, if there is one. The stop bit or bits ensure that the receiver has enough time to get ready for the next character.

- **DTE (Data Terminal Equipment):** The device at the far end of the connection. It is usually a computer or terminal. The DTE uses a male DB-25 connector, and utilizes twenty-two of the twenty-five available pins.

- **DCE (Data Circuit-terminating Equipment):** Refers to the modem or other terminal of the telephone line interface. DCE has a female DB-25 connector, and utilizes the same 22 pins as the DTE for signals and ground. DB-9 connectors are also used.

- **Half-duplex:** A system that allows serial communications in both directions, but only one direction at a time. Half-duplex communications are reminiscent of radio communications where one user says the word "Over" to indicate the end of transmission. In other words, half-duplex is similar to a one-lane road in which traffic controllers at each end can direct flow in either direction, but only in one direction at a time.

- **Full-duplex:** A full-duplex system allows communication in both directions simultaneously. A full-duplex system is reminiscent of a two-lane highway in which traffic can flow in both directions at once.

Serial Bit Stream

In the RS-232-C protocol, the transmission/reception parameters are selected from a range of standard values. The following are the most common ones:

Baud rate: 50, 110, 300, 600, 1200, 2400, 4800, 9600, and 19200

Data bits: 5, 6, 7, or 8

Parity bit: Odd, even, or no parity

Stop bits: 1, 1.5, or 2

RS-232-C defines *DTE (Data Terminal Equipment)* and *DCE (Data Circuit-terminating Equipment)*, sometimes called *Data Communications Equipment.* According to the standard, the DTE designation includes both terminals and computers and DCE refers to modems, transducers, and other devices. The serial port in a computer is defined as a DTE device.

Parity Testing

In RS-232 communications, a bit called a parity bit may optionally be transmitted along with the data. A parity bit provides a simple, but not too reliable, error test to detect data corruption that takes place during transmission. Parity can be even, odd, or none. Even or odd parity refers to the number of 1 bits in each data byte. The parity bit immediately follows the data bits.

If even parity is selected, the parity bit is transmitted with a value of 0 if the number of high bits is even. For example, the binary value

```
0110 0011
```

contains a total of four 1 bits; therefore, the parity bit is 0. By the same token, if even parity is selected, then the binary value

```
0101 0001
```

requires that the parity bit be 1. One way of describing the parity bit is to say that the bit is set to indicate a parity error; therefore, it serves as a *parity error detector.* Another description is that the parity coincides with the number of 1 bits in the data, plus the parity bit. Thus, when even parity is selected the parity bit is added to the number of 1 bits in the data to produce an even number.

Odd parity is the opposite of even parity. If odd parity were selected, then the parity bit in the previous example would be 0. Given odd or even parity, the sender counts the number of 1 bits and sets or clears the parity bit accordingly. The receiver, knowing that the parity is odd or even, can do likewise to determine if the number of 1 bits received matches the required parity setting.

Parity error checking is very primitive. In the first place, the parity error does not identify the bit or bits that cause the error. Furthermore, if an even number of bits are incorrect, then the parity bit would not show the error. On the other hand, over a long transmission, the parity check is likely to detect garbled data.

Connectors and Wiring

The RS-232-C standard requires specific hardware connectors with either twenty-five or nine pins. The twenty-five-pin connector is called a *D-shell connector*, or DB-25. The connector with nine pins is called the *9-pin D-shell connector* or DB-9. In addition, the RJ-45 connector (the name stands for *Registered-Jack 45*) is used for twisted-pair cables. RJ-45 use in RS-232-C serial interface is regulated by the EIA/TIA-561 standard. A common application of RJ-45 connectors is in *Ethernet* cables. Figure 20-2 shows the male DB-25, DB-9, and the female RJ-45 connectors.

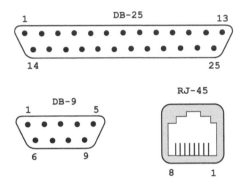

Figure 20-2 DB-25, DB-9, and RJ-45 Connectors.

The function assigned to each pin varies in the common connectors. Table 20.1 lists the assignation of the RS-232-C lines in the different hardware. The cable linking DTE and DCE devices is a parallel straight-through cable with no cross-over or self-connects.

Table 20.1

Definition of Common RS-232-C Lines

DB-25	CONNECTOR DB-9	RJ-45	FUNCTION	CODE NAME	DIRECTION
1		4	Ground	G	
2	3	6	Transmit data	TXD	Output
3	2	5	Receive data	RXD	Input
4	7	8	Request to send	RTS	Output
5	8	7	Clear to send	CTS	Input
6	6		Data set ready	DSR	Input
7	5		Chassis ground	G	
8	1	2	Carrier detect	CD	
20	4	3	Data terminal ready	DTR	Output
22	9	1	Ring indicator	RI	Input

Null Modem

The RS-232-C standards describe the way a computer communicates with a peripheral device, such as a *modem*. In this case, the DTE and DCE lines serve as a communications control. In this context, DTE means data terminal equipment, such as a computer, and DCE is the abbreviation for data communication equipment, such as modems. Often, communications must take place in an environment that does not include a modem; for example, computers communicating with each other or with other devices such as a PIC-based board. In these cases, the use of the DTE/DTE communication lines in flow control is not well defined. The common RS-232-C control and data signals appear in Table 20.2.

Table 20.2
Definition of Common RS-232-C Lines

SIGNAL NAME	DIRECTION	PURPOSE
CONTROL SIGNALS		
Request to Send	DTE -> DCE	DTE wishes to send
Clear to Send	DTE <- DCE	Response to Request to Send
Data Set Ready	DTE <- DCE	DCE ready to operate
Data Terminal Ready	DTE -> DCE	DTE ready to operate
Ring Indicator	DTE <- DCE	DTE receiving telephone ringing signal
Carrier Detect	DTE <- DCE	DTE receiving a carrier signal
DATA SIGNALS		
Transmitted Data	DTE -> DCE	Data generated by DTE
Received Data	DTE <- DCE	Data generated by DCE

The term *null modem* refers to situations in which serial communications take place without the presence of a modem. In this case, the connection between the communicating devices, usually a cable, is wired in such a way so as to allow data transmission without a modem.

In Table 20.1, two pins are used in flow control: *RTS (request to send)* and *CTS (clear to send)*. In conventional RS232 communication (as is the case when a computer communicates with a modem), the RTS signal is an output and DCE an input. Before a character is sent, the sender sets the RTS line high to asks the DTE's permission. Until the DTE grants permission, no data is sent. The DTE grants its permission by setting the CTS line high. If the DCE cannot receive new data, it keeps the CTS signal low. This interface, which provides a simple mechanism for flow control in a single direction, is called a *handshake*.

In full-duplex transmission, the handshake must take place in both directions, that is, both devices must be able to signal their status. The *DTR (data terminal ready)* and *DSR (data set ready)* signals can be used for a second level of flow control. Finally, the *CD (carrier detect)* signal serves as an indication of the state of a modem.

Null Modem Cable

Implementing handshaking without a modem requires that we take into account that two communicating devices can expect to find certain signals on given lines. For example, a device checks the CTS signal for a high value before sending data. If the CTS signal never goes high, transmission does not take place. When a cable is wired so that two devices can communicate without one of them being a modem, the cable is said to be a *null modem*.

One simple approach is to completely eliminate handshaking. In this case, cable wiring interconnects the transmit and the receive lines and the ground wire. The remaining pins are left unconnected, as shown in the null modem cable in Figure 20-3.

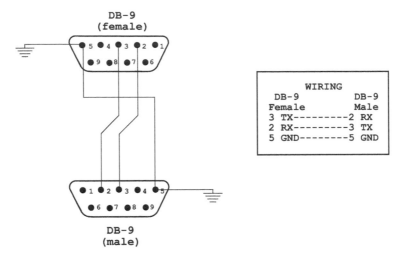

Figure 20-3 Null Modem with No Handshaking.

The three-wire null modem cable can be used to interface devices that do not use modem control signals. However, if one of the devices checks one of the handshake lines, such as RTS/CTS, then the three-wire modem cable fails. To solve this problem, a modem cable can be designed so that the handshake signals are interconnected; for example, DTS to DSR and vice versa. Not knowing which handshake signals are to be used, manufacturers of standard modem cables usually interconnect all handshake lines, as shown in Figure 20-4

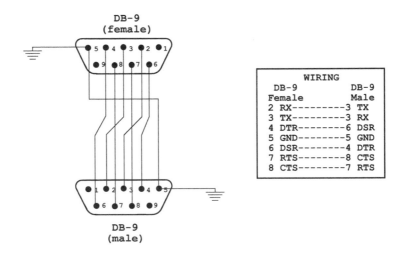

Figure 20-4 Null Modem with Full Handshaking.

Some variations of the full-handshake null modem connect the DTR to the CD line at each end. Pin number 1 (CD) in both male and female connectors is dummied-out to pin number 4 (CDR).

A conventional, straight-through serial cable can be converted to null modem by means of a commercial *null modem adapter* that crosses over the corresponding signal lines. A continuity test is used to determine whether a serial cable is wired as null modem or not. If it is null modem, pin number 2 on one end would show continuity with number 3 pin on the other end.

A *circuit tester* is used to diagnose serial cables. The tester, which is plugged into the port connector, contains an LED for each of the communications lines. When the corresponding LED lights up, the line is active. LED colors indicate positive or negative voltages, with green usually indicating positive and red negative. The light pattern is used to identify different handshakes. Figure 20-5 shows a DB-25 mini tester.

Figure 20-4 DB-25 RS232 Line Tester.

20.2.5 EIA-485 Standard

EIA-485 provides a two-wire, half-duplex serial connection standard, also known as RS-485. This convention provides a multipoint connection with differential signaling. The connection can be made full-duplex using four wires. In this standard, data is conveyed by voltage differences. One polarity represents logic 1 and the reverse one logic 0. The standard requires that the difference of potential be at least 0.2 volts, but any voltage between +12 and – 7 volts allows correct operation.

EIA-485 does not specify a data transmission protocol, making possible the implementation of simple, inexpensive local network and communications links. Its data transmission speeds can reach 35 Mbits/s at distances of up to 10 m, and 100 Kbit/s at distances up to 1200 m. The use of a twisted wire pair and the differential balanced line allows spanning distances of up to 4000 m.

EIA-485 is often used with common UARTs and USARTs to implement low-speed data communications that require minimal hardware. It is also found in programmable logic controllers that are used with proprietary data communications systems. In factories and other electrically charged environments, the differential feature of EIA-485 makes it resistant to electromagnetic interference from motors and other equipment. The standard also finds use in large sound systems, such as those found in theaters and music events. EIA-485 does not specify any connector.

EIA-485 in PIC-based Systems

In PIC-based systems, EIA-485 is often used to provide strong serial signals that can travel up to 4000 m at high baud rates in noisy electrical environments. Only two wires are needed to carry the EIA-485 signals. These are usually labeled the A and B lines. Once the A/B data line is established, up to thirty-two devices can be connected to it. The system is referred to as an EIA-485 network.

Implementing the EIA-485 network requires some way of converting the 485 signal levels to the TTL-levels in the PIC circuit. This is accomplished by means of a dedicated IC, such as the *Texas Instruments Differential Bus Transceiver* chip called the SN75176. The chip actually converts 485 signals to RS-232-C TTL-level signals. This allows devices that traditionally communicate over RS-232-C serial connections to communicate over a two-wire EIA-485 network. Figure 20-6 shows the pin diagram of the SN75176.

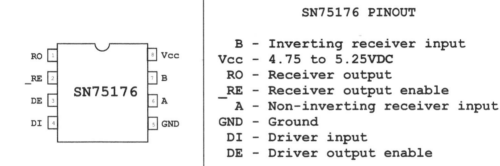

Figure 20-6 Pinout of the SN75176 IC.

In addition to the SN75176, an EIA-485 circuit requires a 485 chip such as the MAX485. In PIC-based systems, the EIA-485 is sometimes used to communicate with multiple devices in a chain. It uses the same 8-bit asynchronous serial communications format as was described previously for RS-232-C.

20.3 Parallel Data Transmission

Parallel communications is the process of sending several bits of data simultaneously over individual data lines. In the computer environment, parallel communications are often associated with a popular printer interface developed by Centronics and sometimes called the *Centronics* or *printer interface*. Originally, the Centronics interface was designed for one-way communications. Later, it was made bi-directional, allowing its use in high-speed data transfers. The Centronics or parallel printer interface is now considered a legacy port.

In PIC-based systems, parallel communications often refer to the general principle rather than to the specific Centronics implementation. For example, wiring an eight-line toggle switch to the eight pins of the 16F85 port-B line provides parallel communications between the switch and the PIC.

PIC circuits that use parallel data transfers offer many advantages. In the first place, parallel transmission is fast and the software is simple to develop. The hardware implementation is straightforward and does not require many additional components. Examples are connecting a multiple toggle switch to each of the lines of a PIC input port, or each of the pins of a Seven-Segment LED to the various pins of a PIC output port. The disadvantages of parallel systems are the distance limitations and the cost in system resources. Furthermore, parallel data transfers do not work well for data transmission over long distances. Many of the circuits and programs covered in previous chapters use parallel data transmission techniques. Because PIC-based systems rarely communicate with parallel printers or use the Centronics standard for data transfer, no further discussion of the Centronics standard is justifiable in this context.

20.3.1 PIC Parallel Slave Port (PSP)

Some PICs are equipped with an 8-bit *Parallel Slave Port* module (PSP). At present, the PSP is multiplexed onto Port D and is found in PICs of the mid-range family, such as the 16F877. The PSP is also called the *microprocessor port*.

The PSP module provides an interface mechanism with one or more microprocessors. The parallel slave port has an operating speed of 200 ns with a clock rate of 20 MHz, as well as several on-chip peripheral functions for implementing real-world interfaces.In PICs equipped with the PSP, the parallel slave port functions are assigned to port D, with some port E bits providing control signals. To initialize PSP mode, data direction bits in the TRISE register that correspond to RD, WR, and CS (TRISE<2:0>) are configured as inputs and the control bit PSPMODE (TRISE) is set. When the PSP mode is active, port D is asynchronously readable and writable through the chip Select (RE2/CS), Read (RE0/RD), and Write (RE1/WR) control inputs. At this time, not many general-purpose applications for the PSP port have been documented, outside of its use as a multi-microprocessor interface. For this reason we have excluded PSP programming from this context.

20.4 PIC "Free-Style" Serial Programming

This section is about PIC serial programming and circuit design that does not follow any specific communications protocol. In this sense, we have used the expression *"free-style"* as opposed to circuits and programs constrained by the requirements of a standard or convention. Many self-contained PIC circuits that do not interface with standardized components can benefit from not having to follow any specific standard.

Later in this chapter, and in other chapters in the book, we present examples of PIC circuits and programs that follow established communications protocols. The titles of the corresponding sections refer to the specific standards or protocols; for example, the section titled "PIC RS-232-C Serial Programming" found in this chapter.

The advantages of so-called "free-style" circuit design and programming are greater ease in development and the use of fewer hardware components. When designer and programmer are not constrained by the specifications of a standard, the circuit can be implemented with a minimal number of hardware components. By the same token, software is simpler and easier to develop.

The following examples of free-style communications systems are presented in the sections that follow:

1. A PIC-to-PIC communications circuit and program. Two programs are required: one for the receiver PIC and one for the sender.

2. Serial-to-parallel and parallel-to-serial circuit and program. Circuit uses 74HC164 and 74HC165 ICs.

20.4.1 PIC-to-PIC Serial Communications

Perhaps the most obvious and straightforward mode of PIC serial communications is one that takes place between two PICs. In this case, one PIC acts as a sender, or master, and the other one as a receiver or slave, although it is also possible for sender and receiver to exchange roles. Consider a circuit in which one PIC polls the state of a bank of switches and then sends the result serially to a second PIC that controls a bank of LEDs to be lighted according to the switch settings. The reason for this circuit is that some PICs may not have a sufficient number of ports to monitor eight switches and control eight LEDs.

PIC-to-PIC Serial Communications Circuits

Actually, the system required for one PIC reading data and serially sending the result to another PIC that outputs the data can be visualized as two separate circuits. One circuit is used to read the state of the eight DIP switches and to send the data serially to another PIC circuit that displays the results. Figure 20-7 shows the two PIC-based circuits.

Structurally, the circuits in Figure 20-7 are quite similar to ones described previously in this book. The bottom circuit contains eight DIP switches wired to ports RB0 to RB7. A pushbutton switch is wired to port RA2 and an LED to port RA3. The serial output is through port RA1. The circuit at the top of Figure 20-7 has eight LEDs wired to ports RB0 to RB7. There is a pushbutton on port RA2 and an LED on port RA3. Input into the circuit is through port RA0. In the remainder of this description we refer to the bottom circuit as the *sender circuit and PIC*, and the one on the top as the *receiver circuit and PIC*.

The pushbuttons are necessary so that sender and receiver are synchronized. In operation, the receiver circuit is first activated by pressing the switch labeled *"receive ready."* The LED on the top circuit lights to indicate the ready state. The sender circuit has an LED labeled "ready" that indicates its state. The user presses the switch labeled *"send ready"* in the sender circuit. At this time, the program in the sender reads the state of the DIP switches and sends the data out, one bit at a time, through the line labeled "serial out" in the diagram. The receiver reads the eight bits in its *"serial in"* line and lights the LEDs accordingly.

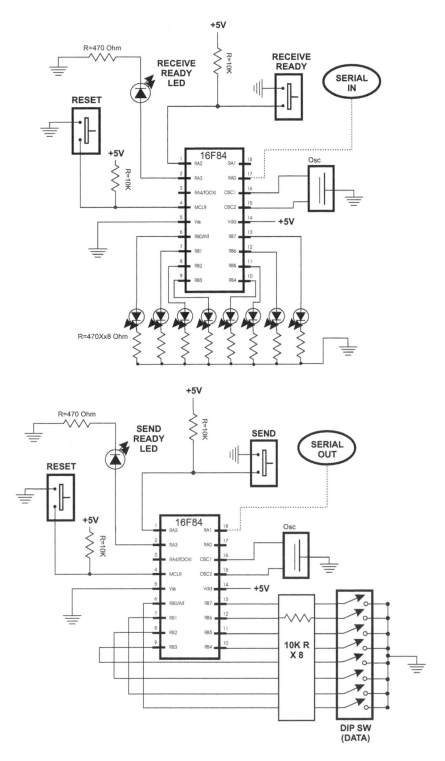

Figure 20-7 PIC-to-PIC Serial Communications Circuits.

PIC-to-PIC Serial Communications Programs

The software consists of two different programs, one to run in the sender PIC and one in the receiver PIC. Asynchronous communications require that sender and receiver operate at the same data speed. Both devices need not run at the same clock speed, but both must synchronize data transmission and reception at the same clock rate. Because the easiest way to accomplish this is to have both PICs use the same oscillator at the same speed, we make this assumption in the programs that follow.

The instruction time and clock rate of a PIC are one-fourth of its clock speed. Thus, a PIC with a 4-MHz clock runs at 1,000,000 cycles per second, and the default timer speed is

$$\frac{1,000,000}{256} = 3,906.25 \, \mu s. \, \text{per bit}$$

Approximately 3,906 µs per clock cycle. Although 3,906 µs is not a standard baud rate, the present application is self-contained and there is no need to conform to RS-232-C or any other protocol.

Because it seems more intuitive to associate a high voltage with a logic 1 and a low voltage with a logic 0, we will adopt this convention in the current application. Nevertheless, we will borrow the character structure from the RS-232-C convention, that is, information will contain a start bit, a series of eight data bits, and a stop bit. No parity is implemented. Figure 20-8 shows the bit structure for one character in our application.

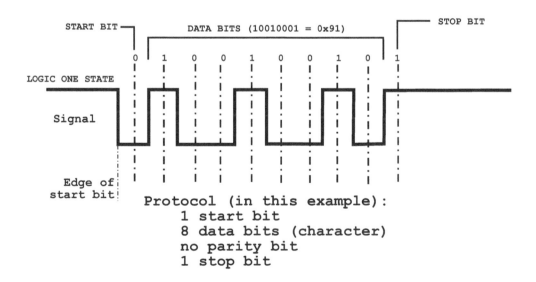

Figure 20-8 Data Structure for PIC-to-PIC Application.

The sender program, named SerialSnd, performs the following initialization operations:

1. Line RA2 is initialized for input because the pushbutton switch is located on this line. Lines RB0 to RB7 are also input, as they are connected to the DIP switch array.

2. The prescaler is assigned to the Watchdog timer so that channel TMR0 runs at full processor speed.

3. Interrupts are disabled.

Initialization code is as follows:

```
; Port-A, bit 2 is input. All others are output
        movlw    b'00000100'       ; Port-A bit 2 is input
                                   ; all others are output
        tris     porta
; Port-B is all input
        movlw    b'11111111'
        tris     portb
        bsf      porta,1        ;Marking bit
; Prepare to set prescaler
        clrf     tmr0
        clrwdt
; Set up OPTION register for full timer speed
        movlw             b'11011000'
;    1  1  0  1  1  0  0  0 <= OPTION bits
;    |  |  |  |  |  |__|__|_____ PS2-PS0 (prescaler bits)
;    |  |  |  |  |  |            Values for Timer0
;    |  |  |  |  |  |            *000 = 1:2    001 = 1:4
;    |  |  |  |  |  |            010 = 1:8    011 = 1:16
;    |  |  |  |  |  |            100 = 1:32  101 = 1:64
;    |  |  |  |  |  |            110 = 1:128 111 = 1:256
;    |  |  |  |  |  |_____ PSA (prescaler assign)
;    |  |  |  |  |               *1 = to WDT
;    |  |  |  |  |               0 = to Timer0
;    |  |  |  |  |_____ TOSE (Timer0 edge select)
;    |  |  |  |               0 = increment on low-to-high
;    |  |  |  |               *1 = increment in high-to-low
;    |  |  |_____ TOCS (TMR0 clock source)
;    |  |               *0 = internal clock
;    |  |               1 = RA4/TOCKI bit source
;    |  |_____ INTEDG (Edge select)
;    |               0 = falling edge
;    |               *1 = rising edge
;    |_____ RBPU pullups
;                  0 = enabled
;                  *1 = disabled
        option
; Disable interrupts
        bcf      intcon,5   ; Timer0 overflow disabled
```

```
        bcf        intcon,7    ; Global interupts disabled
```

Once initialized, the program performs the following functions:

1. The SEND READY LED is turned on.
2. Code monitors the SEND pushbutton switch.
3. Once the switch is pressed, the program turns off the SEND READY LED.
4. The state of the DIP switches is obtained by reading RB0 to RB7.
5. The byte from port B is sent through the serial line.

The following code fragment shows the procedure to send serial data:

```
;================================================================
;                    procedure to send serial data
;================================================================
; ON ENTRY:
;        local variable dataReg holds 8-bit value to be
;        transmitted through port labeled serialLN
; OPERATION:
;        1. The timer at register TMR0 is set to run at
;           maximum clock speed, that is, 256 clock beats.
;           The timer overflow flag in the INTCON register
;           is set when the timer cycles from 0xff to 0x00.
;        2. Each bit (start, data, and stop bits) are sent
;           at a rate of 256 timer beats. That is, each bit is
;           held high or low for one full timer cycle (256
;           clock beats.)
;        3. The procedure tests the timer overflow flag
;           (tmrOVF) to determine when the timer cycle has
;           ended, that is when 256 clock beats have passed.
;
sendData:
        movlw     0x08              ; Setup shift counter
        movwf     bitCount
;========================
;    send START bit
;========================
; Set line low then hold for 256 timer clock beats.
        bcf       PORTA,serialLN    ; Send start bit
; First reset timer
        clrf      TMR0              ; Reset timer counter
        bcf       INTCON,tmrOVF     ; Reset TMR0 overflow flag
; Wait for 256 timer clock beats
startBit:
        btfss     INTCON,tmrOVF     ; timer overflow?
        goto      startBit          ; Wait until set
; At this point timer has cycled. Start bit has ended
        bcf       INTCON,tmrOVF     ; Clear overflow flag
```

```
;=========================
;    send 8 DATA bits
;=========================
; Eight data bits are sent through the serial line
; starting with the high-order bit. The data byte is
; stored in the register named dataReg. The bits are
; rotated left to the carry flag. Code assumes the bit
; is zero and sets the serial line low. Then the carry
; flag is tested. If the carry is set the serial line
; is changed to high. The line is kept low or high for
; 256 timer beats.
send8:
        rlf      dataReg,f         ; Bit into carry flag
        bcf      PORTA,serialLN  ; 0 to serial line
; Code can assume the bit is a zero and set the line
; low because, if low is the wrong state, it will only
; remain for two timer beats. The receiver will not
; check the line for data until 128 timer beats have
; elapsed, so the error will be harmless. In any case,
; there is no assurance that the previous line state is
; the correct one, so leaving the line in its previous
; state could also be wrong.
        btfsc    STATUS,c          ; Test carry flag
        bsf      PORTA,serialLN  ; Bit is set. Fix error.
bitWait:
        btfss    INTCON,tmrOVF   ; Timer cycled?
        goto     bitWait           ; Not yet
; At this point timer has cycled.
; Test for end of byte, if not, send next bit
        bcf      INTCON,tmrOVF   ; Clear overflow flag
        decfsz   bitCount,f        ; Last bit?
        goto     send8             ; not yet
;=========================
;    hold MARKING state
;=========================
; All 8 data bits have been sent. The serial line must
; now be held high (MARKING) for one clock cycle
        bsf      PORTA,serialLN  ; Marking state
markWait:
        btfss    INTCON,tmrOVF   ; Done?
        goto     markWait          ; not yet
;=========================
;   end of transmission
;=========================
        return
```

The code comments explain the routine's operation.

The receiving program, named SerialRcv, runs in the receiver PIC. In this case, the serial line is RA0. Input from the sender program is received through this line. The program performs the following initialization operations:

1. Lines RA0 and RA2 are initialized for input because the pushbutton switch is located on RA2 and RA0 is the serial input line. Lines RB0 to RB7 are output because they are wired to the eight LEDs.

2. The prescaler is assigned to the Watchdog timer so that channel TMR0 runs at full processor speed.

3. Interrupts are disabled.

Once initialized, code performs the following functions:

1. The SEND READY LED is turned on.

2. Code monitors the RECEIVE READY pushbutton switch.

3. Once the switch is pressed, the program turns on the RECEIVE READY LED.

4. Code then monitors the serial line for the first low that indicates the leading edge of the start bit.

5. Once the start bit is detected, code waits for 128 clock cycles to locate the center of the start bit. This synchronizes the receiver with the sender and accommodates small timing errors.

6. The eight data bits are then received and stored.

7. After waiting for the stop bit, code turns off the RECEIVE READY LED and sets the eight LEDs according to the data received through the serial line.

The following code fragment is the procedure *rcvData* from the SerialRcv program:

```
;=============================================================
;                  procedure to receive serial data
;=============================================================
; ON ENTRY:
;         local variable dataReg is used to store 8-bit value
;         received through port (labeled serialLN)
; OPERATION:
;         1. The timer at register TMR0 is set to run at
;            maximum clock speed, that is, 256 clock beats.
;            The timer overflow flag in the INTCON register
;            is set when the timer cycles from 0xff to 0x00.
;         2. When the START signal is received, the code
;            waits for 128 timer beats so as to read data in
;            the middle of the send period.
;         3. Each bit (start, data, and stop bits) is read
;            at intervals of 256 timer beats.
;         4. The procedure tests the timer overflow flag
;            (tmrOVF) to determine when the timer cycle has
```

```
;              ended, that is when 256 clock beats have passed.
;=============================================================
rcvData:
        clrf     TMR0       ; Reset timer
        movlw    0x08       ; Initialize bit counter
        movwf    bitCount
;=========================
;   wait for START bit
;=========================
startWait:
        btfsc    PORTA,0 ; Is port A0 low?
        goto     startWait          ; No. Wait for mark
;=========================
;   offset 128 clock beats
;=========================
; At this point the receiver has found the falling
; edge of the start bit. It must now wait 128 timer
; beats to synchronize in the middle of the sender's
; data rate, as follows:
;              |<========= falling edge of START bit
;              |
;              |-----|<====== 128 clock beats offset
;   -----------.     |     .-------
;              |           |   <== SIGNAL
;              -----------
;              |<---256--->|
;
        movlw    0x80               ; 128 clock beats offset
        movwf    TMR0               ; to TMR0 counter
        bcf      INTCON,tmrOVF      ; Clear overflow flag
offsetWait:
        btfss    INTCON,tmrOVF      ; Timer overflow?
        goto     offsetWait                ; Wait until
        btfsc    PORTA,0            ; Test start bit for error
        goto     offsetWait                ; Recycle if a false
start
;===========================
;      receive data
;===========================
        clrf     TMR0               ; Restart timer
        bcf      INTCON,tmrOVF      ; Clear overflow flag
; Wait for 256 timer cycles for first/next data bit
bitWait:
        btfss    INTCON,tmrOVF      ; Timer cycle end?
        goto     bitWait            ; Keep waiting
; Timer has counter 256 beats
        bcf      INTCON,tmrOVF      ; Reset overflow flag
        movf     PORTA,w            ; Read Port-A into w
```

```
        movwf    temp      ; Store value read
        rrf      temp,f    ; Rotate bit 0 into carry flag
        rlf      rcvReg,f  ; Rotate carry into rcvReg bit 0
        decfsz   bitCount,f          ; 8 bits received
        goto     bitWait             ; Next bit
; Wait for one time cycle at end of reception
markWait:
        btfss    INTCON,tmrOVF       ; Timer overflow flag
        goto     markWait            ; keep waiting
;========================
;   end of reception
;========================
        return
```

Neither the SerialRcv nor the SerialSnd programs contain any handshake signal. The programs rely on the user turning on the receiver before the send function is activated. If this is not the case, the programs fail to communicate. But looking at the circuit diagram in Figure 20-7, we notice that there are available ports in both receiver and sender circuits. The circuit designer could interconnect two ports, one in the receiver and one in the sender, so as to provide a handshake signal.

For example, lines RA4 in both circuits can be interconnected. Then Port A, line 4, in the sender circuit is defined as input and the same line as output in the receiver. The receiver could then set the handshake line high to indicate that it is ready to receive. The sender monitors this same port and does not start the transmission of each character until it reads that the handshake line is high. In this manner, the receiver can suspend transmission at any time and prevent data from being lost. At the same time, the "receiver ready" and "send ready" LEDs can be eliminated.

20.4.2 Program Using Shift Register ICs

The problem of handling multiple input and output lines, which was resolved in the previous example by using two PICs, can also be tackled by means of special-purpose integrated circuits. The term *shift register* refers to the fact that register input and output are connected in a way that data is shifted-down a set of flip-flops when the circuits are activated. Many variations of shift registers ICs are available, the most popular ones being serial-in to serial-out, parallel-in to parallel-out, serial-in to parallel-out, and parallel-in to serial-out. In shift register terminology, the *in* and *out* terms refer to the function in the registers themselves and are not related to the functions that these elements perform in a particular circuit. Figure 20-9 shows an input/output circuit using shift registers.

The circuit in Figure 20-9 shows the use of a parallel-to-serial IC (74HC165) that reads the state of eight input switches, and a serial-to-parallel IC (74HC164) that outputs data to eight LEDs. Without the shift register ICs, the circuit would require sixteen ports, more than those available in the 16F84. Using the shift registers, only six PIC ports are required, leaving eight ports available on the PIC. The demonstration program for the circuit in Figure 20-9 is called Serial6465.

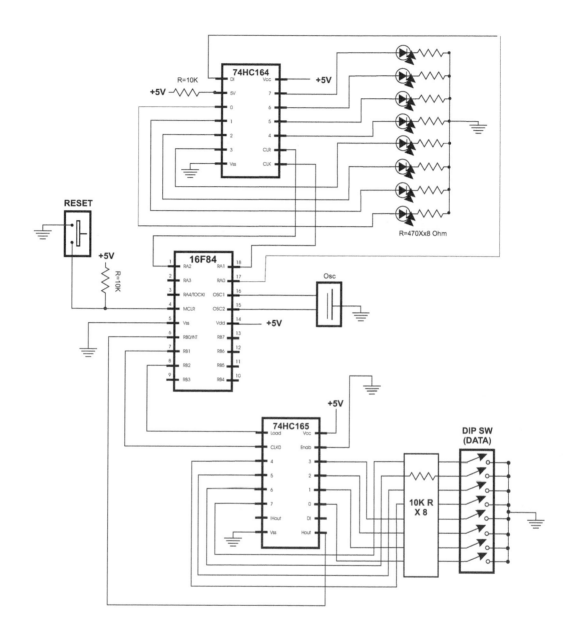

Figure 20-9 Input/output Circuit Using Shift Registers.

74HC165 Parallel-to-Serial Shift Register

The 74HC165 (sometimes called the 165) is a parallel-in, serial-out high-speed 8-bit shift register. Figure 20-10 shows the pin-out of the 74HC165.

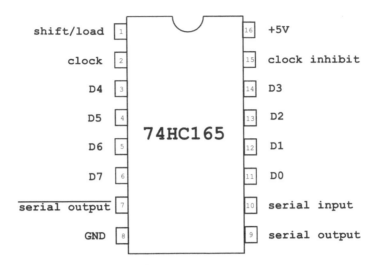

Figure 20-10 74HC165 Pin-Out.

In the 165, pins 3 to 6 and 11 to 14 (labeled D0 to D7) are used as parallel data input lines. Normally these pins are connected to input sources, such as switches or other two-state devices. Serial output takes place through pin number 9, labeled serial output Q. An inverted output is available at pin number 7. The *shift/load control line*, at pin number 1, is used to latch the data into the shift registers of the 165. For example, assume that the 165's input lines are connected to sources that can change state in time. These highs and lows are not recorded internally in the 165 until the shift/load line is pulsed. When this line is pulsed, line values are said to be *latched*. After the data lines are latched, the 165 clock-line is pulsed in order to sequentially shift out each of the eight bits stored internally. Shifting takes place with the most-significant bit first. The actual operations are as follows:

1. A local data storage register is cleared and a local counter is initialized for eight data bits.

2. The 165 shift/load line is pulsed to reset the shift register.

3. The status of the serial output line (165 pin number 9) can now be read to determine the value of the bit shifted out.

4. The bit is stored in a data register and the bit counter is decremented. If the last bit was read, the routine ends.

5. If not, the clock line is pulsed to shift-out the next bit. Execution continues at Step 3.

The wiring of the 165 normally requires at least three interface lines with the PIC. One line connects to the 165 serial output (pin number 9), another one to the clock line (pin number 2), and a third one to the shift/load line (pin number 1). The eight data lines of the 165 are normally wired to the input source.

The following code fragment lists a procedure to interface a 16F84 PIC with a 74HC165 parallel-to-serial shift register:

```
;===============================================================
;            constant definitions from wiring diagram
;===============================================================
#define clk65LN 1    ;| - 74HC165 lines
#define loadLN 2     ;|
.
.
.
;===============================================================
;     74HC165 procedure to read parallel data and send
;                      serially to PIC
;===============================================================
; OPERATION:
;         1. Eight DIP switches are connected to the input
;            ports of a 74HC165 IC. Its output line Hout,
;            and its control lines CLK and load are connected
;            to the PIC's Port-B lines 0, 1, and 2,
;            respectively
;         2. Procedure sets a counter (bitCount) for 8
;            iterations and clears a data holding register
;            (dataReg).
;         3. Port-B bits are read into w. Only the lsb of
;            Port-B is relevant. Value is stored in a working
;            register and the meaningful bit is rotated into
;            the carry flag, then the carry flag bit is
;            shifted into the data register.
;         4. The iteration counter is decremented. If this
;            is the last iteration the routine ends. Otherwise
;            the bitwise read-and-write operation is repeated.

in165:
        clrf    dataReg           ; Clear data register
        movlw   0x08              ; Initialize counter
        movwf   bitCount
        bcf     PORTB,loadLN      ; Reset shift register
        bsf     PORTB,loadLN
nextBit:
        movf    PORTB,w           ; Read Port-B (only LOB is
                                  ; meaningful in this routine)
        movwf   workReg           ; Store value in local
                                  ; register
        rrf     workReg,f         ; Rotate LOB bit into carry
                                  ; flag
        rlf     dataReg,f         ; Carry flag into dataReg
        decfsz  bitCount,f        ; Decrement bit counter
        goto    shiftBits         ; Continue if not zero
```

```
        Return                          ; done
shiftBits:
        bsf        PORTB,clk65LN        ; Pulse clock
        bcf        PORTB,clk65LN
        goto       nextBit              ; Continue
```

The procedure in 165 is in the program Serial6465 listed at the end of this chapter.

74HC164 Serial-to-Parallel Shift Register

The circuit in Figure 20-9 also uses a 74HC164 serial-to-parallel shift register for output to the eight LEDs. Figure 20-11 shows the pin-out of the 74HC164 IC.

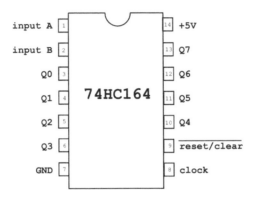

Figure 20-11 74HC164 Pinout.

Serial input into the 164 is through the input A line (pin number 1). Parallel output is through the lines labeled Q0 to Q7. The reset/clear line (on pin 9) and the clock line (on pin 8) provide the control functions. The operations are as follows:

1. A local data storage register holds the 8-bit value that serves as data input. A local counter is initialized for eight data bits.

2. The 164 shift register is cleared by pulsing the reset/clear line.

3. The first/next bit of the data operand is placed on the input line.

4. Bit is shifted in by pulsing the 164 clock line.

5. Bit counter is decremented. If it goes to zero, the routine ends.

6. Otherwise, the bits in the source operand are shifted and execution continues at Step 3.

The following code fragment lists a procedure to interface a 16F84 PIC with a 74HC164 serial-to-parallel shift register:

```
;=========================================================
;       constant definitions from wiring diagram
;=========================================================
#define clockLN 1           ;|
```

```
#define clearLN 2          ;| => 74HC164 lines
#define dataLN 0           ;|
...
;===========================================================
;             74HC164 procedure to send serial data
;===========================================================
; ON ENTRY:
;          local variable dataReg holds 8-bit value to be
;          transmitted through port labeled serialLN
; OPERATION:
;        1. A local counter (bitCount) is initialized to
;           8 bits
;        2. Code assumes that the first bit is zero by
;           setting the data line low. Then the high-order
;           bit in the data register (dataReg) is tested.
;           If set, the data line is changed to high.
;        3. Bits are shifted in by pulsing the 74HC164
;           clock line (CLK).
;        4. Data bits are then shifted left and the bit
;           counter is tested. If all 8 bits have been sent,
;           the procedure returns.
out164:
; Clear 74HC164 shift register
        bcf       PORTA,clearLN    ; 74HC164 CLR clear low
        bsf       PORTA,clearLN    ; then high again
; Init counter
        movlw     0x08             ; Initialize bit counter
        movwf     bitCount
sendBit:
        bcf       PORTA,dataLN     ; Set data line low (assume)
; Using this assumption is possible because the bit is not
; shifted in until the clock line is pulsed.
        btfsc     dataReg,highBit  ; test number bit 7
        bsf       PORTA,dataLN     ; Change assumption if set
;=========================
;    pulse clock line
;=========================
; Bits are shifted in by pulsing the 74HC164 CLK line
        bsf       PORTA,clockLN    ; CLK high
        bcf       PORTA,clockLN    ; CLK low
;=========================
; Rotate data bits left
;=========================
        rlf       dataReg,f        ; Shift left data bits
        decfsz    bitCount,f       ; Decrement bit counter
        goto      sendBit          ; Repeat if not 8 bits
;=========================
;   end of transmission
```

```
;============================
        return
```

It is important to note that serial communications that use shift register ICs are described as synchronous. Synchronous serial transmission requires that the sender and receiver use the same clock signal, or that the sender provide signal or pulse so as to indicate to the receiver when to read the next data element from the line. In the circuits discussed in this section, the shift/load, reset/clear, and clock lines provide this synchronous interface between the PIC and the shift register IC.

The program named Serial6465, in this book's online software, is a demonstration of PIC-to-shift register interfacing.

20.5 PIC Protocol-Based Serial Programming

In the preceding sections we discussed circuits and developed software using PIC serial communications that did not conform to any particular protocol or standard. This style is adequate for stand-alone applications and circuits. On the other hand, PIC-based circuits sometimes communicate with systems that conform to a specific communications standard, for example, with a PC through its RS-232-C serial port. In this case, the PIC software and hardware must conform to the protocol, at least to an operational minimum that ensures satisfactory interfacing with the *protocol-based system*.

In the context of *protocol-based programming*, two situations are possible: either the PIC in use supports the communications standard or protocol, or it does not. In the case of the smaller PICs, such as the 16F84, the software emulates communications protocols because hardware provides no support. The more complex PICs, on the other hand, often contain hardware modules that provide a functionality equivalent to that required by the various standards. In this sense, mid-range and high-range PICs often include hardware support for one or more communication standards and conventions. For instance, the 16F87X PIC family includes an *MSSP* (*Master Synchronous Serial Port*) module and a *USART* (*Universal Synchronous/asynchronous Receiver and Transmitter*) module.

In the sections that follow we develop circuits and programs for cases in which the on-board PIC does not contain hardware support for the standard and for cases in which it does. Examples with PICs that do not provide hardware support for serial communications use the 16F84. Examples with PICs that provide hardware serial communications support use the 16F877, which contains an MSSP and a USART module. The 16F877 circuits and applications in this chapter use the processor's USART module. The 16F877 MSSP module is demonstrated in the chapter on EEPROM programming.

20.5.1 RS-232-C Communications on the 16F84

The *UART* (*Universal Asynchronous Receiver/Transmitter*) controller is a serial communications IC found in computers and other data communication devices. In the PC, the UART was originally National Semiconductor INS8250. With the introduction of the PC AT, IBM changed its serial IC to the NC16450, an improved 8250. Later PCs adopted the NS16550A UART as their serial communications controllers. Other vendors, including Intel and Western Digital, furnish clones of the NS16550A and other UARTs.

The UART-based serial port implementation and circuitry in the PC is compliant with RS/EIA232. For a PIC-based circuit to communicate with a PC's serial port, it must either implement in hardware or emulate in software the RS-232 signals and protocol. One possibility is to include a UART or UART-like IC in the circuit. But this option is not simple to implement because RS-232-C requires voltage levels that are not TTL-compatible.

For PIC-based systems without a UART module, a viable approach is to emulate UART functions in software, at least those required for interfacing with the PC hardware. This is quite feasible due to the availability of dedicated ICs that provide RS-232-C-compatible signals and voltage levels in systems in which a ±12 volt source is not available. These chips, sometimes called *RS-232-C Drivers/Receivers* or *Transceivers*, are especially useful in interfacing UART and USART-based systems with PIC-based hardware.

RS-232-C Transceiver IC

RS-232-C interface ICs are available from several vendors, although the ones from Dallas Semiconductors' Maxim line are probably the most popular. These chips, sometimes called RS-232-C driver/receivers, have in common the use of so-called *charge-pump DC/DC converters* that generate, from the +5- volt TTL power source, the polarities and voltage levels required by RS-232-C.

One of the most popular implementations of the RS-232-C transceiver used in PIC-based systems is the MAX232 and its upgrade, the MAX202. One improvement in the MAX202 is to provide some degree of human-body *electrostatic discharge protection* (ESD), a desirable feature in experimenter boards. Other versions are the MAX233 and MAX203, which do not require external capacitors. Other RS-232-C transceiver ICs with various additional features, such as automatic shutdown, are available. Figure 20-12 is a pin-out of the MAX232 and MAX202 ICs.

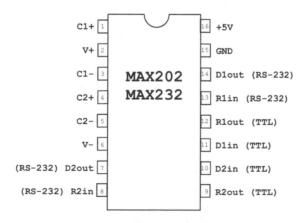

Figure 20-12 MAX202 and MAX232 Transceiver Pinout.

Note that the MAX232 and MAX202 consist of two drivers and two receivers per chip. Lines 14 and 7 (labeled D1out and D2out) provide RS-232-C output. Lines 13 and 8 (labeled R1in and R2in) are RS-232-C input. Lines 10 and 11 (labeled D1in and D2in) are TTL (or CMOS) inputs. Lines 9 and 12 (labeled R2out and R1out) are TTL output. In this designation the letter R stands for receiver and the letter D for driver. The digit 1 indicates the first driver/receiver set and the digit 2 the second one. The lines labeled D are wired to capacitors.

A circuit using the transceiver ICs is simple and easy to build. If a single communication line is required, then the TTL input line can be wired to pin 10 (D2in) and the TTL output to pin 9 (R2out). The RS-232-C input is wired to pin 8 (R2in) and the output to pin 7 (D2out). Later in this section we present a circuit that uses the MAX202 with a 16F84 PIC.

PIC-to-PC Communications

Often, a PIC-based circuit has to communicate with a device that conforms to a standard communications protocol. One of the most common cases is a PIC board that interfaces with a computer, usually a PC or Mac with an RS-232-C port. For example, a PIC board is placed somewhere to collect information, such as temperature, pressure, and humidity. Before the internal storage capacity of the PIC board is exhausted, it is connected to a laptop PC and the data is downloaded from the PIC board to the computer. Once this is done, then the local PIC memory is cleared so that new data can be collected and stored. This application, called a *data logger*, requires some way of transferring data from the PIC-based board to the PC. The RS-232-C line is often available on the PC end and the required interface hardware and programming are uncomplicated.

On the PC end, the communications software can be off-the-shelf applications or specially developed programs. If the purpose is simply to download data to the PC or send simple command to the PIC board, then a standard utility is used. For example, the Windows program named *Hyper Terminal* allows sending and receiving files and commands at various baud rates and RS-232-C communications parameters. *Hyper Terminal* is included with most Windows versions or can be downloaded free from the developer's website.

The PIC board must have a system that conforms to the communications protocol of the device, in this case, the PC. In order to use the PC's serial port, PIC hardware and software must be able to generate required signal levels, baud rate, and other RS-232-C communications parameters. Hardware interfacing is implemented using a transceiver chip, such as the MAX232 or 202 previously described. If the PIC contains a UART or USART module, then the communications software is easy to develop. This case is explored later in this chapter.

RS-232-C TTY Board

The terms *teletype* and *teletypewriter* refer to an obsolete electromechanical typewriter that was used to send and receive information through a simple communication channel. In a modern sense, *TTY* refers to a simple style of communications where the same device sends and receives text messages interactively. The current board is actually a TTY receiver because it does not contain a keyboard that allows sending data. Figure 20-13 shows the circuit diagram for a 16F84-based PC-to-PIC serial communications board.

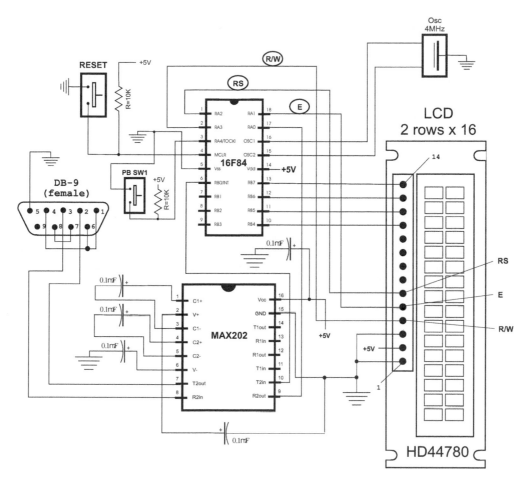

Figure 20-13 PC-to-PIC Serial Communications Circuit.

The circuit in Figure 20-13 contains previously discussed components. The LCD is wired in 4-bit mode, with control lines for RS (reset), E (pulse), and R/W (read/write). The MAX202 provides the TTL-to-RS-232-C conversion and vice versa. The physical connection between the PC and the PIC board is by means of a DB-9 connector and a standard null modem cable. The cable is not shown in the circuit diagram.

16F84A UART Emulation

The 16F84A PIC contains no built-in facilities for RS-232-C communications; therefore, a 16F84A application that communicates through the serial port using the RS-232-C protocol must emulate the protocol in software. The programs previously developed for PIC-to-PIC communications, discussed in Section 20.3.1, serve as a base for the UART emulation application. The major differences between a "free-style" PIC communications program and one that complies with RS-232-C are the following:

1. Data must be transmitted and received at one of the standard RS-232-C baud rates. The most often-used baud rates in this case are 600, 1,200, 2,400, 4,800, 9,600, and 19,200.

2. Data must be formatted according to the protocol's conventions; that is, a start bit, 5, 6, 7, or 8 data bits; the presence or absence of a parity bit; and 1, 1½, or 2 stop bits.

3. RS-232-C communication data is transmitted and received with the least-significant bit first.

The first problem (transmitting and receiving at a standard baud rate) often requires an approximation. The PIC's instructions execute at the rate of its internal clock, which also determines the rate of its timer module.

The time taken by each counter iteration is obtained by dividing the PIC's clock speed by four. For example, a PIC running on a 4-MHz oscillator clock increments the counter every 1 MHz. The counter register is incremented at a rate of 1 µs (assuming no prescaler). If we were to use the unmodified timer rate to measure bit time, the result would be a baud rate of approximately 3,906. Because 3,906 is not a standard baud rate, the timer is adjusted to approximate one of the standard

$$\frac{1}{4,800} = 208.33 \mu s.$$

RS-232-C baud rates. For example, at 4,800 baud, the time per bit is

Because the timer of a PIC with a 4-MHz clock runs at 1 µs per timer iteration, we could count up from 0 to 208 iterations of the counter in order to approximate the bit time of 208 µs needed at 4,800 baud. In addition, we would have to calculate one-half the bit time because synchronization requires offsetting the timer from the edge to the center of the start bit (see Section 20.3.1). In this case, to delay approximately 104 µs we would count up from 0 to 104.

But counting up is inconvenient with the PIC timer/counter because the signal is produced when the counter reaches its maximum. A better solution is to preset the timer counter (TMR0) to a calculated value such that the desired time lapse occurs when the timer register reaches 255. So the actual delays for 4,800 baud are as follows:

```
DELAY            CALCULATION        TMR0 PRESET
208 µs           255 minus 208      47
104 µs           255 minus 104      151
```

Once we have obtained the clock rate for a standard baud rate, it is easy to obtain slower standard rates by slowing down the clock with the prescaler. For example, if the prescaler is assigned to the timer/counter register with a bit value of 000, then the counter rate is one-half the unscaled rate. This would produce a baud rate of 2,400 baud. By the same token, assigning a 1:4 prescaler to the timer produces a baud rate of 1,200 baud using the same preset values previously calculated. Faster baud rates are easily calculated by the same method.

Formatting the data transmission according to the RS-232-C protocol presents no major problem. In fact, the communications programs previously listed in this chapter use a start bit to commence character transmission, followed by eight data bits, and one stop bit to end it, with no parity bit. This same format is compatible with RS-232-C.

The third compatibility issue refers to the bit order in RS-232-C, which requires that the low-order bit be transmitted first. In previous applications, we have sent the high-order bit first by rotating the bits left inside the holding register and testing the carry flag. In the RS-232-C routine, the bits are rotated right into the carry flag and then the carry flag is rotated into the storage variable.

The demonstration program for the circuit in Figure 20-13, named TTYUsart, uses a two-line by sixteen-character LCD to display the characters received from the PC through the serial line. The program initially sends the test string "Ready-" to the PC to test the data transmission routine and to let the PC user know that the PIC board is ready to receive. The program operates at 2,400 baud, one start bit, eight data bits, no parity, and one stop bit. The communications program on the PC must be set to these parameters.

LCD Scrolling Routine

LCDs have limited capacity for data display. A two-line by sixteen-character LCD fills the screen when 32 characters are displayed. For some applications it is convenient to have a procedure that takes some reasonable action when the LCD screen is full. One approach is to detect when the last character in the second LCD line is displayed, then move the second line to the first line, clear the second line, and continue displaying at the start of the second line. This is the standard *screen handling* for a computer program.

An LCD screen scroll routine can be called as each character is displayed. For the scroll to work, the program must keep track of the currently selected LCD line (variable *LCDline* can be 0 for line 1, and 1 for line 2), of the number of characters displayed on that line (variable *LCDcount*), and of the total capacity of the line (constant *LCDlimit*). Given this information, the logic for an LCD line scrolling routine can be as follows:

1. Add current character to *LCDcount*. If *LCDcount* is equal to *LCD limit*, then the end of a line was reached. If not, exit routine.

2. If line end reached is for line 1, set current display address to start of line 2. Reset variable *LCDcount*. Exit routine.

3. If line end reached is for line 2, then copy the characters displayed in line 2 to line 1. Clear line 2. Reset the display address to the start of line 2. Reset *LCDline* variable to line 2. Reset variable *LCDcount*. Exit routine.

Of these operations, copying the characters from the second line to the first one can be the most troublesome. One possibility is to read the data from the LCD directly. This approach requires that the connection between the PIC and the LCD include the R/W line. Another option is to create a buffer in RAM and copy each character displayed to this area. In the case of an LCD with sixteen characters per line, the buffer requires a capacity of 16 bytes. Because the line input is "remembered" in the buffer, the program scrolls a line by copying the contents of the buffer to the other line. This alternative does not require reading the LCD and saves implementing the R/W line.

Storing the characters received in a local buffer first requires reserving a 16-byte area (the buffer) in PIC RAM. There are several ways of accomplishing this. A simple one is using the **cblock** directive, as shown in the following code fragment:

```
;========================================================
;             buffer and variables in PIC RAM
;========================================================
; Create a 16-byte storage area
        cblock  0x0c    ; Start of first data block
        lineBuf                 ; buffer for text storage
        endc
; Leave 16 bytes and continue with local variables;
        cblock  0x1c    ; Second data block
        count1          ; Counter # 1
        count2          ; Counter # 2
. . . other variables can go here
        endc
```

In reality, the buffer is most likely accessed by indirect addressing, so a buffer name (lineBuf in this case) is not really necessary. This is due to the fact that PIC assembly language does not contain a directive for finding the address of a variable. So the buffer address must be hard-coded or defined in a constant. But, in any case, having a buffer name does not cost storage capacity and it may help make the code clearer.

In our design, the scrolling routine depends on finding the characters in the ending line stored in the RAM area mentioned in the preceding paragraph. The buffer locations are accessed directly by referencing the address. For example, the first byte in lineBuf is stored at address 0x0c, the second one at 0xod, and so on. A more effective way of using a buffer is by creating and keeping a *buffer pointer variable* that has the current offset from the start of the buffer. The buffer pointer is then added to the buffer's base address in order to access the current buffer location. *Indirect addressing* using the FSR and the INDF registers simplifies the process, as shown in the following code fragment:

```
; Store character in local line buffer using indirect
; addressing. Byte to store is in rcvData variable.
; 16-byte buffer named lineBuf starts at address 0x0c
; Register variable bufPtr holds offset into buffer
        movlw   0x0c            ; Buffer base address
        addwf   bufPtr,w        ; Add pointer in w
        movwf   FSR             ; Value to index register
        movf    rcvData,w       ; Character into w
        movwf   INDF            ; Store w in [FSR]
        incf    bufPtr,f        ; Bump pointer
```

The manipulation requires loading the *base address* of the buffer (0x0c in this case) in the w register, adding the value stored in the buffer pointer variable (bufPtr), and storing the sum in the FSR register. The character is then loaded into the w register and moved into the INDF register, which has the effect of storing it in the address pointer at by FSR. Conventionally, brackets are used to indicate indirect addressing, so [FSR] means the memory location referenced by the FSR register.

Once the line characters are stored locally, all that is left is the design of a line scrolling routine following the processing steps previously listed. The following procedure performs the necessary operations:

```
;===========================
;   scroll LCD line 2
;===========================
; Procedure to count the number of characters displayed on
; each LCD line. If the number reaches the value in the
; constant LCDlimit, then display is scrolled to the second
; LCD line. If at the end of the second line, then the
; second line is scrolled to the first line and display
; continues at the start of the second line
; reset to the first line.
LCDscroll:
        incf    LCDcount,f              ; Bump counter
; Test for line limit
        movf    LCDcount,w
        sublw   LCDlimit        ; Count minus limit
        btfss   STATUS,z        ; Is count - limit = 0
        goto    scrollExit      ; Go if not at end of line
; At this point the end of the LCD line was reached
; Test if this is also the end of the second line
        movf    LCDline,w
        sublw   0x01            ; Is it line 1?
        btfsc   STATUS,z        ; Is LCDline minus 1 = 0?
        goto    line2End        ; Go if end of second line
; At this point it is the end of the top LCD line
        call    line2           ; Scroll to second line
        clrf    LCDcount        ; Reset counter
        incf    LCDline,f       ; Bump line counter
        goto    scrollExit
; End of second LCD line
line2End:
; Scroll second line to first line. Characters to be
; scrolled are stored in buffer starting at address 0x0c.
; 16 characters are to be moved
; First clear LCD
        call    initLCD
        call    delay_5         ; Make sure not busy
; Set up for data
        bcf     PORTA,E_line    ; E line low
```

```
        bsf       PORTA,RS_line    ; RS line high for data
; Set up counter for 16 characters
        movlw     D'16'                    ; Counter = 16
        movwf     count2
; Get address of storage buffer
        movlw     0x0c
        movwf     FSR                      ; W to FSR
getchar:
        movf      INDF,w   ; get character from display RAM
                           ; location pointed to by file select
                           ; register
        call      send8    ; 4-bit interface routine
; Test for 16 characters displayed
        decfsz    count2,f ; Decrement counter
        goto      nextchar ; Skipped if done
; At this point scroll operation has concluded
        clrf      LCDcount ; Clear counters
; Stay at line 2
        clrf      LCDline
        incf      LCDline,f
        call      line2    ; Set for second line
scrollExit:
        return
nextchar:
        incf      FSR,f    ; Bump pointer
        goto      getchar
;============================
;   clear line buffer
;============================
; Use indirect addressing to store 16 blanks in the
; buffer located at 0x0c
blankBuf:
        Bank0
        movlw     0x0c     ; Pointer to RAM
        movwf     FSR      ; To index register
blank16:
        clrf      INDF     ; Clear memory pointed at by FSR
        incf      FSR,f    ; Bump pointer
        btfss     FSR,4    ; 000x0000 when bit 4 is set
                           ; count reached 16
        goto      blank16
        return
;=======================
; Set address register
;    to LCD line 1
;=======================
; ON ENTRY:
;         Address of LCD line 1 in constant LCD_1
```

```
line1:
        bcf     PORTA,E_line        ; E line low
        bcf     PORTA,RS_line       ; RS line low, set up for
control
        call    delay_5             ; busy?
; Set to second display line
        movlw   LCD_1               ; Address and command bit
        call    send8               ; 4-bit routine
; Set RS line for data
        bsf     PORTA,RS_line       ; Setup for data
        call    delay_5             ; Busy?
; Clear buffer and pointer
        call    blankBuf
        clrf    bufPtr              ; Pointer
        return
;========================
; Set address register
;    to LCD line 2
;========================
; ON ENTRY:
;          Address of LCD line 2 in constant LCD_2
line2:
        bcf     PORTA,E_line        ; E line low
        bcf     PORTA,RS_line       ; RS line low, setup for
control
        call    delay_5             ; Busy?
; Set to second display line
        movlw   LCD_2               ; Address with high-bit set
        call    send8
; Set RS line for data
        bsf     PORTA,RS_line       ; RS = 1 for data
        call    delay_5                     ; Busy?
; Clear buffer and pointer
        call    blankBuf
        clrf    bufPtr              ; Pointer
        return
```

The entire program, named TTYUsart, is found in this book's online software package.

20.5.2 RS-232-C Communications on the 16F87x

The second alternative for *protocol-compliant communications* is using a PIC that provides hardware support for the standard. The 16F84, our workhorse in this book, contains no such facilities. However, other mid-range PICs do provide hardware support for one or several serial communications protocols.

For the examples that follow, we have selected what is perhaps the second most popular PIC of the mid-range family (after the 16F84): the 16F87x. The architecture and basic programming facilities of the 16F87x PIC family were discussed in Chapter 6. At this time, we should recall that 16F87x includes the PIC 16F873, 16F874, 16F876, and 16F877. For our sample programs we have selected the 16F877 because it is the most powerful one in the group. The 16F877 has an operating frequency of up to 20 MHz, 8K of flash program memory, 368 bytes of data memory, 256 bytes of EEPROM, five input/output ports, and two modules for serial communications: a Master Synchronous Serial Port and a Universal Synchronous/Asynchronous Receiver and Transmitter. We focus on the USART module and leave the MSSP for the chapter on EEPROM programming.

16F87x USART Module

The Universal Synchronous Asynchronous Receiver Transmitter (USART) module in the 16F87X family is also known as a *Serial Communications Interface*, or *SCI*. The USART module is useful in communicating with devices and systems that support RS-232-C communications, including computers and terminals. It can be configured as an asynchronous full-duplex device, as a synchronous half-duplex master, or as a synchronous half-duplex slave. In the synchronous mode, the USART module is used mostly in communicating with analog-to-digital and digital-to-analog integrated circuits or for accessing serial EEPROM. Both of these functions were discussed in Chapter 14.

Five registers relate to USART operation in the 16F877: RCSTA, TXREG, RCREG, TXSTA, and SPBRG. The first three are located in bank 0 and the second two in bank 1. TXSTA is the Transmit Status and Control register and RCSTA is the Receive Status and Control register. Figure 20-14 shows the bitmap for the TXSTA register located at address 0x98 in bank 1.

The RCSTA register contains control and status bits for the *receive* function. The register is found at address 0x18 in bank 0. Figure 20-15 is a bitmap of the RCSTA register.

USART Baud Rate Generator

In the USART emulation programs for the 16F84, we were forced to approximate the RS-232-C baud rate with the system clock. The USART module in the 16F87X PICs contains its own baud rate generator, but it is also dependent on the system clock.

Setting the baud rate in the USART module consists of manipulating the *Baud Rate Generator* (BRG) unit. The BRG is a dedicated 8-bit generator that supports both the asynchronous and synchronous modes. The SPBRG is an 8-bit register that controls the rate of a dedicated timer. In the asynchronous mode, the bit labeled BRGH in the TXSTA register (see Figure 20-14) also relates to the baud rate because it allows setting either a slow-speed or a high-speed baud rate. The baud-rate-speed-select bit is inactive in the synchronous mode.

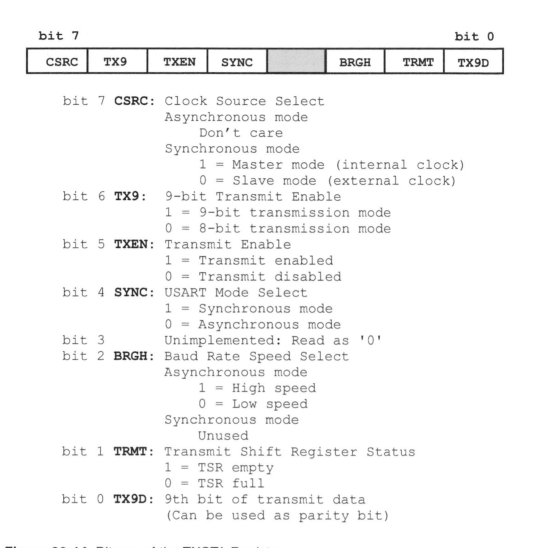

bit 7   **bit 0**

CSRC	TX9	TXEN	SYNC		BRGH	TRMT	TX9D

```
      bit 7 CSRC: Clock Source Select
                  Asynchronous mode
                      Don't care
                  Synchronous mode
                      1 = Master mode (internal clock)
                      0 = Slave mode (external clock)
      bit 6 TX9:  9-bit Transmit Enable
                  1 = 9-bit transmission mode
                  0 = 8-bit transmission mode
      bit 5 TXEN: Transmit Enable
                  1 = Transmit enabled
                  0 = Transmit disabled
      bit 4 SYNC: USART Mode Select
                  1 = Synchronous mode
                  0 = Asynchronous mode
      bit 3       Unimplemented: Read as '0'
      bit 2 BRGH: Baud Rate Speed Select
                  Asynchronous mode
                      1 = High speed
                      0 = Low speed
                  Synchronous mode
                      Unused
      bit 1 TRMT: Transmit Shift Register Status
                  1 = TSR empty
                  0 = TSR full
      bit 0 TX9D: 9th bit of transmit data
                  (Can be used as parity bit)
```

Figure 20-14 Bitmap of the TXSTA Register.

The formula for computing the baud rate takes into account the *system oscillator speed* (Fosc); the setting of the *Baud-Rate-Speed-Select* bit (BRGH), which is set for the high-speed mode and cleared for slow-speed, and also the setting of the SYNC bit in the TXSTA register, which selects either asynchronous or synchronous mode. The formula is as follows:

$$ABR = \frac{Fosc}{S(x+1)},$$

where *ABR* represents the Asynchronous Baud Rate, x is the value in the SPRGB register (range 0 to 255), S is 64 in the high-speed mode (BRGH bit is 1) and 16 in the slow speed mode (BRGH bit is 0). Solving the formula in terms of the value to be placed in the SPRGB register, we get

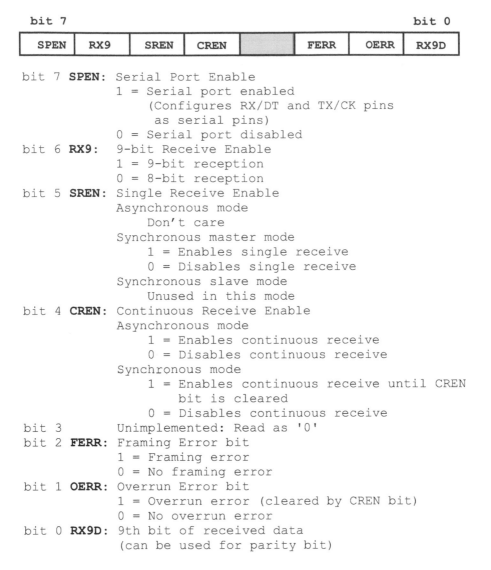

bit 7   bit 0

SPEN	RX9	SREN	CREN		FERR	OERR	RX9D

bit 7 **SPEN**: Serial Port Enable
              1 = Serial port enabled
                  (Configures RX/DT and TX/CK pins
                   as serial pins)
              0 = Serial port disabled
bit 6 **RX9**: 9-bit Receive Enable
              1 = 9-bit reception
              0 = 8-bit reception
bit 5 **SREN**: Single Receive Enable
              Asynchronous mode
                  Don't care
              Synchronous master mode
                  1 = Enables single receive
                  0 = Disables single receive
              Synchronous slave mode
                  Unused in this mode
bit 4 **CREN**: Continuous Receive Enable
              Asynchronous mode
                  1 = Enables continuous receive
                  0 = Disables continuous receive
              Synchronous mode
                  1 = Enables continuous receive until CREN
                     bit is cleared
                  0 = Disables continuous receive
bit 3   Unimplemented: Read as '0'
bit 2 **FERR**: Framing Error bit
              1 = Framing error
              0 = No framing error
bit 1 **OERR**: Overrun Error bit
              1 = Overrun error (cleared by CREN bit)
              0 = No overrun error
bit 0 **RX9D**: 9th bit of received data
              (can be used for parity bit)

Figure 20-15 Bitmap of the RCSTA Register.

$$ABR = \frac{Fosc}{S(x+1)}$$

For example, to calculate the setting of the SPRGB register for 9,600 baud, with a 16-MHz oscillator, at the high-speed rate ($S = 64$) the equation becomes:

$$x = \left(\frac{16,000,000}{9,600 \cdot 64} \right) - 1 = 25.042 \approx 25$$

In this case, the value to store in the SPRGB register is 25. The actual baud rate can now be calculated using the first equation, as follows:

$$ABR = \frac{16,000,000}{64 \cdot (25+1)} = 9615.38$$

The percent error in the baud rate can be estimated by dividing the difference between the desired and the actual baud rate by the desired baud rate. The percent error is 0.16.

16F87x USART Asynchronous Transmitter

The USART in the 16F87x PICs uses a *nonreturn-to-zero format*, consisting of one start bit, eight or nine data bits, no parity, and one stop bit. In compliance with RS-232-C, the USART transmits and receives the least-significant bit first. Transmitter and receiver units are functionally independent but use the same data format and baud rate.

Although parity is not directly supported by the hardware, it can be implemented in software using the ninth data bit. Figure 20-16 shows the 16F87x registers related to asynchronous transmission.

REGISTER NAME	7	6	5	4	3	2	1	0 bits
TXSTA		TX9	TXEN	SYNC		BRGH	TRMT	TX9D
RCSTA	SPEN							
TXREG	TX7	TX6	TX5	TX4	TX3	TX2	TX1	TX0
PIR1				TXIF				
PIE1				TXIE				
SPBRG	(Baud Rate Generator)							
INTCON	GIE	PEIE						

Figure 20-16 16F87x Registers Used in Asynchronous Transmission.

The transmitter function also uses the *Transmit Shift register* (TSR), which is not mapped in memory and is thus not accessible to code. TSR obtains its data from the read/*write transmit buffer*, named *TXREG*, which is loaded in software after the stop bit is received. Then TXREG transfers the data to TSR and becomes empty. At this time the TXIF flag bit is set. An interrupt related to the TXIF bit is enabled/disabled by setting/clearing the TXIE enable bit in the PIE1 register. However, the TXIF flag bit is set regardless of the state of the TXIE enable bit. The TXIF flag is reset automatically when new data is loaded into TXREG.

While the TXIF flag indicates the status of TXREG, the TRMT bit, in TXSTA, reflects the status of TSR. TRMT is set when TSR is empty. This is a read-only bit. No interrupts are linked to the TRMT bit, so the program has to poll this bit to determine if TSR is empty. Transmission is enabled by setting the TXEN bit in TXSTA. The actual transmission does not occur until TXREG is loaded with data and the *baud rate generator* (BRG) has produced a clock beat. Alternatively, transmission can be started by loading TXREG and then setting the TXEN enable bit.

When transmission starts, the (not accessible) TSR register usually is empty. Thereafter, transferring data to TXREG results in a transfer to TSR, which then produces an empty TXREG. This mechanism makes possible the *back-to-back transfer*. Clearing the TXEN enable bit during transmission aborts the transmission. This action also resets the transmitter and sets the TX/CK pin high.

16F87x USART Asynchronous Receiver

When Asynchronous mode is selected by setting the SYNC bit in TXSTA, then reception can be enabled by setting the CREN bit in the RCSTA register. Figure 20-17 shows the registers related to asynchronous reception.

REGISTER NAME	7	6	5	4	3	2	1	0	bits
TXSTA				SYNC		BRGH			
RCSTA	SPEN	RX9		CREN		FERR	OERR	RX9D	
RCREG	RX7	RX6	RX5	RX4	RX3	RX2	RX1	RX0	
PIR1			RCIF						
PIE1			RCIE						
SPBRG	(Baud Rate Generator)								
INTCON	GIE	PEIE							

Figure 20-17 Registers Used in Asynchronous Reception.

The main operational register is the *RSR (Receive Shift Register)*, which, like TSR, is not accessible to application software. As soon as the stop bit is detected in the RX/TX pin, the received data in RSR is transferred to RCREG if it is empty. In this case, the RCIF flag bit is set. The interrupt linked to the RCIF flag is enabled or disabled by means of the RCIE in the PIE1 register. The RCIF flag bit is read-only and can be cleared only by hardware; this happens when the RCREG register has been read and is empty.

RCREG is *double-buffered*, meaning that it is possible for two bytes of data to be started simultaneously while a third byte begins shifting to RSR. If the stop bit is detected while RCREG is not empty, then the *overrun error bit* (OERR) is set in RCSTA. RCREG operates in *first-in-first-out order*. When it is read twice, the two bytes are retrieved in this order.

The *overrun error bit* (OERR) inhibits transfer from RSR into RCREG; therefore, it is important to clear this bit once the error is detected. The *framing error bit* (FERR) in the RCSTA register is set if a stop bit is not detected.

The following steps are followed in initializing and executing asynchronous reception:

1. The SPBRG register is set up for the selected baud rate.
2. Asynchronous reception is enabled by clearing the SYNC bit in the TXSTA register and setting the SPEN bit in the RCSTA register.
3. To enable the receive data interrupt, the RCIE, GIE, and PEIE bits must be set.
4. Reception is activated by setting the CREN bit in RCSTA.
5. When reception has concluded, the RCIF bit in the PIE1 register is set. At that time, an interrupt is generated if the RCIE bit was set.
6. Received data is retrieved by reading RCREG.
7. If any error occurred, the CREN bit must be cleared.

PIC-to-PC RS-232-C Communications Circuit

To demonstrate serial communications with the RS-232-C protocol we developed a circuit consisting of a 4-by-4 keypad and a 2-line by 20-character LCD display. Characters typed on the keypad are converted to ASCII codes for the hexadecimal digit set, that is, the numeral digits and the letters A through F. When a key is pressed, the corresponding ASCII code is displayed in the LCD and transmitted through the serial port to a PC application. Characters received though the serial line are displayed on the LCD. Figure 20-18 is a wiring diagram of the circuit.

The program SerComLCD demonstrates the circuit in Figure 20-18.

16F877 PIC Initialization Code

The following code fragment shows the initialization of the UART module in the 16F877 PIC for 2400 baud, 8 bits, no parity, and one stop bit. No interrupts are used in this example.

```
;=================================================================
;               USART initialization procedure
;=================================================================
; Initialize serial port for 2400 baud, 8 bits, no parity,
; 1 stop
InitSerial:
        Bank1               ; Macro to select bank1
; Bits 6 and 7 of Port C are multiplexed as TX/CK and RX/DT
```

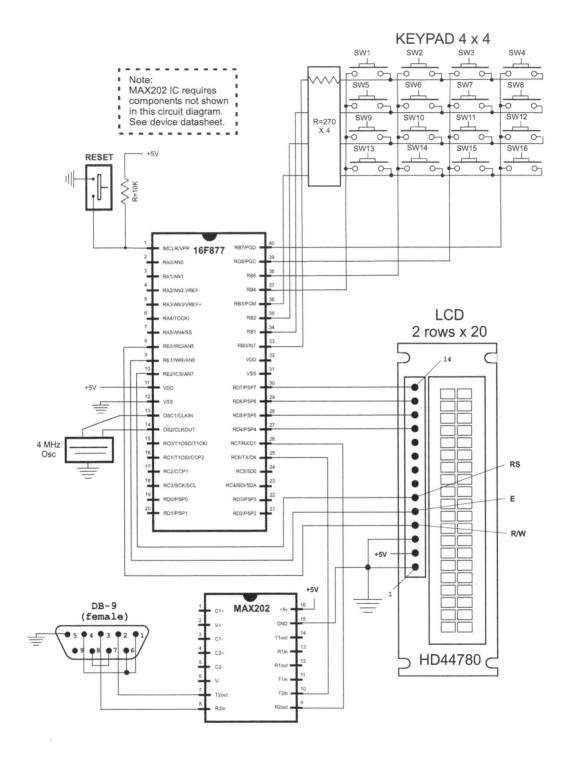

Figure 20-18 USART Communications Circuit with PIC 16F877.

```
; for USART operation. These bits must be set to input in the
; TRISC register
          movlw    b'11000000'       ; Bits for TX and RX
          iorwf    TRISC,f           ; OR into Trisc register
; The asynchronous baud rate is calculated as follows:
;                        Fosc
;              ABR = ---------
;                       S*(x+1)
; Where x is the value in the SPBRG register and S is 64 if the
; high baud rate select bit (BRGH) in the TXSTA control register
; is clear, and 16 if the BRGH bit is set. For setting to 9600
; baud using a 4 MHz oscillator at a high-speed baud rate the
; formula is:
;            4,000,000   4,000,000
;            ---------   --------- = 9,615 baud (0.16% error)
;            16*(25+1)      416
;
; At slow speed (BRGH = 0)
;            4,000,000   4,000,000
;            ---------   --------- = 2,403.85 (0.16% error)
;            64*(25+1)     1,664
;
          movlw    spbrgVal          ; Value in spbrgVal = 25
          movwf    SPBRG             ; Place in baud rate generator
;
; TXSTA (Transmit Status and Control Register) bitmap:
;   7  6  5  4  3  2  1  0  <== bits
;   |  |  |  |  |  |  |  |  |_____ TX9D 9th data bit on
;   |  |  |  |  |  |  |  |          ? (used for parity)
;   |  |  |  |  |  |  |  |_____ TRMT Transmit Shift Register
;   |  |  |  |  |  |  |            1 = TSR empty
;   |  |  |  |  |  |  |          * 0 = TSR full
;   |  |  |  |  |  |  |_____ BRGH High Speed Baud Rate
;   |  |  |  |  |  |               (Asynchronous mode only)
;   |  |  |  |  |  |               1 = high speed (* 4)
;   |  |  |  |  |  |             * 0 = low speed
;   |  |  |  |  |_____ NOT USED
;   |  |  |  |_____ SYNC USART Mode Select
;   |  |  |                1 = syncrhonous mode
;   |  |  |              * 0 = asynchronous mode
;   |  |  |_____ TXEN Transmit Enable
;   |  |                  * 1 = transmit enabled
;   |  |                    0 = transmit disabled
;   |  |_____ TX9 Enable 9-bit Transmit
;   |                        1 = 9-bit transmission mode
;   |                      * 0 = 8-bit mode
;   |_____ CSRC Clock Source Select
;                           Not used in asynchronous mode
```

```
;                                       Synchronous mode:
;                                           1 = Master Mode (internal clock)
;                                         * 0 = Slave mode (external clock)
; Setup value: 0010 0000 = 0x20
        movlw   0x20              ; Enable transmission and high
baud rate
        movwf   TXSTA
        Bank0                 ; Bank 0
; RCSTA (Receive Status and Control Register) bitmap:
;    7  6  5  4  3  2  1  0  <== bits
;    |  |  |  |  |  |  |  |  |_____ RX9D 9th data bit received
;    |  |  |  |  |  |  |  |          ? (can be parity bit)
;    |  |  |  |  |  |  |  |_____ OERR Overrun errror
;    |  |  |  |  |  |  |            ? 1 = error (cleared by software
;    |  |  |  |  |  |  |_____ FERR Framing Error
;    |  |  |  |  |  |               ? 1 = error
;    |  |  |  |  |  |_____ NOT USED
;    |  |  |  |  |_____ CREN Continuous Receive Enable
;    |  |  |  |                   Asynchronous mode:
;    |  |  |  |                   *    1 = Enable continuous receive
;    |  |  |  |                        0 = Disables continuous receive
;    |  |  |  |                   Synchronous mode:
;    |  |  |  |                        1 = Enables until CREN cleared
;    |  |  |  |                        0 = Disables continuous receive
;    |  |  |_____ SREN Single Receive Enable
;    |  |                     ? Asynchronous mode =  don't care
;    |  |                       Synchronous master mode:
;    |  |                         1 = Enable single receive
;    |  |                         0 = Disable single receive
;    |  |_____ RX9 9th-bit Receive Enable
;    |                       1 = 9-bit reception
;    |                     * 0 = 8-bit reception
;    |_____ SPEN Serial Port Enable
;                          * 1 = RX/DT and TX/CK are serial pins

;                          0 = Serial port disabled
; Setup value: 1001 0000 = 0x90
        movlw   0x90     ; Enable serial port and continuous
                         ; reception
        movwf   RCSTA
;
        clrf    errorFlags ; Clear local error flags register
        Return
```

USART Receive and Transmit Routines

The transmit data routine is quite simple. Code checks the TXIF bit in PIR1. If the bit is set, data is transmitted by storing the data byte in TXREG. The following procedure performs the required operations:

```
;===============================
;        transmit data
;===============================
; Test for Transmit Register Empty and transmit data in w
SerialSend:
        Bank0                       ; Select bank 0
busyWait:
        btfss   PIR1,TXIF          ; check if transmitter busy
        goto    busyWait ; wait until transmitter is not busy
        movwf   TXREG              ; and transmit the data
        return
```

Receiving data is more complicated than transmitting it. One of the reasons is that code must test for and handle several possible errors that can occur during reception. The following code fragment shows the local variables and processing required for simple data reception:

```
;=====================================================
;              variables in PIC RAM
;=====================================================
; Local variables
        cblock  0x20                ; Start of block
        .
        .
        .
; Communications variables
        newData                ; not 0 if new data received
        ascVal
        errorFlags
        endc


;=============================================================
;              USART receive data procedure
;=============================================================
; Procedure to test line for data received and return value
; in w. Overrun and framing errors are detected and
; remembered in the variable errorFlags, as follows:
;       7  6  5  4  3  2  1  0   <== errorFlags
;       |-- not used --- |  |___ overrun error
;                        |_____ framing error
SerialRcv:
        clrf    newData ; Clear new data received register
        Bank0           ; Select bank 0
; Bit 5 (RCIF) of the PIR1 Register is clear if the USART
```

```
; receive buffer is empty. If so, no data has been received
        btfss    PIR1,RCIF         ; Check for received data
        return                     ; Exit if no data
; At this point data has been received. First eliminate
; possible errors: overrun and framing.
; Bit 1 (OERR) of the RCSTA register detects overrun
; Bit 2 (FERR) of the RCSTA register detects framing error
        btfsc    RCSTA,OERR        ; Test for overrun error
        goto     OverErr ; Error handler
        btfsc    RCSTA,FERR        ; Test for framing error
        goto     FrameErr ; Error handler
; At this point no error was detected
; Received data is in the USART RCREG register
        movf     RCREG,w ; get received data
        bsf      newData,7   ; Set bit 7 to indicate new data
; Clear error flags
        clrf     errorFlags
        return
;===========================
;     error handlers
;===========================
; Overrun error detected
OverErr:
        bsf      errorFlags,0     ; Bit 0 is overrun error
; Reset system
errExit:
        bcf      RCSTA,CREN        ; Clear continuous receive bit
        bsf      RCSTA,CREN        ; Set to re-enable reception
        return
; Error. FERR framing error bit is set
FrameErr:
        bsf      errorFlags,1     ; Bit 1 is framing error
        movf     RCREG,W          ; Read and throw away bad data
        goto     errExit
```

The procedures listed previously are from the program SerComLCD in the book's online software. The applicable circuit is shown in Figure 20-18.

USART Receive Interrupt

Polled routines for serial communications are adequate when the application does little else but check transmission lines. If the application has other tasks to perform, polled routines can waste processing time and even lose data. In this sense, the *send* function is usually less critical. An application can typically determine when to send data and have available all the data when the send operation activates. This is often not the case in receiving data, especially in applications that execute full-duplex.

A practical solution is to use interrupts for receiving characters through the serial line. The 60F87x includes facilities for implementing interrupt routines by both the send and the receive functions. To enable interrupts for the USART receive operation, the following preparatory steps are necessary:

- • 1. Peripheral and global interrupts must be enabled by setting bits 6 and 7 of the INTCON register.
- • 2. The receive interrupt must be enabled by setting the RCIF bit in the PIE1 register.

The handler for the serial reception interrupt usually performs the following functions:

1. The context is saved. This includes, but is not limited to, the status register, the w register, the PCLATH register, and the FSR register.

2. Code tests for received data by checking the RCIF bit in the PIR1 register. If this bit is clear, the interrupt did not originate in received data.

3. Code can also check if the *interrupt enable bit* (RCIE) is set in the RCIE register. If not enabled, the interrupt is related to serial data.

4. The handler usually checks two possible errors during reception: *overflow* and *framing error*. The first one by checking the OERR bit and the second one by checking the FERR bit, both in the RCSTA register. If reception errors have taken place, the handler takes appropriate action.

5. If no error is detected, then the received data can be retrieved from the RCREG.

6. On exit, the interrupt handler restores the context and issues the **retfie** instruction.

The following code fragment lists the variables and processing routine for an interrupt handler for serial data reception:

```
=========================================================
;                  variables in PIC RAM
;========================================================
; Local variables
        cblock   0x20                ; Start of block
           .
           .
           .
; Communications variables
        errorFlags
; Temporary storage used by interrupt handler
        tempW
        tempStatus
        tempPclath
        tempFsr
        endc
;==============================================================
;==============================================================
;          interrupt handler for received characters
;==============================================================
;==============================================================
IntServ:
```

```
        movwf     tempW    ; Save W
        movf      STATUS,W ; Store STATUS in W
        clrf      STATUS   ; Select bank0
        movwf     tempStatus          ; Save STATUS
        movf      PCLATH,W ; Store PCLATH in W
        movwf     tempPclath          ; Save PCLATH
        clrf      PCLATH   ; Select program memory page 0
        movf      FSR,W    ; Store FSR in W
        movwf     tempFsr  ; Save FSR value
; Test for received data interrupt
        Bank0               ; select bank0
;   7  6  5  4  3  2  1  0  <= PIR1
;           |_____ (RCIF) USART receive interrupt
;                               flag
        Btfsc     PIR1,RCIF          ; Test bit 5
        bsf       STATUS,RP0         ; Bank 1 if RCIF set
;   7  6  5  4  3  2  1  0  <= PIE1
;           |_____ (RCIE) Receive interrupt enable
;                               bit
        btfss     PIE1,RCIE          ; Test if interrupt is enabled
        goto      IntExit ; Go if not enabled
;===============================
;     received data
;===============================
; Routine to handler received data. Overrun and framing
; errors are detected and remembered in the variable
; errorFlags, as follows:
;       7  6  5  4  3  2  1  0   <== errorFlags
;       | -- not used --  |  |___ overrun error
;                            |_____ framing error
        Bank0                         ; Select bank 0
; Test for overrun and framing errors.
; Bit 1 (OERR) of the RCSTA register detects overrun
; Bit 2 (FERR) of the RCSTA register detects framing error
        btfsc     RCSTA,OERR         ; Test for overrun error
        goto      OverErr ; Error handler
        btfsc     RCSTA,FERR         ; Test for framing error
        goto      FrameErr ; Error handler
; At this point no error was detected
; Received data is in the USART RCREG register
        movf      RCREG,w  ; Received data into w
; Clear error flags
        clrf      errorFlags
        goto      IntExit
;=========================
;     error handlers
;=========================
; Errors are returned as bits in the errorFlags register
```

```
;   7  6  5  4  3  2  1  0  <= errorFlags
;   |- not used ---|  |  |____ overrun error
;                     |_____ framing error
; Error responses to be made by main code
OverErr:
        bsf       errorFlags,0      ; Bit 0 is overrun error
; Reset system
        bcf       RCSTA,CREN        ; Clear continuous receive bit
        bsf       RCSTA,CREN        ; Set to re-enable reception
        goto      IntExit
FrameErr:
        bsf       errorFlags,1; Bit 1 is framing error
        movf      RCREG,W  ; Read and throw away bad data
;==============================
;     interrupt handler exit
;==============================
IntExit:
        Bank0
        movf      tempFsr,w         ; Recover FSR value
        movwf     FSR               ; Restore in register
        movf      tempPclath,w      ; Recover PCLATH value
        movwf     PCLATH            ; Restore in register
        movf      tempStatus,W      ; Recover STATUS
        movwf     STATUS            ; Restore in register
        swapf     tempW,F           ; Swap file register in itself
        swapf     tempW,W           ; Restore in register
        retfie
```

The program SerIntLCD in this book's online software, is an Interrupt-Driven demonstration for the circuit in Figure 20-18.

20.6 Demonstration Programs

The sample programs listed in the following sections refer to the programming discussed in this chapter.

20.6.1 SerialSnd Program

```
; File name: SerialSnd.asm
; Date: May 5, 2011
; Authors: Sanchez and Canton
; Processor: 16F84A
;
; Description:
; Two programs to exercise serial communications between
; two PIC 16F84A both running at 4 MHz. One program sends
; data through a single line and the other one receives
; it. This program is the sender.
;
```

```
; Circuit:
;        Port A1 is the serial transmission line.
;        Port A2 is an active-low pushbutton switch that
;                serves to initiate communications.
;        Port A3 is a LED that is ON when the program is
;                ready to send data. Once data starts
;                being sent the LED is turned OFF.
;        Port-B0-B7 is a 8 x toggle switch that provides
;                the data byte to be sent.
;        A pushbutton swtich is in the 16F84 RESET line
;                and serves to restart the program.
;
; Communications parameters:
;        Timer channel TMR0 is used for synchronizing data
;        transmission. The timer runs at the maximum rate of
;        256 cycles per iteration. In a 4 MHz system the
;        timer rate is 1 MHz, thus the bit rate is
;                        1,000,000/256
;        which is approximately 3,906 microseconds per bit.
;
;===========================
;        switches
;===========================
; Switches used in __config directive:
;    _CP_ON          Code protection ON/OFF
; *  _CP_OFF
; *  _PWRTE_ON       Power-up timer ON/OFF
;    _PWRTE_OFF
;    _WDT_ON         Watchdog timer ON/OFF
; *  _WDT_OFF
;    _LP_OSC         Low power crystal oscillator
; *  _XT_OSC         External parallel crystal oscillator
;    _HS_OSC         High speed crystal resonator (8 to 10 MHz)
;                    Resonator: Murate Erie CSA8.00MG = 8 MHz
;    _RC_OSC         Resistor/capacitor oscillator
; |                  (simplest, 20% error)
; |
; |_____ * indicates setup values presently selected

;=========================
; setup and configuration
;=========================
        processor 16f84A
        include   <p16f84A.inc>
        __config  _XT_OSC & _WDT_OFF & _PWRTE_ON & _CP_OFF
;==============================================================
;                       M A C R O S
;==============================================================
```

```
; Macros to select the register banks
Bank0    MACRO                        ; Select RAM bank 0
         bcf      STATUS,RP0
         ENDM


Bank1    MACRO                        ; Select RAM bank 1
         bsf      STATUS,RP0
         ENDM
;=======================================================
;         constant definitions for pin wiring
;=======================================================
#define readySW  2          ; |
#define readyLED 3          ; | => from wiring diagram
#define serialLN 1          ; |
;=======================================================
;              PIC register flag equates
;=======================================================
c        equ      0         ; Carry flag
tmrOVF   equ      2         ; Timer overflow bit
;=======================================================
;              variables in PIC RAM
;=======================================================
         cblock   0x0d      ; Start of block
         bitCount ; Counter for 8 bits
         dataReg            ; Data to send
         endc


;=========================================================
;                            program
;=========================================================
         org      0         ; start at address
         goto     main
; Space for interrupt handlers
         org      0x04

main:
; Port-A, bit 2 is input. Rest is output
         Bank1
         movlw    b'00000100'        ; Port-A bit 2 is input
                                     ; all others are output
         movwf    TRISA
; Port-B is all input
         movlw    b'11111111'
         movwf    TRISB
         Bank0
         bsf      PORTA,1            ;Marking bit
; Prepare to set prescaler
         clrf     TMR0
```

```
        clrwdt
; Setup OPTION register for full timer speed
        movlw    b'11011000'
;   1  1  0  1  1  0  0  0 <= OPTION bits
;   |  |  |  |  |  |__|__|_____ PS2-PS0 (prescaler bits)
;   |  |  |  |  |  |              Values for Timer0
;   |  |  |  |  |  |             *000 = 1:2    001 = 1:4
;   |  |  |  |  |  |              010 = 1:8    011 = 1:16
;   |  |  |  |  |  |              100 = 1:32   101 = 1:64
;   |  |  |  |  |  |              110 = 1:128 111 = 1:256
;   |  |  |  |  |_____ PSA (prescaler assign)
;   |  |  |  |  |                *1 = to WDT
;   |  |  |  |  |                 0 = to Timer0
;   |  |  |  |_____ TOSE (Timer0 edge select)
;   |  |  |                       0 = increment on low-to-high
;   |  |  |                      *1 = increment on high-to-low
;   |  |  |_____ TOCS (TMR0 clock source)
;   |  |                         *0 = internal clock
;   |  |                          1 = RA4/TOCKI bit source
;   |  |_____ INTEDG (Edge select)
;   |                            0 = falling edge
;   |                           *1 = rising edge
;   |_____ RBPU pullups
;                                0 = enabled
;                               *1 = disabled
        option
; Dissable interrupts
        bcf      INTCON,5        ; Timer0 overflow disabled
        bcf      INTCON,7        ; Global interrupts disabled
; Turn on ready LED
        bsf      PORTA,3         ; LED on
;============================
;   wait for READY switch
;       to be pressed
;============================
ready2send:
        btfsc    PORTA,readySW
        goto     ready2send
;============================
;     send serial data
;============================
; At this point program proceeds to send data through
; the serial port line
; Turn off LED
        bcf      PORTA,readyLED
; Read switches and store in local variable
        movf     PORTB,w
        movwf    dataReg
```

```
;===========================
;    call serial output
;        procedure
;===========================
         call     sendData ; call serial output procedure
;===========================
;        wait forever
;===========================
endloop:
         goto     endloop


;=============================================================
;                procedure to send serial data
;=============================================================
; ON ENTRY:
;        local variable dataReg holds 8-bit value to be
;        transmitted through port labeled serialLN
; OPERATION:
;      1. The timer at register TMR0 is set to run at
;         maximum clock speed, that is, 256 clock beats.
;         The timer overflow flag in the INTCON register
;         is set when the timer cycles from 0xff to 0x00.
;      2. Each bit (start, data, and stop bits) is sent
;         at a rate of 256 timer beats. That is, each bit is
;         held high or low for one full timer cycle (256
;         clock beats).
;      3. The procedure tests the timer overflow flag
;         (tmrOVF) to determine when the timer cycle has
;         ended, that is, when 256 clock beats have passed.
;
sendData:
         movlw    0x08              ; Set up shift counter
         movwf    bitCount
;=======================
;    send START bit
;=======================
; Set line low then hold for 256 timer clock beats.
         bcf      PORTA,serialLN    ; Send start bit
; First reset timer
         clrf     TMR0              ; Reset timer counter
         bcf      INTCON,tmrOVF     ; Reset TMR0 overflow flag
; Wait for 256 timer clock beats
startBit:
         btfss    INTCON,tmrOVF     ; timer overflow?
         goto     startBit ; Wait until set
; At this point timer has cycled. Start bit has ended
         bcf      INTCON,tmrOVF     ; Clear overflow flag
;=======================
```

```
;      send 8 DATA bits
;==========================
; Eight data bits are sent through the serial line
; starting with the high-order bit. The data byte is
; stored in the register named dataReg. The bits are
; rotated left to the carry flag. Code assumes the bit
; is zero and sets the serial line low. Then the carry
; flag is tested. If the carry is set the serial line
; is changed to high. The line is kept low or high for
; 256 timer beats.
send8:
          rlf       dataReg,f          ; bit into carry flag
          bcf       PORTA,serialLN     ; 0 to serial line
; Code can assume the bit is a zero and set the line
; low because, if low is the wrong state, it will only
; remain for two timer beats. The receiver will not
; check the line for data until 128 timer beats have
; elapsed, so the error will be harmless. In any case,
; there is no assurance that the previous line state is
; the correct one, so leaving the line in its previous
; state could also be wrong.
          btfsc     STATUS,c           ; test carry flag
          bsf       PORTA,serialLN     ; bit is set. Fix error.
bitWait:
          btfss     INTCON,tmrOVF      ; Timer cycled?
          goto      bitWait            ; not yet
; At this point timer has cycled.
; Test for end of byte, if not, send next bit
          bcf       INTCON,tmrOVF      ; clear overflow flag
          decfsz    bitCount,f         ; Last bit?
          goto      send8              ; not yet
;==========================
;     hold MARKING state
;==========================
; All 8 data bits have been sent. The serial line must
; now be held high (MARKING) for one clock cycle
          bsf                 PORTA,serialLN    ; Marking state
markWait:
          btfss     INTCON,tmrOVF      ; Done?
          goto      markWait           ; not yet
;==========================
;    end of transmission
;==========================
          return                       ; done


;============================================================
;                         end of program
;============================================================
```

```
            end


20.6.2  SerialRcv Program

;  File name: SerialRcv.asm
;  Date: May 6, 2011
;  Authors: Canton and Sanchez
;  Processor: 16F84A
;
;  Description:
;  Two programs to exercise serial communications between
;  two PIC 16F84A both running at 4 MHz. One program sends
;  data through a single line and the other one receives
;  it. This program is the receiver.
;
;  Circuit:
;         Port A0 is the serial transmission line
;         Port A2 is an active-low pushbutton switch that
;                 serves to initiate communications.
;         Port A3 is an LED that is ON when the program is
;                 ready to receive data. Once data starts
;                 being received the LED is turned OFF.
;         Port-B0-B7 are 8 LEDs that display the data bits
;                 that have been received.
;         A pushbutton switch is in the 16F84 RESET line
;                 and serves to restart the program.
;
;  Communications parameters:
;         Timer channel TMR0 is used for synchronizing data
;         transmission. The timer runs at the maximum rate of
;         256 cycles per iteration. In a 4 MHz system the
;         timer rate is 1 MHz, thus the bit rate is
;                         1,000,000/256
;         which is approximately 3,906 microseconds per bit.
;
;         Upon receiving the START bit, the program waits for
;         one half a clock cycle (128 timer beats) to
;         synchronize with the sender.
;============================
;         switches
;============================
;  Switches used in __config directive:
;    _CP_ON        Code protection ON/OFF
;  * _CP_OFF
;  * _PWRTE_ON     Power-up timer ON/OFF
;    _PWRTE_OFF
;    _WDT_ON       Watchdog timer ON/OFF
;  * _WDT_OFF
```

```
;    _LP_OSC          Low power crystal oscillator
; *  _XT_OSC          External parallel crystal oscillator
;    _HS_OSC          High speed crystal resonator (8 to 10 MHz)
;                     Resonator: Murate Erie CSA8.00MG = 8 MHz
;    _RC_OSC          Resistor/capacitor oscillator
;  |                  (simplest, 20% error)
;  |
;  |_____ * indicates setup values presently selected


;==========================
; setup and configuration
;==========================
        processor 16f84A
        include   <p16f84A.inc>
        __config  _XT_OSC & _WDT_OFF & _PWRTE_ON & _CP_OFF
;===================================================================
;                       M A C R O S
;===================================================================
; Macros to select the register banks
Bank0   MACRO                   ; Select RAM bank 0
        bcf     STATUS,RP0
        ENDM


Bank1   MACRO                   ; Select RAM bank 1
        bsf             STATUS,RP0
        ENDM
;=======================================================
;           constant definitions for pin wiring
;=======================================================
#define readySW  2         ;|
#define readyLED 3         ;| => from wiring diagram
#define serialLN 0         ;|
;=======================================================
;           PIC register and flag equates
;=======================================================
c       equ     0       ; Carry flag
tmrOVF  equ     2       ; Timer overflow bit
;
;=======================================================
;           variables in PIC RAM
;=======================================================
        cblock  0x0c    ; Start of block
        bitCount        ; Counter for 8 bits
        rcvReg                  ; Data to send
        temp
        endc
;=======================================================
;                       program
```

```
;===========================================================
          org       0              ; start at address
          goto      main
; Space for interrupt handlers
          org       0x04

main:
          Bank1
; Port-A bits 0 and 2 are input. All others are output
          movlw     b'00000101' ; Port-A setup
          movwf     TRISA
; Port-B is all output
          movlw     b'00000000' ; Port-B setup
          MOVWF     TRISB
          Bank0
; Turn off all Port-B LEDs
          clrf      PORTB
; And receiver register
          clrf      rcvReg
; Prepare to set prescaler
          clrf      TMR0
          clrwdt
; Setup OPTION register for full timer speed
          movlw     b'11011000'
;    1  1  0  1  1  0  0  0 <= OPTION bits
;    |  |  |  |  |  |  |__|__|_____ PS2-PS0 (prescaler bits)
;    |  |  |  |  |  |                Values for Timer0
;    |  |  |  |  |  |                *000 = 1:2    001 = 1:4
;    |  |  |  |  |  |                 010 = 1:8    011 = 1:16
;    |  |  |  |  |  |                 100 = 1:32  101 = 1:64
;    |  |  |  |  |  |                 110 = 1:128 111 = 1:256
;    |  |  |  |  |  |_____ PSA (prescaler assign)
;    |  |  |  |  |                  *1 = to WDT
;    |  |  |  |  |                   0 = to Timer0
;    |  |  |  |  |_____ TOSE (Timer0 edge select)
;    |  |  |  |                       0 = increment on low-to-high
;    |  |  |  |                      *1 = increment on high-to-low
;    |  |  |  |_____ TOCS (TMR0 clock source)
;    |  |  |                          *0 = internal clock
;    |  |  |                           1 = RA4/TOCKI bit source
;    |  |  |_____ INTEDG (Edge select)
;    |  |                               0 = falling edge
;    |  |                              *1 = rising edge
;    |  |_____ RBPU pullups
;    |                                   0 = enabled
;    |                                  *1 = disabled
          option
; Disable interrupts
```

```
        bcf       INTCON,5    ; Timer0 overflow disabled
        bcf       INTCON,7    ; Global interrupts disabled
;==========================
;  wait for READY switch
;      to be pressed
;==========================
ready2rcv:
        btfsc     PORTA,readySW    ; Test switch
        goto      ready2rcv        ; loop
; Turn ON the ready-to-receive LED
        bsf       PORTA,readyLED
;============================
;          receiving
;============================
        call      rcvData          ; Call serial input procedure
;============================
;       data received
;============================
; Turn ready to receive LED off
        bcf       PORTA,readyLED
; Display received data
        movf      rcvReg,w         ; Byte received to w
        movwf     PORTB            ; display in Port-B
;============================
;       wait forever
;============================
endloop:
        goto      endloop
;===============================================================
;                  procedure to receive serial data
;===============================================================
; ON ENTRY:
;        local variable dataReg is used to store 8-bit value
;        received through port (labeled serialLN)
; OPERATION:
;        1. The timer at register TMR0 is set to run at
;           maximum clock speed, that is, 256 clock beats.
;           The timer overflow flag in the INTCON register
;           is set when the timer cycles from 0xff to 0x00.
;        2. When the START signal is received, the code
;           waits for 128 timer beats so as to read data in
;           the middle of the send period.
;        3. Each bit (start, data, and stop bits) is read
;           at intervals of 256 timer beats.
;        4. The procedure tests the timer overflow flag
;           (tmrOVF) to determine when the timer cycle has
;           ended, that is, when 256 clock beats have passed.
```

```
;================================================================
rcvData:
        clrf      TMR0              ; Reset timer
        movlw     0x08              ; Initialize bit counter
        movwf     bitCount
;==========================
;   wait for START bit
;==========================
startWait:
        btfsc     PORTA,0           ; Is port A0 low?
        goto      startWait         ; No. Wait for mark
;==========================
;   offset 128 clock beats
;==========================
; At this point the receiver has found the falling
; edge of the start bit. It must now wait 128 timer
; beats to synchronize in the middle of the sender's
; data rate, as follows:
;                   |<========= falling edge of START bit
;                   |
;                   |-----|<====== 128 clock beats offset
;      ---------    |            .---------
;                   |            |  <== SIGNAL
;                   -----------
;                   |<-- 256--->|
;
        movlw     0x80              ; 128 clock beats offset
        movwf     TMR0              ; to TMR0 counter
        bcf       INTCON,tmrOVF     ; Clear overflow flag
offsetWait:
        btfss     INTCON,tmrOVF     ; timer overflow?
        goto      offsetWait        ; Wait until
        btfsc     PORTA,0           ; Test start bit for error
        goto      offsetWait        ; Recycle if a false start
;==========================
;      receive data
;==========================
        clrf      TMR0              ; Restart timer
        bcf               INTCON,tmrOVF    ; Clear overflow flag
; Wait for 256 timer cycles for first/next data bit
bitWait:
        btfss     INTCON,tmrOVF     ; Timer cycle end?
        goto      bitWait           ; Keep waiting
; Timer has counter 256 beats
        bcf       INTCON,tmrOVF     ; Reset overflow flag
        movf      PORTA,w           ; Read Port-A into w
        movwf     temp              ; Store value read
        rrf       temp,f            ; Rotate bit 0 into carry flag
```

```
        rlf       rcvReg,f           ; Rotate carry into rcvReg 0
decfsz   bitCount,f        ; 8 bits received
        goto      bitWait            ; Next bit
; Wait for one time cycle at end of reception
markWait:
        btfss     INTCON,tmrOVF    ; timer overflow flag
        goto      markWait ; keep waiting
;========================
;   end of reception
;========================
        return


;=========================================================
;                       end of program
;=========================================================
        end
```

20.6.3 Serial6465 Program

```
; File name: Serial6465.asm
; Last update: May 7, 2011
; Authors: Canton and Sanchez
; Processor: 16F84A
;
; Description:
; Program to exercise serial communications using a
; PIC 16F84A and two shift registers: a 74HC164 and a
; 74HC165. The 74HC165 inputs 8 lines from a DIP switch
; and transmits settings to PIC through a serial line.
; PIC sends data serially to an 74HC164 that is wired
; to 8 LEDs that display the received data. A total of
; 6 PIC lines are used in interfacing 8 input switches
; to 8 output LEDs.
;
; Circuit:
;     * Port A0 is the serial transmission line that
;       comes from the 74HC165.
;     * Port A1 is wired to the 74HC164 CLOCK pin
;     * Port A2 is wired to the 74HC164 CLEAR pin
;     * 74HC164 output pins 0 to 7 are wired to LEDs.
;     * Port B0 is wired to the 74HC165 Hout line
;     * Port B1 is wired to the 74HC165 CLK line
;     * Port B2 is wired to the 74HC165 load line
;     * A pushbutton switch is in the 16F84 RESET line
;       and serves to restart the program
; Communications protocol:
;       Communication between PIC and the 74HC164 and
;       74HC165 is synchronous as the shift registers
```

```
;          clock lines serve to shift in and out the data
;          bits.
;
;===========================
;          switches
;===========================
; Switches used in __config directive:
;   _CP_ON         Code protection ON/OFF
; * _CP_OFF
; * _PWRTE_ON      Power-up timer ON/OFF
;   _PWRTE_OFF
;   _WDT_ON        Watchdog timer ON/OFF
; * _WDT_OFF
;   _LP_OSC        Low power crystal oscillator
; * _XT_OSC        External parallel resonator oscillator
;   _HS_OSC        High speed crystal resonator (8 to 10 MHz)
;                  Resonator: Murate Erie CSA8.00MG = 8 MHz
;   _RC_OSC        Resistor/capacitor oscillator
; |                (simplest, 20% error)
; |
; |_____ * indicates setup values presently selected


;=========================
; setup and configuration
;=========================
        processor 16f84A
        include   <p16f84A.inc>
        __config  _XT_OSC & _WDT_OFF & _PWRTE_ON & _CP_OFF


;================================================================
;                         M A C R O S
;================================================================
; Macros to select the register banks
Bank0   MACRO                ; Select RAM bank 0
        bcf       STATUS,RP0
        ENDM

Bank1   MACRO                ; Select RAM bank 1
        bsf       STATUS,RP0
        ENDM
; Note: in the case of the 16F84A the bank select macros
;       do not make the code more efficient, but they
;       do serve to clarify the bank selection operations.
;
;=======================================================
;      constant definitions from wiring diagram
;=======================================================
#define clockLN 1           ; |
```

```
#define clearLN 2          ;| - 74HC164 lines
#define dataLN 0 ;|
;
#define clk65LN 1          ;| - 74HC165 lines
#define loadLN 2           ;|
;========================================================
;              PIC register and flag equates
;========================================================
highBit equ               7                    ; High order bit
;========================================================
;               variables in PIC RAM
;========================================================
        cblock  0x0d      ; Start of block
        bitCount ; Counter for 8 bits
        dataReg           ; Data to send
        workReg           ; Work register for bit shifts
        endc
;========================================================
;                         program
;========================================================
        org     0         ; start at address
        goto    main
; Space for interrupt handlers
        org     0x04

main:
; Port-A is all output
        Bank1
        movlw   b'00000000'
        movwf   TRISA
; Port-B line 0 is input, all others are output
        movlw   b'00000001'
        movwf   TRISB
        Bank0
; Make sure Port-A line 2 (clear line) is high
        movlw   b'00000100'
        movwf   PORTA
;========================
; read input from 165 IC
;========================
        call    in165             ; Local procedure
; dataReg contains input
;==========================
;   call serial output
;       procedure
;==========================
        call    out164            ; Call serial output procedure
;==========================
```

```
;         wait forever
;===========================
endloop:
        goto      endloop
;===============================================================
;             74HC164 procedure to send serial data
;===============================================================
; ON ENTRY:
;         local variable dataReg holds 8-bit value to be
;         transmitted through port labeled serialLN
; OPERATION:
;       1. A local counter (bitCount) is initialized to
;          8 bits.
;       2. Code assumes that the first bit is zero by
;          setting the data line low. Then the high-order
;          bit in the data register (dataReg) is tested.
;          If set, the data line is changed to high.
;       3. Bits are shifted in by pulsing the 74HC164
;          clock line (CLK).
;       4. Data bits are then shifted left and the bit
;          counter is tested. If all 8 bits have been sent,
;          the procedure returns.
out164:
; Clear 74HC164 shift register
        bcf       PORTA,clearLN    ; 74HC164 CLR clear low
        bsf       PORTA,clearLN    ; then high again
; Init counter
        movlw     0x08             ; Initialize bit counter
        movwf     bitCount
sendBit:
        bcf       PORTA,dataLN     ; Set data line low (assume)
; Using this assumption is possible because the bit is not
; shifted in until the clock line is pulsed.
        btfsc     dataReg,highBit  ; test number bit 7
        bsf       PORTA,dataLN     ; Change assumption if set
;=========================
;    pulse clock line
;=========================
; Bits are shifted in by pulsing the 74HC164 CLK line
        bsf       PORTA,clockLN    ; CLK high
        bcf       PORTA,clockLN    ; CLK low
;=========================
; Rotate data bits left
;=========================
        rlf       dataReg,f        ; Shift left data bits
        decfsz    bitCount,f       ; Decrement bit counter
        goto      sendBit          ; Repeat if not 8 bits
;=========================
```

```
;   end of transmission
;=========================
          return


;===============================================================
;      74HC165 procedure to read parallel data and send
;                      serially to PIC
;===============================================================
; OPERATION:
;          1. Eight DIP switches are connected to the input
;             ports of an 74HC165 IC. Its output line Hout
;             and its control lines CLK and load are connected
;             to the PIC's Port-B lines 0, 1, and 2
;             respectively
;          2. Procedure sets a counter (bitCount) for eight
;             iterations and clears a data holding register
;             (dataReg).
;          3. Port-B bits are read into w. Only the lsb of
;             Port-B is relevant. Value is stored in a working
;             register and the meaningful bit is rotated into
;             the carry flag, then the carry flag bit is
;             shifted into the data register.
;          4. The iteration counter is decremented. If this
;             is the last iteration, the routine ends. Otherwise
;             the bitwise read-and-write operation is repeated.

in165:
          clrf     dataReg          ; Clear data register
          movlw    0x08             ; Initialize counter
          movwf    bitCount
          bcf      PORTB,loadLN     ; Reset shift register
          bsf      PORTB,loadLN
nextBit:
          movf     PORTB,w          ; Read Port-B (only LOB is
                                    ; meaningful in this routine)
          movwf    workReg          ; Store value in local
                                    ; register
          rrf      workReg,f        ; Rotate LOB bit into carry
                                    ; flag
          rlf      dataReg,f        ; Carry flag into dataReg
          decfsz   bitCount,f       ; Decrement bit counter
          goto     shiftBits        ; Continue if not zero
          return                    ; done
shiftBits:
          bsf      PORTB,clk65LN    ; Pulse clock
          bcf      PORTB,clk65LN
          goto     nextBit          ; Continue
;===========================================================
```

```
;                          end of program
;============================================================
        end
```

20.6.4 TTYUsart Program

```
; File name: TTYUsart.asm
; Last update: May 1, 2010
; Authors: Sanchez and Canton
; Processor: 16F84A
;
; Description:
; Program to emulate USART operation in PIC code. Uses
; PIC-to-LCD interface. Display has two lines, each with
; sixteen characters.
; Program operation:
; Characters received from the RS232 line are displayed on
; the LCD. LCD lines scroll automatically. A pushbutton
; activates the send operation by transmitting the text
; string: Ready- which is also displayed on the LCD.
;
; Program communications and LCD parameters are stored in
; #define statements. These statements can be edited to
; accommodate a different setup. Program uses delay loops
; for interface timing.
;
; WARNING:
; Code assumes 4 MHz clock. Delay routines must be
; edited for faster clock.
;
; BAUD RATE CALCULATIONS:
; A 4 MHz clock oscillator has a clock frequency of 1 MHz:
; Because the baud rate is the number of clock cycles per
; second, for a 4 MHz clock it is:
;                  1
; bit time = ----- sec. = 208.33 microseconds
;              4,800
; Calculating one-half the baud rate allows resetting the
; clock from the edge to the center of a time pulse:
;
;          |<======== falling edge of start bit
;          |         |<======== center of bit time
;        >|         |< one-half baud rate
;          |         |
;_____.         |         ._____.
;          |_____|            |_____
;            208/2 = 104
; The PIC clock counts up from 0 to 255. So to implement
```

```
; a 104 microsecond delay we must start counting at
; clock beat:
;                 255 - 104 = 151
; plus one microsecond for movlw instruction used to
; initialize the clock:
;                 151 + 1 = 152
; For one full baud rate delay:
;                 255 - 208 = 47 + 1 = 48
; The following two constants are stored in #define
; statements:
;                 halfBaud = 152
;                 fullBaud = 48
; Setting the prescaler to TMR0 reduces the baud rate
; to one-half. Other prescaler values will reduce the
; baud`rate accordingly.
;
; Wiring diagram:
;    RB4-RB7 ===> LCD data lines 4 to 7 (output)
;    RB0 =======> MAX202 T2in line (output)
;    RA0 =======> MAX202 R2out line (input)
;    RA1 =======> LCD E line (output)
;    RA2 =======> LCD RS line (output)
;    RA3 =======> LCD R/W line (output - not used)
;    RA4 =======> Pushbutton switch 1
;                 (input - active low)
;
;============================
;        switches
;============================
; Switches used in __config directive:
;   _CP_ON        Code protection ON/OFF
; * _CP_OFF
; * _PWRTE_ON     Power-up timer ON/OFF
;   _PWRTE_OFF
;   _WDT_ON       Watchdog timer ON/OFF
; * _WDT_OFF
;   _LP_OSC       Low power crystal oscillator
; * _XT_OSC       External parallel resonator/crystal oscillator

;   _HS_OSC       High speed crystal resonator (8 to 10 MHz)
;                 Resonator: Murate Erie CSA8.00MG = 8 MHz
;   _RC_OSC       Resistor/capacitor oscillator
; |               (simplest, 20% error)
; |
; |_____ * indicates setup values presently selected

;=========================
; setup and configuration
```

```
;=========================
        processor 16f84A
        include    <p16f84A.inc>
        __config  _XT_OSC & _WDT_OFF & _PWRTE_ON & _CP_OFF


;============================================================
;                      M A C R O S
;============================================================
; Macros to select the register banks
Bank0   MACRO              ; Select RAM bank 0
        bcf       STATUS,RP0
        ENDM


Bank1   MACRO              ; Select RAM bank 1
        bsf       STATUS,RP0
        ENDM
;============================================================
;                   constant definitions
;       for PIC-to-LCD pin wiring and LCD line addresses
;============================================================
#define E_line 1           ; |
#define RS_line 2          ; | -- from wiring diagram
#define RW_line 3          ; |
; LCD line addresses (from LCD data sheet)
#define LCD_1 0x80         ; First LCD line constant
#define LCD_2 0xc0         ; Second LCD line constant
#define LCDlimit .16; Number of characters per line
; 4800 baud clock countdown values
; Code reduces rate to 2400 baud by entering a minimal
; prescaler to TRM0
#define halfBaud .152      ; For one-half bit time
#define fullBaud .48       ; For one full bit time
;
; Note: The constants that define the LCD display line
;       addresses have the high-order bit set in
;       order to facilitate the controller command
;
;==========================================================
;          buffer and variables in PIC RAM
;==========================================================
; Create a 16-byte storage area
        cblock   0x0c      ; Start of first data block
        lineBuf            ; buffer for text storage
        endc
; Leave 16 bytes and Continue with local variables
        cblock   0x1c      ; Second data block
        count1             ; Counter # 1
        count2             ; Counter # 2
```

```
        J                       ; counter J
        K                       ; counter K
        store1                  ; Local temporary storage
        store2                  ; Storage # 2
; For LCDscroll procedure
        LCDcount                ; Counter for characters per line
        LCDline                 ; Current display line (0 or 1)
        bufPtr                  ; Buffer pointer
; Variables for serial communications
        tempData                ; Temporary storage for bit
                                ; manipulations
        rcvData                 ; Final storage for received character
        bitCount ; Bit counter
        sendData ; Character to send
        endc


;============================================================
;                m a i n     p r o g r a m
;============================================================
        org     0               ; start at address
        goto    main
; Space for interrupt handlers
        org     0x08
main:
        Bank1
        movlw   b'00010001'     ; Port-A lines I/O setup
                                ; RA0 = RS232 input (R2out)
                                ; RA4 = Pushbutton SW # 1
        movwf   TRISA
        movlw   b'00000000' ; Port-B lines as follows:
;    RB4-RB7 ===> LCD data lines 4 to 7 (output)
;    RB0 =======> MAX202 T2in line (output)
        movwf   TRISB
        Bank0
; Clear bits in Port-A output lines
        bcf     PORTA,1
        bcf     PORTA,2
        bcf     PORTA,3
        movlw   b'00000000'     ; All outputs ports low
        movwf   PORTB
; Wait and initialize HD44780
        call    delay_5         ; Allow LCD time to initialize
                                ; itself
        call    delay_5
        call    initLCD         ; Then do forced
initialization
        call    delay_5         ; Wait again
; Set Port-B, line 0 high so start bit is detected
```

```
        bsf       PORTB,0
;=============================
;   wait for start command
;=============================
; Program waits until pushbutton number 1 is pressed
; to continue execution. Pushbutton 1 is active low
; and wired to RA4
pb1Wait:
        btfsc     PORTA,4           ; Test Port-A, line 4
        goto      pb1Wait           ; Loop if not clear
;=============================
;   display and send "Ready-"
;=============================
; Set LCD base address
        call      line1
; Initialize system for UART emulation at 2400 baud
        call      initTTY
; Display on LCD and test serial transmission by sending
; the string "Ready-"
        movlw     'R'
        movwf     sendData ; Store in send register
        call      send8             ; Local LCD display procedure
        call      sendTTY           ; Local send procedure
        movlw     'e'
        movwf     sendData ; Store in send register
        call      send8             ; Local LCD display procedure
        call      sendTTY           ; Local send procedure
        movlw     'a'
        movwf     sendData ; Store in send register
        call      send8             ; Local LCD display procedure
        call      sendTTY           ; Local send procedure
        movlw     'd'
        movwf     sendData ; Store in send register
        call      send8             ; Local LCD display procedure
        call      sendTTY           ; Local send procedure
        movlw     'y'
        movwf     sendData ; Store in send register
        call      send8             ; Local LCD display procedure
        call      sendTTY           ; Local send procedure
        movlw     '-'
        movwf     sendData ; Store in send register
        call      send8             ; Local LCD display procedure
        call      sendTTY           ; Local send procedure
; Init  character counter and line counter variables for
; LCD line scroll procedure
        movlw     0x06              ; 6 characters already
displayed
        movwf     LCDcount
```

```
        clrf     LCDline          ; LCD line counter
;=============================
;    monitor RS232 line
;=============================
nextChar:
        call     rcvTTY           ; Receive character
; Store character in local line buffer using indirect
; addressing
; 16-byte buffer named lineBuf starts at address 0x0c
; Register variable bufPtr holds offset into buffer
        movlw    0x0c             ; Buffer base address
        addwf    bufPtr,w         ; Add pointer in w
        movwf    FSR              ; Value to index register
        movf     rcvData,w        ; Character into w
        movwf    INDF             ; Store w in [FSR]
        incf     bufPtr,f         ; Bump pointer
; Send character (still in w)
        call     send8            ; Display it
        call     LCDscroll        ; Scroll display lines
        goto     nextChar ; Continue


;===============================================================
;                  initialize LCD for 4-bit mode
;===============================================================
initLCD:
; Initialization for Densitron LCD module as follows:
;        4-bit interface
;   2 display lines of 16 characters each
;   cursor on
;   left-to-right increment
;   cursor shift right
;   no display shift
;=====================|
;   set command mode  |
;=====================|
        bcf      PORTA,E_line     ; E line low
        bcf      PORTA,RS_line    ; RS line low
        bcf      PORTA,RW_line    ; Write mode
        call     delay_125        ; delay 125 microseconds
;*********************|
;     FUNCTION SET    |
;*********************|
        movlw    0x28     ; 0 0 1 0 1 0 0 0 (FUNCTION SET)
                          ;       | | | |__ font select:
                          ;       | | |    1 = 5x10 in 1/8 or 1/11
                          ;       | | |    0 = 1/16 dc
                          ;       | | |___ Duty cycle select
                          ;       | |      0 = 1/8 or 1/11
```

```
                          ;      | |        1 = 1/16
                          ;      | |___ Interface width
                          ;      |       0 = 4 bits
                          ;      |       1 = 8 bits
                          ;      |___ FUNCTION SET COMMAND
          call    send8   ; 4-bit send routine

; Set 4-bit mode command must be repeated
          movlw   0x28
          call    send8

;*********************|
; DISPLAY AND CURSOR ON |
;*********************|
          movlw   0x0e    ; 0 0 0 0 1 1 1 0 (DISPLAY ON/OFF)
                          ;             | | | |___ Blink character
                          ;             | | |       1 = on, 0 = off
                          ;             | | |___ Cursor on/off
                          ;             | |       1 = on, 0 = off
                          ;             | |____ Display on/off
                          ;             |         1 = on, 0 = off
                          ;             |_____ COMMAND BIT
          call    send8
;*********************|
;   set entry mode    |
;*********************|
          movlw   0x06    ; 0 0 0 0 0 1 1 0 (ENTRY MODE SET)
                          ;           | | |___ display shift
                          ;           | |       1 = shift
                          ;           | |       0 = no shift
                          ;           | |____ increment mode
                          ;           |         1 = left-to-right
                          ;           |         0 = right-to-left
                          ;           |___ COMMAND BIT
          call    send8

;*********************|
; cursor/display shift |
;*********************|
          movlw   0x14    ; 0 0 0 1 0 1 0 0 (CURSOR/DISPLAY
                          ;         | | | | |  SHIFT)
                          ;         | | | |_|___ don't care
                          ;         | |_|__ cursor/display shift
                          ;         |       00 = cursor shift left
                          ;         |       01 = cursor shift right
                          ;         |       10 = cursor and display
                          ;         |             shifted left
                          ;         |       11 = cursor and display
```

```
                              ;          |                 shifted right
                              ;          |___ COMMAND BIT
        call      send8
;*********************|
;    clear display    |
;*********************|
        movlw     0x01     ; 0 0 0 0 0 0 0 1 (CLEAR DISPLAY)
                           ;               |___ COMMAND BIT
        call      send8
; Per documentation
        call      delay_5  ; Test for busy
        return


;=======================
;  Procedure to delay
;   42 microseconds
;=======================
delay_125:
        movlw     D'42'              ; Repeat 42 machine cycles
        movwf     count1             ; Store value in counter
repeat:
        decfsz    count1,f           ; Decrement counter
        goto      repeat             ; Continue if not 0
        return                       ; End of delay


;=======================
;  Procedure to delay
;   5 milliseconds
;=======================
delay_5:
        movlw     D'41'              ; Counter = 41
        movwf     count2             ; Store in variable
delay:
        call      delay_125          ; Delay
        decfsz    count2,f           ; 40 times = 5 milliseconds
        goto      delay
        return                       ; End of delay
;=======================
;    pulse E line
;=======================
pulseE
        bsf       PORTA,E_line       ; Pulse E line
        nop
        bcf       PORTA,E_line
        return


;============================
;   long delay sub-routine
```

```
;       (for debugging)
;==============================
long_delay
        movlw   D'200'              ; w = 200 decimal
        movwf   J                   ; J = w
jloop:
        movwf   K                   ; K = w
kloop:
        decfsz  K,f                 ; K = K-1, skip next if zero
        goto    kloop
        decfsz  J,f                 ; J = J-1, skip next if zero
        goto    jloop
        return
;=======================
;   send 2 nibbles in
;      4-bit mode
;=======================
; Procedure to send two 4-bit values to Port-B lines
; 7, 6, 5, and 4. High-order nibble is sent first
; ON ENTRY:
;          w register holds 8-bit value to send
send8:
        movwf   store1              ; Save original value
        call    merge4              ; Merge with Port-B
; Now w has merged byte
        movwf   PORTB               ; w to Port-B
        call    pulseE              ; Send data to LCD
; High nibble is sent
        movf    store1,w            ; Recover byte into w
        swapf   store1,w            ; Swap nibbles in w
        call    merge4
        movwf   PORTB
        call    pulseE              ; Send data to LCD
        call    delay_125
        return
;==================
;   merge bits
;==================
; Routine to merge the 4 high-order bits of the
; value to send with the contents of Port-B
; so as to preserve the 4 low-bits in Port-B
; Logic:
;      AND value with 1111 0000 mask
;      AND Port-B with 0000 1111 mask
;      Now low nibble in value and high nibble in
;      PortB are all 0 bits:
;                value = vvvv 0000
;      PortB = 0000 bbbb
```

```
;       OR value and Port-B resulting in:
;                vvvv bbbb
; ON ENTRY:
;       w contains value bits
; ON EXIT:
;       w contains merged bits
merge4:
        andlw    b'11110000'       ; ANDing with 0 clears the
                                   ; bit. ANDing with 1 preserves
                                   ; the original value
        movwf    store2            ; Save result in variable
        movf     PORTB,w           ; Port-B to w register
        andlw    b'00001111'       ; Clear high nibble in Port-B
                                   ; and preserve low nibble
        iorwf    store2,w          ; OR two operands in w
        return
;========================
; Set address register
;    to LCD line 1
;========================
; ON ENTRY:
;        Address of LCD line 1 in constant LCD_1
line1:
        bcf      PORTA,E_line      ; E line low
        bcf      PORTA,RS_line     ; RS line low, set up for
                                   ; control
        call     delay_5           ; busy?
; Set to second display line
        movlw    LCD_1             ; Address and command bit
        call     send8             ; 4-bit routine
; Set RS line for data
        bsf      PORTA,RS_line     ; Set up for data
        call     delay_5           ; Busy?
; Clear buffer and pointer
        call     blankBuf
        clrf     bufPtr            ; Clear
        return
;========================
; Set address register
;    to LCD line 2
;========================
; ON ENTRY:
;        Address of LCD line 2 in constant LCD_2
line2:
        bcf      PORTA,E_line      ; E line low
        bcf      PORTA,RS_line     ; RS line low, set up for
                                   ; control
        call     delay_5           ; Busy?
```

```
; Set to second display line
        movlw   LCD_2               ; Address with high-bit set
        call    send8
; Set RS line for data
        bsf     PORTA,RS_line       ; RS = 1 for data
        call    delay_5             ; Busy?
; Clear buffer and pointer
        call    blankBuf
        clrf    bufPtr
        return


;=========================
;    scroll LCD line 2
;=========================
; Procedure to count the number of characters displayed on
; each LCD line. If the number reaches the value in the
; constant LCDlimit, then display is scrolled to the second
; LCD line. If at the end of the second line, then the
; second line is scrolled to the first line and display
; continues at the start of the second line
; reset to the first line.
LCDscroll:
        incf    LCDcount,f          ; Bump counter
; Test for line limit
        movf    LCDcount,w
        sublw   LCDlimit            ; Count minus limit
        btfss   STATUS,Z            ; Is count minus limit = 0
        goto    scrollExit          ; Go if not at end of line
; At this point the end of the LCD line was reached
; Test if this is also the end of the second line
        movf    LCDline,w
        sublw   0x01                ; Is it line 1?
        btfsc   STATUS,Z            ; Is LCDline minus 1 = 0?
        goto    line2End            ; Go if end of second line
; At this point it is the end of the top LCD line
        call    line2               ; Scroll to second line
        clrf    LCDcount            ; Reset counter
        incf    LCDline,f           ; Bump line counter
        goto    scrollExit
; End of second LCD line
line2End:
; Scroll second line to first line. Characters to be
; scrolled are stored in buffer starting at address 0x0c.
; sixteen characters are to be moved
; First clear LCD
        call    initLCD
        call    delay_5             ; Make sure not busy
; Set up for data
```

```
        bcf     PORTA,E_line      ; E line low
        bsf     PORTA,RS_line     ; RS line high for data
; Set up counter for 16 characters
        movlw   D'16'             ; Counter = 16
        movwf   count2
; Get address of storage buffer
        movlw   0x0c
        movwf   FSR               ; W to FSR
getchar:
        movf    INDF,w    ; get character from display RAM
                          ; location pointed to by file select
                          ; register
        call    send8     ; 4-bit interface routine
; Test for 16 characters displayed
        decfsz  count2,f          ; Decrement counter
        goto    nextchar ; Skipped if done
; At this point scroll operation has concluded
        clrf    LCDcount          ; Clear counters
; Stay at line 2
        clrf    LCDline
        incf    LCDline,f
        call    line2             ; Set for second line
scrollExit:
        return
nextchar:
        incf    FSR,f             ; Bump pointer
        goto    getchar


;============================
;   clear line buffer
;============================
; Use indirect addressing to store 16 blanks in the
; buffer located at 0x0c
blankBuf:
        Bank0
        movlw   0x0c      ; Pointer to RAM
        movwf   FSR       ; To index register
blank16:
        clrf    INDF      ; Clear memory pointed at by FSR
        incf    FSR,f     ; Bump pointer
        btfss   FSR,4     ; 000x0000 when bit 4 is set
                          ; count reached 16
        goto    blank16
        return


;==============================================================
;                     initialize for TTY
;==============================================================
```

```
; Procedure to initialize RS232 reception
; Assumes:
;                     2400 baud
;                     8 data bits
;                     no parity
;                     one stop bit
initTTY:
; First initialize receiver to RS-232-C line parameters
; Disable global and peripheral interrupts
;   7   6   5   4   3   2   1   0   <= INTCON bitmap
;   |   ?   |   ?   ?   ?   ?   ?   (? = unrelated bits)
;   |       |_____ Timer0 interrupt on overflow
;   |_____ Global interrupts
            bcf     INTCON,5        ; Disable TMR0 interrupts
            bcf     INTCON,7        ; Disable global interrupts
            clrf    TMR0            ; Reset timer
            clrwdt                  ; Clear WDT for prescaler
assign
            Bank1
; Set up the OPTION regiser bitmap
;   7   6   5   4   3   2   1   0 <= OPTION bits
;   1   1   0   1   1   0   0   0 <= setup
;   |   |   |   |   |   |   |__|__|_____ PS2-PS0 (prescaler bits)
;   |   |   |   |   |   |               Values for Timer0
;   |   |   |   |   |   |               *000 = 1:2    001 = 1:4
;   |   |   |   |   |   |               010 = 1:8     011 = 1:16
;   |   |   |   |   |   |               100 = 1:32    101 = 1:64
;   |   |   |   |   |   |               110 = 1:128  111 = 1:256
;   |   |   |   |   |   |_____ PSA (prescaler assign)
;   |   |   |   |   |                    1 = to WDT
;   |   |   |   |   |                    *0 = to Timer0
;   |   |   |   |   |_____ TOSE (Timer0 edge select)
;   |   |   |   |                        0 = increment on low-to-high
;   |   |   |   |                        *1 = increment in high-to-low
;   |   |   |   |_____ TOCS (TMR0 clock source)
;   |   |   |                            *0 = internal clock
;   |   |   |                            1 = RA4/TOCKI bit source
;   |   |   |_____ INTEDG (Edge select)
;   |   |                                0 = falling edge
;   |   |                                *1 = rising edge
;   |_____ RBPU (Pullup enable)
;                                        0 = enabled
;                                        *1 = disabled
            movlw   b'11010000'  ; set up timer/counter
            movwf   OPTION_REG
            Bank0
            return
;=============================================================
```

```
;                             receive character
;================================================================
; Receive a single character through the serial port.
; Assumes: 4800 baud, 8 data bits, no parity, 1 stop bit.
; Receiving line is Port-A, line 0
rcvTTY:
        movlw    0x08                 ; Counter for 8 bits
        movwf    bitCount
; The start of character transmission is signaled by
; the sender by setting the line low
startBit:
        btfsc    PORTA,0              ; Test for low on line
        goto     startBit ; Go if not low
;==========================
;   offset to data bit
;==========================
; At this point the receiver has found the falling
; edge of the start bit. It must now wait one and
; one-half the baud rate to synchronize in the center
; of the sender's first data bit, as follows:
;       |<========= falling edge of START bit
;       |       |<========== center of start bit
;       |       |                |<====== center of data bit
;       |-----|-----|
;_____             _____             _____
;       |         |           |           |           <== SIGNAL
;       -----------           ----------
;       |<-- 208 -->|h| <====== ms. for 4800 baud
;
; Clock start count for one-half bit  = 255 - 104 = 151
; Clock start count for one full bit =  255 - 208 = 47
; One clock cycle is added for the movwf instruction:
;    clkHalf = 152 (for one-half bit countdown)
;    clkFull = 48 (for one full bit countdown)
        movlw    halfBaud             ; Skip one-half bit
        movwf    TMR0     ; Initialize tmr0 and start count
        bcf      INTCON,2             ; Clear overflow flag
;==========================
;         start bit
;==========================
wait1:
        btfss    INTCON,2             ; Timer count overflow?
        goto     wait1                ; No, keep waiting
; At this point we are at the center of the start bit
        btfsc    PORTA,0              ; Check to see it is still low
        goto     startBit             ; No, it is high. False start.
; At this point the clock is at the center of the start
; bit. The first data bit must be read one full baud
```

```
; period later
        movlw   fullBaud        ; One full bit delay
        movwf   TMR0            ; Start timer
        bcf     INTCON,2        ; clear tmr0 overflow flag
wait2:
        btfss   INTCON,2        ; End of one full baud period?
        goto    wait2           ; Wait if not end of period
; Timer is now at the center of the first/next data bit
; Timer must be reset immediately so that code will not
; lose synchronization with sender
        movlw   fullBaud ; Skip to next data bit
        movwf   TMR0            ; Restart timer
        bcf     INTCON,2        ; Reset overflow flag
; Now the data bit can be read and and stored
        movf    PORTA,w         ; Read Port-B
        movwf   tempData ; Store in temporary variable
        rrf     tempData,f      ; Rotate bit 0 into carry flag
        rrf     rcvData,f       ; Rotate carry flag into
                                ; storage register high-order
                                ; bit
        decfsz  bitCount,f      ; End of data?
        goto    wait2    ; Continue until 8 bits received
;=============================
;       stop bit
;=============================
stopWait:
        btfss   INTCON,2        ; Test time
        goto    stopWait ; Wait
        return                  ; Exit


;============================================================
;                     send character
;============================================================
; Procedure to send one character through the RS232 line.
; Assumes: 2400 baud, 8 data bits, no parity, one stop bit
; Sending line is Port-B, line 0
; ON ENTRY:
;       variable sendData holds character to send
sendTTY:
        movlw   0x08            ; Init bit counter
        movwf   bitCount
        bcf     PORTB,0         ; Low for start bit
        movlw   fullBaud ; For one baud space
        movwf   TMR0            ; Start timer
        bcf     INTCON,2        ; Clear timer flag
start2snd:
        btfss   INTCON,2        ; Full baud done?
        goto    start2snd       ; No
```

```
            movlw     fullBaud ; Reset for one full bit period
            movwf     TMR0              ; Start timer
            bcf       INTCON,2          ; Clear flag
; At this point the start bit has been sent
; Data follows
sendOut:
            rrf       sendData,f        ; Rotate bit into carry
            bcf       PORTB,0           ; Assume data bit is 0
            btfsc     STATUS,c          ; Test if carry set
            bsf       PORTB,0 ; Change bit to 1 if clear
; Hold bit for 1 baud period
timeBit:
            btfss     INTCON,2          ; Wait for baud period to end
            goto      timeBit           ; Loop if not yet
            movlw     fullBaud          ; Reset timer
            movwf     TMR0              ; Start timer
            bcf       INTCON,2          ; Clear flag
; Test for last bit
            decfsz    bitCount,f        ; Count this bit
            goto      sendOut           ; Continue if not last bit
; Done. Send stop bit
            bsf       PORTB,0           ; High for stop bit
stopBit:
            btfss     INTCON,2          ; Timer done?
            goto      stopBit           ; No
; Set Port-B line 0 high back again
            bsf       PORTB,0
            call      delay_5           ; And hold
            return

            End
```

20.6.5 SerComLCD Program

```
; File name: SerComLCD.asm
; Last revision: May 14, 2011
; Authors: Sanchez and Canton
; Processor: 16F877
;
; Description:
; Decode 4 x 4 keypad, display scan code in LCD, and send
; ASCII character through the serial port. Also receive
; data through serial port and display on LCD. LCD lines
; are scrolled by program.
; Default serial line setting:
;               2400 baud
;               no partity
;               1 stop bit
```

```
;                      8 character bits
;
; Program uses 4-bit PIC-to-LCD interface.
; Code assumes that LCD is driven by Hitachi HD44780
; controller and PIC 16F977. Display supports two lines
; each one with 20 characters. The length, wiring, and base
; address of each display line is stored in #define
; statements. These statements can be edited to accommodate
; a different set up.
; Keypad switch wiring (values are scan codes):
;         -- KEYPAD --
;         0    1    2    3    <= port B0 |
;         4    5    6    7    <= port B1 |-- ROWS = OUTPUTS
;         8    9    A    B    <= port B2 |
;         C    D    E    F    <= port B3 |
;         |    |    |    |
;         |    |    |    |_____ port B4 |
;         |    |    |_____ port B5 |-- COLUMNS = INPUTS
;         |    |_____ port B6 |
;         |_____ port B7 |
;
; Operations:
; 1. Key press action generates a scan code in the range
;    0x0 to 0xf.
; 2. Scan code is converted to an ASCII digit and displayed
;    on the LCD. LCD lines are scrolled as end-of-line is
;    reached.
; 3. Characters typed on the keypad are also transmitted
;    through the serial port.
; 4. Serial port is polled for received characters. These
;    are displayed on the LCD.
;
; WARNING:
; Code assumes 4 MHz clock. Delay routines must be
; edited for faster clock. Clock speed also determines
; values for baud rate setting (see spbrgVal constant).
;
;==========================
;      16F877 switches
;==========================
; Switches used in __config directive:
;    _CP_ON         Code protection ON/OFF
; * _CP_OFF
; * _PWRTE_ON       Power-up timer ON/OFF
;    _PWRTE_OFF
;    _BODEN_ON       Brown-out reset enable ON/OFF
; * _BODEN_OFF
; * _PWRTE_ON       Power-up timer enable ON/OFF
```

```
;   _PWRTE_OFF
;   _WDT_ON          Watchdog timer ON/OFF
; * _WDT_OFF
;   _LPV_ON          Low voltage IC programming enable ON/OFF
; * _LPV_OFF
;   _CPD_ON          Data EE memory code protection ON/OFF
; * _CPD_OFF
; OSCILLATOR CONFIGURATIONS:
;   _LP_OSC          Low power crystal oscillator
;   _XT_OSC          External parallel crystal ocillator
; * _HS_OSC          High speed crystal resonator
;   _RC_OSC          Resistor/capacitor oscillator
; |                  (simplest, 20% error)
; |
; |_____  * indicates setup values presently selected

        processor         16f877              ; Define processor
        #include <p16f877.inc>
        __CONFIG _CP_OFF & _WDT_OFF & _BODEN_OFF & _PWRTE_ON &
_HS_OSC & _WDT_OFF & _LVP_OFF & _CPD_OFF

; __CONFIG directive is used to embed configuration data
; within the source file. The labels following the directive
; are located in the corresponding .inc file.
;=============================================================
;                       M A C R O S
;=============================================================
; Macros to select the register banks
Bank0   MACRO   ; Select RAM bank 0
        bcf     STATUS,RP0
        bcf     STATUS,RP1
        ENDM

Bank1   MACRO   ; Select RAM bank 1
        bsf     STATUS,RP0
        bcf     STATUS,RP1
        ENDM

Bank2   MACRO   ; Select RAM bank 2
        bcf     STATUS,RP0
        bsf     STATUS,RP1
        ENDM

Bank3   MACRO   ; Select RAM bank 3
        bsf     STATUS,RP0
        bsf     STATUS,RP1
        ENDM
;=====================================================
```

```
;                      constant definitions
;   for PIC-to-LCD pin wiring and LCD line addresses
;=======================================================
#define E_line 1          ; |
#define RS_line 0         ; | -- from wiring diagram
#define RW_line 2         ; |
; LCD line addresses (from LCD data sheet)
#define LCD_1 0x80        ; First LCD line constant
#define LCD_2 0xc0        ; Second LCD line constant
#define LCDlimit .20; Number of characters per line
#define spbrgVal .25; For 2400 baud on 4 MHz clock
; Note: The constants that define the LCD display
;       line addresses have the high-order bit set
;       so as to meet the requirements of controller
;       commands.
;
;=======================================================
;               variables in PIC RAM
;=======================================================
; Local variables
        cblock   0x20               ; Start of block
        count1              ; Counter # 1
        count2              ; Counter # 2
        count3              ; Counter # 3
        J                   ; Counter J
        K                   ; Counter K
        store1              ; Local storage
        store2
; For LCDscroll procedure
        LCDcount            ; Counter for characters per line
        LCDline             ; Current display line (0 or 1)
; Keypad processing variables
        keyMask             ; For keypad processing
        rowMask             ; For masking-off key rows
        rowCode             ; Row addend for calculating scan code
        rowCount            ; Counter for key rows (0 to 3)
        scanCode            ; Final key code
        newScan             ; 0 if no new scan code detected
; Communications variables
        newData             ; not 0 if new data received
        ascVal
        errorFlags
        endc


;==============================================================
;                     P R O G R A M
;==============================================================
                org              0          ; start at address
```

```
                    goto    main
; Space for interrupt handlers
        org     0x08
main:
; Wiring:
;     LCD data to Port D, lines 0 to 7
;     E line -> port E, 1
;     RW line -> port E, 2
;     RS line -> port E, 0
; Set PORTE D and E for output
; Data memory bank selection bits:
; RP1:RP0            Bank
;   0:0               0     Ports A,B,C,D, and E
;   0:1               1     Tris A,B,C,D, and E
;   1:0               2
;   1:1               3
; First, initialize Port-B by clearing latches
        clrf    STATUS
        clrf    PORTB
; Select bank 1 to tris Port D for output
        bcf     STATUS,RP1      ; Clear banks 2/3 selector
        bsf     STATUS,RP0      ; Select bank 1 for tris
                                ; registers
; Tris Port D for output. Port D lines 4 to 7 are wired
; to LCD data lines. Port D lines 0 to 4 are wired to LEDs.
        movlw   B'00000000'
        movwf   TRISD           ; and Port D
; By default, Port-A lines are analog. To configure them
; as digital code, must set bits 1 and 2 of the ADCON1
; register (in bank 1)
        movlw   0x06    ; binary 0000 0110  is code to
                        ; make all Port-A lines digital
        movwf   ADCON1
; Port-B lines are wired to keypad swtiches, as follows:
;   7 6 5 4 3 2 1 0
;   | | | | |_|_|_|_____ switch rows (output)
;   |_|_|_|_____ switch columns (input)
; rows must be defined as output and columns as input
        movlw   b'11110000'
        movwf   TRISB
; Tris port E for output
        movlw   B'00000000'
        movwf   TRISE           ; Tris port E
; Enable Port-B pullups for switches in OPTION register
;   7 6 5 4 3 2 1 0 <= OPTION bits
;   | | | | | |__|__|_____ PS2-PS0 (prescaler bits)
;   | | | | |             Values for Timer0
;   | | | | |             000 = 1:2   001 = 1:4
```

```
;      |   |   |   |   |   |                         010 = 1:8    011 = 1:16
;      |   |   |   |   |   |                         100 = 1:32   101 = 1:64
;      |   |   |   |   |   |                         110 = 1:128 *111 = 1:256
;      |   |   |   |   |   |_____ PSA (prescaler assign)
;      |   |   |   |   |                   *1 = to WDT
;      |   |   |   |   |                    0 = to Timer0
;      |   |   |   |   |_____ TOSE (Timer0 edge select)
;      |   |   |   |                       *0 = increment on low-to-high
;      |   |   |   |                        1 = increment on high-to-low
;      |   |   |_____ TOCS (TMR0 clock source)
;      |   |   |                           *0 = internal clock
;      |   |   |                            1 = RA4/TOCKI bit source
;      |   |_____ INTEDG (Edge select)
;      |   |                               *0 = falling edge
;      |_____ RBPU (Pullup enable)
;      |                                   *0 = enabled
;                                           1 = disabled
           movlw    b'00001000'
           movwf    OPTION_REG
; Back to bank 0
           bcf                STATUS,RP0
; Initialize serial port for 9600 baud, 8 bits, no parity
; 1 stop
           call     InitSerial
; Test serial transmission by sending "RDY-"
           movlw    'R'
           call     SerialSend
           movlw    'D'
           call     SerialSend
           movlw    'Y'
           call     SerialSend
           movlw    '-'
           call     SerialSend
           movlw    0x20
           call     SerialSend
; Clear all output lines
           movlw    b'00000000'
           movwf    PORTD
           movwf    PORTE
; Wait and initialize HD44780
           call     delay_5 ; Allow LCD time to initialize itself
           call     initLCD ; Then do forced initialization
           call     delay_5 ; (Wait probably not necessary)
; Clear character counter and line counter variables
           clrf     LCDcount
           clrf     LCDline
; Set display address to start of second LCD line
           call     line1
```

```
;================================================================
;                        scan keypad
;================================================================
; Keypad switch wiring:
;        x   x   x   x   <= port B0 |
;        x   x   x   x   <= port B1 |-- ROWS = OUTPUTS
;        x   x   x   x   <= port B2 |
;        x   x   x   x   <= port B3 |
;        |   |   |   |
;        |   |   |   |_____ port B4 |
;        |   |   |_____ port B5 |-- COLUMNS = INPUTS
;        |   |_____ port B6 |
;        |_____ port B7 |
; Swtiches are connected to Port-B lines
; Clear scan code register
        clrf    scanCode
;============================
;  scan keypad and display
;============================
keyScan:
; Port-B lines are wired to pushbutton switches, as follows:
;   7 6 5 4 3 2 1 0
;   | | | | |_|_|_|_____ switch rows (output)
;   |_|_|_|_____ switch columns (input)
; Keypad processing:
; switch rows are successively grounded (row = 0)
; Then column values are tested. If a column returns 0
; in a 0 row, that switch is down.
; Initialize row code addend
        clrf    rowCode         ; First row is code 0
        clrf    newScan         ; No new scan code detected
; Initialize row count
        movlw   D'4'            ; Four rows
        movwf   rowCount        ; Register variable
        movlw   b'11111110'     ; All set but LOB
        movwf   rowMask
keyLoop:
; Initialize row eliminator mask:
; The row mask is ANDed with the key mask to successively
; mask off each row, for example:
;
;                   |----- row 3
;                   ||---- row 2
;                   |||--- row 1
;                   ||||-- row 0
;           0000 1111 <= key mask
;       AND 1111 1101 <= mask for row 1
;           ---------
```

```
;               0000 1101 <= row 1 is masked off
;
; The row mask, which is initally 1111 1110, is rotated left
; through the carry in order to mask off the next row
        movlw   b'00001111'         ; Mask off all lines
        movwf   keyMask             ; To local register
; Set row mask for current row
        movf    rowMask,w           ; Mask to w
        andwf   keyMask,f           ; Update key mask
        movf    keyMask,w           ; Key mask to w
        movwf   PORTB               ; Mask off Port-B lines
; Read Port-B lines 4 to 7 (columns are input)
        btfss   PORTB,4
        call    col0                ; Key column procedures
        btfss   PORTB,5
        call    col1
        btfss   PORTB,6
        call    col2
        btfss   PORTB,7
        call    col3
; Index to next row by adding 4 to row code
        movf    rowCode,w           ; Code to w
        addlw   D'4'
        movwf   rowCode
;=========================
;     shift row mask
;=========================
; Set the carry flag
        bsf     STATUS,C
        rlf     rowMask,f           ; Rotate mask bits in storage
;=========================
;     end of keypad?
;=========================
; Test for last key row (maximum count is 4)
        decfsz  rowCount,f  ;      Decrement counter
        goto    keyLoop
;=============================================================
;=============================================================
;               display, send, and receive data
;=============================================================
;=============================================================
; At this point all keys have been tested.
; Variable newScan = 0 if no new scan code detected, else
; variable scanCode holds scan code
        movf    newScan,f           ; Copy onto intsef (sets Z
                                    ; flag)
        btfsc   STATUS,Z            ; Is it zero?
        goto    receive
```

```
; At this point a new scan code is detected
        movf      scanCode,w         ; To w
; If scan code is in the range 0 to 9, that is, a decimal
; digit, then ASCII conversion consists of adding 0x30.
; If the scan code represents one of the hex letters
; (0xa to 0xf), then ASCII conversion requires adding
; 0x37
        sublw     0x09               ; 9 - w
; if w from 0 to 9 then 9 - w = positive (C flag = 1)
; if w = 0xa then 9 - 10 = -1 (C flag = 0)
; if w = 0xc then 9 - 12 = -2 (C flag = 0)
        btfss     STATUS,C ; Test carry flag
        goto      hexLetter ; Carry clear, must be a letter
; At this point scan code is a decimal digit in the
; range 0 to 9. Convert to ASCII by adding 0x30
        movf      scanCode,w    ; Recover scan code
        addlw     0x30          ; Convert to ASCII
        goto      displayDig
hexLetter:
        movf      scanCode,w    ; Recover scan code
        addlw     0x37          ; Convert to ASCII
displayDig:
; Store so it can be sent
        movwf     ascVal
        call      send8         ; Display routine
        call      LCDscroll
        call      long_delay    ; Debounce
; Recover ASCII
        movf      ascVal,w
        call      SerialSend
        goto      scanExit
;==========================
;    receive serial data
;==========================
receive:
; Call serial receive procedure
        call      SerialRcv
; HOB of newData register is set if new data
; received
        btfss     newData,7
        goto      scanExit
; At this point, new data was received
        call      send8         ; Display in LCD
        call      LCDscroll     ; Scroll at end of line
scanExit:
        goto      keyScan       ; Continue
;==========================
;    calculate scan code
```

```
;==========================
; The column position is added to the row code (stored
; in rowCode register). Sum is the scan code.
col0:
        movf    rowCode,w       ; Row code to w
        addlw   0x00            ; Add 0 (clearly not
necessary)
        movwf   scanCode ; Final value
        incf    newScan,f                   ; New scan code
        return

col1:
        movf    rowCode,w       ; Row code to w
        addlw   0x01            ; Add 1
        movwf   scanCode
        incf    newScan,f
        return

col2:
        movf    rowCode,w       ; Row code to w
        addlw   0x02            ; Add 2
        movwf   scanCode
        incf    newScan,f
        return

col3:
        movf    rowCode,w       ; Row code to w
        addlw   0x03            ; Add 3
        movwf   scanCode
        incf    newScan,f
        return
;==============================================================
;==============================================================
;               L O C A L    P R O C E D U R E S
;==============================================================
;==============================================================
;==========================
; init LCD for 4-bit mode
;==========================
initLCD:
; Initialization for Densitron LCD module as follows:
;       4-bit interface
;   2 display lines of 16 characters each
;   cursor on
;   left-to-right increment
;   cursor shift right
;   no display shift
;======================|
```

```
;    set command mode     |
;=====================|
        bcf        PORTE,E_line        ; E line low
        bcf        PORTE,RS_line       ; RS line low
        bcf        PORTE,RW_line       ; Write mode
        call       delay_125                    ; delay 125
microseconds
;*********************|
;      FUNCTION SET      |
;*********************|
        movlw      0x28        ; 0 0 1 0 1 0 0 0 (FUNCTION SET)
                               ;       | | | |__ font select:
                               ;       | | |   1 = 5x10 in 1/8 or 1/11
                               ;       | | |   0 = 1/16 dc
                               ;       | | |___ Duty cycle select
                               ;       | |    0 = 1/8 or 1/11
                               ;       | |    1 = 1/16
                               ;       | |___ Interface width
                               ;       |     0 = 4 bits
                               ;       |     1 = 8 bits
                               ;       |___ FUNCTION SET COMMAND
        call       send8       ; 4-bit send routine

; Set 4-bit mode command must be repeated
        movlw      0x28
        call       send8


;*********************|
; DISPLAY AND CURSOR ON |
;*********************|
        movlw      0x0e        ; 0 0 0 0 1 1 1 0 (DISPLAY ON/OFF)
                               ;         | | | |___ Blink character
                               ;         | | |    1 = on, 0 = off
                               ;         | | |___ Cursor on/off
                               ;         | |     1 = on, 0 = off
                               ;         | |____ Display on/off
                               ;         |      1 = on, 0 = off
                               ;         |_____ COMMAND BIT
        call       send8
;*********************|
;    set entry mode      |
;*********************|
        movlw      0x06        ; 0 0 0 0 0 1 1 0 (ENTRY MODE SET)
                               ;           | | |___ display shift
                               ;           | |    1 = shift
                               ;           | |    0 = no shift
                               ;           | |____ increment mode
                               ;           |      1 = left-to-right
```

```
                                        ;              |         0 = right-to-left
                                        ;              |___ COMMAND BIT
          call      send8

;**********************|
; cursor/display shift  |
;**********************|
          movlw     0x14      ; 0 0 0 1 0 1 0 0 (CURSOR/DISPLAY
SHIFT)
                                        ;          | | | |_|___ don't care
                                        ;          | |_|__ cursor/display shift
                                        ;          |        00 = cursor shift left
                                        ;          |        01 = cursor shift right
                                        ;          |        10 = cursor and display
                                        ;          |             shifted left
                                        ;          |        11 = cursor and display
                                        ;          |             shifted right
                                        ;          |___ COMMAND BIT
          call      send8
;**********************|
;    clear display      |
;**********************|
          movlw     0x01      ; 0 0 0 0 0 0 0 1 (CLEAR DISPLAY)
                                        ;                   |___ COMMAND BIT
          call      send8
; Per documentation
          call      delay_5   ; Test for busy
          return


;=====================
;  Procedure to delay
;   42 microseconds
;=====================
delay_125:
          movlw     D'42'               ; Repeat 42 machine cycles
          movwf     count1              ; Store value in counter
repeat
          decfsz    count1,f            ; Decrement counter
          goto      repeat              ; Continue if not 0
          return                        ; End of delay


;=====================
;  Procedure to delay
;   5 milliseconds
;=====================
delay_5:
          movlw     D'42'               ; Counter = 41
          movwf     count2              ; Store in variable
```

```
delay
        call    delay_125         ; Delay
        decfsz  count2,f          ; 40 times = 5 milliseconds
        goto    delay
        return                    ; End of delay
;========================
;    pulse E line
;========================
pulseE
        bsf     PORTE,E_line      ; Pulse E line
        nop
        bcf     PORTE,E_line
        return


;============================
;   long delay sub-routine
;============================
long_delay
                movlw   D'200'    ; w delay count
                movwf   J         ; J = w
jloop:  movwf   K                 ; K = w
kloop:
        decfsz  K,f               ; K = K-1, skip next if zero
        goto    kloop
        decfsz  J,f               ; J = J-1, skip next if zero
        goto    jloop
        return


;========================
;   send 2 nibbles in
;     4-bit mode
;========================
; Procedure to send two 4-bit values to Port-B lines
; 7, 6, 5, and 4. High-order nibble is sent first
; ON ENTRY:
;         w register holds 8-bit value to send
send8:
        movwf   store1            ; Save original value
        call    merge4            ; Merge with Port-B
; Now w has merged byte
        movwf   PORTD             ; w to Port D
        call    pulseE            ; Send data to LCD
; High nibble is sent
        movf    store1,w          ; Recover byte into w
        swapf   store1,w          ; Swap nibbles in w
        call    merge4
        movwf   PORTD
        call    pulseE            ; Send data to LCD
```

```
            call      delay_125
            return
;==========================
;        merge bits
;==========================
; Routine to merge the 4 high-order bits of the
; value to send with the contents of Port-B
; so as to preserve the 4 low-bits in Port-B
; Logic:
;        AND value with 1111 0000 mask
;        AND Port-B with 0000 1111 mask
;        Now low nibble in value and high nibble in
;        Port-B are all 0 bits:
;             value = vvvv 0000
;            Port-B = 0000 bbbb
;        OR value and Port-B resulting in:
;                    vvvv bbbb
; ON ENTRY:
;        w contain value bits
; ON EXIT:
;        w contains merged bits
merge4:
            andlw     b'11110000'     ; ANDing with 0 clears the
                                      ; bit. ANDing with 1 preserves
                                      ; the original value.
            movwf     store2          ; Save result in variable
            movf      PORTD,w         ; Port-B to w register.
            andlw     b'00001111'     ; Clear high nibble in Port-B
                                      ; and preserve low nibble
            iorwf     store2,w        ; OR two operands in w.
            return


;==========================
;   Set address register
;       to LCD line 2
;==========================
; ON ENTRY:
;        Address of LCD line 2 in constant LCD_2
line2:
            bcf       PORTE,E_line    ; E line low
            bcf       PORTE,RS_line   ; RS line low, set up for
control
            call      delay_5         ; Busy?
; Set to second display line
            movlw     LCD_2           ; Address with high-bit set
            call      send8
; Set RS line for data
            bsf       PORTE,RS_line   ; RS = 1 for data
```

```
            call    delay_5              ; Busy?
            return

;===========================
;   Set address register
;       to LCD line 1
;===========================
; ON ENTRY:
;          Address of LCD line 1 in constant LCD_1
line1:
            bcf     PORTE,E_line         ; E line low
            bcf     PORTE,RS_line        ; RS line low, set up for
control
            call    delay_5              ; busy?
; Set to second display line
            movlw   LCD_1                ; Address and command bit
            call    send8                ; 4-bit routine
; Set RS line for data
            bsf     PORTE,RS_line        ; Set up for data
            call    delay_5              ; Busy?
            return

;===========================
;   scroll to LCD line 2
;===========================
; Procedure to count the number of characters displayed on
; each LCD line. If the number reaches the value in the
; constant LCDlimit, then display is scrolled to the second
; LCD line. If at the end of the second line, then LCD is
; reset to the first line.
LCDscroll:
            incf    LCDcount,f           ; Bump counter
; Test for line limit
            movf    LCDcount,w
            sublw   LCDlimit             ; Count minus limit
            btfss   STATUS,Z             ; Is count minu limit = 0?
            goto    scrollExit           ; Go if not at end of line
; At this point the end of the LCD line was reached
; Test if this is also the end of the second line
            movf    LCDline,w
            sublw   0x01                 ; Is it line 1?
            btfsc   STATUS,Z             ; Is LCDline minus 1 = 0?
            goto    line2End             ; Go if end of second line
; At this point it is the end of the top LCD line
            call    line2                ; Scroll to second line
            clrf    LCDcount             ; Reset counter
            incf    LCDline,f            ; Bump line counter
            goto    scrollExit
```

```
; End of second LCD line
line2End:
        call      initLCD           ; Reset
        clrf      LCDcount          ; Clear counters
        clrf      LCDline
        call      line1             ; Display to first line
scrollExit:
        return


;================================================================
;                    communications procedures
;================================================================
; Initizalize serial port for 2400 baud, 8 bits, no parity,
; 1 stop
InitSerial:
        Bank1                ; Macro to select bank1
; Bits 6 and 7 of Port C are multiplexed as TX/CK and RX/DT
; for USART operation. These bits must be set to input in the
; TRISC register
        movlw     b'11000000'       ; Bits for TX and RX
        iorwf     TRISC,f           ; OR into Trisc register
; The asynchronous baud rate is calculated as follows:
;                        Fosc
;              ABR = ---------
;                       S*(x+1)
; where x is value in the SPBRG register and S is 64 if the high
; baud rate select bit (BRGH) in the TXSTA control register is
; clear, and 16 if the BRGH bit is set. For setting to 9600 baud
; using a 4 MHz oscillator at a high-speed baud rate the formula
; is:
;            4,000,000   4,000,000
;            --------- = --------- = 9,615 baud (0.16% error)
;            16*(25+1)      416
;
; At slow speed (BRGH = 0)
;            4,000,000   4,000,000
;            --------- = --------- = 2,403.85 (0.16% error)
;            64*(25+1)     1,664
;
        movlw     spbrgVal          ; Value in spbrgVal = 25
        movwf     SPBRG             ; Place in baud rate generator
; TXSTA (Transmit Status and Control Register) bitmap:
;    7  6  5  4  3  2  1  0  <== bits
;    |  |  |  |  |  |  |  |_____ TX9D 9th data bit on ?
;    |  |  |  |  |  |  |          (used for parity)
;    |  |  |  |  |  |  |_____ TRMT Transmit Shift Register
;    |  |  |  |  |  |             1 = TSR empty
;    |  |  |  |  |  |           * 0 = TSR full
```

```
;   |   |   |   |   |   |   |_____ BRGH High Speed Baud Rate
;   |   |   |   |   |   |                   (Asynchronous mode only)
;   |   |   |   |   |   |                   1 = high speed (* 4)
;   |   |   |   |   |   |                 * 0 = low speed
;   |   |   |   |   |_____ NOT USED
;   |   |   |   |_____ SYNC USART Mode Select
;   |   |   |   |                   1 = syncrhonous mode
;   |   |   |   |                 * 0 = asynchronous mode
;   |   |   |_____ TXEN Transmit Enable
;   |   |   |                 * 1 = transmit enabled
;   |   |   |                   0 = transmit disabled
;   |   |_____ TX9 Enable 9-bit Transmit
;   |   |                   1 = 9-bit transmission mode
;   |   |                 * 0 = 8-bit mode
;   |_____ CSRC Clock Source Select
;   |                   Not used in asynchronous mode
;   |                   Synchronous mode:
;   |                     1 = Master Mode (internal clock)
;   |                   * 0 = Slave mode (external clock)
; Setup value: 0010 0000 = 0x20
          movlw    0x20      ; Enable transmission and high baud
                             ; rate
          movwf    TXSTA
          Bank0              ; Bank 0
; RCSTA (Receive Status and Control Register) bitmap:
;   7   6   5   4   3   2   1   0   <== bits
;   |   |   |   |   |   |   |   |   |_____ RX9D 9th data bit received?
;   |   |   |   |   |   |   |   |               (can be parity bit)
;   |   |   |   |   |   |   |   |_____ OERR Overrun errror?
;   |   |   |   |   |   |   |                   1 = error (cleared by software)
;   |   |   |   |   |   |   |_____ FERR Framing Error?
;   |   |   |   |   |   |                   1 = error
;   |   |   |   |   |_____ NOT USED
;   |   |   |   |_____ CREN Continuous Receive Enable
;   |   |   |                   Asynchronous mode:
;   |   |   |                 *   1 = Enables continuous receive
;   |   |   |                     0 = Disables continuous receive
;   |   |   |                   Synchronous mode:
;   |   |   |                     1 = Enables until CREN cleared
;   |   |   |                     0 = Disables continuous receive
;   |   |   |_____ SREN Single Receive Enable
;   |   |                   ? Asynchronous mode =  don't care
;   |   |                   Synchronous master mode:
;   |   |                     1 = Enable single receive
;   |   |                     0 = Disable single receive
;   |_____ RX9 9th-bit Receive Enable
;   |                   1 = 9-bit reception
;   |                 * 0 = 8-bit reception
```

```
;     |_____ SPEN Serial Port Enable
;                              * 1 = RX/DT and TX/CK are serial pins

;                                0 = Serial port disabled
; Setup value: 1001 0000 = 0x90
           movlw    0x90              ; Enable serial port and
                                      ; continuous reception
           movwf    RCSTA
;
           clrf     errorFlags        ; Clear local error flags
                                      ; register
           return

;===============================
;       transmit data
;===============================
; Test for Transmit Register Empty and transmit data in w
SerialSend:
           Bank0                      ; Select bank 0
           btfss    PIR1,TXIF         ; check if transmitter busy
           goto     $-1               ; wait until transmitter is
                                      ; not busy
           movwf    TXREG             ; and transmit the data
           return
;===============================
;       receive data
;===============================
; Procedure to test line for data received and return value
; in w. Overrun and framing errors are detected and
; remembered in the variable errorFlags, as follows:
;       7  6  5  4  3  2  1  0   <== errorFlags
;       |-- not used - |  |__|_____ overrun error
;                         |_____ framing error
SerialRcv:
           clrf     newData ; Clear new data received register
           Bank0            ; Select bank 0
; Bit 5 (RCIF) of the PIR1 Register is clear if the USART
; receive buffer is empty. If so, no data has been received
           btfss    PIR1,RCIF         ; Check for received data
           return                     ; Exit if no data
; At this point, data has been received. First eliminate
; possible errors: overrun and framing.
; Bit 1 (OERR) of the RCSTA register detects overrun
; Bit 2 (FERR( of the RCSTA register detects framing error
           btfsc    RCSTA,OERR        ; Test for overrun error
           goto     OverErr           ; Error handler
           btfsc    RCSTA,FERR        ; Test for framing error
           goto     FrameErr          ; Error handler
```

```
; At this point no error was detected
; Received data is in the USART RCREG register
        movf      RCREG,w          ; get received data
        bsf       newData,7        ; Set bit 7 to indicate new
                                   ; data
; Clear error flags
        clrf      errorFlags
        return
;==========================
;     error handlers
;==========================
OverErr:
        bsf       errorFlags,0     ; Bit 0 is overrun error
; Reset system
        bcf       RCSTA,CREN       ; Clear continuous receive bit
        bsf       RCSTA,CREN       ; Set to re-enable reception
        return
; error because FERR framing error bit is set
; can do special error handling here - this code simply clears
; and continues
FrameErr:
        bsf       errorFlags,1     ; Bit 1 is framing error
        movf      RCREG,W          ; Read and throw away bad data
        return

        end
```

20.6.6 SerIntLCD Program

```
; File name: SerIntLCD.asm
; Last revision: May 14, 2011
; Authors: Sanchez and Canton
; Processor: 16F877
;
; Interrupt-Driven version of the SerComLCD program
;
; Description:
; Decode 4 x 4 keypad, display scan code in LCD, and send
; ASCII character through the serial port. Also receive
; data through serial port and display on LCD. LCD lines
; are scrolled by program.
; Default serial line setting:
;              2400 baud
;              no partity
;              1 stop bit
;              8 character bits
;
; Program uses 4-bit PIC-to-LCD interface.
```

```
; Code assumes that LCD is driven by Hitachi HD44780
; controller and PIC 16F977. Display supports two lines
; each one with twenty characters. The length, wiring and
; base address of each display line is stored in #define
; statements. These statements can be edited to accommodate
; a different setup.
; Keypad switch wiring (values are scan codes):
;         --- KEYPAD --
;         0   1   2   3    <= port B0 |
;         4   5   6   7    <= port B1 |-- ROWS = OUTPUTS
;         8   9   A   B    <= port B2 |
;         C   D   E   F    <= port B3 |
;         |   |   |   |
;         |   |   |   |_____ port B4 |
;         |   |   |_____ port B5 |-- COLUMNS = INPUTS
;         |   |_____ port B6 |
;         |_____ port B7 |
;
; Operations:
; 1. Key press action generates a scan code in the range
;    0x0 to 0xf.
; 2. Scan code is converted to an ASCII digit and displayed
;    on the LCD. LCD lines are scrolled as end-of-line is
;    reached.
; 3. Characters typed on the keypad are also transmitted
;    through the serial port.
; 4. Received characters generate an interrupt. The interrupt
;    handler displays received characters on the LCD.
;
; WARNING:
; Code assumes 4 MHz clock. Delay routines must be
; edited for faster clock. Clock speed also determines
; values for baud rate setting (see spbrgVal constant).
;
;============================
;       16F877 switches
;============================
; Switches used in __config directive:
;   _CP_ON          Code protection ON/OFF
; * _CP_OFF
; * _PWRTE_ON       Power-up timer ON/OFF
;   _PWRTE_OFF
;   _BODEN_ON       Brown-out reset enable ON/OFF
; * _BODEN_OFF
; * _PWRTE_ON       Power-up timer enable ON/OFF
;   _PWRTE_OFF
;   _WDT_ON         Watchdog timer ON/OFF
; * _WDT_OFF
```

```
;   _LPV_ON          Low voltage IC programming enable ON/OFF
; * _LPV_OFF
;   _CPD_ON          Data EE memory code protection ON/OFF
; * _CPD_OFF
; OSCILLATOR CONFIGURATIONS:
;   _LP_OSC          Low power crystal oscillator
;   _XT_OSC          External parallel resonator/crystal oscillator

; * _HS_OSC          High speed crystal resonator
;   _RC_OSC          Resistor/capacitor oscillator
; |                  (simplest, 20% error)
; |
; |_____   * indicates setup values presently selected

        processor        16f877              ; Define processor
        #include <p16f877.inc>
        __CONFIG _CP_OFF & _WDT_OFF & _BODEN_OFF & _PWRTE_ON &
_HS_OSC & _WDT_OFF & _LVP_OFF & _CPD_OFF


; __CONFIG directive is used to embed configuration data
; within the source file. The labels following the directive
; are located in the corresponding .inc file.
;===============================================================
;                         M A C R O S
;===============================================================
; Macros to select the register banks
Bank0   MACRO                ; Select RAM bank 0
        bcf      STATUS,RP0
        bcf      STATUS,RP1
        ENDM

Bank1   MACRO                ; Select RAM bank 1
        bsf      STATUS,RP0
        bcf      STATUS,RP1
        ENDM

Bank2   MACRO                ; Select RAM bank 2
        bcf      STATUS,RP0
        bsf      STATUS,RP1
        ENDM

Bank3   MACRO                ; Select RAM bank 3
        bsf      STATUS,RP0
        bsf      STATUS,RP1
        ENDM
;=======================================================
;                   constant definitions
;   for PIC-to-LCD pin wiring and LCD line addresses
```

```
;========================================================
#define E_line 1          ;|
#define RS_line 0         ;| -- from wiring diagram
#define RW_line 2         ;|
; LCD line addresses (from LCD data sheet)
#define LCD_1 0x80        ; First LCD line constant
#define LCD_2 0xc0        ; Second LCD line constant
#define LCDlimit .20; Number of characters per line
#define  spbrgVal .25; For 2400 baud on 4 MHz clock
; Note: The constants that define the LCD display
;       line addresses have the high-order bit set
;       so as to meet the requirements of controller
;       commands.
;
;========================================================
;                variables in PIC RAM
;========================================================
; Local variables
        cblock   0x20     ; Start of block
        count1            ; Counter # 1
        count2            ; Counter # 2
        count3            ; Counter # 3
        J                 ; Counter J
        K                 ; Counter K
        store1            ; Local storage
        store2
; For LCDscroll procedure
        LCDcount ; Counter for characters per line
        LCDline           ; Current display line (0 or 1)
; Keypad processing variables
        keyMask           ; For keypad processing
        rowMask           ; For masking-off key rows
        rowCode           ; Row addend for calculating scan code
        rowCount ; Counter for key rows (0 to 3)
        scanCode ; Final key code
        newScan           ; 0 if no new scan code detected
; Communications variables
        ascVal
        errorFlags
; Temporary storage used by interrupt handler
        tempW
        tempStatus
        tempPclath
        tempFsr
        endc


;============================================================
;                         P R O G R A M
```

```
;================================================================
        org     0                       ; start at address
        goto    main
; Space for interrupt handlers
        org     0x04
InterruptCode:
        goto    IntServ                 ; Interrupt service routine


;================================================================
;                         main program
;================================================================
main:
; Wiring:
;     LCD data to Port D, lines 0 to 7
;     E line -> port E, 1
;     RW line -> port E, 2
;     RS line -> port E, 0
; Set port D and E for output
; Data memory bank selection bits:
; RP1:RP0              Bank
;   0:0                 0      Ports A,B,C,D, and E
;   0:1                 1      Tris A,B,C,D, and E
;   1:0                 2
;   1:1                 3
; First, initialize Port-B by clearing latches
        clrf    STATUS
        clrf    PORTB
; Select bank 1 to tris Port D for output
        Bank1
; Tris Port D for output. Port D lines 4 to 7 are wired
; to LCD data lines. Port D lines 0 to 4 are wired to LEDs.
        movlw   B'00000000'
        movwf   TRISD                   ; and Port D
; By default Port-A lines are analog. To configure them
; as digital code must set bits 1 and 2 of the ADCON1
; register (in bank 1)
        movlw   0x06        ; binary 0000 0110  is code to
                            ; make all Port-A lines digital
        movwf   ADCON1
; Port-B lines are wired to keypad switches as follows:
;   7 6 5 4 3 2 1 0
;   | | | | |_|_|_|_____ switch rows (output)
;   |_|_|_|_____ switch columns (input)
; rows must be defined as output and columns as input
        movlw   b'11110000'
        movwf   TRISB
; Tris port E for output
        movlw   B'00000000'
```

```
            movwf    TRISE             ; Tris port E
; Enable Port-B pullups for switches in OPTION register
;    7  6  5  4  3  2  1  0 <= OPTION bits
;    |  |  |  |  |  |  |__|__|_____ PS2-PS0 (prescaler bits)
;    |  |  |  |  |  |                Values for Timer0
;    |  |  |  |  |  |                000 = 1:2    001 = 1:4
;    |  |  |  |  |  |                010 = 1:8    011 = 1:16
;    |  |  |  |  |  |                100 = 1:32   101 = 1:64
;    |  |  |  |  |  |                110 = 1:128 *111 = 1:256
;    |  |  |  |  |_____ PSA (prescaler assign)
;    |  |  |  |                      *1 = to WDT
;    |  |  |  |                      0 = to Timer0
;    |  |  |  |_____ TOSE (Timer0 edge select)
;    |  |  |                         *0 = increment on low-to-high
;    |  |  |                         1 = increment on high-to-low
;    |  |  |_____ TOCS (TMR0 clock source)
;    |  |                            *0 = internal clock
;    |  |                            1 = RA4/TOCKI bit source
;    |  |_____ INTEDG (Edge select)
;    |                               *0 = falling edge
;    |_____ RBPU (Pullup enable)
;                                     *0 = enabled
;                                     1 = disabled
        movlw    b'00001000'
        movwf    OPTION_REG
; Back to bank 0
        Bank0
; Initialize serial port for 9600 baud, 8 bits, no parity
; 1 stop
        call     InitSerial
; Test serial transmission by sending RDY-
        movlw    'R'
        call     SerialSend
        movlw    'D'
        call     SerialSend
        movlw    'Y'
        call     SerialSend
        movlw    '-'
        call     SerialSend
        movlw    0x20
        call     SerialSend
; Clear all output lines
        movlw    b'00000000'
        movwf    PORTD
        movwf    PORTE
; Wait and initialize HD44780
        call     delay_5 ; Allow LCD time to initialize itself
        call     initLCD ; Then do forced initialization
```

```
        call    delay_5 ; (Wait probably not necessary)
; Clear character counter and line counter variables
        clrf    LCDcount
        clrf    LCDline
; Set display address to start of second LCD line
        call    line1
;=============================================================
;                       scan keypad
;=============================================================
; Keypad switch wiring:
;           x   x   x   x    <= Port B0 |
;           x   x   x   x    <= Port B1 |-- ROWS = OUTPUTS
;           x   x   x   x    <= port B2 |
;           x   x   x   x    <= port B3 |
;           |   |   |   |
;           |   |   |   |_____ port B4 |
;           |   |   |_____ port B5 |-- COLUMNS = INPUTS
;           |   |_____ port B6 |
;           |_____ port B7 |
; Switches are connected to Port-B lines
; Clear scan code register
        clrf    scanCode
;=============================
;   scan keypad and display
;=============================
keyScan:
; Port-B lines are wired to pushbutton switches as follows:
;   7 6 5 4 3 2 1 0
;   | | | | |_|_|_|_____ switch rows (output)
;   |_|_|_|_____ switch columns (input)
; Keypad processing:
; switch rows are successively grounded (row = 0)
; Then column values are tested. If a column returns 0
; in a 0 row, that switch is down.
; Initialize row code addend
        clrf    rowCode         ; First row is code 0
        clrf    newScan         ; No new scan code detected
; Initialize row count
        movlw   D'4'            ; Four rows
        movwf   rowCount        ; Register variable
        movlw   b'11111110'     ; All set but LOB
        movwf   rowMask
keyLoop:
; Initialize row eliminator mask:
; The row mask is ANDed with the key mask to successively
; mask off each row, for example:
;
;                       |----- row 3
```

```
;                        ||---- row 2
;                        |||--- row 1
;                        ||||-- row 0
;              0000 1111 <= key mask
;       AND    1111 1101 <= mask for row 1
;              ----------
;              0000 1101 <= row 1 is masked off
;
; The row mask, which is initally 1111 1110, is rotated left
; through the carry in order to mask off the next row
        movlw   b'00001111'       ; Mask off all lines
        movwf   keyMask           ; To local register
; Set row mask for current row
        movf    rowMask,w         ; Mask to w
        andwf   keyMask,f         ; Update key mask
        movf    keyMask,w         ; Key mask to w
        movwf   PORTB             ; Mask off Port-B lines
; Read Port-B lines 4 to 7 (columns are input)
        btfss   PORTB,4
        call    col0              ; Key column procedures
        btfss   PORTB,5
        call    col1
        btfss   PORTB,6
        call    col2
        btfss   PORTB,7
        call    col3
; Index to next row by adding 4 to row code
        movf    rowCode,w         ; Code to w
        addlw   D'4'
        movwf   rowCode
;=========================
;     shift row mask
;=========================
; Set the carry flag
        bsf     STATUS,C
        rlf     rowMask,f         ; Rotate mask bits in storage
;=========================
;     end of keypad?
;=========================
; Test for last key row (maximum count is 4)
        decfsz  rowCount,f  ; Decrement counter
        goto    keyLoop
;=============================================================
;=============================================================
;                  display and send data
;=============================================================
;=============================================================
; At this point all keys have been tested.
```

```
; Variable newScan = 0 if no new scan code detected, else
; variable scanCode holds scan code
        movf    newScan,f       ; Copy onto itself
        btfsc   STATUS,Z ; Is it zero
        goto    ScanExit
; At this point a new scan code is detected
        movf    scanCode,w      ; To w
; If scan code is in the range 0 to 9, that is, a decimal
; digit, then ASCII conversion consists of adding 0x30.
; If the scan code represents one of the hex letters
; (0xa to 0xf), then ASCII conversion requires adding
; 0x37
        sublw   0x09            ; 9 - w
; if w from 0 to 9 then 9 - w = positive (C flag = 1)
; if w = 0xa then 9 - 10 = -1 (C flag = 0)
; if w = 0xc then 9 - 12 = -2 (C flag = 0)
        btfss   STATUS,C ; Test carry flag
        goto    hexLetter       ; Carry clear, must be a
letter
; At this point scan code is a decimal digit in the
; range 0 to 9. Convert to ASCII by adding 0x30
        movf    scanCode,w      ; Recover scan code
        addlw   0x30            ; Convert to ASCII
        goto    displayDig
hexLetter:
        movf    scanCode,w      ; Recover scan code
        addlw   0x37            ; Convert to ASCII
displayDig:
; Store so it can be sent
        movwf   ascVal
        call    send8           ; Display routine
        call    LCDscroll
        call    long_delay      ; Debounce
; Recover ASCII
        movf    ascVal,w
        call    SerialSend
ScanExit:
        goto    keyScan         ; Continue
;===========================
;   calculate scan code
;===========================
; The column position is added to the row code (stored
; in rowCode register). Sum is the scan code
col0:
        movf    rowCode,w       ; Row code to w
        addlw   0x00            ; Add 0
        movwf   scanCode        ; Final value
        incf    newScan,f       ; New scan code
```

```
        return
col1:
        movf    rowCode,w           ; Row code to w
        addlw   0x01                ; Add 1
        movwf   scanCode
        incf    newScan,f
        return

col2:
        movf    rowCode,w           ; Row code to w
        addlw   0x02                ; Add 2
        movwf   scanCode
        incf    newScan,f
        return

col3:
        movf    rowCode,w           ; Row code to w
        addlw   0x03                ; Add 3
        movwf   scanCode
        incf    newScan,f
        return
;===============================================================
;===============================================================
;                 L O C A L    P R O C E D U R E S
;===============================================================
;===============================================================
;==========================
; init LCD for 4-bit mode
;==========================
initLCD:
; Initialization for Densitron LCD module as follows:
;    4-bit interface
;    2 display lines of 16 characters each
;    cursor on
;    left-to-right increment
;    cursor shift right
;    no display shift
;=====================|
;   set command mode  |
;=====================|
        bcf             PORTE,E_line        ; E line low
        bcf             PORTE,RS_line       ; RS line low
        bcf             PORTE,RW_line       ; Write mode
        call    delay_125                   ; delay 125
microseconds
;*********************|
;     FUNCTION SET    |
;*********************|
```

```
          movlw    0x28      ; 0 0 1 0 1 0 0 0 (FUNCTION SET)
                             ;         | | | |__ font select:
                             ;         | | |    1 = 5x10 in 1/8 or 1/11
                             ;         | | |    0 = 1/16 dc
                             ;         | | |___ Duty cycle select
                             ;         | |      0 = 1/8 or 1/11
                             ;         | |      1 = 1/16)
                             ;         | |___ Interface width
                             ;         |      0 = 4 bits
                             ;         |      1 = 8 bits
                             ;         |___ FUNCTION SET COMMAND
          call     send8     ; 4-bit send routine

; Set 4-bit mode command must be repeated
          movlw    0x28
          call     send8

;***********************|
; DISPLAY AND CURSOR ON |
;***********************|
          movlw    0x0e      ; 0 0 0 0 1 1 1 0 (DISPLAY ON/OFF)
                             ;           | | | |___ Blink character
                             ;           | | |     1 = on, 0 = off
                             ;           | | |___ Cursor on/off
                             ;           | |       1 = on, 0 = off
                             ;           | |____ Display on/off
                             ;           |        1 = on, 0 = off
                             ;           |_____ COMMAND BIT
          call     send8
;***********************|
;    set entry mode     |
;***********************|
          movlw    0x06      ; 0 0 0 0 0 1 1 0 (ENTRY MODE SET)
                             ;             | | |___ display shift
                             ;             | |      1 = shift
                             ;             | |      0 = no shift
                             ;             | |____ increment mode
                             ;             |       1 = left-to-right
                             ;             |       0 = right-to-left
                             ;             |___ COMMAND BIT
          call     send8

;***********************|
; cursor/display shift  |
;***********************|
          movlw    0x14      ; 0 0 0 1 0 1 0 0 (CURSOR/DISPLAY
SHIFT)
                             ;         | | | |_|___ don't care
```

```
                          ;              |  |_|__ cursor/display shift
                          ;              |          00 = cursor shift left
                          ;              |          01 = cursor shift right
                          ;              |          10 = cursor and display
                          ;              |                shifted left
                          ;              |          11 = cursor and display
                          ;              |                shifted right
                          ;              |___ COMMAND BIT
            call    send8
;*********************|
;    clear display    |
;*********************|
            movlw   0x01      ; 0 0 0 0 0 0 0 1 (CLEAR DISPLAY)
                          ;                 |___ COMMAND BIT
            call    send8
; Per documentation
            call    delay_5  ; Test for busy
            return


;=====================
;  Procedure to delay
;   42 microseconds
;=====================
delay_125:
            movlw   D'42'            ; Repeat 42 machine cycles
            movwf   count1           ; Store value in counter
repeat:
            decfsz  count1,f         ; Decrement counter
            goto    repeat           ; Continue if not 0
            return                   ; End of delay


;=====================
;  Procedure to delay
;   5 milliseconds
;=====================
delay_5:
            movlw   D'42'            ; Counter = 41
            movwf   count2           ; Store in variable
delay:
            call    delay_125        ; Delay
            decfsz  count2,f         ; 40 times = 5 milliseconds
            goto    delay
            return                   ; End of delay
;=====================
;    pulse E line
;=====================
pulseE
            bsf     PORTE,E_line     ; Pulse E line
```

```
        nop
        bcf         PORTE,E_line
        return


;==============================
;   long delay sub-routine
;==============================
long_delay
        movlw       D'200'              ; w delay count
        movwf       J                   ; J = w
jloop:
        movwf       K                   ; K = w
kloop:
        decfsz      K,f                 ; K = K-1, skip next if zero
        goto        kloop
        decfsz      J,f                 ; J = J-1, skip next if zero
        goto        jloop
        return


;========================
;   send 2 nibbles in
;     4-bit mode
;========================
; Procedure to send two 4-bit values to Port-B lines
; 7, 6, 5, and 4. High-order nibble is sent first
; ON ENTRY:
;         w register holds 8-bit value to send
send8:
        movwf       store1              ; Save original value
        call        merge4              ; Merge with Port-B
; Now w has merged byte
        movwf       PORTD               ; w to Port D
        call        pulseE              ; Send data to LCD
; High nibble is sent
        movf        store1,w            ; Recover byte into w
        swapf       store1,w            ; Swap nibbles in w
        call        merge4
        movwf       PORTD
        call        pulseE              ; Send data to LCD
        call        delay_125
        return
;==========================
;       merge bits
;==========================
; Routine to merge the 4 high-order bits of the
; value to send with the contents of Port-B
; so as to preserve the 4 low-bits in Port-B
; Logic:
```

```
;          AND value with 1111 0000 mask
;          AND Port-B with 0000 1111 mask
;          Now low nibble in value and high nibble in
;          Port-B are all 0 bits:
;               value = vvvv 0000
;               Port-B = 0000 bbbb
;          OR value and Port-B resulting in:
;                        vvvv bbbb
;  ON ENTRY:
;       w contain value bits
;  ON EXIT:
;       w contains merged bits
merge4:
          andlw    b'11110000'         ; ANDing with 0 clears the
                                       ; bit. ANDing with 1 preserves
                                       ; the original value
          movwf    store2              ; Save result in variable
          movf     PORTD,w             ; Port-B to w register
          andlw    b'00001111'         ; Clear high nibble in Port-B
                                       ; and preserve low nibble
          iorwf    store2,w            ; OR two operands in w
          return


;==========================
;   Set address register
;       to LCD line 2
;==========================
; ON ENTRY:
;          Address of LCD line 2 in constant LCD_2
line2:
          bcf      PORTE,E_line        ; E line low
          bcf      PORTE,RS_line       ; RS line low, setup for
control
          call     delay_5             ; Busy?
; Set to second display line
          movlw    LCD_2               ; Address with high-bit set
          call     send8
; Set RS line for data
          bsf      PORTE,RS_line       ; RS = 1 for data
          call     delay_5             ; Busy?
          return


;==========================
;   Set address register
;       to LCD line 1
;==========================
; ON ENTRY:
;          Address of LCD line 1 in constant LCD_1
```

```
line1:
        bcf         PORTE,E_line        ; E line low
        bcf         PORTE,RS_line       ; RS line low, set up for
                                        ; control
        call        delay_5             ; busy?
; Set to second display line
        movlw       LCD_1               ; Address and command bit
        call        send8               ; 4-bit routine
; Set RS line for data
        bsf         PORTE,RS_line       ; Set up for data
        call        delay_5             ; Busy?
        return


;===========================
;   scroll to LCD line 2
;===========================
; Procedure to count the number of characters displayed on
; each LCD line. If the number reaches the value in the
; constant LCDlimit, then display is scrolled to the second
; LCD line. If at the end of the second line, then LCD is
; reset to the first line.
LCDscroll:
        incf        LCDcount,f          ; Bump counter
; Test for line limit
        movf        LCDcount,w
        sublw       LCDlimit            ; Count minus limit
        btfss       STATUS,Z            ; Is count minus limit = 0?
        goto        scrollExit          ; Go if not at end of line
; At this point the end of the LCD line was reached
; Test if this is also the end of the second line
        movf        LCDline,w
        sublw       0x01                ; Is it line 1?
        btfsc       STATUS,Z            ; Is LCDline minus 1 = 0?
        goto        line2End            ; Go if end of second line
; At this point it is the end of the top LCD line
        call        line2               ; Scroll to second line
        clrf        LCDcount ; Reset counter
        incf        LCDline,f           ; Bump line counter
        goto        scrollExit
; End of second LCD line
line2End:
        call        initLCD             ; Reset
        clrf        LCDcount ; Clear counters
        clrf        LCDline
        call        line1               ; Display to first line
scrollExit:
        return
```

```
;================================================================
;                   communications procedures
;================================================================
; Initizalize serial port for 2400 baud, 8 bits, no parity,
; 1 stop
InitSerial:
        Bank1                       ; Macro to select bank1
; Bits 6 and 7 of Port C are multiplexed as TX/CK and RX/DT
; for USART operation. These bits must be set to input in the
; TRISC register
        movlw    b'11000000'        ; Bits for TX and RX
        iorwf    TRISC,f            ; OR into Trisc register
; The asynchronous baud rate is calculated as follows:
;                      Fosc
;             ABR = ---------
;                     S*(x+1)
; where x is value in the SPBRG register and S is 64 if the high
; baud rate select bit (BRGH) in the TXSTA control register is
; clear, and 16 if the BRGH bit is set. For setting to 9600 baud
; usign a 4-MHz oscillator at a high-speed baud rate the formula
; is:
;            4,000,000   4,000,000
;            --------- = --------- = 9,615 baud (0.16% error)
;            16*(25+1)      416
;
; At slow speed (BRGH = 0)
;            4,000,000   4,000,000
;            --------- = ---------- = 2,403.85 (0.16% error)
;            64*(25+1)     1,664
;
        movlw    spbrgVal ; Value in spbrgVal = 25
        movwf    SPBRG               ; Place in baud rate generator
; TXSTA (Transmit Status and Control Register) bitmap:
;   7  6  5  4  3  2  1  0  <== bits
;   |  |  |  |  |  |  |  |  |_____ TX9D 9nth data bit on
;   |  |  |  |  |  |  |  |          ? (used for parity)
;   |  |  |  |  |  |  |  |_____ TRMT Transmit Shift Register
;   |  |  |  |  |  |  |             1 = TSR empty
;   |  |  |  |  |  |  |             * 0 = TSR full
;   |  |  |  |  |  |  |_____ BRGH High Speed Baud Rate
;   |  |  |  |  |  |               (Asynchronous mode only)
;   |  |  |  |  |  |               1 = high speed (* 4)
;   |  |  |  |  |  |               * 0 = low speed
;   |  |  |  |  |_____ NOT USED
;   |  |  |  |_____ SYNC USART Mode Select
;   |  |  |               1 = syncrhonous mode
;   |  |  |               * 0 = asynchronous mode
;   |  |  |_____ TXEN Transmit Enable
```

```
;   |   |                        * 1 = transmit enabled
;   |   |                          0 = transmit disabled
;   |   |_____ TX9 Enable 9-bit Transmit
;   |                          1 = 9-bit transmission mode
;   |                        * 0 = 8-bit mode
;   |_____ CSRC Clock Source Select
;                             Not used in asynchronous mode
;                             Synchronous mode:
;                               1 = Master Mode (internal clock)
;                             * 0 = Slave mode (external clock)
; Setup value: 0010 0000 = 0x20
        movlw   0x20      ; Enable transmission and low baud rate
        movwf   TXSTA
        Bank0                                     ; Bank 0
; RCSTA (Receive Status and Control Register) bitmap:
;   7  6  5  4  3  2  1  0  <== bits
;   |  |  |  |  |  |  |  |  |_____ RX9D 9th data bit received?
;   |  |  |  |  |  |  |  |          (can be parity bit)
;   |  |  |  |  |  |  |  |_____ OERR Overrun errror?
;   |  |  |  |  |  |  |             1 = error (cleared by software)
;   |  |  |  |  |  |  |_____ FERR Framing Error?
;   |  |  |  |  |  |               1 = error
;   |  |  |  |  |  |_____ NOT USED
;   |  |  |  |  |_____ CREN Continuous Receive Enable
;   |  |  |  |                   Asynchronous mode:
;   |  |  |  |                 *   1 = Enable continuous receive
;   |  |  |  |                     0 = Disables continuous receive
;   |  |  |  |                   Synchronous mode:
;   |  |  |  |                     1 = Enables until CREN cleared
;   |  |  |  |                     0 = Disables continous receive
;   |  |  |  |_____ SREN Single Receive Enable
;   |  |  |                   ? Asynchronous mode =  don't care
;   |  |  |                     Synchronous master mode:
;   |  |  |                       1 = Enable single receive
;   |  |  |                       0 = Disable single receive
;   |  |_____ RX9 9th-bit Receive Enable
;   |  |                     1 = 9-bit reception
;   |  |                   * 0 = 8-bit reception
;   |_____ SPEN Serial Port Enable
;                           * 1 = RX/DT and TX/CK are serial pins

;                             0 = Serial port disabled
; Setup value: 1001 0000 = 0x90
        movlw   0x90      ; Enable serial port and
                          ; continuous reception
        movwf   RCSTA
; Enable global and peripheral interrupts
;   7  6  5  4  3  2  1  0  <= INTCON bitmap
```

```
;   |   |   - unrelated --
;   |   |_____ Peripheral interrupts enable
;   |_____ Global interrupts enable
        movlw   b'11000000'
        movwf   INTCON
; Enable receive interrupt in PIE1 register
;  7  6  5  4  3  2  1  0  <= PIE1 bitmap
;         |_____ USART receive interrupt enable
        Bank1
        movlw   b'00100000'
        movwf   PIE1
; Clear error flags register
        Bank0
        clrf    errorFlags
        return


;==============================
;        transmit data
;==============================
; Test for Transmit Register Empty and transmit data in w
SerialSend:
        Bank0           ; Select bank 0
        btfss   PIR1,TXIF       ; check if transmitter busy
        goto    $-1     ;wait until transmitter is not busy
        movwf   TXREG   ;and transmit the data
        return


;===============================================================
;===============================================================
;        interrupt handler for received characters
;===============================================================
;===============================================================
IntServ:
        movwf   tempW           ; Save W
        movf    STATUS,W        ; Store STATUS in W
        clrf    STATUS          ; Select bank0
        movwf   tempStatus      ; Save STATUS
        movf    PCLATH,W        ; Store PCLATH in W
        movwf   tempPclath      ; Save PCLATH
        clrf    PCLATH          ; Select program memory page 0
        movf    FSR,W           ; Store FSR in W
        movwf   tempFsr         ; Save FSR value

; Test for received data interrupt
        Bank0                           ; select bank0
;  7  6  5  4  3  2  1  0  <= PIR1
;         |_____ (RCIF) USART receive interrupt
;                             flag
```

```
        btfsc    PIR1,RCIF            ; Test bit 5
        bsf               STATUS,RP0          ; Bank 1 if RCIF set
; 7  6  5  4  3  2  1  0  <= PIE1
;       |_____ (RCIE) Receive interrupt enable
;                            bit
        btfss    PIE1,RCIE           ; Test if interrupt is enabled
        goto     IntExit            ; Go if not enabled
;==============================
;       received data
;==============================
; Routine to handle received data. Overrun and framing
; errors are detected and remembered in the variable
; errorFlags, as follows:
;       7  6  5  4  3  2  1  0   <== errorFlags
;       |- not used -- |  |  |___ overrun error
;                         |_____ framing error
        Bank0                               ; Select bank 0
; Test for overrun and framing errors.
; Bit 1 (OERR) of the RCSTA register detects overrun
; Bit 2 (FERR) of the RCSTA register detects framing error
        btfsc    RCSTA,OERR          ; Test for overrun error
        goto     OverErr            ; Error handler
        btfsc    RCSTA,FERR          ; Test for framing error
        goto     FrameErr ; Error handler
; At this point no error was detected
; Received data is in the USART RCREG register
        movf     RCREG,w             ; Received data into w
        call     send8               ; Display in LCD
        call     LCDscroll           ; Scroll at end of line
; Clear error flags
        clrf     errorFlags
        goto     IntExit
;==========================
;       error handlers
;==========================
; Errors are returned as bits in the errorFlags register
;       7  6  5  4  3  2  1  0   <== errorFlags
;       |- not used -- |  |  |___ overrun error
;                         |_____ framing error
; Error responses to be made by main code
OverErr:
        bsf      errorFlags,0        ; Bit 0 is overrun error
; Reset system
        bcf      RCSTA,CREN          ; Clear continuous receive bit
        bsf      RCSTA,CREN          ; Set to re-enable reception
        goto     IntExit
FrameErr:
        bsf      errorFlags,1; Bit 1 is framing error
```

```
            movf      RCREG,W                ; Read and throw away bad data
;==============================
;    interrupt handler exit
;==============================
IntExit:
            Bank0
            movf      tempFsr,w              ; Recover FSR value
            movwf     FSR                          ; Restore in register
            movf      tempPclath,w           ; Recover PCLATH value
            movwf     PCLATH                 ; Restore in register
            movf      tempStatus,W           ; Recover STATUS
            movwf     STATUS                 ; Restore in register
            swapf     tempW,F                ; Swap file register in itself
            swapf     tempW,W                ; Restore in register
            retfie
            end
```

Resistor Color Codes

The resistor color coding system applies to carbon film, metal oxide film, fusible, precision metal film, and wirewound resistors of the axial lead type. This system is employed when the surface area is not sufficient to print the actual resistance value. Several color codes are used, the most common ones are the 4-band and 5-band codes. In the 4-band code, the first two bands represent the magnitude of the resistance, the third band is a multiplier for this value, and the fourth band encodes the error tolerance. In the 5-band code the first three bands represent the magnitude, the fourth band serves as a multiplier, and the fifth band is the error tolerance.

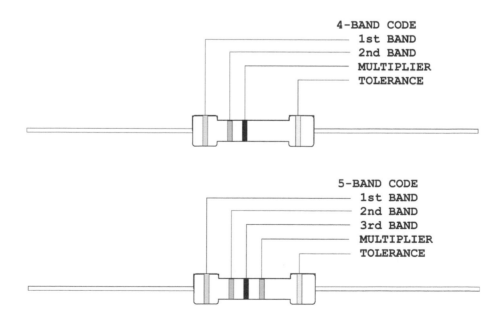

The color codes for the various bands are as follows:

COLOR	MAGNITUDE	MULTIPLIER	TOLERANCE
Black	0	1	
Brown	1	10	1%
Red	2	100	2%
Orange	3	1K	
Yellow	4	10K	
Green	5	100K	0.5%
Blue	6	1M	0.25%
Violet	7	10M	0.10%
Grey	8		0.05%
White	9		
Gold		0.1	5%
Silver		0.01	10%

To read the resistance value, first determine if it is a 4-band or a 5-band encoding. Then proceed to identify the tolerance band, which is usually either gold or silver. Starting at the opposite end, read the two or three magnitude bands and multiply this value by the multiplier band. For example, a resistor with four color bands, red, orange, brown, and gold, is a 230-Ohm resistor with a 5% error tolerance.

There are several online calculators that allow you to easily find the resistance value. You can locate these calculators by searching for the keywords: resistor color codes.

Appendix B

Essential Electronics

B.1 Atom

Until the end of the nineteenth century it was assumed that matter was composed of small, indivisible particles called *atoms*. The work of J.J. Thompson, Rutherford, and Bohr proved that atoms were complex structures that contained both positive and negative particles. The positive ones were called *protons* and the negative ones *electrons*.

Several models of the atom were proposed: the one by Thompson assumed that there were equal numbers of protons and electrons inside the atom and that these elements were scattered at random, as in the leftmost drawing in Figure B-1. Later, in 1913, Rutherford's experiments led him to believe that atoms contained a heavy central positive nucleus with the electrons scattered randomly. So he modified Thompson's model as shown in the center drawing. Finally, Neils Bohr theorized that electrons had different energy levels, as if they moved around the nucleus in different orbits, like planets around a sun. The rightmost drawing represents this orbital model.

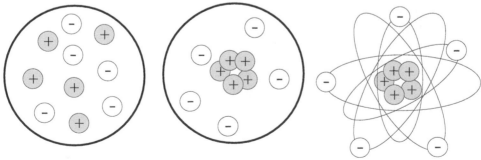

Figure B-1 Models of the Atom.

Investigations also showed that the normal atom is *electrically neutral*. Protons (positively charged particles) have a mass of 1.673 times 10^{-24} grams. Electrons (negatively charged particles) have a mass of 9.109 times 10^{-28} grams. Furthermore, the orbital model of the atom is not actually valid because orbits have little meaning at the atomic level. A more accurate representation is based on concentric spherical shells about the nucleus. An active area of research deals with atomic and sub-atomic structures.

The number of protons in an atom determines its *atomic number*; for example, the hydrogen atom has a single proton and an atomic number of 1, helium has two protons, carbon has six, and uranium has ninety-two. But when we compare the ratio of mass to electrical charge in different atoms we find that the nucleus must be made up of more than protons. For example, the helium nucleus has twice the charge of the hydrogen nucleus, but four times the mass. The additional mass is explained by assuming that there is another particle in the nucleus, called a neutron, which has the same mass as the proton but no electrical charge. Figure B-2 shows a model of the helium atom with two protons, two electrons, and two neutrons.

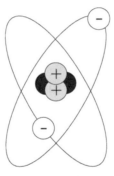

Figure B-2 Model of the Helium Atom.

B.2 Isotopes and Ions

But nature is not always consistent with such neat models. Whereas in a neutral atom, the number of protons in the atomic nucleus exactly matches the number of electrons, the number of protons need not match the number of neutrons. For example, most hydrogen atoms have a single proton, but no neutrons, while a small percentage have one neutron, and an even smaller one has two neutrons. In this sense, atoms of an element that contains different numbers of neutrons are *isotopes* of the element; for example water (H_2O) containing hydrogen atoms with two neutrons (deuterium) is called "heavy water."

An atom that is electrically charged due to an excess or deficiency of electrons is called an *ion*. When the dislodged elements are one or more electrons, the atom takes a positive charge. In this case it is called a *positive ion*. When a stray electron combines with a normal atom, the result is called a *negative ion*.

B.3 Static Electricity

Free electrons can travel through matter or remain at rest on a surface. When electrons are at rest, the surface is said to have a static electrical charge that can be positive or negative. When electrons are moving in a stream-like manner, we call this movement an *electrical current*. Electrons can be removed from a surface by means of friction, heat, light, or a chemical reaction. In this case the surface becomes positively charged.

The ancient Greeks discovered that when amber was rubbed with wool the amber became electrically charged and would attract small pieces of material. In this case, the charge is a positive one. Friction can cause other materials, such as hard rubber or plastic, to become negatively charged. Observing objects that have positive and negative charges we note that like charges repel and unlike charges attract each other, as shown in Figure B-3.

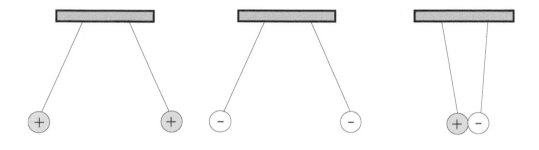

Figure B-3 Like and Unlike Charges.

Friction causes loosely held electrons to be transferred from one surface to the other. This results in a net negative charge on the surface that has gained electrons, and a net positive charge on the surface that has lost electrons. If there is no path for the electrons to take to restore the balance of electrical charges, these charges remain until they gradually leak off. If the electrical charge continues building, it eventually reaches the point where it can no longer be contained. In this case it discharges itself over any available path, as is the case with lightning.

Static electricity does not move from one place to another. While some interesting experiments can be performed with it, it does not serve the practical purpose of providing energy to do sustained work.

Static electricity certainly exists, and under certain circumstances we must allow for it and account for its possible presence, but it will not be the main theme of these pages.

B.4 Electrical Charge

Physicists often resort to models and theories to describe and represent some force that can be measured in the real world. But very often these models and representations are no more than concepts that fail to physically represent the object. In this sense, no one knows exactly what gravity is, or what an electrical charge is. Gravity, which can be felt and measured, is the force between masses.

By the same token, bodies in "certain electrical conditions" also exert measurable forces on one another. The term "electrical charge" was coined to explain these observations.

Three simple postulates or assumptions serve to explain all electrical phenomena:

1. Electrical charge exists and can be measured. Charge is measured in Coulombs, a unit named for the French scientist Augustin de Coulomb.

2. Charge can be positive or negative.

3. Charge can neither be created nor destroyed. If two objects with equal amounts of positive and negative charge are combined on some object, the resulting object will be electrically neutral and will have zero net charge.

B.4.1 Voltage

Objects with opposite charges attract, that is, they exert a force upon each other that pulls them together. In this case, the magnitude of the force is proportional to the product of the charge on each mass. Like gravity, electrical force depends inversely on the distance squared between the two bodies; the closer the bodies, the greater the force. Consequently, it takes energy to pull apart objects that are positively and negatively charged, in the same manner that it takes energy to raise a big mass against the pull of gravity.

The potential that separate objects with opposite charges have for doing work is called *voltage*. Voltage is measured in units of *volts* (V). The unit is named for the Italian scientist *Alessandro Volta*.

The greater the charge and the greater the separation, the greater the stored energy, or voltage. By the same token, the greater the voltage, the greater the force that drives the charges together.

Voltage is always measured between two points that represent the positive and negative charges. In order to compare voltages of several charged bodies, a common reference point is necessary. This point is usually called "ground."

B.4.2 Current

Electrical charge flows freely in certain materials, called *conductors*, but not in others, called *insulators*. Metals and a few other elements and compounds are good conductors, while air, glass, plastics, and rubber are insulators. In addition, there is a third category of materials called semiconductors; sometimes they seem to be good conductors but much less so at other times. Silicon and germanium are two such semiconductors. We discuss semiconductors in the context of integrated circuits in Section B-7 of this appendix.

Figure B-4 shows two connected, oppositely charged bodies. The force between them has the potential for work; therefore, there is voltage. If the two bodies are connected by a conductor, as in the illustration, the positive charge moves along the wire to the other sphere. On the other end, the negative charge flows out on the wire toward the positive side. In this case, positive and negative charges combine to neutralize each other until there are no charge differences between any points in the system.

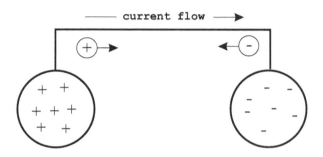

Figure B-4 Connected Opposite Charges.

The flow of an electrical charge is called a current. Current is measured in *amperes* (A), also called *amps*, after Andre Ampere, a French mathematician and physicist. An ampere is defined as the flow of one Coulomb of charge in one second.

Electrical current is directional; therefore, a positive current is the flow current from a positive point A to a negative point B. However, most current results from the flow of negative-to-positive charges.

B.4.3 Power

Current flowing through a conductor produces heat. The heat is the result of the energy that comes from the charge traveling across the voltage difference. The work involved in producing this heat is *electrical power*. Power is measured in units of *watts* (W), named after the Englishman James Watt, who invented the steam engine.

B.4.4 Ohm's Law

The relationship between voltage, current, and power is described by *Ohm's Law*, named after the German physicist Georg Simon Ohm. Using equipment of his own *creation*, Ohm determined that the current that flows through a wire is proportional to its cross-sectional area and inversely proportional to its length. This allowed defining the relationship between voltage, current, and power, as expressed by the equation:

$$P = V \times I$$

where P represents the power in watts, V is the voltage in volts, and I is the current in amperes. Ohm's Law can also be formulated in terms of voltage, current, and resistance as shown later in this chapter.

B.5 Electrical Circuits

An *electrical network* is an interconnection of electrical elements. An *electrical circuit* is a network in a closed-loop, giving a return path for the current. A network is a connection of two or more simple elements, and may not necessarily be a circuit.

Although there are several types of electrical circuits, they all have some of the following elements:

1. A power source, which can be a battery, alternator, etc., produces an electrical potential.
2. Conductors, in the form of wires or circuit boards, provide a path for the current.
3. Loads, in the form of devices such as lamps, motors, etc., use the electrical energy to produce some form of work.
4. Control devices, such as potentiometers and switches, regulate the amount of current flow or turn it on and off.
5. Protection devices, such as fuses or circuit breakers, prevent damage to the system in case of overload.
6. A common ground.

Figure B-5 shows a simple circuit that contains all these elements.

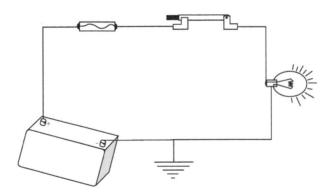

Figure B-5 Simple Circuit.

B.5.1 Types of Circuits

There are three common types of circuits: series, parallel, and series-parallel. The circuit type is determined by how the components are connected. In other words, by how the circuit elements, power source, load, and control and protection devices are interconnected. The simplest circuit is one in which the components offer a single current path. In this case, although the loads may be different, the amount of current flowing through each one is the same. Figure B-6 shows a series circuit with two light bulbs.

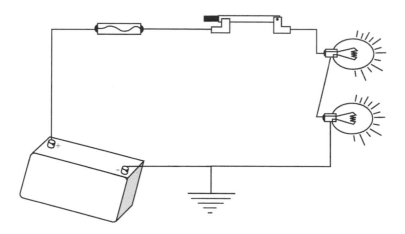

Figure B-6 Series Circuit.

In the *series circuit* in Figure B-6, if one of the light bulbs burn out, the circuit flow is interrupted and the other one will not light. Some Christmas lights are wired in this manner, and if a single bulb fails the whole string will not light.

In a *parallel circuit* there is more than one path for current flow. Figure B-7 shows a circuit wired in parallel.

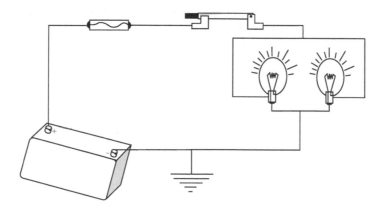

Figure B-7 Parallel Circuit.

In the circuit of Figure B-7, if one of the light bulbs burns out, the other one will still light. Also, if the load is the same in each circuit branch, so is the current flow in that branch. By the same token, if the load in each branch is different, so is the current flow in each branch.

The series-parallel circuit has some components wired in series and others in parallel. Therefore, the circuit shares the characteristics of both series and parallel circuits. Figure B-8 shows the same parallel circuit to which a series *rheostat* (dimmer) has been added in series.

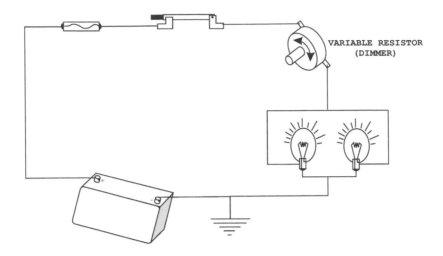

Figure B-8 Series-Parallel Circuit.

In the circuit of Figure B-8 the two light bulbs are wired in parallel, so if one fails the other one will not. However, the rheostat (dimmer) is wired in series with the circuit, so its action affects both light bulbs.

B.6 Circuit Elements

So far we have represented circuits using a pictorial style. Circuit diagrams are more often used because they achieve the same purpose with much less artistic effort and are easier to read. Figure B-9 is a diagrammatic representation of the circuit in Figure B-8.

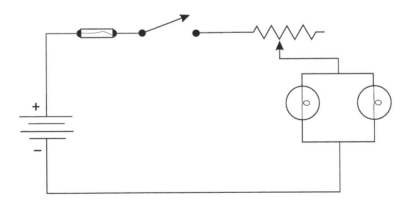

Figure B-9 Diagram of a Series-Parallel Circuit.

Certain components are commonly used in electrical circuits. These include power sources, resistors, capacitors, inductors, and several forms of semiconductor devices.

B.6.1 Resistors

If the current flow from, say a battery, is not controlled, a *short-circuit* takes place and the wires can melt or the battery may even explode. *Resistors* provide a way of controlling the flow of current from a source. A resistor is to current flow in an electrical circuit as a valve is to water flow: both elements "resist" flow. Resistors are typically made of materials that are poor conductors. The most common ones are made from powdered carbon and some sort of binder. Such carbon composition resistors usually have a dark-colored cylindrical body with a wire lead on each end. Color bands on the body of the resistor indicate its value, measured in Ohms and represented by the Greek letter Ω. The color code for resistor bands can be found in Appendix A.

The potentiometer and the rheostat are variable resistors. When the knob of a potentiometer or rheostat is turned, a slider moves along the resistance element and reduces or increases the resistance. A potentiometer is used as a dimmer in the circuits of Figure B-8 and Figure B-9. The photoresistor or photocell is composed of a light sensitive material whose resistance decreases when exposed to light. Photoresistors can be used as light sensors.

B.6.2 Revisiting Ohm's Law

We have seen how Ohm's Law describes the relationship between voltage, current, and power. The law is reformulated in terms of resistance so as to express the relationship between voltage, current, and resistance, as follows:

In this case V represents voltage, I is the current, and R is the resistance in the circuit. Ohm's Law equation can be manipulated in order to find current or resistance in terms of the other variables, as follows

$$V = I \times R$$

Note that the voltage value in Ohm's Law refers to the voltage across the resistor, in other words, the voltage between the two terminal wires. In this sense the voltage is actually produced by the resistor, as the resistor is restricting the flow of charge much as a valve or nozzle restricts the flow of water. It is the restriction created by the resistor that forms an excess of charge with respect to the other side of the circuit. The charge difference results in a voltage between the two points. Ohm's Law is used to calculate the voltage if we know the resistor value and the current flow.

$$I = \frac{V}{R}$$

$$R = \frac{V}{I}$$

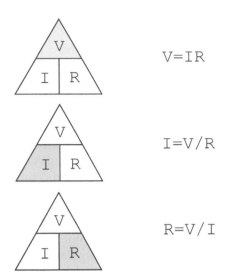

Figure B-10 Ohm's Law Pyramid.

A popular mnemonic for Ohm's Law consists of drawing a pyramid with the voltage symbol at the top and current and resistance in the lower level. Then, it is easy to solve for each of the values by observing the position of the other two symbols in the pyramid, as shown in Figure B-10.

B.6.3 Resistors in Series and Parallel

When resistors are in series, the total resistance equals the sum of the individual resistances. The diagram in Figure B-11 shows two resistors (R1 and R2) wired in series in a simple circuit.

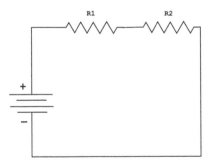

Figure B-11 Resistors in Series.

In Figure B-11 the total resistance (RT) is calculated by adding the resistance values of R1 and R2; thus, RT = R1 + R2.

In terms of water flow, a series of partially closed valves in a pipe add up to slow the flow of water.

Resistors can also be connected in parallel, as shown in Figure B-12.

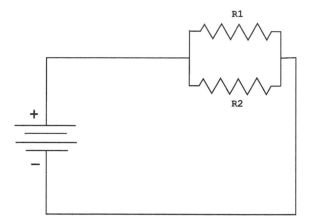

Figure B-12 Resistors in Parallel.

When resistors are placed in parallel, the combination has less resistance than any one of the resistors. If the resistors have different values, then more current flows through the path of least resistance. The total resistance in a parallel circuit is obtained by dividing the product of the individual resistors by their sum, as in the formula:

$$RT = \frac{R1 \times R2}{R1 + R2}$$

If more than two resistors are connected in parallel, then the formula can be expressed as follows:

$$RT = \frac{1}{\dfrac{1}{R1} + \dfrac{1}{R2} + \dfrac{1}{R3} \cdots}$$

Also note that the diagram representation of resistors in parallel can have different appearances. For example, the circuit in Figure B-13 is electrically identical to the one in Figure B-12.

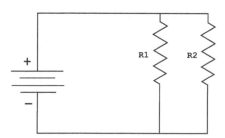

Figure B-13 Alternative Circuit of Parallel Resistors.

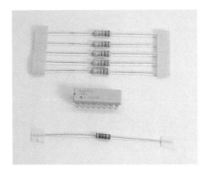

Figure B-14 Resistors.

Figure B-14 shows several commercial resistors. The integrated circuit at the center of the image combines eight resistors of the same value. These devices are convenient when the circuit design calls for several identical resistors. The color-coded cylindrical resistors in the image are made of carbon

Appendix A contains the color codes used in identifying resistors whose surface area does not allow printing its value.

B.6.4 Capacitors

An element often used in the control of the flow of an electrical charge is a *capacitor*. The name originated in the notion of a "capacity" to store charge. In that sense a capacitor functions as a small battery. Capacitors are made of two conducting surfaces separated by an insulator. A wire lead is usually connected to each surface. Two large metal plates separated by air would perform as a capacitor. More frequently, capacitors are made of thin metal foils separated by a plastic film or another form of solid insulator. Figure B-15 shows a circuit that contains both a capacitor and a resistor.

In Figure B-15, charge flows from the battery terminals, along the conductor wire, and onto the capacitor plates. Positive charges collect on one plate and negative charges on the other plate. The initial current is limited only by the resistance of the wires and by the resistor in the circuit. As charge builds up on the plates, charge repulsion resists the flow, and the current is reduced. At some point the repulsive force from the charge on the plates is strong enough to balance the force from the charge on the battery, and the current stops.

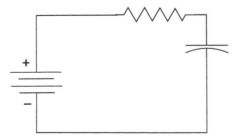

Figure B-15 Capacitor Circuit.

The existence of charges on the capacitor plates means there must be a voltage between the plates. When the current stops this voltage is equal to the voltage in the battery. Because the points in the circuit are connected by conductors, they have the same voltage, even if there is a resistor in the circuit. If the current is zero, there is no voltage across the resistor, according to Ohm's law.

The amount of charge on the plates of the capacitor is a measure of the value of the capacitor. This "capacitance" is measured in *farads* (f), named in honor of the English scientist Michael Faraday.

The relationship is expressed by the equation

$$C = \frac{Q}{V}$$

where C is the capacitance in farads, Q is the charge in Coulombs, and V is the voltage. Capacitors of 1 farad or more are rare. Generally capacitors are rated in microfarads (μf), one-millionth of a farad, or picofarads (pf), one-trillionth of a farad.

Consider the circuit of Figure B-15 after the current has stabilized. If we now remove the capacitor from the circuit, it still holds a charge on its plates. That is, there is a voltage between the capacitor terminals. In one sense, the charged capacitor appears somewhat like a battery. If we were to short-circuit the capacitor's terminals, a current would flow as the positive and negative charges neutralize each other. But unlike a battery, the capacitor does not replace its charge. So the voltage drops, the current drops, and finally there is no net charge and no voltage difference anywhere in the circuit.

B.6.5 Capacitors in Series and in Parallel

Like resistors, capacitors can be joined together in series and in parallel. Connecting two capacitors in parallel results in a bigger capacitance value, because of the larger plate area. Thus, the formula for total capacitance (CT) in a parallel circuit containing capacitors C1 and C2 is

$$CT = C1 + C2$$

Note that the formula for calculating capacitance in parallel is similar to the one for calculating series resistance. By the same token, where several capacitors are connected in series, the formula for calculating the total capacitance is

$$CT = \frac{1}{\dfrac{1}{C1} + \dfrac{1}{C2} + \dfrac{1}{C3} \ldots}$$

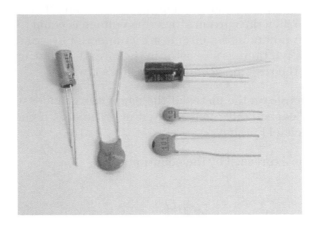

Figure B-16 Assorted Commercial Capacitors.

Note that the total capacitance of a connection in series is lower than for any capacitor in the series, considering that for a given voltage across the entire group, there is less charge on each plate.

There are several types of commercial capacitors, including mylar, ceramic, disk, and electrolytic. Figure B-16 shows several commercial capacitors.

B.6.6 Inductors

Inductors are the third type of basic circuit components. An inductor is a coil of wire with many windings. The wire windings are often made around a core of a magnetic material, such as iron. The properties of inductors are derived from magnetic rather than electric forces.

When current flows through a coil, it produces a magnetic field in the space outside the wire. This makes the coil behave just like a natural, permanent magnet. Moving a wire through a magnetic field generates a current in the wire, and this current will flow through the associated circuit. Because it takes mechanical energy to move the wire through the field, it is the mechanical energy that is transformed into electrical energy. A generator is a device that converts mechanical to electrical energy by means of induction. An electric motor is the opposite of a generator. In the motor, electrical energy is converted to mechanical energy by means of induction.

The current in an inductor is similar to the voltage across a capacitor. In both cases it takes time to change the voltage from an initially high current flow. Such induced voltages can be very high and can damage other circuit components, so it is common to connect a resistor or a capacitor across the inductor to provide a current path to absorb the induced voltage. In combination, inductors behave just like resistors: inductance adds in series. By the same token, parallel connection reduces induction. Induction is measured in henrys (h), but more commonly in mh, and μh.

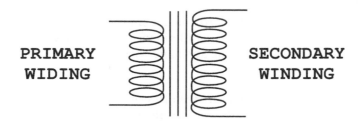

PRIMARY WIDING   **SECONDARY WINDING**

Figure B-17 Transformer Schematics.

B.6.7 Transformers

The *transformer* is an induction device that changes voltage or current levels. The typical transformer has two or more windings wrapped around a core made of laminated iron sheets. One of the windings, called the primary, receives a fluctuating current. The other winding, called the secondary, produces a current induced by the primary. Figure B-17 shows the schematics of a transformer.

The device in Figure B-17 is a step-up transformer. This is determined by the number of windings in the primary and secondary coils. The ratio of the number of turns in each winding determines the voltage increase. A transformer with an equal number of turns in the primary and secondary transfers the current unaltered. This type of device is sometimes called an isolation transformer. A transformer with less turns in the secondary than in the primary coil is a step-down transformer and its effect is to reduce the primary voltage at the secondary coil.

Transformers require an alternating or fluctuating current because it is the fluctuations in the current flow in the primary that induce a current in the secondary. The ignition coil in an automobile is a transformer that converts the low-level battery voltage to the high-voltage level necessary to produce a spark.

B.7 Semiconductors

The word *semiconductor* stems from the property of some materials that act either as a conductor or as an insulator, depending on certain conditions. Several elements are classified as semiconductors, including silicon, zinc, and germanium. Silicon is the most widely used semiconductor material because it is easily obtained.

In the ultra-pure form of silicon, the addition of minute amounts of certain impurities (called *dopants*) alters the atomic structure of the silicon. This determines whether the silicon can then be made to act as a conductor or as a nonconductor, depending upon the polarity of an electrical charge applied to it.

In the early days of radio, receivers required a device called a rectifier to detect signals. Ferdinand Braun used the rectifying properties of the galena crystal, a semiconductor material composed of lead sulfide, to create a "cat's whisker" diode that served this purpose. This was the first semiconductor device.

B.7.1 Integrated Circuits

Until 1959, electronic components performed a single function; therefore, many of them had to be wired together to create a functional circuit. Transistors were individually packaged in small cans. Packaging and hand wiring the components into circuits was extremely inefficient.

In 1959, at Fairchild Semiconductor, Jean Hoerni and Robert Noyce developed a process which made it possible to diffuse various layers onto the surface of a silicon wafer, while leaving a layer of protective oxide on the junctions. By allowing the metal interconnections to be evaporated onto the flat transistor surface the process replaced hand wiring. By 1961, nearly 90% of all the components manufactured were integrated circuits.

B.7.2 Semiconductor Electronics

To understand the workings of semiconductor devices, we need to reconsider the nature of the electrical charge. Electrons are one of the components of atoms, and atoms are the building blocks of all matter. Atoms bond with each other to form molecules. Molecules of just one type of atom are called elements. In this sense, gold, oxygen, and plutonium are elements because they all consist of only one type of atom. When a molecule contains more than one atom, it is known as a compound. Water, which has both hydrogen and oxygen atoms, is a compound. Figure B-18 represents an orbital model of an atom with five protons and three electrons.

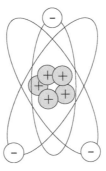

Figure B-18 Orbital Model of the Boron Atom with Its Valence Electrons.

In Figure B-18, protons carry positive charge and electrons carry negative charge. Neutrons, not represented in the illustration, are not electrically charged. Atoms that have the same number of protons and electrons have no net electrical charge.

Electrons that are far from the nucleus are relatively free to move around because the attraction from the positive charge in the nucleus is weak at large distances. In fact, it takes little force to completely remove an outer electron from an atom, leaving an ion with a net positive charge. A free electron can move at speeds approaching the speed of light (approximately 186,282 miles per second).

Electric current takes place in metal conductors due to the flow of free electrons. Because electrons have negative charge, the flow is in a direction opposite to the positive current. Free electrons traveling through a conductor drift until they hit other electrons attached to atoms. These electrons are then dislodged from their orbits and replaced by the formerly free electrons. The newly freed electrons then start the process anew.

B.7.3 P-Type and N-Type Silicon

Semiconductor devices are made primarily of silicon. Pure silicon forms rigid crystals because of its four outermost electrons. Because it contains no free electrons, it is not a conductor. But silicon can be made conductive by combining it with other elements (doping) such as boron and phosphorus. The boron atom has three outer valence electrons (Figure B-18) and the phosphorus atom has five. When three silicon atoms and one phosphorus atom bind together, creating a structure of four atoms, there is an extra electron and a net negative charge.

The combination of silicon and phosphorous, with the extra phosphorus electron, is called an N-type silicon. In this case, the N stands for the extra negative electron. The extra electron donated by the phosphorus atom can easily move through the crystal; therefore N-type silicon can carry an electrical current.

When a boron atom combines with a cluster of silicon atoms, there is a deficiency of one electron in the resulting crystal. Silicon with a deficient electron is called P-type silicon (P stands for positive). The vacant electron position is sometimes called a "hole." An electron from another nearby atom can "fall" into this hole, thereby moving the hole to a new location. In this case, the hole can carry a current in the P-type silicon.

B.7.4 Diode

Both P-type and N-type silicon conduct electricity. In either case, the conductivity is determined by the proportion of holes or the surplus of electrons. By forming some P-type silicon in a chip of N-type silicon, it is possible to control electron flow so that it takes place in a single direction. This is the principle of the diode, and the p-n action is called a pn-junction.

A diode is said to have a forward bias if it has a positive voltage across it from the P- to N-type material. In this condition, the diode acts rather like a good conductor and current can flow, as in Figure B-19.

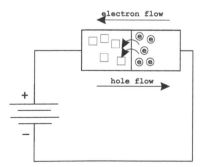

Figure B-19 A Forward Biased Diode.

If the polarity of the voltage applied to the silicon is reversed, then the diode is *reverse-biased* and appears nonconducting. This nonsymmetric behavior is due to the properties of the *pn-junction*. The fact that a diode acts like a one-way valve for current is a very useful characteristic. One application is to convert *alternating current* (AC) into *direct current* (DC). Diodes are so often used for this purpose that they are sometimes called rectifiers.

Appendix C

Numeric Data

C.1 Numbers in Computing

In order to perform more efficient digital operations on numeric data, mathematicians have devised systems and structures that differ from those used traditionally. This chapter presents the background material necessary for understanding and using the number systems and numeric data storage structures employed in digital devices.

C.1.1 Counting

The fundamental application of a number system is counting. A Stone-Age hunter uses his or her fingers to show other members of the tribe how many mammoths were spotted at the bottom of the ravine. In this manner the hunter is able to transmit a unique type of information that does not relate to the species, size, or color of the animals, but to their numbers. Our minds have the ability to capture this notion of "oneness" independently of other properties of objects.

The most primitive method of counting consists of using objects to represent degrees of oneness. The Stone-Age hunter used fingers to represent individual mammoth. Alternatively, the hunter could have resorted to pebbles, sticks, lines on the ground, or scratches on the cave wall to show how many units there were of the object.

C.1.2 Tally System

The tally system probably originated from notches on a stick or scratches on a cave wall. In its simplest form, each scratch, notch, or line, represents an object. The method is so simple and intuitive that we still resort to it occasionally. Tallying requires no knowledge of quantity and no elaborate symbols. Had there been twelve mammoth in the ravine the cave wall would have appeared as follows:

| | | | | | | | | | | | |

A logical evolution of the tally system consists of grouping marks. Because we have five fingers on each hand, the twelve mammoth could be grouped as follows:

| | | | | | | | | | | |

Perhaps a primitive mathematical genius added one final sophistication to the tally system. By drawing one tally line diagonally the visualization is further improved, as in this familiar style:

C.1.3 Roman Numerals

Roman numerals show how a simple graphical tally system evolved into a symbolic numeric representation. The first five digits were encoded with the symbols:

```
I, II, III, IIII, and V
```

The Roman symbol V is conceivably a simplification of the tally encoding using a diagonal line to complete the grouping, as shown in Table C.1.

Table C.1

Symbols in the Roman Numeration System

ROMAN	DECIMAL
I	1
V	5
X	10
L	50
C	100
D	500
M	1000

The Roman numeral system is based on an add-subtract rule whereby the elements of a number, read left-to-right, are either added to or subtracted from the previous sum according to its value. Thereby, the decimal number 1994 is represented in Roman numerals as follows:

```
MCMXCIV = M + (C - M) + (X - C) + (I - V)

        = 1000 + (1000 - 100) + (100 - 10) + (5 - 1)

        = 1000 + 900 + 90 + 4

        = 1994
```

The uncertainty in the positional value of each digit, the absence of a symbol for zero, and the fact that some numbers require either one or two symbols (I, IV, V, IX, and X) complicate the rules of arithmetic using Roman numerals.

C.2 Origins of the Decimal System

The one element of our civilization that has transcended all cultural and social differences is our decimal system of numbers. While mankind is yet to agree on the most desirable political order, on generally acceptable rules of moral behavior, or on a universal language, the Hindu-Arabic numerals have been adopted by practically all the nations and cultures of the world.

By the ninth century A.D. the Arabs were using a ten-symbol positional system of numbers that included the special symbol for 0. The Latin title of the first book on the subject of "Indian numbers" is *Liber Algorismi de Numero Indorum.* The author is the Arab mathematician al-Khowarizmi.

Despite of the evident advantages of this number system, its adoption in Europe took place only after considerable debate and controversy. Many scholars of the time still considered Roman numerals to be easier to learn and more convenient for operations on the *abacus.* The supporters of the Roman numeral system, called abacists, engaged in intellectual combat with the algorists, who were in favor of the Hindu-Arabic numerals as described by al-Khowarizmi. For several centuries abacists and algorists debated about the advantages of their systems, with the Catholic church often siding with the abacists. This controversy explains why the Hindu-Arabic numerals were not accepted into general use in Europe until the beginning of the sixteenth century.

It is sometimes said that the reason for there being ten symbols in the Hindu-Arabic numerals is related to the fact that we have ten fingers. However, if we make a one-to-one correlation between the Hindu-Arabic numerals and our fingers, we find that the last finger must be represented by a combination of two symbols, 10. Also, one Hindu-Arabic symbol, 0, cannot be matched to an individual finger. In fact, the decimal system of numbers, as used in a positional notation that includes a zero digit, is a refined and abstract scheme that should be considered one of the greatest achievements of human intelligence. We will never know for certain if the Hindu-Arabic numerals are related to the fact that we have ten fingers, but its profoundness and usefulness clearly transcend this biological fact.

The most significant feature of the Hindu-Arabic numerals is the presence of a special symbol, 0, which by itself represents no quantity. Nevertheless, the special symbol 0 is combined with the other ones. In this manner the nine other symbols are reused to represent larger quantities. Another characteristic of decimal numbers is that the value of each digit depends on its position in a digit string. This positional characteristic, in conjunction with the use of the special symbol 0 as a placeholder, allow the following representations:

```
   1 = one
  10 = ten
 100 = hundred
1000 = thousand
```

The result is a counting scheme where the value of each symbol is determined by its column position. This positional feature requires the use of the special symbol, 0, which does not correspond to any unit-amount, but is used as a placeholder in multicolumn representations. We must marvel at the intelligence, capability for abstraction, and even the sense of humor of the mind that conceived a counting system that has a symbol that represents nothing. We must also wonder about the evolution of mathematics, science, and technology had this system not been invented. One intriguing question is whether a positional counting system that includes the zero symbol is a natural and predictable step in the evolution of our mathematical thought, or whether its invention was a stroke of genius that could have been missed for the next two thousand years.

C.2.1 Number Systems for Digital-Electronics

The computers built in the United States during the early 1940s operated on decimal numbers. However, in 1946, von Neumann, Burks, and Goldstine published a trend-setting paper titled "Preliminary Discussion of the Logical Design of an Electronic Computing Instrument," in which they state:

> *"In a discussion of the arithmetic organs of a computing machine one is naturally led to a consideration of the number system to be adopted. In spite of the long-standing tradition of building digital machines in the decimal system, we must feel strongly in favor of the binary system for our device."*

In their paper, von Neumann, Burks, and Goldstine also consider the possibility of a computing device that uses binary-coded decimal numbers. However, the idea is discarded in favor of a pure binary encoding. The argument is that binary numbers are more compact than binary-coded decimals. Later in this book you will see that binary-coded decimal numbers (called BCD) are used today in some types of computer calculations.

In 1941, Konrad Zuse, a German who had done pioneering work in computing machines, released the first programmable computer designed to solve complex engineering equations. The machine, called the Z3, was controlled by perforated strips of discarded movie film and used the binary number system.

The use of the binary number system in digital calculators and computers was made possible by previous research on number systems and on numerical representations, starting with an article by G.W. Leibnitz published in Paris in 1703. Researchers concluded that it is possible to count and perform arithmetic operations using any set of symbols as long as the set contains at least two symbols, one of which must be zero.

In digital electronics the binary symbol 1 is equated with the electronic state ON, and the binary symbol 0 with the state OFF. The two symbols of the binary system can also represent conducting and nonconducting states, positive or negative, or any other bi-valued condition. It was the binary system that presented the Hindu-Arabic decimal number system with the first challenge in 800 years. In digital-electronics, two steady states are easier to implement and more reliable than a ten-digit encoding.

C.2.2 Positional Characteristics

All modern number systems, including decimal, hexadecimal, and binary, are positional and include the digit zero. It is the positional feature that is used to determine the total value of a multi-digit representation. For example, the digits in the decimal number 4359 have the following positional weights:

```
4  3  5  9

|  |  |  |_____      units

|  |  |_____      ten units

|  |_____      hundred units

|_____      thousand units
```

The total value is obtained by adding the column weights of each unit:

```
        4000  ---  4 thousand units

         300  ---  3 hundred units

   +      50  ---  5 ten units

           9  ---  9 unit

        ----

        4359
```

C.2.3 Radix or Base of a Number System

In any positional number system, the weight of each column is determined by the total number of symbols in the set, including zero. This is called the base or radix of the system. The base of the decimal system is 10 and the base of the binary system is 2. The positional value or weight (P) of a digit in a multi-digit number is determined by the formula

$$P = d \times B^c$$

where d is the digit, B is the base or radix, and c is the zero-based column number, starting from right to left. Note that the increase in column weight from right to left is purely conventional. You could construct a number system in which the column weights increase in the opposite direction. In fact, in the original Hindu notation, the most significant digit was placed at the right.

In radix-positional terms, a decimal number can be expressed as a sum of digits by the formula

$$\sum_{i=-m}^{n} d_i \times 10^i$$

where i is the system's range and n is its limit.

C.3 Types of Numbers

By the adoption of special representations for different types of numbers, the usefulness of a positional number system can be extended beyond the simple counting function.

C.3.1 Whole Numbers

The digits of a number system, called the positive integers or *natural numbers*, are an ordered set of symbols. The notion of an *ordered set* means that the numerical symbols are assigned a predetermined sequence. A positional system of numbers also requires the special digit zero that, by itself, represents the absence of oneness, or nothing, and thus is not included in the set of natural numbers. However, 0 assumes a cardinal function when it is combined with other digits, for instance, 10 or 30. The *whole numbers* are the set of natural numbers, including the number zero.

C.3.2 Signed Numbers

A number system can also encode direction. We generally use the + and − signs to represent opposite numerical directions. The typical illustration for a set of signed numbers is as follows:

```
 -9 -8 -7 -6 -5 -4 -3 -2 -1   0 +1 +2 +3 +4 +5 +6 +7 +8 +9

 negative numbers  <-          zero          -> positive numbers
```

The number zero, which separates the positive and the negative numbers, has no sign of its own; although in some binary encodings we can end up with a negative and a positive zero.

C.3.3 Rational, Irrational, and Imaginary Numbers

A number system also represents parts of a whole. For example, when a carpenter cuts one board into two boards of equal length, we can represent the result with the fraction 1/2; the fraction 1/2 represents one of the two parts that make up the object. Rational numbers are those expressed as a ratio of two integers, for example, 1/2, 2/3, 5/248. Note that this use of the word "rational" is related to the mathematical concept of a ratio, and not to reason.

The denominator of a rational number expresses the number of potential parts. In this sense 2/5 indicates two of five possible parts. There is no reason why the number 1 cannot be used to indicate the number of potential parts, for example 2/1, 128/1. In this case the ratio x/1 indicates x elements of an undivided part. Therefore, it follows that x/1 = x. The implication is that the set of rational numbers includes the integers, because an integer can be expressed as a ratio using a unit denominator.

But not all non-integer numbers can be written as an exact ratio of two integers. The discovery of the first irrational number is usually associated with the investigation of a right triangle by the Greek mathematician *Pythagoras* (approximately 600 B.C.). The *Pythagorean Theorem* states that in any right triangle the square of the longest side (hypotenuse) is equal to the sum of the squares of the other two sides.

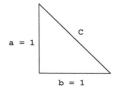

For this triangle, the Pythagorean theorem states that

$$a^2 + b^2 = c^2$$
$$2 = c^2$$
$$2 = c \times c$$
$$c = \sqrt{2}$$

Therefore, the length of the hypotenuse in a right triangle with unit sides is a number, that when multiplied by itself, gives 2. This number (approximately 1.414213562) cannot be expressed as the exact ratio of two integers. Other irrational numbers are the square roots of 3 and 5, as well as the mathematical constants π and e.

The set of numbers that includes the natural numbers, the whole numbers, and the rational and irrational numbers is called the real numbers. Most common mathematical problems are solved using real numbers. However, during the investigation of squares and roots, we notice that there can be no real number whose square is negative. Mathematicians of the eighteenth century extended the number system to include operations with roots of negative numbers. They did this by defining an imaginary unit as follows:

$$i = \sqrt{-1}$$

The imaginary unit makes possible new set of numbers, called complex numbers, that consist of a real part and an imaginary part. One of the uses of complex numbers is in finding the solution of a quadratic equation. Complex numbers are also useful in vector analysis, graphics, and in solving many engineering, scientific, and mathematical problems.

C.4 Radix Representations

The radix of a number system is the number of symbols in the set, including zero. Thus, the radix of the decimal system is 10, and the radix of the binary system is 2. Digital electronics is based on circuits that can be in one of two stable states. Therefore, a number system based on two symbols is better suited for work in digital electronics, because each state can be represented by a digit.

C.4.1 Decimal versus Binary Numbers

The binary system of numbers uses two symbols, 1 and 0. It is the simplest possible set of symbols with which we can count and perform arithmetic. Most of the difficulties in learning and using the binary system arise from this simplicity. Figure C.1 shows sixteen groups of four electronic cells each in all possible combinations of two states.

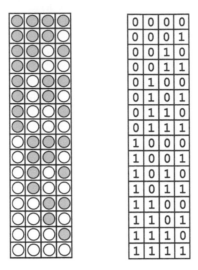

Figure C-1 Electronic Cells and Binary Numbers

It is interesting to note that binary numbers match the physical state of each electronic cell. If we think of each cell as a miniature light bulb, then the binary number 1 can be used to represent the state of a charged cell (light ON) and the binary number 0 to represent the state of an uncharged cell (light OFF).

C.4.2 Hexadecimal and Octal

Binary numbers are convenient in digital electronics; however, one of their drawbacks is the number of symbols required to encode a large value. For example, the number 9134 is represented in four decimal digits. However, the binary equivalent 10001110101110 requires fourteen digits. In addition, large binary numbers are difficult to remember.

One possible way of compensating for these limitations of binary numbers is to use individual symbols to represent groups of binary digits. For example, a group of three binary numbers allow eight possible combinations. In this case, we can use the decimal digits 0 to 7 to represent each possible combination of three binary digits. This grouping of three binary digits gives rise to the following table:

```
        binary          octal

        0 0 0             0

        0 0 1             1

        0 1 0             2

        0 1 1             3

        1 0 0             4

        1 0 1             5

        1 1 0             6

        1 1 1             7
```

The *octal* encoding serves as a shorthand representation for groups of three-digit binary numbers.

Hexadecimal numbers (base 16) are used for representing values encoded in four binary digits. Because there are only ten decimal digits, the hexadecimal system borrows the first six letters of the alphabet (A, B, C, D, E, and F). The result is a set of sixteen symbols, as follows:

0 1 2 3 4 5 6 7 8 9 A B C D E F

Most modern computers are desgined with memory cells, registers, and data paths in multiples of four binary digits. Table C.2 lists some common units of memory storage.

Table C.2

Units of Memory Storage

UNIT	BITS	HEX DIGITS	HEX RANGE
Nibble	4	1	0 to F
Byte	8	2	0 to FF
Word	16	4	0 to FFFF
Doubleword	32	8	0 to FFFFFFFF

In most digital-electronic devices memory addressing is organized in multiples of four binary digits. Here again, the hexadecimal number system provides a convenient way to represent addresses. Table C.3 lists some common memory addressing units and their hexadecimal and decimal range.

Table C.3

Units of Memory Addressing

UNIT	DATA PATH IN BITS	ADDRESS RANGE DECIMAL	HEX
1 paragraph	4	0 to 15	0-F
1 page	8	0 to 255	0-FF
1 kilobyte	16	0 to 65,535	0-FFFF
1 megabyte	20	0 to 1,048,575	0-FFFFF
4 gigabytes	32	0 to 4,294,967,295	0-FFFFFFFF

C.5 Number System Conversions

We use decimal numbers in our everyday life because they meaningfully represent common units used in the real world. To state that a certain historical event took place in the year 7C6 hexadecimal would convey little information to the average person. However, in computer systems based on two-state electronic cells, binary representations are more convenient. Also note that hexadecimal and octal numbers are handy shorthand for representing groups of binary digits.

Numerical conversions between positional systems of different radices are based on the number of symbols in the respective sets and on the positional value (weight) of each column. But methods used for manual conversions are not always suitable for machine conversions, as we will see in the forthcoming sections.

C.5.1 Binary-to-ASCII-Decimal

To manually convert a binary number to its decimal equivalent we take into account
the positional weight of each binary digit, as shown in Figure C-2.

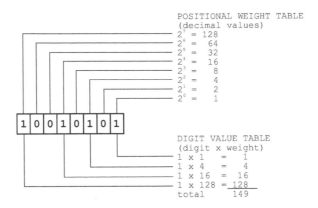

Figure C-2 Binary to ASCII Decimal Conversion Example

The positional weight table in Figure C-2 lists the decimal value of each binary
column. These weights are powers of the system's base (2 in the binary system). In
the digit value table, also in Figure C-2, the decimal values of the binary columns
holding a 1 digit are added. The sum of the weights of all the 1-digits in the operand
is the decimal equivalent of the binary number. In this case, 10010101 binary = 149
decimal.

The method in Figure C-2, although useful in manual conversions, is not an algorithm for computer conversions. Figure C-3 is a flowchart of a low-level bi-
nary-to-decimal conversion routine.

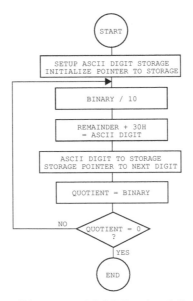

Figure C-3 Flowchart for a Binary to ASCII Decimal Conversion

The algorithm for the processing in Figure C-3 can be written as follows:

1. Set up and initialize a string storage area (sometimes called a buffer) to hold the ASCII decimal digits of the result. Set up the buffer pointer to the rightmost digit position of the result.

2. Obtain the remainder of the value divided by 10.

3. Add 30H to remainder digit to convert to ASCII representation.

4. Store remainder digit in buffer and index the buffer pointer to the preceding digit.

5. Quotient of division by 10 becomes the new binary value.

6. End conversion routine if quotient is equal to 0. Otherwise, continue at Step 2.

Note that the numerical digits are located from 30H to 39H in the ASCII table. This makes is easy to convert a binary digit to ASCII simply by adding 30H. Likewise, an ASCII digit is converted to binary by subtracting 30H.

C.5.2 Binary-to-Hexadecimal Conversion

The method described in Section 2.4.1 for a binary-to-ASCII decimal conversion can be adapted to other radices by representing the positional weight of each binary digit in the number system to which the conversion is to be made. In the case of a binary-to-ASCII hexadecimal conversion the positional weight of each binary digit is a hexadecimal value. Figure C-4 shows the conversion of the binary value 10010101 into hexadecimal using the corresponding positional weights.

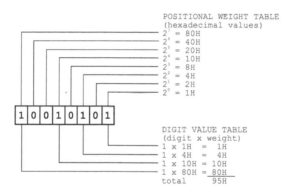

Figure C-4 Binary to ASCII Hexadecimal Conversion Example

The machine conversion binary-to-ASCII hexadecimal is similar to the binary-to-ASCII decimal algorithm described previously. In the case of the conversion into ASCII hexadecimal digits, the buffer need only hold four ASCII characters, because a 16-bit binary cannot exceed the value FFFFH. In the case of binary-to-ASCII hex, the divisor for obtaining the digits is 16 instead of 10.

C.5.3 Decimal-to-Binary Conversion

Longhand conversion of decimal into binary can be performed using the positional weights to find the binary 1 digits and then subtracting this positional weight from the decimal value. The process is shown in Figure C-5.

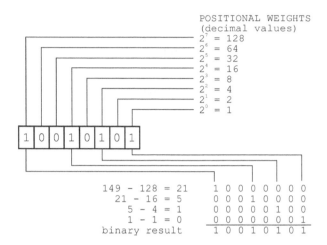

Figure C-5 Example of Decimal-to-Binary Conversion

In the example of Figure C-5, we start with the decimal value 149. Because the highest power of 2 smaller than 149 is 128, which corresponds to bit 7, we set bit 7 in the result and perform the subtraction:

$$149 - 128 = 21$$

At this point the highest positional weight smaller that 21 is 16, which corresponds to bit 4. Therefore we set bit 4 and perform the subtraction:

$$21 - 16 = 5$$

The remaining steps in the conversion can be seen in the illustration. The conversion is finished when the result of the subtraction is 0.

Suppose there is a numerical value in the form of a string of ASCII decimal, octal, or hexadecimal digits. In order for a processor to perform simple arithmetic operations on such data, the data must first be converted to binary. The binary value is then loaded into machine registers or memory cells. However, methods suited for manual conversion do not always make a good computer algorithm. Figure C.6 shows two decimal-to-binary conversion algorithms that are suited for machine coding.

Using the first method of Figure C-6, the individual decimal digits are multiplied by their corresponding positional values. The final result is obtained by adding all the partial products. Although this method is frequently used, it has the disadvantage that a different multiplier is used during each iteration (1, 10, 100, 1000). The second method in Figure C-6 starts with the high-order ASCII-decimal digit. The calculations consist of multiplying an accumulated value by 10. Initially, this accumulated value is set to 0. After multiplication by 10, the value of the digit is added to the accumulated value. The following algorithm is based on the second method in Figure C-6

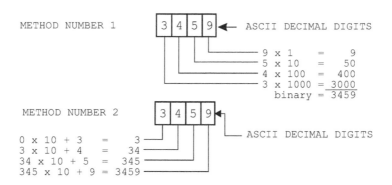

Figure C-6 Machine Conversion of ASCII Decimal to Binary.

1. Set up and initialize to binary zero a storage location for holding the value accumulated during conversion. Set up a pointer to the highest-order ASCII digit in the source string.

2. Test the ASCII digit for a value in the range 0 to 9. End of routine if the ASCII digit is not in this range.

3. Subtract 30H from ASCII decimal digit.

4. Multiply accumulated value by 10.

5. Add digit to accumulated value.

6. Increment the pointer to the next digit and continue at Step 2.

Figure C-7 is a flowchart of the conversion algorithm

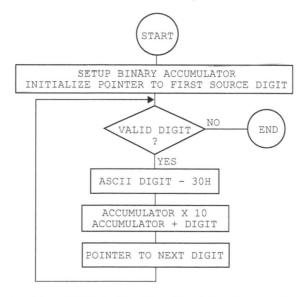

Figure C-7 Flowchart for ASCII to Machine Register Conversion.

Appendix D

Character Data

D.1 Character Representations

In this appendix we review the various encodings and formats used for representing character and numeric data in digital systems. The numeric formats allow representing binary numbers as signed and unsigned integers in several forms, binary floating-point numbers, and decimal floating-point numbers, usually called *binary-coded decimals* or *BCD*.

D.1.1 Electronic-Digital Machines

The mechanization of arithmetic is often traced back to the abacus, slide rule, mechanical calculators, and punch card machines. The work of John von Neumann at Princeton's Institute for Advanced Study and Research marks the first highlight in the design and construction of a digital-electronic calculating machine. In von Neumann's design, data and instructions are stored in a common memory area. An alternative approach, known as Harvard architecture, was discarded at first but has recently been revalidated and is in use in several microcontroller families.

The calculating power of the first computer was approximately 2,000 operations per second, while previous electromechanical devices were capable of performing only three or four operations. Today's digital machines can execute more than a billion instructions per second. Technological advances and miniaturization techniques have reduced the cost and size of computing machinery.

D.2 Character Representations

Over the years, data representation issues have often been determined by the various conventions used by the different hardware manufacturer. Machines have had different word lengths and different character sets and have used various schemes for storing character and data. Fortunately, in microprocessor and microcontroller design, the encoding of character data has not been subject to major disagreements.

Historically, the methods used to represent characters have varied widely, but the basic approach has always been to choose a fixed number of bits and then map the various bit combinations to the various characters. Clearly, the number of bits of the storage format limits the total number of distinct characters that can be represented. In this manner, the 6-bit codes used on a number of earlier computing machines allow representing 64 characters. This range allows including the uppercase letters, the decimal digits, some special characters, but not the lowercase letters. Computer manufacturers that used the 6-bit format often argued that their customers had no need for lowercase letters. Nowadays, 7- and 8-bit codes that allow representing the lowercase letters have been adopted almost universally.

Most of the world (except IBM) has standardized character representations using the *ISO (International Standards Organization)* code. ISO exists in several national variants; the one used in the United States is called ASCII, which stands for *American Standard Code for Information Interchange.* All microcomputers and microcontrollers use ASCII as the code for character representation.

D.2.1 ASCII

ASCII is a character encoding based on the English alphabet. ASCII was first published as a standard in 1967 and was last updated in 1986. The first 33 codes, referred to as non-printing codes, are mostly obsolete control characters. The remaining 95 printable characters (starting with the space character) include the common characters found in a standard keyboard, the decimal digits, and the upper- and lowercase characters of the English alphabet. Table D.1 lists the ASCII characters in decimal, hexadecimal, and binary.

Table D.1
ASCII Character Representation

DECIMAL	HEX	BINARY	VALUE	
000	000	00000000	annual	(Null character)
001	001	00000001	SOH	(Start of Header)
002	002	00000010	STX	(Start of Text)
003	003	00000011	ETX	(End of Text)
004	004	00000100	EOT	(End of Transmission)
005	005	00000101	ENQ	(Enquiry)
006	006	00000110	ACK	(Acknowledgment)
007	007	00000111	BEL	(Bell)
008	008	00001000	BS	(Backspace)
009	009	00001001	HT	(Horizontal Tab)
010	00A	00001010	LF	(Line Feed)
011	00B	00001011	VT	(Vertical Tab)
012	00C	00001100	FF	(Form Feed)
013	00D	00001101	CR	(Carriage Return)
014	00E	00001110	SO	(Shift Out)
015	00F	00001111	SI	(Shift In)
016	010	00010000	DLE	(Data Link Escape)
017	011	00010001	DC1	(XON)(Device Control 1)
018	012	00010010	DC2	(Device Control 2)
019	013	00010011	DC3	(XOFF)(Device Control 3)
020	014	00010100	DC4	(Device Control 4)
021	015	00010101	NAK	(- Acknowledge)
022	016	00010110	SYN	(Synchronous Idle)

(continues)

Table D.1

ASCII Character Representation (conitnued)

DECIMAL	HEX	BINARY		VALUE
000	000	00000000	annual	(Null character)
023	017	00010111	ETB	(End of Trans. Block)
024	018	00011000	CAN	(Cancel)
025	019	00011001	EM	(End of Medium)
026	01A	00011010	SUB	(Substitute)
027	01B	00011011	ESC	(Escape)
028	01C	00011100	FS	(File Separator)
029	01D	00011101	GS	(Group Separator)
030	01E	00011110	RS	(Request to Send)
031	01F	00011111	US	(Unit Separator)
032	020	00100000	SP	(Space)
033	021	00100001	!	(exclamation mark)
034	022	00100010	"	(double quote)
035	023	00100011	#	(number sign)
036	024	00100100	$	(dollar sign)
037	025	00100101	%	(percent)
038	026	00100110	&	(ampersand)
039	027	00100111	'	(single quote)
040	028	00101000	(	(left/opening parenthesis)
041	029	00101001	)	(right/closing parenthesis)
042	02A	00101010	*	(asterisk)
043	02B	00101011	+	(plus)
044	02C	00101100	,	(comma)
045	02D	00101101	-	(minus or dash)
046	02E	00101110	.	(dot)
047	02F	00101111	/	(forward slash)
048	030	00110000	0	(decimal digits ...)
049	031	00110001	1	
050	032	00110010	2	
051	033	00110011	3	
052	034	00110100	4	
053	035	00110101	5	
054	036	00110110	6	
055	037	00110111	7	
056	038	00111000	8	
057	039	00111001	9	
058	03A	00111010	:	(colon)
059	03B	00111011	;	(semi-colon)
060	03C	00111100	<	(less than)
061	03D	00111101	=	(equal sign)
062	03E	00111110	>	(greater than)
063	03F	00111111	?	(question mark)
064	040	01000000	@	(AT symbol)
065	041	01000001	A	
066	042	01000010	B	
067	043	01000011	C	
. . .				
090	05A	01011010	Z	
091	05B	01011011	[	(left/opening bracket)
092	05C	01011100	\	(back slash)
093	05D	01011101	]	(right/closing bracket)

(continues)

Table D.1

ASCII Character Representation (conitnued)

DECIMAL	HEX	BINARY		VALUE
094	05E	01011110	^	(circumflex)
095	05F	01011111	_	(underscore)
096	060	01100000	'	
097	061	01100001	a	
098	062	01100010	b	
099	063	01100011	c	
...				
122	07A	01111010	z	
123	07B	01111011	{	(left/opening brace)
124	07C	01111100	|	(vertical bar)
125	07D	01111101	}	(right/closing brace)
126	07E	01111110	~	(tilde)
127	07F	01111111	DEL	(delete)

D.2.2 EBCDIC and IBM

In spite of ASCII's general acceptance, IBM continues to use EBCDIC (*Extended Binary Coded Decimal Interchange Code*) for character encoding. IBM mainframes and mid-range systems such as the AS/400 use a wholly incompatible character set primarily designed for punched cards.

EBCDIC uses the full eight bits available to it, so there is no place left to implement parity checking. On the other hand, EBCDIC has a wider range of control characters than ASCII.

EBCDIC character encoding is based on Binary Coded Decimal (BCD), which we discuss in Section D.5. There are four main blocks in the EBCDIC code page:

1. The range 0000 0000 to 0011 1111 is reserved for control characters.

2. The range 0100 0000 to 0111 1111 is for punctuation.

3. The range 1000 0000 to 1011 1111 is for lowercase characters.

4. The range 1100 0000 to 1111 1111 is for uppercase characters and numbers.

Actually, microprocessor and microcontroller design need not address how character data is encoded. Usually a set of instructions allows manipulating 8-bit quantities, but the processor need not be concerned with what the encodings represent. On the other hand, some mainframe processors do have instructions that manipulate character codes. For example, the EDIT instruction on the IBM 370 implements the kind of picture conversion that appears in COBOL programs.

D.2.3 Unicode

One of the limitations of the ASCII code is that eight bits are not enough for representing characters sets in languages such as Japanese or Chinese which use large character sets. This has led to the development of encodings that allow representing large character sets. *Unicode* has been proposed as a universal character encoding standard that can be used for representation of text for computer processing.

Unicode attempts to provide a consistent way of encoding multilingual text and thus make it possible to exchange text files internationally. The design of Unicode is based on the ASCII code, but goes beyond the Latin alphabet to which ASCII is limited. The Unicode Standard provides the capacity to encode all the characters used for the written languages of the world. Like ASCII, Unicode assigns each character a unique numeric value and name. Unicode uses three encoding forms that use a common repertoire of characters. These forms allow encoding as many as a million characters.

The three encoding forms of the Unicode Standard allow the same data to be transmitted in a byte, word, or double-word format, that is, in 8-, 16-, or 32-bits per character.

- UTF-8 is a way of transforming all Unicode characters into a variable length encoding of bytes. In this format the Unicode characters corresponding to the familiar ASCII set have the same byte values as ASCII. By the same token, Unicode characters transformed into UTF-8 can be used with existing software.

- UTF-16 is designed to balance efficient access to characters with economical use of storage. It is reasonably compact and all the heavily used characters fit into a single 16-bit code unit, while all other characters are accessible via pairs of 16-bit code units.

- UTF-32 is used where memory space is no concern, but fixed width, single code unit access to characters is desired. In UTF-32, each Unicode character is represented by a single 32-bit code.

D.3 Storage and Encoding of Integers

The Indian mathematician Pingala first described binary number in the fifth century B.C. The modern system of binary numbers first appeared in the work of Gottfried Leibniz during the seventeenth century. During the mid-nineteenth century, the British logician George Boole described a logical system that used binary numbers to represent logical true and false. In 1937, Claude Shannon published his master's thesis that used Boolean algebra and binary arithmetic to implement electronic relays and switches. The thesis paper entitled *A Symbolic Analysis of Relay and Switching Circuits* is usually considered the origin of modern digital circuit design.

Also in 1937, George Stibitz completed a relay-based computer that could perform binary addition. The Bell Labs *Complex Number Computer*, also designed by Stibitz, was completed in January 1940. The system was demonstrated to the *American Mathematical Society* in September 1940. The attendants included John Von Neumann, John Mauchly, and Norbert Wiener. In 1945, von Neumann wrote a seminal paper in which he stated that binary numbers were the ideal computational format.

D.3.1 Signed and Unsigned Representations

For unsigned integers there is little doubt that the binary representation is ideal. Successive bits indicate powers of 2, with the most-significant bit at the left and the least-significant one on the right, as is customary in decimal representations. Figure D-1 shows the digit weights and conventional bit numbering in the binary encoding.

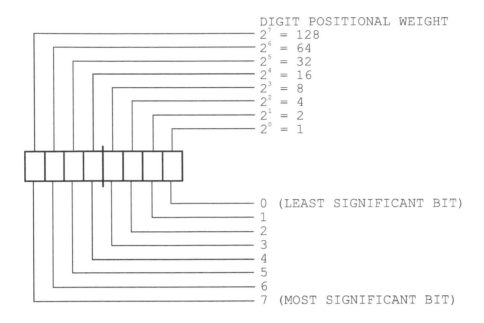

Figure D-1 Binary Digit Weights and Numbering.

In order to perform arithmetic operations, the digital machine must be capable of storing and retrieving numerical data. Numerical data is stored in standard formats, designed to minimize space and optimize processing. Historically, numeric data was stored in data structures devised to fit the characteristics of a specific machine, or the preferences of its designers. It was in 1985 that the Institute of Electrical and Electronics Engineers (IEEE) and the American National Standards Institute (ANSI) formally approved mathematical standards for encoding and storing numerical data in digital devices.

The electronic and physical mechanisms used for storing data have evolved with technology. One common feature of many devices, from punched tape to integrated circuits, is that the encoding is represented in two possible states. In paper tape the two states are holes and no holes, while in electronic media they are usually the presence and absence of an electrical charge.

Data stored in processor registers, in magnetic media, in optical devices, or in punched tape is usually encoded in binary. Thus, the programmer and the operator can usually ignore the physical characteristics of the storage medium. In other words, the bit pattern 10010011 can be encoded as holes in a strip of paper tape, as magnetic charges on a mylar-coated disk, as positive voltages in an integrated circuit memory cell, or as minute craters on the surface of a CD. In all cases, 10010011 represents the decimal number 147.

D.3.2 Word Size

In electronic digital devices the bistable states are represented by a *binary digit*, or *bit*. Circuit designers group several individual cells to form a unit of storage that holds several bits. In a particular machine the basic unit of data storage is called the word size. Word size in computers often ranges from 8 to 128 bits, in powers of 2. Microcontrollers and other digital devices sometimes use word sizes that are determined by their specific architectures. For example, some PIC microcontrollers use a 14-bit word size.

In most digital machines the smallest unit of storage individually addressable is eight bits (one *byte*). Individual bits are not directly addressable and must be manipulated as part of larger units of data storage.

D.3.3 Byte Ordering

The storage of a single-byte integer can be done according to the scheme in Figure D-1. However, the maximum value that can be represented in eight bits is the decimal number 255. To represent larger binary integers requires additional storage area. Because memory is usually organized in byte-size units, any decimal number larger than 255 requires more than one byte of storage. In this case the encoding is padded with the necessary leading zeros. Figure D-2 is a representation of the decimal number 21,141 stored in two consecutive data bytes.

Figure D-2 Representation of an Unsigned Integer.

One issue related to using multiple memory bytes to encode binary integers is the successive layout of the various byte-size units. In other words, does the representation store the most-significant byte at the lowest numbered memory location, or viceversa. For example, when a 32-bit binary integer is stored in a 32-bit storage area, we can follow the conventional pattern of placing the low-order bit on the right-hand side and the high-order bit on the left, as we did in Figure D-1. However, if the 32-bit number is to be stored into four byte-size memory cells, then two possible storage schemes are possible, as shown in Figure D-3.

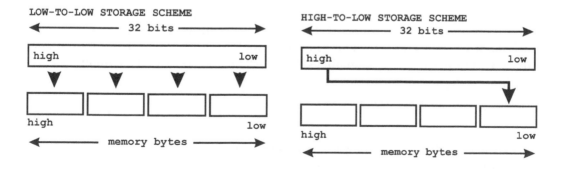

Figure D-3 Byte Ordering Schemes.

In the low-to-low storage scheme, the low-order 8-bits of the operand are stored in the low-order memory byte, the next group of 8-bits are moved to the following memory byte in low-to-high order, and so on. Conceivably, this scheme can be described by saying that the "little end" of the operand is stored first, that is, in lowest memory. According to this notion, the storage scheme is described as the *little-endian* format. If the "big-end" of the operand, that is, the highest valued bits, is stored in the low memory addresses, then the byte ordering is said to be in *big-endian* format. Some Intel processors (like those of the 80x86 family) follow the little-endian format. Some Motorola processors (like those of the 68030 family) follow the big-endian format, while others (such as the MIPS 2000) can be configured to store data in either format.

In many situations the programmer needs to be aware of the byte-ordering scheme; for example, to retrieve memory data into processor registers so as to perform multi-byte arithmetic, or to convert data stored in one format to the other one. This last operation is a simple byte-swap. For example, if the hex value 01020304 is stored in four consecutive memory cells in low-to-high order (little-endian format), it appears in memory (low-to-high) as the values 04030201. Converting this data to the big-endian format consists of swapping the individual bytes so that they are stored in the order 01010304. Figure D-4 is a diagram of a byte-swap operation.

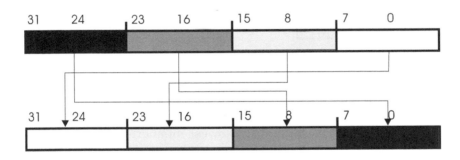

Figure D-4 Data Format Conversion by Byte Swapping.

D.4.4 Sign-Magnitude Representation

Representing signed numbers requires differentiating between positive and negative magnitudes. One possible scheme is to devote one bit to represent the sign. Typically the high-order bit is set (1) to denote negatives and reset (0) to denote positives. Using this convention, the decimal numbers 93 and – 93 are represented as follows:

```
01011101 binary = 93 decimal
11011101 binary = -93 decimal
|
|---------- sign bit
```

This way of designating negative numbers, called a *sign-magnitude* representation, corresponds to the conventional way in which we write negative and positive numbers longhand, that is, we precede the number by its sign. Sign-magnitude representation has the following characteristics:

1. The absolute value of positive and negative numbers is the same.

2. Positive numbers can be distinguished from negative numbers by examining the high-order bit.

3. There are two possible representations for zero, one negative (10000000B) and one positive (00000000B).

But a major limitation of sign-magnitude representation is that the processing required to perform addition is different from that for subtraction. Complicated rules are required for the addition of signed numbers. For example, considering two operands labeled x and y, the following rules must be observed for performing signed addition:

1. If x and y have the same sign, they are added directly and the result is given the common sign.

2. If x is larger than y, then y is subtracted from x and the result is given the sign of x.

3. If y is larger than x, then x / is subtracted from y and the result is given the sign of y.

4. If either x or y is 0 or -0, the result is the non-zero element.

5. If both x and y are -0, then the sum is 0.

However, there are other numeric representations that avoid this situation. A consequence of sign-magnitude representation is that, in some cases, it is necessary to take into account the magnitude of the operands in order to determine the sign of the result. Also, the presence of an encoding for negative zero reduces the numerical range of the representation and is, for most practical uses, an unnecessary complication. An important limitation of using the high-order bit for representing the sign is the resulting halving of the numerical range.

D.3.5 Radix Complement Representation

The *radix complement* of a number is defined as the difference between the number and the next integer power of the base that is larger than the number. In decimal numbers the radix complement is called the *ten's complement*. In the binary system the radix complement is called the *two's complement*. For example, the radix complement of the decimal number 89 (ten's complement) is calculated as follows:

```
100   = higher power of 10
-  89
 ----
   11  = ten's complement of 89
```

The use of radix complements to simplify machine subtraction operations can best be seen in an example. The operation $x = a - b$ with the following values:

```
a = 602
b = 353
            602
        -   353
            ---
x =         249
```

Note that in the process of performing longhand subtraction, we had to perform two borrow operations. Now consider that the radix complement (ten's complement) of 353 is:

```
1000 - 353 = 647
```

Using complements we can reformulate subtraction as the addition of the ten's complement of the subtrahend, as follows:

```
      602
+     647
    _____
     1249
    |_____ discarded digit
```

The result is adjusted by discarding the digit that overflows the number of digits in the operands.

In performing longhand decimal arithmetic there is little advantage in replacing subtraction with ten's complement addition. The work of calculating the ten's complement cancels out any other possible benefit. However, in binary arithmetic the use of radix complements entails significant computational advantages because binary machines can calculate complements efficiently.

The two's complement of a binary number is obtained in the same manner as the ten's complement of a decimal number, that is, by subtracting the number from an integer power of the base that is larger than the number. For example, the two's complement of the binary number 101 is

```
    1000B  =  2^3 = 8 decimal (higher power of 2)
-    101B  =        5 decimal
   _____        _____
    011B   =        3 decimal
```

While the two's complement of 10110B is calculated as follows:

```
   100000B  =  2^5 = 32 decimal (higher power of 2)
-   10110B  =        22 decimal
   _____        _____
   01010B          10 decimal
```

You can perform the binary subtraction of 11111B (31 decimal) – 10110B (22 decimal) by finding the two's complement of the subtrahend, adding the two operands, and discarding any overflow digit, as follows:

```
      11111B  =  31 decimal
+     01010B  =  10 decimal (two's complement of 22)
     _____
     101001B
discard_____|
      01001B  =   9 decimal (31 minus 22 = 9)
```

In addition to the radix complement representation, there is a diminished radix representation that is often useful. This encoding, sometimes called the *radix-minus-one form*, is created by subtracting 1 from an integer power of the base that is larger than the number, then subtracting the operand from this value. In the decimal system, the diminished radix representation is sometimes called the *nine's complement*. This is due to the fact that an integer power of ten, minus one, results in one or more 9-digits. In the binary system, the diminished radix representation is called the one's complement. The nine's complement of the decimal number 76 is calculated as follows:

```
     100  = next highest integer power of 10

      99  = 100 minus 1
 -    76
      ----
      23  = nine's complement of 89
```

The one's complement of a binary number is obtained by subtracting the number from an integer power of the base that is larger than the number, minus one. For example, the one's complement of the binary number 101 (5 decimal) can be calculated as follows:

```
    1000B  =  2^3 = 8 decimal

     111B  =  1000B minus 1 =  7 decimal
 -   101B                      5 decimal
     ------                    ---------
     010B  =                   2 decimal
```

An interesting feature of one's complement is that it can be obtained by changing every 1 binary digit to a 0 and every 0 binary digit to a 1. In this example, 010B is the one's complement of 101B. In this context the 0 binary digit is often said to be the complement of the 1 binary digit, and vice versa. Most modern computers contain an instruction that inverts all the digits of a value by changing all 1 digits into 0, and all 0 digits into 1. The operation is also known as *logical negation*.

Furthermore, the two's complement can be obtained by adding 1 to the one's complement of a number. Therefore, instead of calculating

```
      100000B
 -     10110B
      -------
      01010B
```

we can find the two's complement of 10110B as follows:

```
    10110B  = number
    01001B  = change 0 to 1 and 1 to 0 (one's complement)
 +      1B    then add 1
    ---------
    01010B  = two's complement
```

This algorithm provides a convenient way of calculating the two's complement in a machine equipped with a complement instruction. Finally, the two's complement can be obtained by subtracting the operand from zero and discarding the overflow.

The radix complement of a number is the difference between the number and an integer power of the base that is larger than the number. Following this rule, we calculate the radix complement of the binary number 10110 as follows:

```
    100000B  =  2^5 = 32 decimal
 -   10110B  =        22 decimal
    -------             ----------
     01010B             10 decimal
```

However, the machine calculation of the two's complement of the same value often produces a different result; for example,

```
100000000B  =  28 = 256 decimal
-  00010110B  =         22 decimal
   _____            _____
   11101010B             234 decimal
```

The difference is due to the fact that in the longhand method we have used the next-higher integer power of the base compared to the value of the subtrahend (in this case, 100000B) while the machine calculations use the next-higher integer power of the base compared to the operand's word size, which is normally either 8 or 16 bits. In this example the operand's word size is eight bits and the next-highest integer power of 2 is 100000000B. In either case, the results from two's complement subtraction are valid as long as the minuend is an integer power of the base that is larger than the subtrahend.

For example, to perform the binary subtraction of 00011111B (31 decimal) minus 00010110B (22 decimal) we can find the two's complement of the subtrahend and add, discarding any overflow digit, as follows:

```
         00011111B  =  31 decimal
     +   11101010B  =  234 decimal (two's complement of 22)
         _____
         100001001B
discard____|
         00001001B  =   9 decimal (31 minus 22 = 9)
```

In addition to the simplification of subtraction, two's complement arithmetic has the advantage that there is no representation for negative 0. It can be argued that there are cases in which a negative zero notation could be useful, but in fact this is usually unnecessary. While both the two's complement and the one's complement schemes can be used to implement binary arithmetic, system designers usually prefer the two's complement.

D.4 Encoding of Fractional Numbers

In any positional number system, the weight of each integer digit is determined by the formula

$$P = d * BC$$

where d is the digit, B is the base or radix, and C is the zero-based column number, starting from right to left. Therefore, the value of a multi-digit positive integer to n digits can be expressed as a sum of the digit values:

$$dn*Bn + dn-1*Bn-1 + dn-2*Bn-2 + ... + d0*B0$$

where d is the value of the digit and B is the base or radix of the number system. This representation can be extended to represent fractional values. Recalling that we can extend the sequence to the right of the radix point, as follows

$$x^{-n} = \frac{1}{x^n}$$

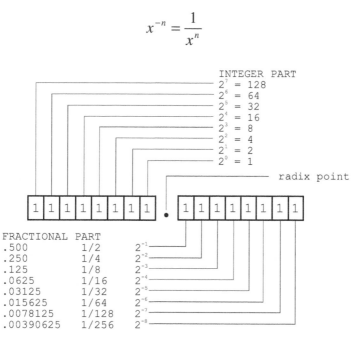

Figure D-5 Positional Weights in a Binary Fraction.

In the decimal system, the value of each digit to the right of the decimal point is calculated as 1/10, 1/100, 1/1000, and so on. The value of each successive digit of a binary fraction is the reciprocal of a power of 2; therefore, the sequence is 1/2, 1/4, 1/8, 1/16, ... Figure D-5 shows the positional weight of the integer and fractional digits in a binary number.

In Chapter 2 we used the positional weights of the binary digits to convert a binary number to its decimal equivalent. A similar method can be used to convert the fractional part of a binary number. Using the decimal equivalents shown in Figure D-5 we convert the binary fraction .10101 to a decimal fraction as follows:

D.4.1 Fixed-Point Representations

The encoding and storage of fractional numbers (also called real numbers) in binary form presents several difficulties. The first one is related to the representation of the *radix point*. Because there are only two symbols in the binary set, and both are used to represent the numerical value of the number, there is no other symbol available for the decimal point.

Figure D-6 Binary Fixed-Point Representation.

One possible solution is to predefine the digit field that represents the integer part and the one that represents the fractional part. For example, if a real number is to be encoded in two data bytes, we can assign the high-order byte to encode the integer part and the low-order byte for the fractional part. In this case, the positive decimal number 58.125 could be encoded as shown in Figure D-6.

In Figure D-6 we assumed that the binary point is positioned between the eighth and the ninth digit of the encoding. Fixed-point representations assume that whatever distribution of digits is selected for the integer and the fractional part of the representation is maintained in all cases. This is the greatest limitation of the fixed-point formats.

Suppose we want to store the value 312.250. This number is represented in binary as follows:

```
312 = 100111000
.250 = .01
```

In this case, the total number of binary digits required for the binary encoding is 11. The number can be physically stored in a 16-digit structure (as the one in Figure D-6), leaving five cells to spare. However, because the fixed-point format we have adopted assigns eight cells to the integer part of the number, 312.250 cannot be encoded because the integer part requires nine binary digits. In spite of this limitation, the-fixed point format was the only one used in early computers.

D.4.2 Floating-Point Representations

An alternative to fixed-point is not to assume that the radix point has a fixed position in the encoding, but to allow it to float, hence the name *floating-point*. The idea of separately encoding the position of the radix point originated in scientific notation, where a number is written as a base greater than or equal to 1 and smaller than 10, multiplied by a power of 10. For example, the value 310.25 in scientific notation is written:

$$3.1025 \times 10^2$$

A number in scientific notation has a real part and an exponent part. Using the terminology of logarithms, these two parts are sometimes called the *mantissa* and the *characteristic*. The following simplification of scientific notation is often used in computer work:

```
3.1025 E2
```

In the computer version of scientific notation, the multiplication symbol and the base are implied. The letter E, which is used to signal the start of the exponent part of the representations, accounts for the name "exponential form." Numbers smaller than 1 can be represented in scientific notation or in exponential form using negative powers. For example, the number .0004256 can be written:

$$4.256 \times 10^{-4}$$

or as

```
4.256 E-4
```

Floating-point representations provide a more efficient use of the machine's storage space. For example, the numerical range of the fixed-point encoding shown in Figure D-6 is from 255.99609375 to 0.00390625. To improve this range we can reassign the sixteen bits of storage so that four bits are used for encoding the exponent and twelve bits for the fractional part, called the significand. In this case the encoded number appears as follows:

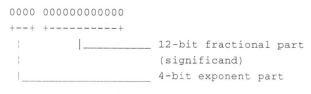

```
0000 000000000000
+--+ +----------+
 ¦          |_____  12-bit fractional part
 ¦                        (significand)
 |_____    4-bit exponent part
```

If we were to use the first bit of the exponent to indicate the sign of the exponent, then the range of the remaining three digits would be 0 to 7. Note that the sign of the exponent indicates the direction in which the decimal point is to be moved; this is unrelated to the sign of the number. In this example, the fractional part (or significand) could hold values in the range 1,048,575 to 1. The combined range of exponent and significand allows representing decimal numbers in the range 4095 to 0.00000001 that considerably exceeds the range in the same storage space in fixed-point format.

D.4.3 Standardized Floating-Point Representations

Both the significand and the exponent of a floating-point number can be stored as an integer, in sign-magnitude, or in radix complement form. The number of bits assigned to each field varies according to the range and the precision required. For example, the computers of the CDC 6000, 7000, and CYBER series used a 96-digit significand with an 11-digit exponent, while the PDP 11 series used 55-digit significands and 8-digit exponents in their extended precision formats.

Variations, incompatibilities, and inconsistencies in floating-point formats led to the development of a standard format. In March and July 1985, the Computer Society of the Institute of Electric and Electronic Engineers (IEEE) and the American National Standards Institute (ANSI) approved a standard for binary floating-point arithmetic (ANSI/IEEE Standard 754-1985). This standard establishes four formats for encoding binary floating-point numbers. Table D.2 summarizes the characteristics of these formats.

Table D.2
ANSI/IEEE Floating Point Formats

PARAMETER	SINGLE	SINGLE EXTENDED	DOUBLE	DOUBLE EXTENDED
total bits	32	43	64	79
significand bits	24	32	53	64
maximum exponent	+127	1023	1023	16383
minimum exponent	−126	1022	−1022	16382
exponent width	8	11	11	15
exponent bias	+127	---	+1023	---

D.4.4 IEEE 754 Single Format

Figure D-7 shows the IEEE floating-point single format.

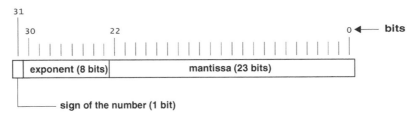

Figure D-7 IEEE Floating-Point Single Format.

If a floating-point encoding is to allow the representation of signed numbers, it must devote one binary digit to encode the number's sign. In the IEEE 754 single format in Figure D-7, the high-order bit represents the sign of the number. A value of 1 indicates a negative number.

The exponent of a binary floating-point number represents the integer power of the base with which the significand must be multiplied. The exponent can be stored in integer, sign magnitude, or radix complement representations. The IEEE 754 standard for floating-point arithmetic establishes that the exponent be stored in biased form, although the bias is not defined in all formats defined in the standard.

The word *bias*, in this context, means a constant that is added to the exponent in order to determine its final value. The term *excess-n notation* has also been used in this context. The constant is usually calculated to be approximately one-half the numerical range of the exponent field. For example, the IEEE single format devotes eight digits for the exponent field (see Figure D-7). The numerical range of eight binary digits is 0 to 255 decimal and one-half of this range is approximately 127. Adding the constant 127 to all positive exponents places them in the range 127 to 255. The lower half of the range (1 to 126) is used for negative exponents. A 0-value in the exponent field is reserved to encode zero and *denormals*. Denormals are a special type of number discussed in the following paragraph. Table D.3 shows the values of the exponent and the biased representation in the IEEE single format for floating-point numbers.

Table D.3

Interpretation of Exponent in the IEEE Single Format

BIASED EXPONENT	SIGN OF NUMBER	TRUE EXPONENT	SIGNIFICAND	CLASS
0000 0000	+	−	00 ... 00	positive zero
	−	−	00 ... 00	negative zero
			11 ... 11	
			to	
			00 ... 01	denormals
0000 0001	−	−126	00 ... 00	normals
to		to	to	
0111 1111		0	11 ... 11	
1000 0000	−	1	00 ... 00	normals
to		to	to	
1111 1110		127	11 ... 11	
1111 1111	+	−	00 ... 00	+ infinity
	−		00 ... 00	− infinity
	−		10 ... 00	Indefinite
	−		00 ... 01	
			to	
			11 ... 11	Not-a-number

Note in Table D.3 that the exponent value 00000000B is used to represent zero and denormal numbers. Denormals, or denormalized numbers, occur when the exponent of the number is too small to represent in the corresponding floating-point format. On the other hand, the exponent 11111111B is used to encode numbers that are too large for the single format, or to represent error conditions. The exponent range 00000001B to 11111110B (decimal values 1 to 254) is used to represent *normal* numbers, that is, numbers that are within the range of the format.

In IEEE 754 floating-point formats the high bit of the exponent field does not encode the sign, as is the case in the sign-magnitude form. Instead, the bias 127 scheme, mentioned previously, is used to represent negative and positive exponents. Negative exponents are in the range 1 to 127 (see Table D.3) and positive exponents are in the range 128 to 254. In contrast with fixed-point conventions, the high bit of the exponent is set to indicate a positive exponent, and is zero to indicate a negative exponent. The main advantage of a biased exponent is that the numbers can be compared bitwise, from left to right, to determine the larger one. The number's true exponent is obtained by subtracting the bias.

The third field of the floating-point representation is known by several names: fractional part, *mantissa*, *characteristic*, and significand (see Figure D-7). The word *significand* is the one most commonly used the literature. Like the exponent, the significand can be stored as an integer, or in sign-magnitude or radix complement representations.

A floating-point binary number is said to be in *normalized form* when the first digit of its significand is 1. An un-normalized binary floating-point number can be normalized by successively shifting the digits of the significand to the left, while simultaneously subtracting one from the exponent. This process is continued until the high-order bit of the significand is a binary 1. The process does not change the value of the number, as shifting the significand bits to the left effectively multiplies the number by 2, while subtracting one from the exponent divides the number by 2. Clearly, the value of a number does not change if it is multiplied and divided by the same value. Also, note that normalization applies to the entire encoded number because it requires adjustments of both the exponent and the significand. Therefore, it is not correct to speak of a normalized significand or a normalized mantissa; we should refer to the significand of a normalized floating-point number.

One advantage of the normalized form is that the significand contains a maximum number of significant bits. However, addition and subtraction of floating-point numbers require that both operands have the same exponent. Therefore, before performing these operations, it is often necessary to shift the significand digits to the right or to the left so that the exponents are equal.

The IEEE standard takes advantage of the fact that a normalized significand of a binary floating point starts with a 1 digit. In the single- and double-precision formats, this leading bit of the significand is assumed, in effect doubling the range of the representation. Not so in the extended formats, in which the digit must be explicitly coded. Note that this assumption is not valid if the exponent is all zeros. A zero exponent and a non-zero significand indicate a *denormal*, as shown in Table D.3. Also, the use of an implicit bit makes necessary a special representation for zero (see Table D.3). This special zero must be handled separately during arithmetic operations.

D.4.5 Encoding and Decoding Floating-Point Numbers

The formats in the IEEE 754 standard for binary floating-point arithmetic were designed to provide maximum storage capacity and processing efficiency. For example, the exponent in the IEEE single format, stored in biased form, takes up eight bits; however, these eight bits do not fall on a byte boundary. The exponent bits take up seven bit positions in the high-order byte, and one bit position in the next byte, as shown in Figure D-7. In the same IEEE single encoding, the significand takes up seven bits of the second byte as well as the third and fourth bytes. The sign of the number is the high-order bit of the high-order byte. Figure D-8 shows the number 127.375 stored in the IEEE floating-point single format.

The encoding in Figure D-8 is interpreted as follows:

```
sign of number = 0 (positive)
biased exponent = 10000101B = 133 decimal
real exponent = 133 - bias = 133 - 127 = 6
significand = 1.1111110 11000000 00000000 (adding explicit digit)
significand is adjusted by moving the radix point six places
to the right
new significand = 1111111.01100...000
```

The significand bits are intepreted as follows::

```
integer part = 1111111 = 127
fractional part = .01100..00 = .375
```

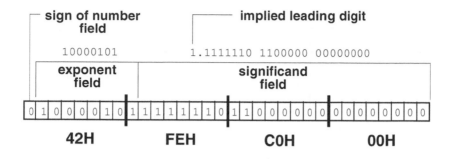

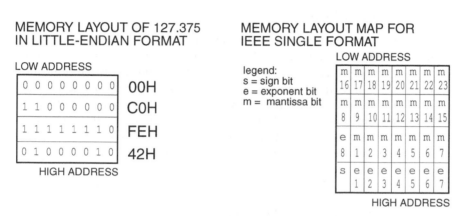

Figure D-8 Encoding of the Number 127.375 in IEEE Single Format.

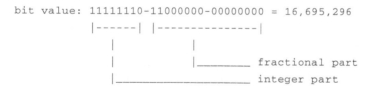

D.5 Binary-Coded Decimals (BCD)

Floating-point encodings are the most efficient format for storing numerical data in a digital device, and binary arithmetic is the fastest way to perform numerical calculations. But other representations are also useful. BCD (binary-coded decimal) is a way of representing decimal digits in binary form. There are two common ways of encoding decimal digits in binary format. One is known as the *packed BCD* format and the other one as *unpacked*. In the unpacked format, each BCD digit is stored in one byte. In packed form, two BCD digits are encoded per byte. The unpacked BCD format does not use the four high-order bits of each byte, which is wasted storage space. On the other hand, the unpacked format facilitates conversions and arithmetic operations on some machines. Figure D.9 shows the memory storage of a packed and unpacked BCD number.

UNPACKED BCD

0 0 0 0	0 0 1 0	2
0 0 0 0	0 0 1 1	3
0 0 0 0	0 1 1 1	7
0 0 0 0	1 0 0 1	9

PACKED BCD

| 0 0 1 0 | 0 0 1 1 | 23 |
| 0 1 1 1 | 1 0 0 1 | 79 |

Figure D-9 Packed and Unpacked BCD.

D.5.1 Floating-Point BCD

Unlike the floating-point binary numbers, binary-coded decimal representations and BCD arithmetic have not been explicitly described in a formal standard. Each machine or software package stores and manipulates BCD numbers in a unique and often incompatible way. Some machines include packed decimal formats, which are sign-magnitude BCD representations. These integer formats are useful for conversions and input-output operations. For performing arithmetic calculations, a floating-point BCD encoding is required. This approach provides all the advantages of floating point as well as the accuracy of decimal encodings.

The BCD floating-point format that we call BCD12 is shown Figure D-10.

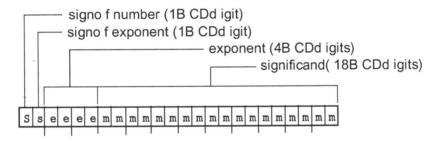

Figure D-10 Map of the BCD12 Format.

BCD12 requires 12 bytes of storage and is described as follows:

1. The sign of the number (S) is encoded in the leftmost packed BCD digit. Therefore, the first four bits are either 0000B (positive number) or 0001B (negative number).

2. The sign of the exponent is represented in the four low-order bits of the first byte. The sign of the exponent is also encoded in one packed BCD digit. As is the case with the sign of the number field, the sign of the exponent is either 0000B (positive exponent) or 0001B (negative exponent)

3. The following two bytes encode the exponent in four packed BCD digits. The decimal range of the exponent is 0000 to 9999.

4. The remaining nine bytes are devoted to the significand field, consisting of eighteen packed BCD digits. Positive and negative numbers are represented with a significand normalized to the range 1.00...00 to 9.00...99. The decimal point follow-

ing the first significand digit is implied. The special value 0 has an all-zero significand.

5. The special value FF hexadecimal in the number's sign byte indicates an invalid number.

The structure of the BCD12 format is described in Table D.4.

Table D.4

Field Structure of the BCD12 Format

CODE	FIELD NAME	BITS WIDE	BCD DIGITS	RANGE
S	Sign of number	4	1	0 – 1 (BCD)
S	Sign of exponent	4	1	0 – 1 (BCD)
E	Exponent	16	4	0 – 9999
M	Significand	72	18	0 – 99..99 (18 digits)
	Format size	96 (12 bytes)		

Notes:
1. The significand is scaled (normalized) to a number in the range 1.00..00 to 9.99..99.
2. The encoding for the value zero (0.00..00) is a special case.
3. The special value FFH in the sign byte indicates an invalid number.

The BCD12 format, as is the case in all BCD encodings, does not make ideal use of the available storage space. In the first place, each packed BCD digit requires four bits, which in binary could serve to encode six additional combinations. At a byte level, the wasted space is of 100 encodings (BCD 0 to 99) out of a possible 256 (0 to FFH). The sign field in the BCD12 format is wasteful because only one binary digit is actually required for storing the sign. Regarding efficient use of storage, BCD formats cannot compete with floating-point binary encodings. The advantages of BCD representations are a greater ease of conversion into decimal forms, and the possibility of using the processors' BCD arithmetic instructions.

Appendix E

Digital Arithmetic and Conversions

E.1 Microcontroller Arithmetic

Microcontrollers are not designed for intensive numeric processing; therefore, they are not equipped with many arithmetic operators usually found in microprocessors. A typical mid-range microcontroller has instructions to add and subtract integers and perhaps to increment and decrement. Hardware multiplication is rarely available and even more rare is division. Likewise, there is usually no hardware support for decimal and floating-point arithmetic. For this reason the microcontroller programmer is often challenged to provide most arithmetic and data processing operations in software.

In this discussion we assume a mid-range microcontroller, such as the PIC 16f8x. These devices contain primitives for adding and subtracting integers, shifting and rotating bits, incrementing and decrementing machine registers, some support for decimal operations and conversions, as well as the basic logic primitives AND, OR, XOR, and NOT. Multiplication and division operators, as well as floating-point operators, are not available in the mid-range devices.

E.2 Unsigned and Two's Complement Arithmetic

In Chapter 3 we discussed the various representations for signed and unsigned binary and decimal numbers. Arithmetic operations of unsigned operands are the simplest. In this case we assume that the encoding always represents a positive number and that all bits relate to the number's magnitude.

Unsigned arithmetic can be binary or decimal. In a machine with 8-bit words, binary arithmetic on unsigned numbers uses the entire range of the format. This is true even when the primitive operations are valid in two's complement form; in fact, it is one of the great advantages of two's complement representation. Table E.1 shows a 4-bit binary in several numeric formats.

Table E.1

Interpretations of 4-bit Binary Numbers

BINARY	1'S COMPLEMENT	DECIMAL VALUES 2'S COMPLEMENT	UNSIGNED
0111	7	7	7
0110	6	6	6
0101	5	5	5
0100	4	4	4
0011	3	3	3
0010	2	2	2
0001	1	1	1
0000	0	0	0
1111	-0	-1	15
1110	-1	-2	14
1101	-2	-3	13
1100	-3	-4	12
1011	-4	-5	11
1010	-5	-6	10
1001	-6	-7	9
1000	-7	-8	8

Assume a machine with a 4-bit word size and consider addition of two unsigned numbers:

```
        BINARY      DECIMAL
        0111        7
      + 0110        6
        ------      ----
        1101        13
```

Note, in the previous example, that if the encoding were in two's complement form, the addition of the positive values 6 plus 7 would produce a result that overflows the capacity of the representation. In 4-bit two's complement representation, there is no way of encoding the value 13.

The question that arises is: In a device that performs two's complement addition, must we always assume that the operands are in two's complement form? The answer is: No. Signed addition of two's complement operands and the unsigned addition of integer operands can be performed with identical processing and by the same electronic circuitry. It is the software that must take into account the encoding of the operands in order to interpret the results. For example, in the 4-binary digit device previously considered, the two's complement addition of the values 6 and 7 produce an overflow, which can be detected by observing the change in the high-order bit (the sign bit) of the result. Therefore, in this case, the result of the addition operation is invalid. However, if the same decimal values represent unsigned operands, then the addition of 7 plus 6 produces the valid result 13. In either case, the binary values of the operands, as well as the result, are the same.

Microcontrollers usually support the fundamental operations of addition and subtraction on signed and unsigned integer operands with a single primitive operation. The addition and subtraction operators in low- and mid-range devices allow two operands. The more powerful microcontrollers support addition and subtraction of three operands, which is useful in implementing multi-digit routines. In either case, the software determines if the result is signed or unsigned by interpreting the changes in the high-order bit of the operands and by evaluating the status flags if these are available.

E.2.1 Operations on Decimal Numbers

Although microcontrollers are binary devices, the instruction set often includes operations for performing arithmetic on binary coded decimal numbers. In Chapter 3 we saw that BCD numbers can be stored in packed or unpacked form. In packed format, two BCD digits are contained in each byte. The low-order BCD digit takes up bits 0 to 3 and the high-order BCD digit takes up bits 4 to 7. Unpacked BCD digits are stored one digit per byte; in this case the high-order nibble is unused. The packed and unpacked binary coded decimal formats can be seen in Figure 3-9.

Microcontroller designers usually adopt the packed BCD format for representing decimal operands. One advantage of packed BCDs is that the two decimal digits encoded in a single byte can be represented as hexadecimal digits. For example, the values H24 and H99 represent the packed BCD digits 24 and 99, respectively. Note that each hex digit is preceded by the letter H to indicate radix 16. In actual microcontroller programming, other ways are often used for representing numbers in hexadecimal notation.

The addition and subtraction of decimal numbers represented in packed BCD can be performed with binary primitive operations, complemented with some additional adjustments. In some cases the addition of two BCD numbers in packed format may produce a valid result; for example

```
      H23          H31          H56
+     H12          H38          H22
      ----         ----         ----
      H35          H69          H78
```

In the previous examples the results are valid because the sum of each digit does not exceed the range of the BCD format. However, the following additions do not produce valid BCD results:

```
      H33          H31          H56
+     H27          H59          H27
      ----         ----         ----
      H5A          H8A          H7D
```

In the case of the first operation, the valid BCD result would be 33 + 27 = 60; in the second one, 31 + 59 = 90; and in the third one, 56 + 27 = 83. A simple adjustment corrects the error, as follows:

```
        H33              H31              H56
   +    H27              H59              H27
        ---              ---              ---
        H5A              H8A              H7D
   +    H 6              H 6              H 6
        ---              ---              ---
        H60              H90              H83
```

In all three cases, adding 6 to the previous sum produces the expected result. The logic for deciding when the value 6 must be added is simple: If the sum of the low-order digit is greater than 9 or if the sum produced a carry out of the low-order nibble, then add 6 to the sum to perform the decimal adjustment. Some high-end microcontrollers contain a primitive instruction that executes the decimal adjustment automatically, that is, without having to test the sum. However, this instruction is not available in low- and mid-range devices.

Also note that the largest number that can be encoded in packed BCD format is the decimal 99. When adding two BCD digits, the high-order digit of the sum cannot be greater than 9. If so, then the capacity of the format has been exceeded and the result cannot be adjusted by the simple addition of 6. Here again, a multi-byte processing routine can be developed in order to accommodate the result of BCD addition when the sum exceeds a single byte.

Many microcontrollers are equipped with a flag that indicates overflow from binary digit number 3. This flag, sometimes called the *digit carry* or the *half carry flag*, can be used to detect that a calculation has overflowed the storage capacity of four binary digits. The availability of this flag simplifies the logic necessary for adjusting binary addition of decimal operands because the value 6 must be added when the digit in the low-order nibble is larger than 9, or when there has been a carry to the next digit. The flowchart in Figure E-1 shows this processing.

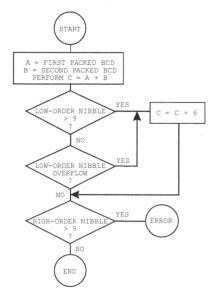

Figure E-1 Flowchart for Two-Byte BCD Addition.

E.3 Bit Manipulations and Auxiliary Operations

In addition to basic logic and arithmetic, microcontrollers contain primitive operators to manipulate individual bits, to compare operands, to make decisions based on the state of individual bits and flags, and to convert data to other formats. As always, the presence or absence of some of these operations, as well as their degree of power and sophistication, varies with the individual microcontroller. In the following subsections we describe the most commonly available primitives.

E.3.1 Bit Shift and Rotate

The fundamental operators to shift and rotate are useful in developing BCD and binary arithmetic routines. One interesting use of bit shifting is in implementing binary multiplication and division routines.

Shift operations consist of transposing to the left or right all the bits in the operand. In microcontrollers the operand is usually a processor register. For example, after a right shift operation, all the bits in the value 01110101B (75H) are moved one position to the right, resulting in the value 00111010B (3AH). Note that on a right shift, the rightmost bit disappears and a zero comes into the high-order bit. By the same token, in a left shift, the high-order bit disappears and a zero comes into the low-order bit. Figure E-2 shows the action of a left-shift operation.

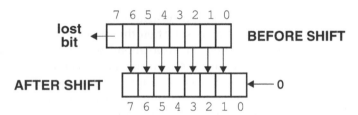

Figure E-2 Left Shift Operation.

The rotate operation differs from the shift in that in the rotate, the low-order bit is either a copy of the high-order bit or of the carry flag. In the first case, the operation is a pure rotate; in the second case, the rotate is referred to as *rotate-through-carry*. Figure E-3 shows the action of a *left-rotate-through-carry* flag.

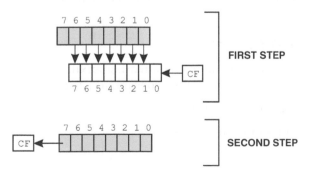

Figure E-3 Rotate-through-carry Left Operation.

Note in Figure E-3 that the contents of the carry flag are first copied to the low-order bit of the destination operand; then the individual bits of the source (in gray in the illustration) are shifted left and moved to the destination. Finally the high-order bit of the source is copied to the carry flag.

There are several possible variations of the rotate operation. The Intel microprocessors distinguish between arithmetic and logic rotates. In the arithmetic rotation the high-order bit is preserved in order to maintain the sign of the operand. The rotate shown in Figure E-3 is the one most common in microcontroller hardware. Clearing the carry flag before the rotate takes place makes the operation identical to a shift.

E.3.2 Comparison Operations

An interesting property of subtraction is its use in finding the relative size of two operands. This interesting action of subtraction is based on the following logic:

1. If the result of a subtraction is zero, then both operands were of the same size.

2. If the result of a subtraction is a positive number, then the subtrahend was smaller than the minuend.

3. If the result of a subtraction is a negative number, the subtrahend was larger than the minuend.

 · In a binary/digital device the result of a subtraction operation can be determined by observing the flags. If the zero flag is set, then the operands were the same (case 1, above). If the carry flag is set, then the subtrahend was larger than the minuend (case 3, above). If neither the carry nor the zero flag is set, then the resulting subtrahend was smaller than the minuend (case 2, above). Because all microcontrollers offer some mechanism for redirecting execution according to the state of the flags, a program can use subtraction to make these decisions.

The one objection to the use of subtraction in comparing the size of two operands is that the process will change one of them. To use subtraction in comparison operations, the programmer has to find some way of preserving the minuend. Alternatively, some devices contain a comparison operator that sets the flags as if a subtraction had taken place but without changing the operands. High-end microcontrollers are equipped with dedicated comparison operators but the middle- and low-range devices usually are not.

E.3.3 Other Support Operations

Mid- and high-range microcontrollers contain other auxiliary bitwise, arithmetic, and logic operators that can be useful to the programmer. These include instructions to

1. Increment and decrement operands

2. Clear registers or storage locations

3. Swap nibbles

4. Clear and set individual bits

5. Test individual bits

Usually, instructions to increment and decrement and to test individual bits are also capable of redirecting execution according to the result. For example, a special decrement can be followed by a jump if decrementing sets the zero flag. Or a bit test instruction can include a jump that is taken if the tested bit is set or reset.

E.4 Unsigned Binary Arithmetic

Because microcontrollers are not used in data processing, microcontroller programming does not usually require the development of powerful or sophisticated numerical routines. At the same time, because microcontrollers often lack primitive support for even the most essential calculations, the programmer makes up for this deficiency. For example, mid-range PIC microcontrollers contain primitive instructions for signed and unsigned addition and subtraction of byte-size operands. Unsigned addition and subtraction operations that exceed one byte, as well as unsigned multiplication and division, must be provided in software.

In unsigned arithmetic, all bits of the binary encoding are interpreted as magnitude bits and all numbers are positive. Addition of unsigned binary numbers is limited by the machine's word size. For example, a mid-range PIC microcontroller performs unsigned addition on 8-bit operands. An overflow of the sum is reported by the carry flag set. In this case the carry flag clear indicates that the sum is within the storage capacity of the format. In unsigned arithmetic processing, routines for extending operations to multiple bytes are straightforward and relatively simple.

E.4.1 Multi-Byte Unsigned Addition

Many microcontrollers are one-byte machines, so operands and results for arithmetic operations must be contained within eight bits. The largest unsigned value that can be represented in a single byte is the decimal number 255. But often applications require adding operands that are larger than a single byte and storing results that exceed this limit. In these cases, multi-byte routines become necessary.

The simplest case is the addition of two unsigned byte-size operands whose sum exceeds 255 decimal. This case requires storing the result in a two-byte area and detecting those cases in which there is a carry into the high-order byte. In this case the largest possible operands for byte addition are the hexadecimal numbers FF. Addition is as follows:

```
        Binary:
      1 1 1 1 1 1 1 1
  +   1 1 1 1 1 1 1 1
      ---------------
      1 1 1 1 1 1 1 0
  C <=
```

In this example the symbol C <= represents a carry out of the high-order bit, the case when the sum exceeds the capacity of a single byte. In hexadecimal, the sum of HFF + HFF = H1FE. You can add two byte-size operands into a two-byte storage area using byte addition to determine the low-order byte of the result and testing for a carry out of the high-order bit. If there is a carry, then the high-order byte of the result is 1; otherwise the high-order byte is 0.

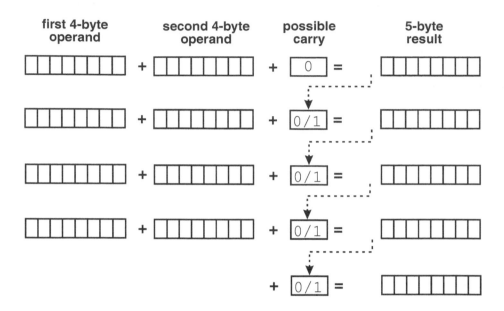

Figure E-4 Unsigned Multi-Byte Addition.

The same logic can be generalized to add more than two byte-size operands as long as the storage area for the result exceeds the size of the operands by one byte. For example, two word-size operands (16 bits each) can be added into a 3-byte (24-bit) storage area, or two double-word operands (32 bits) into a 5-byte storage area. The general algorithm for multi-byte addition is shown in Figure E-4.

The case shown in Figure E-4 consists of adding two 4-byte operands into a 5-byte sum. The addition of the first two operands assumes that there is no carry. In the remaining stages there can be a possible carry from the previous stage if the sum of the two byte-size operands, plus the previous carry, exceed the storage capacity of eight bits. The last byte of the result is determined solely by the possible carry from the previous stage.

In Figure E-4 we see that multi-byte addition requires the sum of three values in all stages except the first and the last one. Some high-end microcontrollers have addition operators that accept a three-byte operand. Others have special addition opcodes that automatically add-in the carry flag. The latter operators are referred to as *add-with-carry*. However, in most low- and mid-range devices, the software must take care of incrementing the sum if there is a carry from the previous stage. The actual multi-byte addition routines are developed in the context of programming the various microcontrollers, discussed later in this book.

E.4.2 Unsigned Multiplication

The case for multiplication cannot be generalized because high-end microcontrollers usually contain one or more multiplication operators; this is not the case in low- and mid-range devices. In the first case implementation is simply by using the corresponding operator. This section explains multiplication in devices that lack a dedicated multiplication operation code.

Arithmetically, multiplication is performed by repeated addition. The multiplier represents the number of times that the multiplicand must be added to itself. Therefore, 3 times 4 is the same as 3 + 3 + 3 + 3. This fact allows implementing multiplication routines in software as long as the device contains an addition operator. The logic is based on using the multiplier as a counter. This counter is decremented each time the multiplicand is added to itself. The routine ends when the counter is exhausted, as shown in the flowchart in Figure E-5.

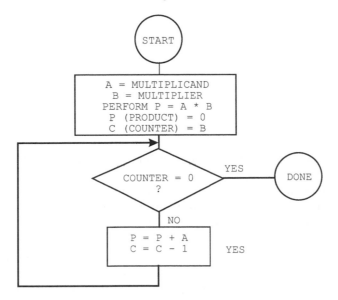

Figure E-5 Unsigned Multiplication Flowchart.

The beauty of the repeated addition algorithm is its simplicity, and its main shortcoming is its slowness. An alternative way of performing multiplication is by shifting the bits of the operand. This method is based on the properties of a binary positional system in which the value of each digit is a successive power of 2. Therefore, by shifting all digits to the left, the value 0001B (1 decimal) successively becomes 0010B (2 decimal), 0100B (4 decimal), 1000B (8 decimal), and so on.

Binary multiplication by means of bit shifting has the downside that the multiplier must be a power of 2. Otherwise, the software must shift by a power of 2 that is smaller than the multiplier and then add the multiplier as many times as necessary to complete the product. In this manner, to multiply by 5, we can shift left twice and add once the value of the multiplicand. To multiply by 7, we would shift left twice and then add three times the value of the multiplicand. As the multiplier gets larger and more distant from the smaller power of 2, the number of addition operations required is also larger, and the effectiveness of the algorithm diminishes.

A third approach is based on the manipulations performed during longhand multiplication. For example, the multiplication of 00101101B (45 decimal) by 01101101B (109 decimal) can be expressed as a series of products and shifts, in the following manner:

```
                    0 0 1 0 1 1 0 1 B = 45 decimal
        times       0 1 1 0 1 1 0 1 B = 109 decimal
                    -----------------
                    0 0 1 0 1 1 0 1
                  0 0 0 0 0 0 0 0
                0 0 1 0 1 1 0 1
              0 0 1 0 1 1 0 1
            0 0 0 0 0 0 0 0
          0 0 1 0 1 1 0 1
        0 0 1 0 1 1 0 1
      0 0 0 0 0 0 0 0
      ---------------------------------
      0 0 1 0 0 1 1 0 0 1 0 1 0 0 1 B = 4905 decimal
```

The actual calculations in this method of binary multiplication are quite simple because the product by a 0 digit is zero and the product by a 1 digit is the multiplicand itself. Consequently, the multiplication routine simply tests each digit in the multiplier. If the digit is zero, no action need be performed; if the digit is one, the multiplicand is shifted left and added into an accumulator.

The storage allocation to hold the product of a multiplication operation is not the same as that to hold the sum. In multi-byte addition, one additional byte is required to hold the sum. In multiplication, the storage allocation must be twice the size of the operands. For example, byte multiplication requires a two-byte storage, while multiplying two double-byte operands requires a four-byte storage allocation.

E.4.3 Unsigned Division

If multiplication can be reduced to repeated addition, then division can be conceptualized as repeated subtraction. In the case of division, the quotient (result) is the number of times the divisor must be subtracted from the dividend before zero or a negative value results from the subtraction. The flowchart in Figure E-6 shows the logic steps in unsigned division.

In Figure E-6 note that the logic tests for a zero divisor, as division by zero is mathematically undefined. Also, because the operation is unsigned, the result cannot be negative; therefore, the divisor must be larger than the dividend. Finally, the logic must consider the case in which subtracting the divisor from the remainder produces a negative value, in which case an adjustment is necessary to produce a valid quotient. This adjustment avoids the need for searching for a trial divisor, as in the case of the common longhand division algorithm. In machine code, the negative result is detected as an overflow (carry flag set) from the subtraction.

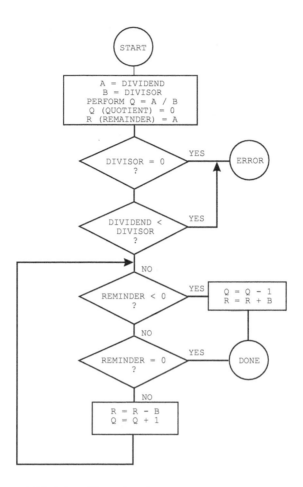

Figure E-6 Unsigned Division Flowchart.

E.5 Signed Binary Arithmetic

In two's complement and sign-magnitude representations, the high-order bit represents the sign of the operand, while its magnitude is represented in the remaining bits. Therefore, in the case of signed numbers, a carry out of the high-order bit is meaningless because the high-order bit is not a magnitude bit. For example, consider the following operation in an 8-bit device that performs unsigned and two's complement addition:

```
      80   =   0101 0000B
  +   90   =   0101 1010B
      -----------------
     170   =   1010 1010B
```

If the operands are assumed to be in unsigned binary format, the result is valid. However, if the operands (the decimal values 80 and 90) are assumed to be positive numbers in two's complement form, then the result is invalid because the positive number 170 cannot be represented in an 8-bit two's complement encoding.

Clearly, multi-byte operations on signed representations cannot be performed identically as with unsigned operands. Table E.2 shows the unsigned and two's complement representations of one-byte numbers.

Table E.2

Signed and Unsigned Representations of One-Byte Numbers

BINARY	2'S COMPLEMENT	UNSIGNED
0000 0000	0	0
0000 0001	1	1
0000 0010	2	2
0000 0011	3	3
.	.	.
.	.	.
.	.	.
0111 1111	127	127
1000 0000	−128	128
1000 0001	−127	129
1000 0010	−126	130
1000 0011	−125	131
.	.	.
.	.	.
1111 1110	−2	254
1111 1111	−1	255

E.5.1 Overflow Detection in Signed Arithmetic

In unsigned addition, the carry flag is magnitude related. It is set when there is a carry out of the high-order bit of the destination operand, which takes place when its capacity has been exceeded. This is usually described as an overflow condition. However, a carry out of the high-order bit of the result is not always meaningful in signed arithmetic. For example, suppose the following two's complement addition:

```
      Decimal          binary
      127            0111 1111
 +    127            0111 1111
      ----           ---------
      ??             1111 1110
```

In this case the sum clearly exceeds the capacity of the format, because the largest positive value that can be represented in a two's complement 8-bit format is 127 (see Table E.2). However, the operation did not generate a carry out of the high-order bit. Therefore, the carry flag could not have been used to detect the overflow error in this case.

Now consider the addition of two negative numbers in two's complement form:

```
      Decimal          binary
      -4             1111 1100
 +    -5             1111 1011
      ----           ---------
      -9        C <= 1111 0111
```

In this case, the addition operation generated a carry out of the high-order bit; however, the sum is arithmetically correct. In fact, any addition of negative operands in two's complement notation generates a carry out of the most-significant bit.

These two examples show that the carry flag, by itself, cannot be used to detect an error or no-error condition in two's complement arithmetic. Detecting an overflow condition in two's complement representations requires observing the carry into the high-order bit of the encoding as well as the carry out. In both previous examples we note that there was a carry into the high-order bit of the result. However, in the first case, there was no carry out. The general rule is: *two's complement overflow takes place when the carry into and the carry out of the high-order bit have opposite values.* Figure E-7 is a flowchart to detect overflow in signed arithmetic.

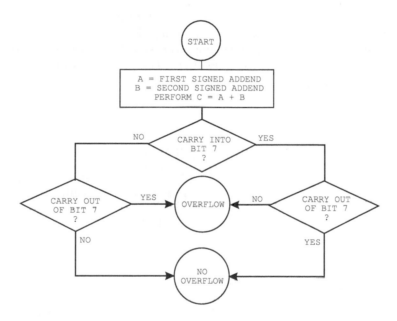

Figure E-7 Detecting Overflow in Two's Complement Arithmetic.

Most microprocessors and some high-end microcontrollers contain hardware facilities for detecting signed arithmetic overflow. In some cases the hardware support consists of a single overflow flag that is set whenever the result of an arithmetic operation exceeds the capacity of the format. In other cases, as in the PIC 18CXX2 family, the status register contains a negative bit flag that indicates a 1-bit in the sign bit position, as well as an overflow bit that is set whenever there is an overflow from the magnitude bits (0 to 6) into the sign bit (bit 7) of the destination operand. In this device, software can test one or both of these flags to detect two's complement overflow.

In low- and mid-range devices, with no hardware support for signed arithmetic, detecting a two's complement overflow is by no means simple. Without a hardware flag to report a carry condition into a particular bit position, software is confronted with several possible alternatives, but none is simple or straightforward.

E.5.2 Sign Extension Operations

Observing the carry into and the carry out of the most-significant bit is a valid way of detecting overflow of a two's complement arithmetic operation. In theory, the logic described in the flowchart of Figure E-7 can be implemented in devices without hardware support for signed overflow; however, the processing is complicated and therefore costly in execution time. An alternative approach is to ensure that the format has sufficient capacity to store the arithmetic result. The rule developed previously lets us determine that, for addition and subtraction, the destination format must have at least one more byte than the operands. In multiplication, the destination operand must be at least twice the size of the source operands.

A simple mechanism for extending the capacity of two's complement encoding is called *sign extension*. The process consists of copying the sign bit into the high-order bit positions of the extended encoding. For example, to extend a two's complement 8-bit number into 16 bits, copy the sign bit of the original value (bit number 7) into all the bits of the extended byte. The process is shown in Figure E-8 for both positive and negative operands.

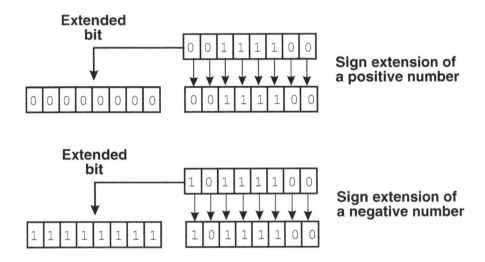

Figure E-8 Sign Extension of Two's Complement Numbers.

E.5.3 Multi-byte Signed Operations

Signed operations on two's complement numbers encoded in multiple bytes can be performed using the processor's arithmetic primitives. Consider the addition of the numbers −513 and −523, each one encoded in 16-bit two's complement form:

```
decimal            binary
                   HOB        LOB
-513               1111 1101  1111 1111
-523               1111 1101  1111 0101
-----              ---------------------
```

                    -1036   1111 1011   1111 0100

In the preceding example, adding the low-order bytes produces the sum shown, plus a carry. Adding the high-order bytes plus the carry, and discarding the overflow, produces the sum of high-order bytes shown above. The result is the correct value in two's complement form. The fact that the result did not overflow the capacity of the 16-bit format can be ascertained by observing that there was a carry into the fifteenth digit, and also a carry out. Carry in and carry out of the sign bit is one of the conditions for no overflow in the flowchart of Figure E-7.

E.6 Data Format Conversions

Quite often, code needs to convert data into and from different numeric formats, for example, to display ASCII digits in an output device, or to convert numeric keyboard input in ASCII into binary or BCD encodings for processing. In this section we consider the logic for the following cases:

1. BCD digits to ASCII decimal

2. Binary to string of ASCII decimal digits

3. String of ASCII decimal digits to binary

4. Binary to string of ASCII hexadecimal digits

As in the previous cases, implementation of these conversions is device dependent and varies in the different hardware.

E.6.1 BCD Digits to ASCII Decimal

Packed BCD digits are encoded in one digit per nibble, as shown in Section E.2.1. Thus, each digit is a binary value in the range 0 to 9. Converting each digit to ASCII consists of isolating each nibble and then changing the binary into an ASCII representation. Note in Table E.1 that the numeric ASCII digits start at 30H for the digit zero and extend to 39H for the digit 9. For this reason, converting a numeric digit from binary to ASCII consists of adding 30H. By the same token, subtracting 30H converts a single ASCII digit to binary.

Assume four packed BCD digits in two consecutive memory bytes, labeled A and B, where A holds the two low-order digits; also, a four-digit storage buffer to which the variable P is a pointer. The conversion algorithm can be described as follows:

1. Initialize buffer pointer P to the first storage location.

2. Copy digit A to temporary digit T.

3. Mask out four high-order bits of T.

4. Add 30H to value in T and store in buffer by pointer P.

5. Bump buffer pointer to next digit storage.

6. Copy digit A to temporary digit T.

7. Mask out four low-order bits in T.

8. Shift four high-order bits to the right by 4 bits.

9. Add 30H to value in *T* and store in buffer by pointer *P*.

10. Bump buffer pointer to next digit.

11. Proceed with digit *B* in the same manner.

E.6.2 Unsigned Binary to ASCII Decimal Digits

Often we hold an unsigned binary number in memory or a machine register and need to display its value to some ASCII-based output device. The process requires converting the binary value to a string of ASCII decimal digits. The number of decimal digits depends on the number of bits in the binary representation. A one-byte unsigned binary requires three ASCII decimal digits because the value ranges from 0 to 255. A two-byte unsigned binary requires a string of five ASCII decimal digits because the range of a two-byte representation is from 0 to 65,535, and so on. The storage area for the ASCII digits is sometimes referred to as a *buffer*.

The process of converting binary to ASCII decimal consists of dividing the binary by 10 to obtain each decimal digit, then adding 30H to the remainder in order to turn the digit into ASCII. The process continues until the original dividend is reduced to zero, as shown in the flowchart of Figure E-9.

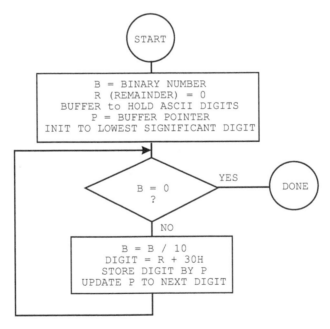

Figure E-9 Unsigned Binary to ASCII Decimal String.

E.6.3 ASCII Decimal String to Unsigned Binary

Another conversion operation frequently needed in software is the transformation of a string of ASCII decimal digits into binary. This type of conversion typically arises when the program needs to receive input that must later be processed by the device. For example, the user enters a numeric value from a keyboard and the application must process this data in binary form.

In designing the conversion routine we must first delimit the value range of the input data so as to allocate a sufficiently large binary format to store the result. For example, code can store in a single unsigned byte a binary in the range 0 to 255, but it requires two bytes to store one in the range 0 to 65,535. Once the binary storage size is determined, the conversion logic is based on converting each ASCII digit to binary, high-to-low, and adding its value to a previous sum multiplied by 10. The flowchart in Figure E-10 describes the conversion logic.

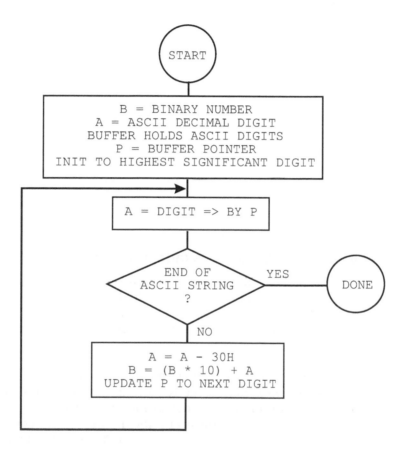

Figure E-10 Decimal String to Unsigned Binary.

The logic in the flowchart of Figure E-10 assumes that there is some way of detecting the end of the string of ASCII digits. This could be a terminator character embedded in the string or a counter for the number of digits. Here again we use a buffer pointer that is initialized to the least-significant digit in the ASCII string. The ASCII digit is converted to binary by subtracting 30H and then added to the previous sum, multiplied by 10. For example, assume the ASCII string of the decimal digits 564.

```
STRING = '564'

FIRST ITERATION:
STEP 1: B = 0
```

```
           P => HIGH DIGIT IN STRING '564'
STEP 2:  A = '5'
STEP 3:  END OF STRING?
         NO
STEP 4:  A = 4 ('5' - 30H = 5)
         B = (0 * 10) + A = 5
         P TO NEXT LOWER DIGIT
SECOND ITERATION:
STEP 2:  A = '6'
STEP 3:  END OF STRING?
         NO
STEP 4:  A = 6 ('6' - 30H = 6)
         B = (5 * 10) + A = 56
         P TO NEXT LOWER DIGIT
THIRD ITERATION:
STEP 2:  A = '4'
STEP 3:  END OF STRING?
         NO
STEP 4:  A = 4 ('4' - 30H = 4)
         B = (56 * 10) + A = 564
         P TO NEXT LOWER DIGIT
FOURTH ITERATION:
STEP 2:  A = ??
STEP 3:  END OF STRING?
         YES
RESULT:  B = 564
```

E.6.4 Unsigned Binary to ASCII Hexadecimal Digits

Converting a binary number to a string of ASCII hex digits is quite similar to converting from binary to an ASCII decimal string, as described in Section E.6.2. Here again, the digit space to allocate for the ASCII string depends on the size of the binary operand. An 8-bit binary is represented in two ASCII hex digits, a 16-bit binary in four ASCII hex digits, and so on.

The process of converting binary to ASCII hexadecimal consists of dividing the binary by 16 to obtain each hex digit. If the remaining hexadecimal digit is in the range 0 to 9 we add 30H to turn it into the corresponding ASCII digit. If it is in the range A to F, then we must add 40H to convert to ASCII. The process continues until the original dividend is reduced to zero, as shown in the flowchart of Figure E-11.

E.6.5 Signed Numerical Conversions

Conversion routines that use signed operands are usually a variation of the unsigned ones described in previous sections. Although logic can be developed that directly encodes to and from two's complement format, the more convenient approach is to determine the sign of the operand and then use unsigned conversion for the digit values. For example, to convert a signed binary in two's complement form into a string of ASCII decimal digits, the logic first determines if the binary operand is negative or positive. If a positive number, then the unsigned conversion routine can be used directly. If the binary operand is a negative number, the – sign is placed in the storage buffer. Then the two's complement binary is converted to an unsigned number so that the ASCII digits can be obtained with the conversion routine described in Section E.6.2.

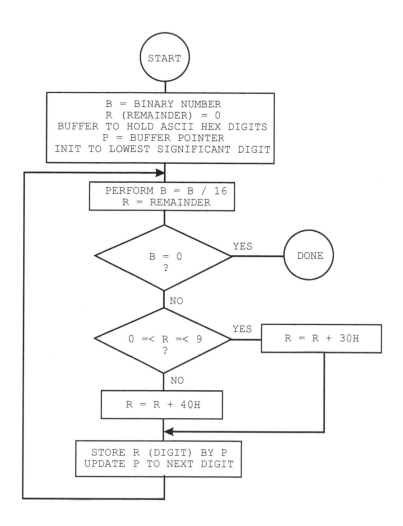

Figure E-11 Unsigned Binary to ASCII Hexadecimal String.

Appendix F

Mid-Range Instruction Set

This appendix describes the instructions in the PIC mid-range family. Not all instructions are implemented in all devices but all of them work in the specific PICs discussed in the text, that is, the 16F84A and the 16F877.

Table F.1

Mid-Range PIC Instruction Set

MNEMONIC	OPERAND	DESCRIPTION	CYCLES	BITS AFFECTED
		BYTE-ORIENTED OPERATIONS:		
ADDWF	f,d	Add w and f	1	C,DC,Z
ANDWF	f,d	AND w with f	1	Z
CLRF	f	Clear f	1	Z
CLRW	-	Clear w	1	Z
COMF	f,d	Complement f	1	Z
DECF	f,d	Decrement f	1	Z
DECFSZ	f,d	Decrement, skip if 0	1(2)	-
INCF	f,d	Increment f	1	Z
INCFSZ	f,d	Increment, skip if 0	1(2)	-
IORWF	f,d	Inclusive OR w and f	1	Z
MOVF	f,d	Move f	1	Z
MOVWF	f	Move w to f	1	-
NOP	-	No operation	1	-
RLF	f,d	Rotate left through carry	1	C
RRF	f,d	Rotate right through carry	1	C
SUBWF	f,d	Subtract w from f	1	C,DC,Z
SWAPF	f,d	Swap nibbles in f	1	-
XORWF				
		BIT-ORIENTED OPERATIONS		
BCF	f,b	Bit clear in f	1	-
BSF	f,b	Bit set in f	1	-
BTFSC	f,b	Bit test, skip if clear	1	-
BTFSS	f,b	Bit test, skip if set	1	-
		LITERAL AND CONTROL OPERATIONS		
ADDLW	k	Add literal and w	1	C,DC,Z

(continues)

Table F.1

Mid-Range PIC Instruction Set (continued)

MNEMONIC	OPERAND	DESCRIPTION	CYCLES	BITS AFFECTED
		LITERAL AND CONTROL OPERATIONS		
ANDLW	k	AND literal and w	1	Z
CALL	k	Call procedure	2	-
CLRWDT	-	Clear Watchdog timer	1	TO,PD
GOTO	k	Go to address	2	-
IORLW	k	Inclusive OR literal with w	1	Z
MOVLW	k	Move literal to w	1	-
RETFIE	-	Return from interrupt	2	-
RETLWk	-	Return literal in w	2	-
RETURN	-	Return from procedure	2	-
SLEEP	-	Go into SLEEP mode	1	TO,PD
SUBLW	k	Subtract literal and w	1	C,DC,Z
XORLW	k	Exclusive OR literal with w	1	Z

Legend:
   f = file register
   d = destination: 0 = w register
                       1 = file register

   b = bit position
   k = 8-bit constant

Table F.2

Conventions used in Instruction Descriptions

FIELD	DESCRIPTION
f	Register file address (0x00 to 0x7F)
w	Working register (accumulator)
b	Bit address within an 8-bit file register (0 to 7)
k	Literal field, constant data or label (may be either an 8-bit or an 11-bit value)
x	Don't care (0 or 1)

d	Destination select;
	    d = 0: store result in w,
	    d = 1: store result in file register f.

dest	Destination either the w register or the specified register file location
label	Label name
TOS	Top of Stack
PC	Program Counter
PCLATH	Program Counter High Latch
GIE	Global Interrupt Enable bit
WDT	Watchdog Timer
!TO	Time-out bit
!PD	Power-down bit
[ ]	Optional element
[XXX]	Contents of memory location pointed at by XXX register
( )	Contents
->	Assigned to
< >	Register bit field
italics	User defined term

ADDLW   Add Literal and w

Syntax: [label] ADDLW   k
Operands: k in range 0 to 255
Operation: (w) + k -> w
Status Affected: C, DC, Z
Description: The contents of the w register are added to
                     the eight bit literal k and the result is
                     placed in the w register
Words: 1
Cycles: 1

Example1:

```
ADDLW   0x15
Before Instruction:        w = 0x10
After Instruction:         w = 0x25
```

Example 2:

```
ADDLW   var1
Before Instruction:        w = 0x10
var1 is data memory variable
var1 = 0x37
After Instruction:         w = 0x47
```

ADDWF   Add w and f

Syntax:	[label] ADDWF f,d
Operands:	f in range 0 to 127
	d = 0 / 1
Operation:	(w) + (f) -> destination
Status Affected:	C, DC, Z
Description:	Add the contents of the w register with register f. If d is 0 the result is stored in the w register. If d is 1 the result is stored back in register f.
Words:	1
Cycles:	1

Example 1:

```
ADDWF   FSR,0
Before Instruction:
        w = 0x17
        FSR = 0xc2
After Instruction:
        W = 0xd9
        FSR = 0xc2
```

Example 2:

```
ADDWF   INDF, 1
Before Instruction:
        w = 0x17
        FSR = 0xC2
        Contents of Address (FSR) = 0x20
After Instruction:
        W = 0x17
        FSR = 0xC2
        Contents of Address (FSR) = 0x37
```

BCF   Bit Clear f

Syntax: [label] BCF f,b
Operands: f in range 0 to 127
                     b in range 0 to 7
Operation: 0 ->f<b>
Status Affected: None
Description: Bit 'b' in register f is cleared.
Words: 1
Cycles: 1

Example 1:

```
BCF     reg1,7
Before Instruction: reg1 = 0xc7 (1100 0111)
After Instruction: reg1 = 0x47 (0100 0111)
```

Example 2:

```
BCF     INDF,3
Before Instruction:   w = 0x17
                      FSR = 0xc2
                      [FSR]= 0x2f
After Instruction:

                      w = 0x17
                      FSR = 0xc2
                      [FSR] = 0x27
```

BSF

Bit Set f

Syntax:	[label] BSF f,b
Operands:	f in range 0 to 127
	b in range 0 to 7
Operation:	1-> f<b>
Status Affected:	None
Description:	Bit 'b' in register f is set.
Words:	1
Cycles:	1
Example 1:	

```
BSF       reg1,6
Before Instruction:        reg1 = 0011 1010
After Instruction:         reg1 = 0111 1010
```

Example 2:

```
BSF       INDF,3
Before Instruction:     w = 0x17
                        FSR = 0xc2
                        [FSR] = 0x20

After Instruction:

                        w = 0x17
                        FSR = 0xc2
                        [FSR] = 0x28
```

BTFSC   Bit Test f, Skip if Clear

Syntax: [label] BTFSC   f,b
Operands: f in range 0 to 127
                     b in range 0 to 7
Operation: skip next instruction if (f<b>) = 0
Status Affected: None
Description: If bit b in register f is 0 then the next
                     instruction is skipped. If bit b is 0 then the
                     next instruction (fetched during the current
                     instruction execution) is discarded, and a NOP
                     is executed instead, making this a 2-cycle
                     instruction.
Words: 1
Example :

```
repeat:
        btfsc   reg1,4
        goto    repeat
Case 1: Before Instruction
                PC = $
                reg1 = xxx0 xxxx
        After Instruction
                because reg1<4>= 0,
                PC = $ + 2 (goto skipped)
Case 2: Before Instruction
                PC = $
                reg1= xxx1 xxxx
        After Instruction
                because FLAG<4>=1,
                PC = $ + 1 (goto executed)
```

BTFSS

Bit Test f, Skip if Set

Syntax:	[label] BTFSC f,b
Operands:	f in range 0 to 127
	b in range 0 to 7
Operation:	skip next instruction if (f<b>) = 1
Status Affected:	None
Description:	If bit b in register f is 1 then the next instruction is skipped. If bit b is 0 then the next instruction (fetched during the current instruction execution) is discarded, and a NOP is executed instead, making this a 2-cycle instruction.
Words:	1
Cycles:	1(2)

```
repeat:
        btfss   reg1,4
        goto    repeat
Case 1: Before Instruction
                PC = $
                Reg1 = xxx1 xxxx
        After Instruction
                because Reg1<4>= 1,
                PC = $ + 2 (goto skiped)
Case 2: Before Instruction
                PC = $
                Reg1 = xxx0 xxxx
        After Instruction
                because Reg1<4>=0,
                PC = $ + 1 (goto executed)
```

CALL   Call Sub-Routine

Syntax:	[label] CALL k
Operands:	k in range 0 to 2047
Operation:	(PC) + -> TOS,
	k-> PC<10:0>,
	(PCLATH<4:3>)-> PC<12:11>
Status Affected:	None
Description:	Call Subroutine. First, the 13-bit return address (PC+1) is pushed onto the stack. The eleven-bit immediate address is loaded into PC bits <10:0>. The upper bits of the PC are loaded from PCLATH<4:3>. CALL is a 2-cycle instruction.
Words:	1
Cycles:	2
Example 1:	

```
Here:
        call    There
Before Instruction:
        PC = AddressHere
After Instruction:
        TOS = Address Here + 1
        PC = Address There
```

CLRF   Clear f

Syntax: [label] CLRF   f
Operands: f in range 0 to 127
Operation: 00h ->f
                        1-> Z
Status Affected: Z
Description: The contents of register f are cleared and
                        the Z bit is set.
Words: 1
Cycles: 1
Example 1:

```
         clrf    reg1
         Before Instruction:     reg1 = 0x5a
         After Instruction:      reg1 = 0x00
                                 Z = 1
```

Example 2:

```
         Clrf    INDF
         Before Instruction:     FSR = 0xc2
                                 [FSR]= 0xAA
         After Instruction:      FSR = 0xc2
                                 [FSR] = 0x00
                                 Z = 1
```

CLRW   Clear w

Syntax: [label] CLRW
Operands: None
Operation: 00h -> w
                     1-> Z
Status Affected: Z
Description: w register is cleared. Zero bit (Z) is set.
Words: 1
Cycles: 1
Example 1:

```
            CLRW
            Before Instruction:      w = 0x5A
            After Instruction:       w = 0x00
                                     Z = 1
```

CLRWDT   Clear Watchdog Timer

Syntax:	[label] CLRWDT
Operands:	None
Operation:	00h -> WDT
	0 -> WDT prescaler count,
	1 -> TO
	1 -> PD
Status Affected:	TO, PD
Description:	CLRWDT instruction clears the Watchdog Timer. It also clears the prescaler count of the WDT. Status bits TO and PD are set. The instruction does not change the assignment of the WDT prescaler.
Words:	1
Cycles:	1
Example 1:	

```
CLRWDT
Before Instruction:        WDT counter= x
                           WDT prescaler = 1:128
After Instruction:         WDT counter=0x00
                           WDT prescaler count=0
                           TO = 1
                           PD = 1
                           WDT prescaler = 1:128
```

COMF   Complement f

Syntax:	[label] COMF f,d
Operands:	f in range 0 to 127
	d is 0 or 1
Operation:	(f) -> destination
Status Affected:	Z
Description:	The contents of register f are 1's complemented. If d is 0 the result is stored in w. If d is 1 the result is stored back in register f.
Words:	1
Cycles:	1

Example 1:

```
comf    reg1,0
Before Instruction:         reg1 = 0x13
After Instruction:          reg1 = 0x13
                            w = 0xEC
```

Example 2:

```
comf    INDF,1
Before Instruction:         FSR = 0xc2
                            [FSR]= 0xAA
After Instruction:          FSR = 0xc2
                            [FSR] = 0x55
```

Example 3:

```
comf    reg1,1
Before Instruction:         reg1= 0xff
After Instruction:          reg1 = 0x00
```

DECF   Decrement f

Syntax: [label] DECF f,d
Operands: f in range 0 to 127
                     d is either 0 or 1
Operation: (f) - 1 -> destination
Status Affected: Z
Description: Decrement register f. If d is 0 the result is
                     stored in the w register. If d is 1 the
                     result is stored back in register f.
Words: 1
Cycles: 1
Example 1:

```
decf    count,1
Before Instruction:          count = 0x01
                             Z = 0
After Instruction:           count = 0x00
                             Z = 1
```

Example 2:

```
decf    INDF,1
Before Instruction:          FSR = 0xc2
                             [FSR] = 0x01
                             Z = 0
After Instruction:           FSR = 0xc2
                             [FSR] = 0x00
                             Z = 1
```

Example 3:

```
decf    count,0
Before Instruction:          count = 0x10
                             w = x
                             Z = 0
After Instruction:           count = 0x10
                             w = 0x0f
```

DECFSZ   Decrement f, Skip if 0

Syntax:	[label] DECFSZ f,d
Operands:	f in the range 0 to 127
	d is either 0 or 1
Operation:	(f) - 1 -> destination; skip if result = 0
Status Affected:	None
Description:	The contents of register f are decremented. If d is 0 the result is placed in the w register. If d is 1 the result is placed back in register f.
	If the result is 0, then the next instruction (fetched during the current instruction execution) is discarded and a NOP is executed instead, making this a 2-cycle instruction.
Words:	1
Cycles:	1(2)

Example

```
here:
        decfsz  count,1
        goto    here
Case 1:
Before Instruction:      PC = $
                         count = 0x01
After Instruction:       count = 0x00
                         PC = $ + 2 (goto skipped)
Case 2:
Before Instruction:      PC = $
                         count = 0x04
After Instruction:       count = 0x03
                         PC = $ + 1 (goto executed)
```

GOTO Unconditional Branch

Syntax:	[label] GOTO k
Operands:	0 £ k £ 2047
Operation:	k -> PC<10:0>
	PCLATH<4:3> ->PC<12:11>
Status Affected:	None
Description:	GOTO is an unconditional branch. The eleven-bit immediate value is loaded into PC bits <10:0>. The upper bits of PC are loaded from PCLATH<4:3>.
	GOTO is a 2-cycle instruction.
Words:	1
Cycles:	2
Example	

```
goto    There
After Instruction:      PC = address of There
```

INCF   Increment f

Syntax:	[label] INCF f,d
Operands:	f in the range 0 to 127
	d is either 0 or 1
Operation:	(f) + 1 -> destination
Status Affected:	Z
Description:	The contents of register f are incremented. If d is 0 the result is placed in the w register. If d is 1 the result is placed back in register f.
Words:	1
Cycles:	1
Example 1:	

```
              incf    count,1
              Before Instruction:        count = 0xff
                                         Z = 0
              After Instruction:         count = 0x00
                                         Z = 1
```

Example 2:

```
              incf    INDF,1
              Before Instruction:        FSR = 0xC2
                                         [FSR] = 0xff
                                         Z = 0
              After Instruction:         FSR = 0xc2
                                         [FSR] = 0x00
                                         Z = 1
```

Example 3:

```
              incf    count,0
              Before Instruction:        count = 0x10
                                         w = x
                                         Z = 0
              After Instruction:         count = 0x10
                                         w = 0x11
                                         Z = 0
```

INCFSZ   Increment f, Skip if 0

Syntax:	[label] INCFSZ f,d
Operands:	f in the range 0 to 127
	d is either 0 or 1
Operation:	(f) + 1 -> destination, skip if result = 0
Status Affected:	None
Description:	The contents of register f are incremented. If d is 0 the result is placed in the w register. If d is 1 the result is placed back in register f. If the result is 0, then the next instruction (fetched during the current instruction execution) is discarded and a NOP is executed instead, making this a 2-cycle instruction.
Words:	1
Cycles:	1(2)
Example	

```
Here:
        incfsz  count,1
        goto    Here
Case 1:
    Before Instruction:     PC = $
                            count = 0x10
    After Instruction:      count = 0x11
                            PC = $ + 1 (goto executed)
Case 2:
    Before Instruction:     PC = $
                            count = 0x00
    After Instruction:      count = 0x01
                            PC = $ + 2 (goto skipped)
```

IORLW   Inclusive OR Literal with w

Syntax: [label] IORLW k
Operands: k is in range 0 to 255
Operation: (w).OR. k -> w
Status Affected: Z
Description: The contents of the w register is ORed with
                        the eight- bit literal k. The result is placed in
                        the w register.
Words: 1
Cycles: 1

Example 1:

```
iorlw   0x35
Before Instruction:          w = 0x9a
After Instruction:           w = 0xbfF
                             Z = 0
```

Example 2:

```
iorlw   myreg
Before Instruction:          w = 0x9a
Myreg is a variable representing a location
in PIC RAM.                  [Myreg] = 0x37
After Instruction:           w = 0x9F
                             Z = 0
```

Example 3:

```
iorlw   0x00
Before Instruction:          w = 0x00
After Instruction:           w = 0x00
```

IORWF   Inclusive OR w with f

Syntax: [label] IORWF f,d
Operands: f is in the range 0 to 127
                     d is either 0 or 1
Operation: (W).OR. (f) -> destination
Status Affected: Z
Description: Inclusive OR the w register with register f. If d
                     is 0 the result is placed in the w register. If d is
                     1 the result is placed back in register f.
Words: 1
Cycles: 1

Example 1:

```
          iorwf   result,0
          Before Instruction:        result = 0x13
                                      w = 0x91
          After Instruction:         result = 0x13
                                      w = 0x93
                                      z = 0
```

Example 2:

```
          iorwf   INDF,1
          Before Instruction:        w = 0x17
                                      FSR = 0xc2
                                      [FSR] = 0x30
          After Instruction:         w = 0x17
                                      FSR = 0xc2
                                      [FSR] = 0x37
                                      z = 0
```

Example 3:

```
          iorwf   result,1
   Case 1: Before Instruction:        result = 0x13
                                      w = 0x91
           After Instruction:         result = 0x93
                                      w = 0x91
                                      z = 0
   Case 2: Before Instruction:        result = 0x00
                                      w = 0x00
           After Instruction:         result = 0x00
                                      w = 0x00
                                      z = 1
```

MOVLW   Move Literal to w

Syntax: [label] MOVLW k
Operands: k in range 0 to 255
Operation: k- > w
Status Affected: None
Description: The eight bit literal k is loaded into w register.
                   The don't cares will assemble as 0s.
Words: 1
Cycles: 1

Example 1:

```
movlw   0x5a
After Instruction:        w = 0x5A
```
Example 2:

```
movlw   myreg
Before Instruction:       w = 0x10
                          [myreg] = 0x37
After Instruction:        w = 0x37
```

MOVF   Move f

Syntax:	[label] MOVF f,d
Operands:	f is in the range 0 to 127
	d is either 0 or 1
Operation:	(f) -> destination
Status Affected:	Z
Description:	The contents of register f is moved to a destination dependent upon the status of d. If d = 0, destination is w register. If d = 1, the destination is file register f itself. d = 1 is useful to test a file register because status flag Z is affected.
Words:	1
Cycles:	1

Example 1:

```
        movf    FSR,0
        Before Instruction:        w = 0x00
                                   FSR = 0xc2
        After Instruction:         w = 0xc2
                                   Z = 0
```

Example 2:

```
        movf    INDF,0
        Before Instruction:        w = 0x17
                                   FSR = 0xc2
                                   [FSR] = 0x00
        After Instruction:         w = 0x17
                                   FSR = 0xc2
                                   [FSR] = 0x00
                                   Z = 1
```

Example 3:

```
        movf    FSR,1
Case 1: Before Instruction:        FSR = 0x43
        After Instruction:         FSR = 0x43
                                   Z = 0
Case 2: Before Instruction:        FSR = 0x00
        After Instruction:         FSR = 0x00
                                   Z = 1
```

MOVWF   Move w to f

Syntax: [label] MOVWF f
Operands: f in range 0 to 127
Operation: (w) -> f
Status Affected: None
Description: Move data from w register to register f.
Words: 1
Cycles: 1

Example 1:

```
movwf   OPTION_REG
Before Instruction:          OPTION_REG = 0xff
                             w = 0x4f
After Instruction:           OPTION_REG = 0x4f
                             w = 0x4f
```

Example 2:

```
movwf   INDF
Before Instruction:          w = 0x17
                             FSR = 0xC2
                             [FSR] = 0x00
After Instruction:           w = 0x17
                             FSR = 0xC2
                             [FSR] = 0x17
```

NOP   No Operation

Syntax:	[label] NOP
Operands:	None
Operation:	No operation
Status Affected:	None
Description:	No operation
Words:	1
Cycles:	1
Example	

```
nop
Before Instruction:     PC = $
After Instruction:      PC = $ + 1
```

OPTION   Load Option Register

Syntax:	[label] OPTION
Operands:	None
Operation:	(w) -> OPTION_REG
Status Affected:	None
Description:	The contents of the w register are loaded in the OPTION_REG register. This instruction is supported for code compatibility with PIC16C5X products. Because OPTION_REG is a readable/writable register, code can directly address it without using this instruction.

Words: 1
Cycles: 1
Example:

```
movlw   b'01011100'
option
```

RETFIE   Return from Interrupt

Syntax:	[label] RETFIE
Operands:	None
Operation:	TOS -> PC,
	1 -> GIE
Status Affected:	None
Description:	Return from Interrupt. The 13-bit address at the Top of Stack (TOS) is loaded in the PC. The Global Interrupt Enable bit, GIE (INTCON<7>), is automatically set, enabling Interrupts. This is a 2-cycle instruction.
Words:	1
Cycles:	2
Example:	

```
retfie
After Instruction:        PC = TOS
                          GIE = 1
```

RETLW   Return with Literal in w

Syntax: [label] RETLW k
Operands: k in the range 0 to 255
Operation: k -> w;
                     TOS -> PC
Status Affected: None
Description: The w register is loaded with the eight bit literal
                     k. The program counter is loaded from the
                     13-bit address at the Top of Stack (the return
                     address). This is a 2-cycle instruction.
Words: 1
Cycles: 2
Example:

```
        movlw   2         ; Load w with desired
                          ; Table offset
        call    table     ; When call returns w
                          ; contains value stored
                          ; in table
Table:
        addwf   pc        ; w = offset
        retlw   .22       ; First table entry
        retlw   .23       ; Second table entry
        retlw   .24
          .
          .
          .
        retlw   .29       ; Last table entry
        Before Instruction:      w = 0x02
        After Instruction:       w = .24
```

RETURN   Return from Sub-Routine

Syntax:	[label] RETURN
Operands:	None
Operation:	TOS -> PC
Status Affected:	None
Description:	Return from sub-routine. The stack is POPed and the Top of Stack (TOS) is loaded into the program counter. This is a 2-cycle instruction.
Words:	1
Cycles:	2
Example:	

```
return
After Instruction:      PC = TOS
```

RLF   Rotate Left f through Carry

Syntax: [label] RLF f,d
Operands: f in the range 0 to 127
                     d is either 0 or 1
Operation: See description below
Status Affected: C
Description: The contents of register f are rotated one bit to
                     the left through the Carry Flag. If d is 0 the
                     result is placed in the w register. If d is 1 the
                     result is stored back in register f.
Words: 1
Cycles: 1

Example 1:

```
            rlf     reg1,0
            Before Instruction:        reg1 = 1110 0110
                                       C = 0
            After Instruction:         reg1 = 1110 0110
                                       w =1100 1100
                                       C =1
```

Example 2:

```
            rlf     INDF,1
    Case 1: Before Instruction:        w = xxxx xxxx
                                       FSR = 0xc2
                                       [FSR] = 0011 1010
                                       C = 1
            After Instruction:         w = 0x17
                                       FSR = 0xc2
                                       [FSR] = 0111 0101
                                       C = 0
    Case 2: Before Instruction:        w = xxxx xxxx
                                       FSR = 0xC2
                                       [FSR] = 1011 1001
                                       C = 0
            After Instruction:         w = 0x17
                                       FSR = 0xC2
                                       [FSR] = 0111 0010
                                       C = 1
```

RRF   Rotate Right f through Carry

Syntax: [label] RRF f,d
Operands: f in the range 0 to 127
                     d is either 0 or 1
Operation: See description below
Status Affected: C
Description: The contents of register f are rotated one bit to
                     the right through the Carry Flag. If d is 0 the
                     result is placed in the w register. If d is 1 the
                     result is placed back in register f.
Words: 1
Cycles: 1

Example 1:

```
          rrf     reg1,0
          Before Instruction:        reg1= 1110 0110
                                      w = xxxx xxxx
                                      C = 0
          After Instruction:         reg1= 1110 0110
                                      w = 0111 0011
                                      C = 0
```

Example 2:

```
          rrf     INDF,1
  Case 1: Before Instruction:        w = xxxx xxxx
                                      FSR = 0xc2
                                      [FSR] = 0011 1010
                                      C = 1
          After Instruction:         w = 0x17
                                      FSR = 0xC2
                                      [FSR] = 1001 1101
                                      C = 0
  Case 2: Before Instruction:        w = xxxx xxxx
                                      FSR = 0xC2
                                      [FSR] = 0011 1001
                                      C = 0
          After Instruction:         w = 0x17
                                      FSR = 0xc2
                                      [FSR] = 0001 1100
                                      C = 1
```

SLEEP

Syntax:	[label] SLEEP
Operands:	None
Operation:	00h -> WDT,
	0 -> WDT prescaler count,
	1 -> TO,
	0 -> PD
Status Affected:	TO, PD
Description:	The power-down status bit PD is cleared. Time-out status bit TO is set. Watchdog Timer and its prescaler count are cleared. The processor is put into SLEEP mode with the oscillator stopped. The SLEEP instruction does not affect the assignment of the WDT prescaler.
Words:	1
Cycles:	1
Example:	

```
SLEEP
```

SUBLW   Subtract w from Literal

Syntax: [label] SUBLW k
Operands: k in range 0 to 255
Operation: k - (W) -> W
Status Affected: C, DC, Z
Description: The w register is subtracted (2's complement
                     method) from the eight bit literal k. The result
                     is placed in the w register.
Words: 1
Cycles: 1

Example 1:

```
            sublw   0x02
    Case 1: Before Instruction:        w = 0x01
                                       C = x
                                       Z = x
            After Instruction:         w = 0x01
                                       C = 1 if result +
                                       Z = 0
    Case 2: Before Instruction:        w = 0x02
                                       C = x
                                       Z = x
            After Instruction:         w = 0x00
                                       C = 1 ; result = 0
                                       Z = 1
    Case 3: Before Instruction:        w = 0x03
                                       C = x
                                       Z = x
            After Instruction:         w = 0xff
                                       C = 0 ; result -
                                       Z = 0
```

Example 2:

```
            sublw   myreg
            Before Instruction:        w = 0x10
                                       [myreg] = 0x37
            After Instruction          w = 0x27
                                       C = 1 ; result +
```

SUBWF   Subtract w from f

Syntax: [label] SUBWF   f,d
Operands: f in the range 0 to 127
                      d is either 0 or 1
Operation: (f) - (W) -> destination
Status Affected: C, DC, Z
Description: Subtract (2's complement method) w register
                      from register f. If d is 0 the result is stored in
                      the w register. If d is 1 the result is stored
                      back in register f.
Words: 1
Cycles: 1

Example :

```
              subwf    reg1,1
     Case 1:  Before Instruction:       reg1 = 3
                                        w = 2
                                        C = x
                                        Z = x
              After Instruction:        reg1 = 1
                                        w = 2
                                        C = 1 ; result +
                                        Z = 0
     Case 2:  Before Instruction:       reg1 = 2
                                        w = 2
                                        C = x
                                        Z = x
              After Instruction:        reg1 = 0
                                        w = 2
                                        C = 1 ; result = 0
                                        Z = 1
     Case 3:  Before Instruction:       reg1 = 1
                                        w = 2
                                        C = x
                                        Z = x
              After Instruction:        reg1 = 0xff
                                        w = 2
                                        C = 0 ; result is -
                                        Z = 0
```

SWAPF   Swap Nibbles in f

Syntax:	[label] SWAPF f,d
Operands:	f in the range 0 to 127
	d is either 0 or 1
Operation:	(f<3:0>) -> destination<7:4>,
	(f<7:4>) -> destination<3:0>
Status Affected:	None
Description:	The upper and lower nibbles of register f are exchanged. If d is 0 the result is placed in w register. If d is 1 the result is placed in register f'.
Words:	1
Cycles:	1

Example 1:

```
swapf   reg,0
Before Instruction:        reg1 = 0xa5
After Instruction:         reg1 = 0xa5
                           W = 0x5a
```

Example 2:

```
Swapf   INDF,1
Before Instruction:        w = 0x17
                           FSR = 0xc2
                           [FSR] = 0x20
After Instruction:         w = 0x17
                           FSR = 0xC2
                           [FSR] = 0x02
```

Example 3:

```
swapf   reg,1
Before Instruction:        reg1 = 0xa5
After Instruction:         reg1 = 0x5a
```

TRIS   Load TRIS Register

Syntax:	[label] TRIS f
Operands:	f in the range 5 to 7
Operation:	(w) -> TRIS register f;
Status Affected:	None
Description:	The instruction is supported for code compatibility with the PIC16C5X products. Because TRIS registers are readable and writable, code can address these registers directly.
Words:	1
Cycles:	1

Example:

```
movlw   B'00000000'
tris    PORTB
```

XORLW   Exclusive OR Literal with w

Syntax:	[label] XORLW k
Operands:	k in the range 0 to 255
Operation:	(w).XOR. k -> W
Status Affected:	Z
Description:	The contents of the w register are XORed with the eight bit literal k. The result is placed in the w register.
Words:	1
Cycles:	1

Example 1:

```
xorlw   b'10101111'
Before Instruction:        w = 1011 0101
After Instruction:         w = 0001 1010
                           Z = 0
```

Example 2:

```
xorlw   myreg
Before Instruction:        w = 0xaf
                           [Myreg] = 0x37
After Instruction:         w = 0x18
                           Z = 0
```

XORWF   Exclusive OR w with f

Syntax: [label] XORWF f,d
Operands: f in range 0 to 127
                    d is either 0 or 1
Operation: (W).XOR. (f) -> destination
Status Affected: Z
Description: Exclusive OR the contents of the w register
                    with register f. If d is 0 the result is stored in
                    the w register. If d is 1 the result is stored back
                    in register f.
Words: 1
Cycles: 1

Example 1:

```
xorwf   reg,1
Before Instruction:        w = 1011 0101
                           reg = 1010 1111
After Instruction:         reg = 0001 1010
                           w = 1011 0101
```

Example 2:

```
xorwf   reg,0
Before Instruction         w =  1011 0101
                           reg = 1010 1111
After Instruction:         reg = 1010 1111
                           w = 0001 1010
```

Example 3:

```
xorwf   INDF,1
Before Instruction:        w = 1011 0101
                           FSR = 0xc2
                           [FSR] = 1010 1111
After Instruction:         w = 1011 0101
                           FSR = 0xc2
                           [FSR] = 0001 1010
```

Appendix G

Printed Circuit Boards

G.1 Introduction

For students in courses in electrical, electronics, and computer engineering (or for anyone interested in these fields), hardware boards and programmers are a valuable learning tool. Many such boards are offered online at different prices and levels of complexity. But making your own demo boards is even more valuable than purchasing one off the shelf. In this appendix we provide information and resources to facilitate making hardware versions of three simple demo boards mentioned in the text.

The reader should note that these hardware boards are neither cheap nor easy to make. One of the problems is that companies that produce printed circuit boards often require orders for more than one board. For example, the company that we have often used in making printed circuit boards (ExpressPCB) provides a ProtoPro service of 4 boards per order. At the time of this writing the cost of a minimal order for each of the book's demo boards was approximately \$180.00. Which results in a cost of approximately \$45.00 per board. Also in this appendix we provide information on producing home-made boards but this option has its own difficulties and drawbacks.

G.2 Printed Circuit Boards (PCBs)

PCBs are a nonconductive substrates (usually of fiberglass or resin) on which conductive traces and pathways are etched on one or more copper layers that are laminated onto board body. The result is an inexpensive and reliable media to which the electronic components can be attached, usually by welding.

Printed circuit boards are often created using a computer program that facilitates drawing the traces and placing the components onto one or more layers. Figure G-1 is a screen snapshot of the ExpressPCB program (furnished free on-line by ExpressPCB).

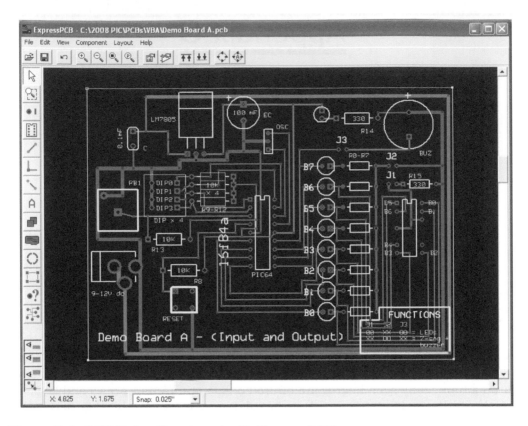

Figure G-1 PCB Being Developed with ExpressPCB.

A PCB editor program (such as ExpressPCB) generates a file from which the printed circuit board can be fabricated. Often the programs are proprietary of the firm that manufactures the boards. In the case of ExpressPCB, the file with the extension .pcb can be submitted online and the manufactured boards are returned to you in a few days.

In this book's software package we include three files in .pcb format for the hardware version of each of the boards used in the text. The location and filenames are:

- PCBs/VBA/Demo Board A.pcb

- PCBs/VBB/Demo Board B.pcb

- PCBs/VBI/Demo Board I.pcb

These files can be used in having printed circuit boards made by ExpressPCB. In order to find out how to go about it you may log on to the company's web site at www.expresspcb.com.

G.3 Parts Lists

Once the printed circuits boards have been made, they must be populated with the electronic components. In order to facilitate this step, we are including a parts list for each of the boards in this book's software package. The location and filenames are as follows:

- PCBs/VBA/parts demo a
- PCBs/VBB/parts demo b
- PCBs/VBI/parts demo c

The files are in .txt format and can be loaded and viewed in any word processing program or in Microsoft notepad.

In addition to the parts listed, the boards require a 9 to 12 volts AC/DC wall or table-top adapter. The power plug on the adapter should be female, 2.5 x 5.5 mm., and center positive polarity. Jameco part number 189238CJ is suitable. Similar ones are available at Radio Shack and from other sources.

G.4 Building Your Own Circuit Boards

Several methods have been developed for making printed circuit boards on a small scale, as would be convenient for the experimenter and prototype developer. If you look through the pages of any electronics supply catalog, you will find kits and components based on different technologies of various levels of complexity. The methods we describe in the following sections are perhaps the simplest because they do not require a photographic process.

The process consists of the following steps:

1. The circuit diagram is drawn on the PC using a general-purpose or specialized drawing program.
2. A printout is made of the circuit drawing on photographic paper.
3. The printout is transferred to a copper-clad circuit board blank by ironing over the paper's backside with a household clothes iron.
4. The resulting board is placed in a ferric chloride etching bath that eats away all the copper, except the circuit image ironed onto the board surface.
5. The board is washed of the etching fluid, cleaned, drilled, and the components soldered to it in the conventional manner.
6. Optionally, another image can be ironed onto the front side of a double-sided board to provide a second conductive layer with or without components.
7. Images containing text, identification, or logos can be ironed onto either side of the board but are usually on the front side.

In this appendix we offer two examples: the first one consists of a simple demo board made on a single-sided board with text on the nonconductive side. The second one is the home-made version of Demo Board A that has two conductive sides and text on the front side.

G.4.1 Tools and Materials

The following tools and materials are required for fabricating the printed circuit board:

- A general-purpose drawing program or a specialized printed circuit drawing application.

- A laser printer such as LaserJet or equivalent or a laser copier. Inkjet printers are not suitable.

- Household clothes iron.

- Scotch-Brite™ (green) abrasive pads.

- Acetone solvent.

- Copper-plated circuit board blank, single- or double-side according to project.

- Ferric chloride etching solution for copper-plated boards.

- Plastic or glass container for holding board during etching.

After the board has been etched, it must be drilled and the components soldered. The following tools and materials are necessary for this phase:

- Small electric drill capable of high revolutions, such as a Dremmel tool.

- Drill bits 0.035" and 0.040".

- Soldering iron or soldering gun suitable for electronic components.

- Light duty rosin-core solder.

G.4.2 Single-Sided Demo Board

We have mentioned that hardware demonstration (or demo) boards are a useful tool in mastering embedded systems. Constructing your own demo boards and circuits is not a difficult task. The components can be placed on a breadboard, wire-wrapped onto a special circuit board. A printed circuit boards can be home-made, or one ordered through the Internet. The book's online resource contains descriptions of commercial PCBs as well as instructions and resources for obtaining commercially made boards for the demo circuits used in this book.

We first illustrate the process with a simple board that contains a 16F84A-PIC, Seven-Segment LED, buzzer, pushbutton switch, and a bank of four toggle switches. Figure G-2 shows the schematics for the board and the power supply.

G.4.3 PCB Images for Demo Board

Commercial PCBs contain circuit etchings on two, four, or more layers. Although multi-layered boards appear complex, most circuit designers will agree that the more layers available the easier it is to place the components without interference. Double-sided boards (two conductive layers) usually have the power and interface connections on one layer and the ground plane on another one. That is the method we followed with the demo boards described in the online documentation and in this appendix. In this case, two circuit board images are required, one for each conductive plane on the board. In addition, most boards contain a text image that includes component placements, company logos, model numbers, and other textual information. In commercially made boards, the text image is silk-screened onto the board.

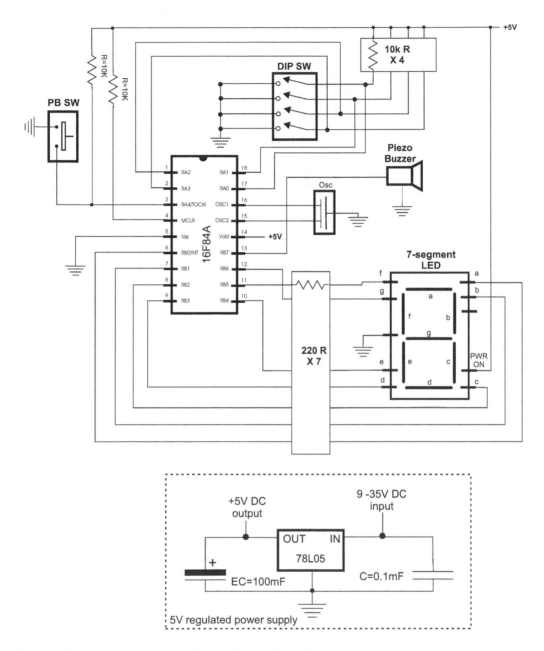

Figure G-2 A Simple PIC 16F87A Demo Board.

The first home-made board example in this appendix uses a single-sided board with an etched image on the conductive side (bronze layer) and a text image on the board's nonconductive side. Both images can be created with a conventional drawing program, such as Corel Draw, Adobe Illustrator, or Windows Paint, or with a specialized application, such as the one furnished by ExpressPCB and discussed earlier in this appendix. Figure G-3 shows the images used for making the PCB for the circuit in Figure G-2.

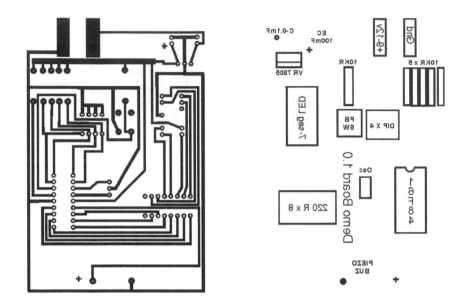

Figure G-3 Bottom-Side and Top-Side Layer Images for a PCB.

In Figure G-3 the image on the left side is etched on the conductive side of the board, that is, the one with the copper layer. This is usually referred to as the bottom layer. In this example the components and the text are placed on the top or nonconductive layer. Notice that the top-side image must be horizontally mirrored before it is transferred to the photographic paper. This is necessary so that the text and graphics coincide with the circuit etchings.

Drawing the Circuit Diagram

Any computer drawing program serves this purpose. We have used CorelDraw™, but there are several specialized PCB drawing programs available on the Internet. Notice in the drawing in Figure G-3 that the circuit locations where the components are to be soldered consist of small circular pads, usually called solder pads. Figure G-4 zooms into a PCB drawing to show the details of the solder pads.

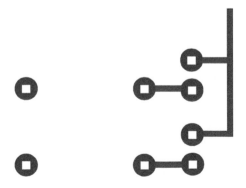

Figure G-4 Detail of Circuit Board Pads.

Quite often it is necessary for a circuit line to cross between two solder pads. In this case the pads can be made smaller or modified so as to allow passing a trace between them. Modified pads are shown in Figure G-5.

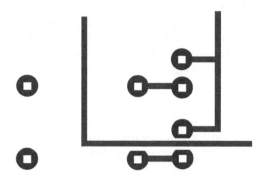

Figure G-5 Modified Circuit Boards Pads.

Printing the PCB Diagram

In the method presently described the circuit diagram is printed using a laser printer. Inkjet toners do not produce an image that resist the action of the etching liquid. Although in our experiments we used LaserJet printers, it is well documented that virtually any laser printer will work. Laser copiers have also been used successfully for creating the PCB circuit image.

However, with this method the width of the traces can become an issue. The traces in the PCB image of Figure G-5 are 2 points, which is 0.027". Traces half that width and less have been used successfully; but as the traces become thinner the entire process turns out more critical. For most simple circuits, 0.020" traces should be a useful limit. Be careful not to touch the glossy side of the photographic paper or the printed image with your fingers.

Selecting the Paper

Practitioners of this method affirm that one of the most critical elements is the paper used in printing the circuit. Pinholes in some papers can degrade the image to the point that the circuit lines (especially if they are very thin) do not etch correctly. Another problem relates to removing the ironed-on paper from the board without damaging the board surface.

Glossy, coated inkjet-printer paper works well (yes, Inkjet paper on a LaserJet printer). Even better results can be obtained with glossy photo paper. We use a common high-gloss photographic paper available from Staples and sold under the name of "picture paper." The 30 sheets, 8-by-10 size, has the Staples number B031420197 1713. The UPC barcode is: 7 18103 02238 5.

Transferring the PCB Image

Transferring the image onto the board blank is done by applying heat from a common clothes iron, set on the hottest setting, onto the paper/board sandwich. In most irons the hottest setting is labeled "linen." After going over the back of the paper several times with the hot iron, the paper becomes fused to the copper side of the blank board. The board/paper sandwich is then allowed to soak in water for about ten minutes, after which the paper can be removed by peeling or light scrubbing with a toothbrush. It has been mentioned that Hewlett-Packard toner cartridges with so-called microfine particles work better than store-brand toner cartridges.

Etching the Board

Once the paper has been removed and the board washed, it is time to prepare the board for etching. The preliminary operations consist of rubbing the copper surface of the board with a Scotchbrite abrasive pad and then scrubbing the surface with a paper towel soaked with acetone solvent.

Once the board is rubbed and clean, you can proceed to etch the circuit. The etching solution contains ferric chloride and is available from Radio Shack as a solution and from Jameco Electronics as a powder to be mixed by the user. PCB ferric chloride etchant should be handled with rubber gloves and rubber apron because it stains the skin and utensils. Also, concentrated acid fumes from ferric chloride solution are toxic and can cause severe burns. These chemicals should be handled according to cautions and warnings posted on the containers.

The ferric chloride solution should be used in a plastic or glass container, never in metal. Faster etching is accomplished if the etching solution is first warmed by placing the bottle in a tub of hot water. Once the board is in the solution, face up, the container is rocked back and forward. It is also possible to aid copper removal by rubbing the surface with a rubber-gloved finger.

Finishing the Board

The etched board should be washed well, first in water and then in lacquer thinner or acetone, either solvent works. It is better to just rub the board surface with a paper towel soaked in the solvent. Keep in mind that most solvents are flammable and explosive, and also toxic.

After the board is clean, the mounting holes can be drilled using the solder pads as a guide. A small electric drill at high revolutions, such as a Dremmel tool, works well for this operation. The standard drill size for the mounting holes is 0.035". A #60 drill (0.040") also works. Online one can purchase a set of special PCB drill bits of various sizes. Once all the holes are drilled, the components can be mounted from the front side and soldered at the pads.

Component-Side Image

The component side (front side) of the PCB can be printed with an image of the elements to be mounted or with logos or other text. A single-sided blank board has no copper coating on the front side, so the image is just ironed on without etching. Probably the best time to print the text image is after the board has been etched and drilled but before mounting the components. Double-sided boards must be etched and washed before the text image is transferred.

G.5 Caveats

Assembling and soldering components onto PCBs require some manual and technical skills, materials, and special equipment. We do not recommend that you undertake this task unless you have previous experience in this area. Neither do we provide information or assistance on how to do it.

The parts listed were available at the mentioned sources and the listed prices at the time of writing. Availability and prices often change, so it may not be the case when you place your order.

The sources recommended for the PVBs and the electronic components are ones that we have used in the past and are popular with experimenters and hobbyists. But there are many other sources that can supply equivalent parts at the same or perhaps better prices. The authors and publishers of this book have no association with or commercial interest in any of the sources mentioned.

Appendix H

Additional Code

In this appendix we list several programs that were developed while writing this book and for some reason were not used in the text. They are provided to the reader as a code grab-bag in the hope that some may find a useful fragment or routine among those listed. Each program contains a description of its purpose and functionality. The code for the supplementary programs is available in this book's online software package.

```
; File: SecondCnt.ASM
; Date: April 29, 2011
; Authors: Canton and Sanchez
;
; Description:
; Using timer0 to delay one second at a signal
; rate of 1,000,000 beats per second
;============================
;          switches
;============================
; Switches used in __config directive:
;    _CP_ON        Code protection ON/OFF
; *  _CP_OFF
; *  _PWRTE_ON     Power-up timer ON/OFF
;    _PWRTE_OFF
;    _WDT_ON       Watchdog timer ON/OFF
; *  _WDT_OFF
;    _LP_OSC       Low power crystal oscillator
; *  _XT_OSC       External parallel crystal oscillator
;    _HS_OSC       High speed crystal resonator (8 to 10 MHz)
;                  Resonator: Murate Erie CSA8.00MG = 8 MHz
;    _RC_OSC       Resistor/capacitor ocillator
;                  (simplest, 20% error)
; |
; |_____ * indicates setup values
```

```
        processor 16f84A
        include   <p16f84A.inc>
        __config  _XT_OSC & _WDT_OFF & _PWRTE_ON & _CP_OFF


;======================================================
;                  PIC register equates
;======================================================
porta   equ                  0x05
portb   equ        0x06
status  equ                  0x03
z                  equ                  0x02
c                  equ                  0x00
tmr0    equ                  0x01
; countL  equ                0x01      ; Alias for tmr0
;
;======================================================
;             variables in PIC RAM
;======================================================
; Local variables
        cblock  0x0d                 ; Start of block
        J                            ; counter J
        K                            ; counter K
; 3-byte auxiliary counter. Low order byte is kept
; in the timer0 register
        countM        ; Medium byte
        countH    ; High byte

        endc
;=========================================================
;                 m a i n    p r o g r a m
;=========================================================
        org     0          ; start at address 0
        goto    main
;
;=============================
;       interrupt handler
;=============================
        org                0x04
;       goto    IntServ
;=============================
;       main program
;=============================
main:
; Clear the Watchdog timer and reset prescaler
        clrf    tmr0
        clrwdt
; Set up the OPTION regiser bitmap
        movlw   b'11011000'
```

```
;     7   6   5   4   3   2   1   0  <=  OPTION bits
;     |   |   |   |   |   |   |__|__|_____  PS2-PS0 (prescaler bits)
;     |   |   |   |   |   |                  Values for Timer0
;     |   |   |   |   |   |                  *000 = 1:2    001 = 1:4
;     |   |   |   |   |   |                   010 = 1:8    011 = 1:16
;     |   |   |   |   |   |                   100 = 1:32   101 = 1:64
;     |   |   |   |   |   |                   110 = 1:128 *111 = 1:256
;     |   |   |   |   |   |_____  PSA (prescaler assign)
;     |   |   |   |   |                    1 = to WDT
;     |   |   |   |   |                    *0 = to Timer0
;     |   |   |   |   |_____  TOSE (Timer0 edge select)
;     |   |   |   |                        0 = increment on low-to-high
;     |   |   |   |                        *1 = increment in high-to-low
;     |   |   |   |_____  TOCS (TMR0 clock source)
;     |   |   |                          *0 = internal clock
;     |   |   |                           1 = RA4/TOCKI bit source
;     |   |   |_____  INTEDG (Edge select)
;     |   |                             *0 = falling edge
;     |   |_____  RBPU (Pullup enable)
;     |                                   0 = enabled
;     |                                   *1 = disabled
          option
; Setup ports
          movlw    0x00                  ; Set port B to output
          tris     portb
          clrf     portb                 ; All port B to 0
; Port A is not used in this program
mloop:
          bsf      portb,0
          call     TM0delay
          bcf      portb,0
          call     TM0delay
          goto     mloop
;*******************************
;   one second delay sub-routine
;          using Timer0
;*******************************
; Routine logic:
; The prescaler is assigned to timer0 and set up so
; that the timer runs at 1:2 rate. This means that
; every time the counter reaches 128 (0x80), a total
; of 256 machine cycles have elapsed. The value 0x80
; is detected by testing bit 7 of the counter
; register. This method gives the routine a total of
; 128 machine cycles before the next counter beat must
; be acknowledged.
TM0delay:
; Timer is designed to count from 0 to 1,000,000
```

```
; 1,000,000 = 0x0f 0x42 0x40
;              ---- ---- ----
;               |    |    |___ (see note)
;               |    |_____ countM
;               |_____ countH
; Note:
;     The initial count of 0x40 (64 decimal) is ensured
; by initializing the tmr0 register to count 32 timer
; beats at the 1:2 prescaler rate. 128 - 32 = 96 = 0x60
; Initialize the counters:
          movlw     0x0f
          movwf     countH
          movlw     0x42
          movwf     countM
          movlw     0x60
          movwf     tmr0
; Routine tests timer overflow by testing bit 7 of
; the tmr0 register.
cycle:
          movlw     3
          subwf     tmr0,w
          btfsc     status,c
          goto      cycle
; Subtract 256 from beat counter by decrementing the
; mid-order byte
          decfsz    countM,f
          goto      cycle              ; Continue if mid-byte not
zero
; At this point the mid-order byte has overflowed.
; High-order byte must be decremented.
          decfsz    countH,f
          goto      cycle
; At this point one second has elapsed
          return
          end
```

```
;  File name: SevenSeg.asm
;  Date: April 9, 2011
;  Authors: Canton and Sanchez
;
;  Reference: SevenSeg Circuit and Board
;
;  Description:
;  Test program for reading four toggle switches and
;  displaying the represented hex number on Seven-Segment
;  LED. Also contains a pushbutton switch to activate a
;  piezo buzzer. Switches are wired active low.
;
;  Switches used in __config directive:
;    _CP_ON          Code protection ON/OFF
;  * _CP_OFF
;  * _PWRTE_ON       Power-up timer ON/OFF
;    _PWRTE_OFF
;    _WDT_ON         Watchdog timer ON/OFF
;  * _WDT_OFF
;    _LP_OSC         Low power crystal oscillator
;  * _XT_OSC         External parallel crystal oscillator
;    _HS_OSC         High speed crystal resonator (8 to 10 MHz)
;                    Resonator: Murate Erie CSA8.00MG = 8 MHz
;    _RC_OSC         Resistor/capacitor oscillator
;                    (simplest, 20% error)
;  |
;  |_____ * indicates setup values
;
;==========================
; setup and configuration
;==========================
        processor 16f84A
        include    <p16f84A.inc>
        __config  _XT_OSC & _WDT_OFF & _PWRTE_ON & _CP_OFF
;
;=====================================================
;                   constant definitions
;              (per circuit wiring diagram)
;=====================================================
#define Pb_sw  4 ; Port A line 4 to pushbutton switch
;
;=====================================================
;                   PIC register equates
;=====================================================
Porta    equ       0x05
Portb    equ       0x06
;==========================
;       local variables
```

```
;==============================
        cblock   0x0c              ; Start of block
        J                          ; counter J
        K                          ; counter K
        endc
;=============================================================
;                              program
;=============================================================
        org      0         ; start at address 0
        goto     main
;
; Space for interrupt handlers
        org      0x08

main:
; Port A. Five low-order lines set for input
        movlw    B'00011111'       ; w = 00011111 binary
        tris     porta             ; port A (lines 0 to 4) to
input
; Port B. All eight lines for output
        movlw    B'00000000'       ; w := 00000000 binary
        tris     portb             ; port B to output
;
;================================
; Pushbutton switch processing
;================================
pbutton:
; Push button switch on demo board is wired to port A bit 4
; Switch logic is active low
        btfss    porta,Pb_sw       ; Test and skip if switch bit
                                   ; set
        goto     buzzit            ; Buzz if switch ON,
; At this point port A bit 4 is set (switch is off)
        call     buzoff            ; Buzzer off
        goto     readdip           ; Read DIP switches
buzzit:
        call     buzon             ; Turn on buzzer
        goto     pbutton
;
;============================
;   dip switch monitoring
;============================
readdip:
; Read port A switches
        movf     porta,w           ; Port A bits to w
; because board is wired active low then all switch bits
; must be negated.  This is done by XORing with 1-bits
        xorlw    b'11111111'       ; Invert all bits in w
```

```
; Mask off 4 high-order bits
        andlw   b'00001111'                 ; And with mask
; At this point the w register contains a 4-bit value
; in the range 0 to 0xf. Use this value (in w) to
; obtain Seven-Segment display code
        call    segment
        movwf   portb               ; Display switch bits
        goto    pbutton
;================================
;  routine to returns 7-segment
;           codes
;================================
segment:
                addwf   PCL,f   ; PCL is program counter latch
                retlw   0x3f    ; 0 code
                retlw   0x06    ; 1
                retlw   0x5b    ; 2
                retlw   0x4f    ; 3
                retlw   0x66    ; 4
                retlw   0x6d    ; 5
                retlw   0x7d    ; 6
                retlw   0x07    ; 7
                retlw   0x7f    ; 8
                retlw   0x6f    ; 9
                retlw   0x77    ; A
                retlw   0x7c    ; B
                retlw   0x39    ; C
                retlw   0x5b    ; D
                retlw   0x79    ; E
                retlw   0x71    ; F
                retlw   0x7f    ; Just in case all on


;============================
;   piezo buzzer ON
;============================
; Routine to turn on piezo buzzer on port B bit 7
buzon:
        bsf     portb,7         ; Tune on bit 7, port B
        return
;
;============================
;   piezo buzzer OFF
;============================
; Routine to turn off piezo buzzer on port B bit 7
buzoff:
        bcf     portb,7         ; Bit 7 port b clear
        return
;============================
```

```
;    long delay sub-routine
;      (for code testing)
;==============================
long_delay
                  movlw    D'200'   ; w = 200 decimal
                  movwf    J                    ; J = w
jloop:   movwf    K                    ; K = w
kloop:   decfsz   K,f                  ; K = K-1, skip next if zero
                  goto     kloop
                  decfsz   J,f                   ; J = J-1, skip next
if zero
                  goto     jloop
                  return
                  end
```

```
;  File name: TestStr.asm
;  Date: May 1, 2011
;  Authors: Sanchez and Canton
;
;  Description:
;  Program to test sending strings to LCD memory directly
;  Program uses delay loops for interface timing.
;  WARNING:
;  Code assumes 4 MHz clock. Delay routines must be
;  edited for faster clock

;  Displays: Minnesota State, Mankato
;
;============================
;         switches
;============================
;  Switches used in __config directive:
;    _CP_ON          Code protection ON/OFF
;  * _CP_OFF
;  * _PWRTE_ON       Power-up timer ON/OFF
;    _PWRTE_OFF
;    _WDT_ON         Watchdog timer ON/OFF
;  * _WDT_OFF
;    _LP_OSC         Low power crystal oscillator
;  * _XT_OSC         External parallel crystal oscillator
;    _HS_OSC         High speed crystal resonator (8 to 10 MHz)
;                    Resonator: Murate Erie CSA8.00MG = 8 MHz
;    _RC_OSC         Resistor/capacitor oscillator
;                    (simplest, 20% error)
;  |
;  |_____ * indicates setup values

;=========================
;  setup and configuration
;=========================
        processor 16f84A
        include   <p16f84A.inc>
        __config  _XT_OSC & _WDT_OFF & _PWRTE_ON & _CP_OFF


;=======================================================
;                 constant definitions
;   for PIC-to-LCD pin wiring and LCD line addresses
;=======================================================
#define E_line 1            ;|
#define RS_line 2           ;| -- from wiring diagram
#define RW_line 3           ;|
;
;  LCD line addresses (from LCD datasheet)
```

```
#define LCD_1 0x80          ; First LCD line constant
#define LCD_2 0xc0          ; Second LCD line constant
; Note: The constants that define the LCD display line
;       addresses have the high-order bit set in
;       order to faciliate the controller command
;
;========================================================
;                   PIC register equates
;========================================================
        porta     equ       0x05
        Portb     equ       0x06
        fsr       equ       0x04
        Status    equ       0x03
        indf      equ       0x00
        z         equ       2
;========================================================
;                 variables in PIC RAM
;========================================================
; Reserve 16 bytes for string buffer
        cblock    0x0c
        strData
        endc
; Leave 16 bytes and Continue with local variables
        cblock    0x1d              ; Start of block
        count1    ; Counter # 1
        count2    ; Counter # 2
        count3    ; Counter # 3
        pic_ad    ; Storage for start of text area
                  ; (labeled strData) in PIC RAM
        J         ; counter J
        K         ; counter K
        index     ; Index into text table (also used
                  ; for auxiliary storage)
        endc
;============================================================
;                          program
;============================================================
                org               0         ; start at address
                goto      main
; Space for interrupt handlers
        org               0x08

main:
        movlw     b'00000000' ; All lines to output
        tris      porta                 ; in port A
        tris      portb                 ; and port B
        movlw     b'00000000' ; All outputs ports low
        movwf     porta
```

```
        movwf   portb
; Wait and initialize HD44780
        call    delay_5         ; Allow LCD time to initialize
itself
        call    initLCD         ; Then do forced
initialization
        call    delay_5         ; (Wait probably not
necessary)
; Store base address of text buffer in PIC RAM
        movlw   0x0c            ; Start address of text buffer
        movwf   pic_ad          ; to local variable
;=========================
;       test routine
;=========================
; Set DDRAM address to start of first line
        call    line1
; Store characters and send directly
        movlw   'H'
        movwf   portb
        call    pulseE
        movlw   'e'
        movwf   portb
        call    pulseE
        movlw   'l'
        movwf   portb
        call    pulseE
        movlw   'l'
        movwf   portb
        call    pulseE
        movlw   'o'
        movwf   portb
        call    pulseE
        call    delay_5
;======================
;       done!
;======================
loopHere:
        goto    loopHere  ;done


;************************************************************
;                   INITIALIZE LCD PROCEDURE
;************************************************************
initLCD
; Initialization for Densitron LCD module as follows:
;       8-bit interface
;    2 display lines of 16 characters each
;    cursor on
;    left-to-right increment
```

```
;    cursor shift right
;    no display shift
;***********************|
;      COMMAND MODE      |
;***********************|
        bcf       porta,E_line      ; E line low
        bcf       porta,RS_line     ; RS line low for command
        bcf       porta,RW_line    ; Write mode
        call      delay_125                    ;delay 125
microseconds
;***********************|
;      FUNCTION SET      |
;***********************|
        movlw     0x38      ; 0 0 1 1 1 0 0 0 (FUNCTION SET)
                            ;        | | | |__ font select:
                            ;        | | |    1 = 5x10 in 1/8 or 1/11
                            ;        | | |    0 = 1/16 dc
                            ;        | | |___ Duty cycle select
                            ;        | |      0 = 1/8 or 1/11
                            ;        | |      1 = 1/16
                            ;        | |___ Interface width
                            ;        |      0 = 4 bits
                            ;        |      1 = 8 bits
                            ;        |___ FUNCTION SET COMMAND
        movwf     portb   ;0011 1000
        call      pulseE  ;pulseE and delay

;***********************|
;    DISPLAY OFF         |
;***********************|
        movlw     0x08      ; 0 0 0 0 1 0 0 0 (DISPLAY ON/OFF)
                            ;          | | | |___ Blink character
                            ;          | | |     1 = on, 0 = off
                            ;          | | |___ Cursor on/off
                            ;          | |       1 = on, 0 = off
                            ;          | |____ Display on/off
                            ;          |        1 = on, 0 = off
                            ;          |____ COMMAND BIT

        movwf     portb
        call      pulseE  ;pulseE and delay

;***********************|
; DISPLAY AND CURSOR ON |
;***********************|
        movlw     0x0e      ; 0 0 0 0 1 1 1 0 (DISPLAY ON/OFF)
                            ;          | | | |___ Blink character
                            ;          | | |     1 = on, 0 = off
```

```
                              ;               | | |___ Cursor on/off
                              ;               | |    1 = on, 0 = off
                              ;               | |____ Display on/off
                              ;               |     1 = on, 0 = off
                              ;               |____ COMMAND BIT
         movwf    portb
         call     pulseE   ;pulseE and delay

;*********************|
;    ENTRY MODE SET   |
;*********************|
         movlw    0x06     ; 0 0 0 0 0 1 1 0 (ENTRY MODE SET)
                           ;            | | |___ display shift
                           ;            | |    1 = shift
                           ;            | |    0 = no shift
                           ;            | |____ cursor increment
                           ;            |     1 = left-to-right
                           ;            |     0 = right-to-left
                           ;            |____ COMMAND BIT
         movwf    portb    ;00000110
         call     pulseE

;*********************|
; CURSOR/DISPLAY SHIFT |
;*********************|
         movlw    0x14     ; 0 0 0 1 0 1 0 0 (CURSOR/DISPLAY
                           ;                     SHIFT)
                           ;         | | | |_|___ don't care
                           ;         | |_|__ cursor/display shift
                           ;         |     00 = cursor shift left
                           ;         |     01 = cursor shift right
                           ;         |     10 = cursor and display
                           ;         |          shifted left
                           ;         |     11 = cursor and display
                           ;         |          shifted right
                           ;         |___ COMMAND BIT
;
         movwf    portb    ;0001 1111
         call     pulseE

;*********************|
;   CLEAR DISPLAY     |
;*********************|
         movlw    0x01     ; 0 0 0 0 0 0 0 1 (CLEAR DISPLAY)
                           ;               |___ COMMAND BIT
         movwf    portb    ;0000 0001
;
         call     pulseE
```

```
        call     delay_5   ;delay 5 milliseconds after init
        return
;*************************************************************
;                  DELAY AND PULSE PROCEDURES
;*************************************************************
;=========================
;   Procedure to delay
;     42 microseconds
;=========================
delay_125
        movlw    D'42'     ; Repeat 42 machine cycles
        movwf    count1    ; Store value in counter
repeat
        decfsz   count1,f  ; Decrement counter
        goto     repeat    ; Continue if not 0
        return             ; End of delay
;-------------------------------------------------------------
;=========================
;   Procedure to delay
;     5 milliseconds
;=========================
delay_5
        movlw    D'41'     ; Counter = 41
        movwf    count2    ; Store in variable
delay
        call     delay_125          ; Delay
        decfsz   count2,f           ; 40 times = 5 milliseconds
        goto     delay
        return                      ; End of delay
;=========================
;      pulse E line
;=========================
pulseE
        bsf      porta,E_line      ;pulse E line
        bcf      porta,E_line
        call     delay_125          ;delay 125 microseconds
        return
;============================
;   long delay sub-routine
;       (for debugging)
;============================
long_delay
        movlw    D'200'   ; w = 200 decimal
        movwf    J                  ; J = w
jloop:  movwf    K                  ; K = w
kloop:  decfsz   K,f                ; K = K-1, skip next if zero
             goto     kloop
        decfsz   J,f                ; J = J-1, skip next if zero
```

```
            goto     jloop
            return
;=============================
;   LCD display procedure
;=============================
; Sends 16 characters from PIC buffer with address stored
; in variable pic_ad to LCD line previously selected
display16:
; Set up for data
            bcf      porta,E_line        ; E line low
            bsf      porta,RS_line       ; RS line low for control
            call     delay_125                   ; Delay
; Set up counter for 16 characters
            movlw    D'16'                       ; Counter = 16
            movwf    count3
; Get display address from local variable pic_ad
            movf     pic_ad,w ; First display RAM address to W
            movwf    fsr      ; W to FSR
getchar:
            movf     indf,w   ; get character from display RAM
                              ; location pointed to by file select
                              ; register

            movwf    portb
            call     pulseE   ;send data to display
; Test for 16 characters displayed
            decfsz   count3,f           ; Decrement counter
            goto     nextchar           ; Skipped if done
            return
nextchar:
            incf     fsr,f              ; Bump pointer
            goto     getchar
;========================
;     blank buffer
;========================
; Procedure to store 16 blank characters in PIC RAM
; buffer starting at address stored in the variable
; pic_ad
blank16:
            movlw    D'16'    ; Setup counter
            movwf    count1
            movf     pic_ad,w ; First PIC RAM address
            movwf    fsr      ; Indexed addressing
            movlw    0x20     ; ASCII space character
storeit:
            movwf    indf     ; Store blank character in PIC RAM
                              ; buffer using fsr register
            decfsz   count1,f ; Done?
            goto     incfsr             ; no
```

```
        return                   ; yes
incfsr:
        incf     fsr,f           ; Bump fsr to next buffer
space
        goto     storeit
;=======================
; Set address register
;    to LCD line 1
;=======================
; ON ENTRY:
;         Address of LCD line 1 in constant LCD_1
line1:
        bcf      porta,E_line    ; E line low
        bcf      porta,RS_line   ; RS line low, set up for
                                 ; control
        call     delay_125       ; delay 125 microseconds
; Set to second display line
        movlw    LCD_1           ; Address and command bit
        movwf    portb
        call     pulseE          ; Pulse and delay
; Set RS line for data
        bsf      porta,RS_line   ; Setup for data
        call     delay_125       ; Delay
        return
;=======================
; Set address register
;    to LCD line 2
;=======================
; ON ENTRY:
;         Address of LCD line 2 in constant LCD_2
line2:
        bcf      porta,E_line    ; E line low
        bcf      porta,RS_line   ; RS line low, setup for
                                 ; control
        call     delay_125       ; delay
; Set to second display line
        movlw    LCD_2           ; Address with high-bit set
        movwf    portb
        call     pulseE          ; Pulse and delay
; Set RS line for data
        bsf      porta,RS_line   ; RS = 1 for data
        call     delay_125       ; delay
        return

;===============================
;   first text string procedure
;===============================
storeMSU:
```

```
; Procedure to store in PIC RAM buffer the message
; contained in the code area labeled msg1
; ON ENTRY:
;          variable pic_ad holds address of text buffer
;          in PIC RAM
;          w register holds offset into storage area
;          msg1 is routine that returns the string characters
;          and a zero terminator
;          index is local variable that holds offset into
;          text table. This variable is also used for
;          temporary storage of offset into buffer
; ON EXIT:
;          Text message stored in buffer
;
; Store offset into text buffer (passed in the w register)
; in temporary variable
          movwf    index              ; Store w in index
; Store base address of text buffer in fsr
          movf     pic_ad,w; first display RAM address to W
          addwf    index,w            ; Add offset to address
          movwf    fsr                         ; W to FSR
; Initialize index for text string access
          movlw    0        ; Start at 0
          movwf    index    ; Store index in variable
; w still = 0
get_msg_char:
          call     msg1     ; Get character from table
; Test for zero terminator
          andlw    0x0ff
          btfsc    status,z ; Test zero flag
          goto     endstr1  ; End of string
; ASSERT: valid string character in w
;         store character in text buffer (by fsr)
          movwf    indf     ; store in buffer by fsr
          incf     fsr,f    ; increment buffer pointer
; Restore table character counter from variable
          movf     index,w  ; Get value into w
          addlw    1        ; Bump to next character
          movwf    index    ; Store table index in variable
          goto     get_msg_char     ; Continue
endstr1:
          return

; Routine for returning message stored in program area
msg1:
          addwf    PCL,f              ; Access table
          retlw    'M'
          retlw    'i'
```

```
        retlw   'n'
        retlw   'n'
        retlw   'e'
        retlw   's'
        retlw   'o'
        retlw   't'
        retlw   'a'
        retlw   0


;==================================
;   second text string procedure
;==================================
storeUniv:
; Processing identical to procedure StoreMSU
        movwf   index    ; Store w in index
; Store base address of text buffer in fsr
        movf    pic_ad,0 ; first display RAM address to W
        addwf   index,0  ; Add offset to address
        movwf   fsr      ; W to FSR
; Initialize index for text string access
        movlw   0        ; Start at 0
        movwf   index    ; Store index in variable
; w still = 0
get_msg_char2:
        call    msg2     ; Get character from table
; Test for zero terminator
        andlw   0x0ff
        btfsc   status,z ; Test zero flag
        goto    endstr2  ; End of string
; ASSERT: valid string character in w
;         store character in text buffer (by fsr)
        movwf   indf     ; Store in buffer by fsr
        incf    fsr,f    ; Increment buffer pointer
; Restore table character counter from variable
        movf    index,w  ; Get value into w
        addlw   1        ; Bump to next character
        movwf   index    ; Store table index in variable
        goto    get_msg_char2    ; Continue
endstr2:
        return

; Routine for returning message stored in program area
msg2:
        addwf   PCL,f              ; Access table
        retlw   'S'
        retlw   't'
        retlw   'a'
        retlw   't'
```

```
        retlw    'e'
        retlw    ','
        retlw    0x20
        retlw    'M'
        retlw    'a'
        retlw    'n'
        retlw    'k'
        retlw    'a'
        retlw    't'
        retlw    'o'
        retlw    0

        end
```

```
; File: Timer0.ASM
; Date: April 27, 2006
; Author: Julio Sanchez
; Processor: a6F84A
;
; Description:
; Program to demonstrate programming of the 16F84A
; TIMER0 module. Program flashes eight LEDs in sequence
; counting from 0 to 0xff. Timer0 is used to delay
; the count.
;============================
;          switches
;============================
; Switches used in __config directive:
;   _CP_ON          Code protection ON/OFF
; * _CP_OFF
; * _PWRTE_ON       Power-up timer ON/OFF
;   _PWRTE_OFF
;   _WDT_ON         Watchdog timer ON/OFF
; * _WDT_OFF
;   _LP_OSC         Low power crystal oscillator
; * _XT_OSC         External parallel crystal oscillator
;   _HS_OSC         High speed crystal resonator (8 to 10 MHz)
;                   Resonator: Murate Erie CSA8.00MG = 8 MHz
;   _RC_OSC         Resistor/capacitor oscillator
;                   (simplest, 20% error)
; |
; |_____ * indicates setup values

        processor 16f84A
        include   <p16f84A.inc>
        __config _XT_OSC & _WDT_OFF & _PWRTE_ON & _CP_OFF
;=========================================================
;                variables in PIC RAM
;=========================================================
; None in this application
;
;=========================================================
;                m a i n   p r o g r a m
;=========================================================
        org     0              ; start at address 0
        goto    main
;
;============================
;      interrupt handler
;============================
        org             0x08
;============================
```

```
;          main program
;==============================
main:
; Clear the Watchdog timer and reset prescaler
          clrwdt
; Set up the OPTION regiser bitmap
          movlw     b'11010111'
;     7   6   5   4   3   2   1   0  <= OPTION bits
;     |   |   |   |   |   |   |__|__|_____ PS2-PS0 (prescaler bits)
;     |   |   |   |   |   |                 Values for Timer0
;     |   |   |   |   |   |                 000 = 1:2     001 = 1:4
;     |   |   |   |   |   |                 010 = 1:8     011 = 1:16
;     |   |   |   |   |   |                 100 = 1:32    101 = 1:64
;     |   |   |   |   |   |                 110 = 1:128 *111 = 1:256
;     |   |   |   |   |_____ PSA (prescaler assign)
;     |   |   |   |                         1 = to WDT
;     |   |   |   |                        *0 = to Timer0
;     |   |   |   |_____ TOSE (Timer0 edge select)
;     |   |   |                         0 = increment on low-to-high
;     |   |   |                        *1 = increment on high-to-low
;     |   |   |_____ TOCS (TMR0 clock source)
;     |   |                         *0 = internal clock
;     |   |                         1 = RA4/TOCKI bit source
;     |   |_____ INTEDG (Edge select)
;     |                         *0 = falling edge
;     |_____ RBPU (Pullup enable)
;                           0 = enabled
;                          *1 = disabled
          option
; Setup ports
          movlw     0x00                ; Set port B to output
          tris      PORTB
          clrf      PORTB               ; All port B to 0
; Port A is not used in this program
mloop:
          incf      PORTB,f             ; Add 1 to register value
          call      TM0delay
          goto      mloop
;*****************************
;     delay sub-routine
;         uses Timer0
;*****************************
TM0delay:
; Initialize the timer register
          clrf      TMR0      ; Clear SFR for Timer0
; Routine tests the value in the TMR0 register by
; subtracting 0xff from the value in TMR0. The zero flag
; is set if TMR0 = 0xff
```

```
cycle:
        movf    TMR0,w              ; Timer to w
; w has TMR0 register value
        sublw   0xff                ; Subtract max value
; Zero flag is set if value in TMR0 = 0xff
        btfss   STATUS,Z ; Test for zero
        goto    cycle               ; Repeat
        return

                End
```

```
; File: TimerTst.ASM
; Date: June 7, 2010
; Authors: Sanchez and Canton
;
; Description:
; Using the timer to generate a signal at 1 MHz
;
;===========================
;         switches
;===========================
; Switches used in __config directive:
;   _CP_ON          Code protection ON/OFF
; * _CP_OFF
; * _PWRTE_ON       Power-up timer ON/OFF
;   _PWRTE_OFF
;   _WDT_ON         Watchdog timer ON/OFF
; * _WDT_OFF
;   _LP_OSC         Low power crystal oscillator
; * _XT_OSC         External parallel crystal oscillator
;   _HS_OSC         High speed crystal resonator (8 to 10 MHz)
;                   Resonator: Murate Erie CSA8.00MG = 8 MHz
;   _RC_OSC         Resistor/capacitor oscillator
;                   (simplest, 20% error)
; |
; |_____ * indicates setup values

        processor 16f84A
        include   <p16f84A.inc>
        __config  _XT_OSC & _WDT_OFF & _PWRTE_ON & _CP_OFF


;=======================================================
;                  PIC register equates
;=======================================================
        porta    equ     0x05
        Portb    equ     0x06
        Status   equ     0x03
        z        equ     0x02
        tmr0     equ     0x01
;
;=======================================================
;               variables in PIC RAM
;=======================================================
; Local variables
        cblock  0x0d    ; Start of block
        J               ; counter J
        K               ; counter K
        countL          ; Auxiliary counter
        countH          ; ISR counter
```

```
        endc
;===========================================================
;                m a i n    p r o g r a m
;===========================================================
        org     0               ; start at address 0
        goto    main
;
;=============================
;       interrupt handler
;=============================
        org             0x04
;       goto    IntServ
;=============================
;       main program
;=============================
main:
; Clear the Watchdog timer and reset prescaler
        clrwdt
; Set up the OPTION regiser bitmap
        movlw   b'11010011'
;   7  6  5  4  3  2  1  0 <= OPTION bits
;   |  |  |  |  |  |  |__|__|_____ PS2-PS0 (prescaler bits)
;   |  |  |  |  |  |               Values for Timer0
;   |  |  |  |  |  |               000 = 1:2    001 = 1:4
;   |  |  |  |  |  |               010 = 1:8    011 = 1:16
;   |  |  |  |  |  |               100 = 1:32   101 = 1:64
;   |  |  |  |  |  |               110 = 1:128 *111 = 1:256
;   |  |  |  |  |_____ PSA (prescaler assign)
;   |  |  |  |                      1 = to WDT
;   |  |  |  |                     *0 = to Timer0
;   |  |  |  |_____ TOSE (Timer0 edge select)
;   |  |  |                         0 = increment on low-to-high
;   |  |  |                        *1 = increment on high-to-low
;   |  |  |_____ TOCS (TMR0 clock source)
;   |  |                           *0 = internal clock
;   |  |                            1 = RA4/TOCKI bit source
;   |  |_____ INTEDG (Edge select)
;   |                              *0 = falling edge
;   |_____ RBPU (Pullup enable)
;                                   0 = enabled
;                                  *1 = disabled
        option
; Set up ports
        movlw   0x00                    ; Set port B for output
        tris    portb
        clrf    portb                   ; All port B to 0
; Port A is not used in this program
mloop:
```

```
            bsf                     portb,0
            call     TM0delay
            bcf                     portb,0
            call     TM0delay
            goto     mloop
;*****************************
;     delay sub-routine
;          uses Timer0
;*****************************
TM0delay:
; Initialize the timer register
            clrf     tmr0      ; Clear SFR for Timer0
; Routine tests the value in the tmr0 register by
; xoring with a mask of all ones. The operation sets
; the zero flag if tmr0 is zero.
cycle:
            movf     tmr0,w              ; Timer to w
; w has tmr0 register value
            sublw    0xff                ; Subtract max value
; Zero flag is set if value in tmr0 = 0xff
            btfss    status,z ; Test for zero
            goto     cycle               ; Repeat
            return

            end
```

```
; File name: TTYUsart.asm
; Last update: May 12, 2009
; Authors: Sanchez and Canton
; Processor: 16F84A
;
; Description:
; Program to emulate USART operation in PIC code. Uses
; PIC-to-LCD interface. Display has two lines, each with
; sixteen characters.
; Program operation:
; Characters received from the RS232 line are displayed on
; the LCD. LCD lines scroll automatically. A pushbutton
; activates the send operation by transmitting the text
; string "Ready-", which is also displayed on the LCD.
;
; Program communications and LCD parameters are stored in
; #define statements. These statements can be edited to
; accommodate a different set up. Program uses delay loops
; for interface timing.
;
; WARNING:
; Code assumes 4 MHz clock. Delay routines must be
; edited for faster clock
;
; BAUD RATE CALCULATIONS:
; A 4 MHz clock oscillator has a clock frequency of 1 MHz:
; because the baud rate is the number of clock cycles per
; second, for a 4 MHz clock it is:
;                 1
; bit time = ------ sec. = 208.33 microseconds
;               4,800
; Calculating one half the baud rate allows resetting the
; clock from the edge to the center of a time pulse:
;
;           |<======== falling edge of start bit
;           |       |<======== center of bit time
;        >|        |< one-half baud rate
;           |       |
;_____.      |       ._____.
;           |_____|            |_____
;              208/2 = 104
; The PIC clock counts up from 0 to 255. So to implement
; a 104-microsecond delay, we must start counting at
; clock beat:
;                 255 - 104 = 151
; plus one microsecond for movlw instruction used to
; initialize the clock:
;                 151 + 1 = 152
```

```
; For one full baud rate delay:
;                 255 - 208 = 47 + 1 = 48
; The following two constants are stored in #define
; statements:
;                 halfBaud = 152
;                 fullBaud = 48
; Setting the prescaler to TMR0 reduces the baud rate
; to one-half. Other prescaler values will reduce the
; baud rate accordingly.
;
; Wiring diagram:
;    RB4-RB7 ===> LCD data lines 4 to 7 (output)
;    RB0 =======> MAX202 T2in line (output)
;    RA0 =======> MAX202 R2out line (input)
;    RA1 =======> LCD E line (output)
;    RA2 =======> LCD RS line (output)
;    RA3 =======> LCD R/W line (output - not used)
;    RA4 =======> Pushbutton switch 1
;                     (input - active low)
;
;===========================
;        switches
;===========================
; Switches used in __config directive:
;   _CP_ON          Code protection ON/OFF
; * _CP_OFF
; * _PWRTE_ON       Power-up timer ON/OFF
;   _PWRTE_OFF
;   _WDT_ON         Watchdog timer ON/OFF
; * _WDT_OFF
;   _LP_OSC         Low power crystal oscillator
; * _XT_OSC         External parallel crystal oscillator
;   _HS_OSC         High speed crystal resonator (8 to 10 MHz)
;                   Resonator: Murate Erie CSA8.00MG = 8 MHz
;   _RC_OSC         Resistor/capacitor oscillator
; |                 (simplest, 20% error)
; |
; |_____ * indicates setup values presently selected

;========================
; setup and configuration
;========================
        processor 16f84A
        include   <p16f84A.inc>
        __config  _XT_OSC & _WDT_OFF & _PWRTE_ON & _CP_OFF

;=============================================================
;                    M A C R O S
```

```
;==============================================================
; Macros to select the register banks
Bank0   MACRO                           ; Select RAM bank 0
        bcf     STATUS,RP0
        ENDM

Bank1   MACRO                           ; Select RAM bank 1
        bsf     STATUS,RP0
        ENDM
;==============================================================
;                   constant definitions
;       for PIC-to-LCD pin wiring and LCD line addresses
;==============================================================
#define E_line 1        ; |
#define RS_line 2       ; | -- from wiring diagram
#define RW_line 3       ; |
; LCD line addresses (from LCD datasheet)
#define LCD_1 0x80      ; First LCD line constant
#define LCD_2 0xc0      ; Second LCD line constant
#define LCDlimit .16; Number of characters per line
; 4800 baud clock countdown values
; Code reduces rate to 2400 baud by entering a minimal
; presclaer to TRM0
#define halfBaud .152 ; For one-half bit time
#define fullBaud .48  ; For one full bit time
;
; Note: The constants that define the LCD display line
;       addresses have the high-order bit set in
;       order to faciliate the controller command.
;
;==========================================================
;           PIC register and flag equates
;==========================================================
z       equ     2       ; Zero flag
c       equ     0       ; Carry flag
;==========================================================
;          buffer and variables in PIC RAM
;==========================================================
; Create a 16-byte storage area
        cblock  0x0c    ; Start of first data block
        lineBuf                 ; buffer for text storage
        endc
; Leave 16 bytes and Continue with local variables
        cblock  0x1c    ; Second data block
        count1          ; Counter # 1
        count2          ; Counter # 2
        J                       ; counter J
        K               ; counter K
```

```
            store1              ; Local temporary storage
            store2              ; Storage # 2
; For LCDscroll procedure
            LCDcount ; Counter for characters per line
            LCDline             ; Current display line (0 or 1)
            bufPtr              ; Buffer pointer
; Variables for serial communications
            tempData ; Temporary storage for bit manipulations
            rcvData             ; Final storage for received character
            bitCount ; Bit counter
            sendData ; Character to send
            endc


;============================================================
;                m a i n     p r o g r a m
;============================================================
                org             0         ; start at address
                goto    main
; Space for interrupt handlers
        org     0x08

main:
        Bank1
        movlw   b'00010001' ; Port A lines I/O setup
                        ; RA0 = RS232 input (R2out)
                        ; RA4 = Pushbutton SW # 1
        movwf   TRISA
        movlw   b'00000000' ; Port B lines as follows:
;    RB4-RB7 ===> LCD data lines 4 to 7 (output)
;    RB0 =======> MAX202 T2in line (output)
; RB0 =
        movwf   TRISB
        Bank0
; Clear bits in port A output lines
        bcf     PORTA,1
        bcf     PORTA,2
        bcf     PORTA,3
        movlw   b'00000000' ; All outputs ports low
        movwf   PORTB
; Wait and initialize HD44780
        call    delay_5 ; Allow LCD time to initialize
                        ; itself
        call    delay_5
        call    initLCD ; Then do forced initialization
        call    delay_5 ; Wait again
; Set port B, line 0 high so start bit is detected
        bsf             PORTB,0
;============================
```

```
;   wait for start command
;=============================
; Program waits until pushbutton number 1 is pressed
; to continue execution. Pushbutton 1 is active low
; and wired to RA4
pb1Wait:
        btfsc   PORTA,4             ; Test port A, line 4
        goto    pb1Wait             ; Loop if not clear
;=============================
;  display and send "Ready-"
;=============================
; Set LCD base address
        call    line1
; Initialize system for UART emulation at 2400 baud
        call    initTTY
; Display on LCD and test serial transmission by sending
; the string "Ready-"
        movlw   'R'
        movwf   sendData ; Store in send register
        call    send8    ; Local LCD display procedure
        call    sendTTY  ; Local send procedure
        movlw   'e'
        movwf   sendData ; Store in send register
        call    send8    ; Local LCD display procedure
        call    sendTTY  ; Local send procedure
        movlw   'a'
        movwf   sendData ; Store in send register
        call    send8    ; Local LCD display procedure
        call    sendTTY  ; Local send procedure
        movlw   'd'
        movwf   sendData ; Store in send register
        call    send8    ; Local LCD display procedure
        call    sendTTY  ; Local send procedure
        movlw   'y'
        movwf   sendData ; Store in send register
        call    send8    ; Local LCD display procedure
        call    sendTTY  ; Local send procedure
        movlw   '-'
        movwf   sendData ; Store in send register
        call    send8    ; Local LCD display procedure
        call    sendTTY  ; Local send procedure
; Init  character counter and line counter variables for
; LCD line scroll procedure
        movlw   0x06               ; 6 characters already
displayed
        movwf   LCDcount
        clrf    LCDline            ; LCD line counter
;=============================
```

```
;      monitor RS232 line
;==============================
nextChar:
        call      rcvTTY                        ; Receive character
; Store character in local line buffer using indirect
; addressing
; 16-byte buffer named lineBuf starts at address 0x0c
; Register variable bufPtr holds offset into buffer
        movlw     0x0c            ; Buffer base address
        addwf     bufPtr,w        ; Add pointer in w
        movwf     FSR             ; Value to index register
        movf      rcvData,        ; Character into w
        movwf     INDF            ; Store w in [FSR]
        incf      bufPtr,f        ; Bump pointer
; Send character (still in w)
        call      send8           ; Display it
        call      LCDscroll       ; Scroll display lines
        goto      nextChar        ; Continue


;================================================================
;                    initialize LCD for 4-bit mode
;================================================================
initLCD:
; Initialization for Densitron LCD module as follows:
;      4-bit interface
;   2 display lines of 16 characters each
;   cursor on
;   left-to-right increment
;   cursor shift right
;   no display shift
;======================|
;   set command mode   |
;======================|
        bcf       PORTA,E_line       ; E line low
        bcf       PORTA,RS_line      ; RS line low
        bcf       PORTA,RW_line      ; Write mode
        call      delay_125                    ; delay 125
microseconds
;*********************|
;     FUNCTION SET    |
;*********************|
        movlw     0x28      ; 0 0 1 0 1 0 0 0 (FUNCTION SET)
                            ;       | | | |__ font select:
                            ;       | | |     1 = 5x10 in 1/8 or 1/11
                            ;       | | |     0 = 1/16 dc
                            ;       | | |___ Duty cycle select
                            ;       | |       0 = 1/8 or 1/11
```

```
                                ;        | |            1 = 1/16
                                ;        | |___ Interface width
                                ;        |        0 = 4 bits
                                ;        |        1 = 8 bits
                                ;        |___ FUNCTION SET COMMAND
         call      send8        ; 4-bit send routine

; Set 4-bit mode command must be repeated
         movlw     0x28
         call      send8

;********************|
; DISPLAY AND CURSOR ON |
;********************|
         movlw     0x0e        ; 0 0 0 0 1 1 1 0 (DISPLAY ON/OFF)
                                ;              | | | |___ Blink character
                                ;              | | |       1 = on, 0 = off
                                ;              | | |___ Cursor on/off
                                ;              | |       1 = on, 0 = off
                                ;              | |____ Display on/off
                                ;              |         1 = on, 0 = off
                                ;              |____ COMMAND BIT
         call      send8
;********************|
;    set entry mode      |
;********************|
         movlw     0x06        ; 0 0 0 0 0 1 1 0 (ENTRY MODE SET)
                                ;            | | |___ display shift
                                ;            | |       1 = shift
                                ;            | |       0 = no shift
                                ;            | |____ increment mode
                                ;            |         1 = left-to-right
                                ;            |         0 = right-to-left
                                ;            |___ COMMAND BIT
         call      send8

;********************|
; cursor/display shift   |
;********************|
         movlw     0x14        ; 0 0 0 1 0 1 0 0 (CURSOR/DISPLAY
                                ;                      SHIFT)
                                ;          | | | |_|___ don't care
                                ;          | |_|__ cursor/display shift
                                ;          |         00 = cursor shift left
                                ;          |         01 = cursor shift right
                                ;          |         10 = cursor and display
                                ;          |             shifted left
                                ;          |         11 = cursor and display
```

```
                                   ;          |                 shifted right
                                   ;          |___ COMMAND BIT
        call      send8
;*********************|
;    clear display    |
;*********************|
        movlw     0x01      ; 0 0 0 0 0 0 0 1 (CLEAR DISPLAY)
                            ;                |___ COMMAND BIT
        call      send8
; Per documentation
        call      delay_5  ; Test for busy
        return


;======================
;  Procedure to delay
;   42 microseconds
;======================
delay_125
        movlw     D'42'    ; Repeat 42 machine cycles
        movwf     count1   ; Store value in counter
repeat
        decfsz    count1,f ; Decrement counter
        goto      repeat   ; Continue if not 0
        return             ; End of delay


;======================
;  Procedure to delay
;   5 milliseconds
;======================
delay_5
        movlw     D'41'    ; Counter = 41
        movwf     count2   ; Store in variable
delay
        call      delay_125         ; Delay
        decfsz    count2,f ; 40 times = 5 milliseconds
        goto      delay
        return             ; End of delay
;======================
;    pulse E line
;======================
pulseE
        bsf       PORTA,E_line      ; Pulse E line
        nop
        bcf       PORTA,E_line
        return
;==========================
;   long delay sub-routine
;     (for debugging)
```

```
;==============================
long_delay
        movlw   D'200'   ; w = 200 decimal
        movwf   J                        ; J = w
jloop:  movwf   K                        ; K = w
kloop:  decfsz  K,f                      ; K = K-1, skip next if zero
        goto    kloop
        decfsz  J,f                      ; J = J-1, skip next if zero
        goto    jloop
        return
;========================
;   send 2 nibbles in
;      4-bit mode
;========================
; Procedure to send two 4-bit values to port B lines
; 7, 6, 5, and 4. High-order nibble is sent first
; ON ENTRY:
;         w register holds 8-bit value to send
send8:
        movwf   store1   ; Save original value
        call    merge4   ; Merge with port B
; Now w has merged byte
        movwf   PORTB    ; w to port B
        call    pulseE   ; Send data to LCD
; High nibble is sent
        movf    store1,w ; Recover byte into w
        swapf   store1,w ; Swap nibbles in w
        call    merge4
        movwf   PORTB
        call    pulseE   ; Send data to LCD
        call    delay_125
        return
;==================
;   merge bits
;==================
; Routine to merge the four high-order bits of the
; value to send with the contents of port B
; so as to preserve the four low-bits in port B
; Logic:
;      AND value with 1111 0000 mask
;      AND port B with 0000 1111 mask
;      Now low nibble in value and high nibble in
;      port B are all 0 bits:
;          value = vvvv 0000
;          port B = 0000 bbbb
;      OR value and port B resulting in:
;                  vvvv bbbb
; ON ENTRY:
```

```
;       w contain value bits
; ON EXIT:
;       w contains merged bits
merge4:
        andlw   b'11110000'         ; ANDing with 0 clears the
                                    ; bit. ANDing with 1 preserves
                                    ; the original value
        movwf   store2              ; Save result in variable
        movf    PORTB,w             ; port B to w register
        andlw   b'00001111' ; Clear high nibble in port B
                                    ; and preserve low nibble
        iorwf   store2,w ; OR two operands in w
        return


;========================
; Set address register
;    to LCD line 1
;========================
; ON ENTRY:
;       Address of LCD line 1 in constant LCD_1
line1:
        bcf     PORTA,E_line        ; E line low
        bcf     PORTA,RS_line       ; RS line low, set up for
                                    ; control
        call    delay_5             ; busy?
; Set to second display line
        movlw   LCD_1               ; Address and command bit
        call    send8               ; 4-bit routine
; Set RS line for data
        bsf     PORTA,RS_line       ; Setup for data
        call    delay_5             ; Busy?
; Clear buffer and pointer
        call    blankBuf
        clrf    bufPtr              ; Clear
        return
;========================
; Set address register
;    to LCD line 2
;========================
; ON ENTRY:
;       Address of LCD line 2 in constant LCD_2
line2:
        bcf     PORTA,E_line        ; E line low
        bcf     PORTA,RS_line       ; RS line low, setup for
control
        call    delay_5             ; Busy?
; Set to second display line
        movlw   LCD_2               ; Address with high-bit set
```

```
        call      send8
; Set RS line for data
        bsf       PORTA,RS_line      ; RS = 1 for data
        call      delay_5            ; Busy?
; Clear buffer and pointer
        call      blankBuf
        clrf      bufPtr
        return


;============================
;   scroll LCD line 2
;============================
; Procedure to count the number of characters displayed on
; each LCD line. If the number reaches the value in the
; constant LCDlimit, then display is scrolled to the second
; LCD line. If at the end of the second line, then the
; second line is scrolled to the first line and display
; continues at the start of the second line
; reset to the first line.
LCDscroll:
        incf      LCDcount,f         ; Bump counter
; Test for line limit
        movf      LCDcount,w
        sublw     LCDlimit           ; Count minus limit
        btfss     STATUS,z           ; Is count - limit = 0
        goto      scrollExit         ; Go if not at end of line
; At this point the end of the LCD line was reached
; Test if this is also the end of the second line
        movf      LCDline,w
        sublw     0x01               ; Is it line 1?
        btfsc     STATUS,z           ; Is LCDline minus 1 = 0?
        goto      line2End           ; Go if end of second line
; At this point it is the end of the top LCD line
        call      line2              ; Scroll to second line
        clrf      LCDcount           ; Reset counter
        incf      LCDline,f          ; Bump line counter
        goto      scrollExit
; End of second LCD line
;
line2End:
; Scroll second line to first line. Characters to be
; scrolled are stored in buffer starting at address 0x0c.
; 16 characters are to be moved.
; First clear LCD
        call      initLCD
        call      delay_5            ; Make sure not busy
; Set up for data
        bcf       PORTA,E_line       ; E line low
```

```
        bsf       PORTA,RS_line    ; RS line high for data
; Set up counter for 16 characters
        movlw     D'16'            ; Counter = 16
        movwf     count2
; Get address of storage buffer
        movlw     0x0c
        movwf     FSR      ; W to FSR
getchar:
        movf      INDF,w   ; get character from display RAM
                           ; location pointed to by file select
                           ; register
        call      send8    ; 4-bit interface routine
; Test for 16 characters displayed
        decfsz    count2,f ; Decrement counter
        goto      nextchar ; Skipped if done
; At this point scroll operation has concluded
        clrf      LCDcount ; Clear counters
; Stay at line 2
        clrf      LCDline
        incf      LCDline,f
        call      line2    ; Set for second line
scrollExit:
        return
nextchar:
        incf      FSR,f    ; Bump pointer
        goto      getchar

;============================
;   clear line buffer
;============================
; Use indirect addressing to store 16 blanks in the
; buffer located at 0x0c
blankBuf:
        Bank0
        movlw     0x0c     ; Pointer to RAM
        movwf     FSR      ; To index register
blank16:
        clrf      INDF     ; Clear memory pointed at by FSR
        incf      FSR,f    ; Bump pointer
        btfss     FSR,4    ; 000x0000 when bit 4 is set
                           ; count reached 16
        goto      blank16
        return

;================================================================
;                    initialize for TTY
;================================================================
; Procedure to initialize RS232 reception
```

```
; Assumes:
;                            2400 baud
;                            8 data bits
;                            no parity
;                            one stop bit
initTTY:
; First initialize receiver to RS-232 line parameters
; Disable global and peripheral interrupts
;  7  6  5  4  3  2  1  0   <= INTCON bitmap
;  |  ?  |  ?  ?  ?  ?  ?   (? = unrelated bits)
;  |     |_____ Timer0 interrupt on overflow
;  |_____ Global interrupts
        bcf     INTCON,5         ; Disable TMR0 interrupts
        bcf     INTCON,7         ; Disable global interrupts
        clrf    TMR0             ; Reset timer
        clrwdt                   ; Clear WDT for prescaler
                                 ; assign
        Bank1
; Set up the OPTION register bitmap
;  7  6  5  4  3  2  1  0 <= OPTION bits
;  1  1  0  1  1  0  0  0 <= setup
;  |  |  |  |  |  |__|__|_____ PS2-PS0 (prescaler bits)
;  |  |  |  |  |              Values for Timer0
;  |  |  |  |  |              *000 = 1:2   001 = 1:4
;  |  |  |  |  |              010 = 1:8    011 = 1:16
;  |  |  |  |  |              100 = 1:32   101 = 1:64
;  |  |  |  |  |              110 = 1:128 111 = 1:256
;  |  |  |  |  |_____ PSA (prescaler assign)
;  |  |  |  |                 1 = to WDT
;  |  |  |  |                 *0 = to Timer0
;  |  |  |  |_____ TOSE (Timer0 edge select)
;  |  |  |                    0 = increment on low-to-high
;  |  |  |                    *1 = increment in high-to-low
;  |  |  |_____ TOCS (TMR0 clock source)
;  |  |                       *0 = internal clock
;  |  |                       1 = RA4/TOCKI bit source
;  |  |_____ INTEDG (Edge select)
;  |                          0 = falling edge
;  |                          *1 = rising edge
;  |_____ RBPU (Pullup enable)
;                             0 = enabled
;                             *1 = disabled
        movlw   b'11010000'  ; set up timer/counter
        movwf   OPTION_REG
        Bank0
        return
;=============================================================
;                    receive character
```

```
;=================================================================
; Receive a single character through the serial port.
; Assumes: 4800 baud, 8 data bits, no parity, 1 stop bit.
; Recieving line is Port A, line 0
rcvTTY:
        movlw    0x08                  ; Counter for 8 bits
        movwf    bitCount
; The start of character transmission is signaled by
; the sender by setting the line low
startBit:
        btfsc    PORTA,0               ; Test for low on line
        goto     startBit ; Go if not low
;==========================
;   offset to data bit
;==========================
; At this point the receiver has found the falling
; edge of the start bit. It must now wait one and
; one-half the baud rate to synchronize in the center
; of the sender's first data bit
;, as follows:
;        |<========= falling edge of START bit
;        |         |<========== center of start bit
;        |         |             |<====== center of data bit
;        |-----------|-----|
;_____              _____             _____
;        |           |             |           |          <== SIGNAL
;        -----------              ----------
;        |<-- 208 -->|<104>| <====== ms. for 4800 baud
;
; Clock start count for one-half bit  = 255 - 104 = 151
; Clock start count for one full bit  =  255 - 208 = 47
; One clock cycle is added for the movwf intruction:
;    clkHalf = 152 (for one-half bit countdown)
;    clkFull = 48 (for one full bit countdown)
        movlw    halfBaud ; Skip one-half bit
        movwf    TMR0     ; Initialize tmr0 and start count
        bcf      INTCON,2           ; Clear overflow flag
;=============================
;         start bit
;=============================
wait1:
        btfss    INTCON,2           ; Timer count overflow?
        goto     wait1              ; No, keep waiting
; At this point we are at the center of the start bit
        btfsc    PORTA,0            ; Check to see it is still low
        goto     startBit ; No, it is high. False start
; At this point the clock is at the center of the start
; bit. The first data bit must be read one full baud
```

```
; period later
        movlw   fullBaud            ; One full bit delay
        movwf   TMR0                ; Start timer
        bcf     INTCON,2            ; clear tmr0 overflow flag
wait2:
        btfss   INTCON,2            ; End of one full baud period?
        goto    wait2               ; Wait if not end of period
; Timer is now at the center of the first/next data bit
; Timer must be reset immediately so that code will not
; lose synchronization with sender
        movlw   fullBaud ; Skip to next data bit
        movwf   TMR0                ; Restart timer
        bcf     INTCON,2            ; Reset overflow flag
; Now the data bit can be read and stored
        movf    PORTA,w             ; Read port B
        movwf   tempData ; Store in temporary variable
        rrf     tempData,f          ; Rotate bit 0 into carry flag
        rrf     rcvData,f   ; Rotate carry flag into storage
                                    ; register high-order bit
        decfsz  bitCount,f          ; End of data?
        goto    wait2    ; Continue until 8 bits received
;============================
;         stop bit
;============================
stopWait:
        btfss   INTCON,2            ; Test time
        goto    stopWait ; Wait
        return                      ; Exit


;============================================================
;                     send character
;============================================================
; Procedure to send one character through the RS232 line.
; Assumes: 2400 baud, 8 data bits, no parity, one stop bit
; Sending line is Port B, line 0
; ON ENTRY:
;       variable sendData holds character to send
sendTTY:
        movlw   0x08                ; Init bit counter
        movwf   bitCount
        bcf     PORTB,0             ; Low for start bit
        movlw   fullBaud            ; For one baud space
        movwf   TMR0                ; Start timer
        bcf     INTCON,2            ; Clear timer flag
start2snd:
        btfss   INTCON,2            ; Full baud done?
        goto    start2snd           ; No
        movlw   fullBaud            ; Reset for one full bit
```

```
                                            ; period
            movwf     TMR0                  ; Start timer
            bcf       INTCON,2              ; Clear flag
; At this point the start bit has been sent
; Data follows
sendOut:
            rrf       sendData,f  ; Rotate bit into carry
            bcf       PORTB,0 ; Assume data bit is 0
            btfsc     STATUS,c ; Test if carry set
            bsf       PORTB,0  ; Change bit to 1 if clear
; Hold bit for 1 baud period
timeBit:
            btfss     INTCON,2              ; Wait for baud period to end
            goto      timeBit               ; Loop if not yet
            movlw     fullBaud              ; Reset timer
            movwf     TMR0                  ; Start timer
            bcf       INTCON,2              ; Clear flag
; Test for last bit
            decfsz    bitCount,f            ; Count this bit
            goto      sendOut               ; Continue if not last bit
; Done. Send stop bit
            bsf       PORTB,0               ; High for stop bit
stopBit:
            btfss     INTCON,2              ; Timer done?
            goto      stopBit               ; No
; Set port B line 0 high back again
            bsf       PORTB,0
            call      delay_5               ; And hold
            return

            End
```

```
; File name: I2CEEP.asm
; Last revision: May 28, 2010
; Authors: Sanchez and Canton
; Processor: 16F877
;
; Description:
; Receive character data through RS-232 line and store in
; 24LC04B EEPROM IC, using the I2C serial protocol in the
; PIC's MSSP module. Received characters are echoed on
; the second LCD line. When <Enter> key is detected (code
; 0x0d) the text stored in EEPROM memory is retrieved and
; displayed on the LCD. On startup the top LCD line displays
; the prompt: "Receiving:". At that time a message "Rdy- " is
; sent through the serial line so as to test the connection.
;
; Default serial line setting:
;               2400 baud
;               no parity
;               1 stop bit
;               8 character bits
;
; Wiring:
; 24LC04B SDA line is wired to PIC RC4 (MSSP SDA)
; 24LC04B SCL line is wired to PIC RC3 (MSSP SCL)
; 24LC04B A0-A2 and WP lines are not used (GND)
;
; Program to uses 4-bit PIC-to-LCD interface.
; Code assumes that LCD is driven by Hitachi HD44780
; controller and PIC 16F877. Display supports two lines
; each one with 20 characters. The length, wiring, and base
; address of each display line is stored in #define
; statements. These statements can be edited to acommodate
; a different set up.
;
; WARNING:
; Code assumes 10-MHz clock. Delay routines must be
; edited for a different clock. Clock speed also determines
; values for baud rate setting (see spbrgVal constant).
;
;============================
;       16F877 switches
;============================
; Switches used in __config directive:
;   _CP_ON          Code protection ON/OFF
; * _CP_OFF
; * _PWRTE_ON       Power-up timer ON/OFF
;   _PWRTE_OFF
;   _BODEN_ON       Brown-out reset enable ON/OFF
```

```
;  *  _BODEN_OFF
;  *  _PWRTE_ON       Power-up timer enable ON/OFF
;     _PWRTE_OFF
;     _WDT_ON         Watchdog timer ON/OFF
;  *  _WDT_OFF
;     _LPV_ON         Low voltage IC programming enable ON/OFF
;  *  _LPV_OFF
;     _CPD_ON         Data EE memory code protection ON/OFF
;  *  _CPD_OFF
; OSCILLATOR CONFIGURATIONS:
;     _LP_OSC         Low power crystal oscillator
;     _XT_OSC         External parallel crystal oscillator
;  *  _HS_OSC         High speed crystal resonator
;     _RC_OSC         Resistor/capacitor oscillator
;  |                  (simplest, 20% error)
;  |
;  |_____  * indicates setup values presently selected

        processor        16f877              ; Define processor
        #include <p16f877.inc>
        __CONFIG _CP_OFF & _WDT_OFF & _BODEN_OFF & _PWRTE_ON &
_HS_OSC & _WDT_OFF & _LVP_OFF & _CPD_OFF

; __CONFIG directive is used to embed configuration data
; within the source file. The labels following the directive
; are located in the corresponding .inc file.
        errorlevel -302
; Suppress bank-related warning
;===============================================================
;                      M A C R O S
;===============================================================
; Macros to select the register banks
Bank0   MACRO                           ; Select RAM bank 0
        bcf      STATUS,RP0
        bcf      STATUS,RP1
        ENDM

Bank1   MACRO                           ; Select RAM bank 1
        bsf      STATUS,RP0
        bcf      STATUS,RP1
        ENDM

Bank2   MACRO                           ; Select RAM bank 2
        bcf      STATUS,RP0
        bsf      STATUS,RP1
        ENDM

Bank3   MACRO                           ; Select RAM bank 3
```

```
        bsf       STATUS,RP0
        bsf       STATUS,RP1
        ENDM
;========================================================
;                 constant definitions
;   for PIC-to-LCD pin wiring and LCD line addresses
;========================================================
#define E_line 1            ; |
#define RS_line 0           ; | -- from wiring diagram
#define RW_line 2           ; |
; LCD line addresses (from LCD data sheet)
#define LCD_1 0x80          ; First LCD line constant
#define LCD_2 0xc0          ; Second LCD line constant
#define LCDlimit .20; Number of characters per line
#define spbrgVal .64; For 2400 baud on 10 MHz clock
; Note: The constant that define the LCD display
;        line addresses have the high-order bit set
;        so as to meet the requirements of controller
;        commands.
;============================================================
;           constants for I2C initialization
;============================================================
; I2C connected to 24LC04B EEPROM.
; The MSSP module is in I2C MASTER mode.
#define LC04READ 0xa0      ; I2C value for read control byte
#define LC04WRITE 0xa1     ; I2C value for write control byte


;============================================================
;                 General Purpose Variables
;============================================================
; Local variables
; Reserve 20 bytes for string buffer
        cblock    0x20
        strData
        endc
; Other data
        cblock    0x34                 ; Start of block
        count1              ; Counter # 1
        count2              ; Counter # 2
        count3              ; Counter # 3
        J                            ; counter J
        K                            ; counter K
        bufAdd
        index
        store1              ; Local storage
        store2
; For LCDscroll procedure
        LCDcount ; Counter for characters per line
```

```
              LCDline            ; Current display line (0 or 1)
          endc
;===============================
;       Common RAM area
;===============================
; These GPRs can be accessed from any bank.
; 15 bytes are available, from 0x70 to 0x7f
          cblock   0x70
; Communications variables
          newData            ; not 0 if new data received
          ascVal
          errorFlags
; EEPROM-related variables
          EEMemAdd ; EEPROM address to access
          EEByte             ; Data byte to write
          endc


;================================================================
;                        P R O G R A M
;================================================================
          org      0         ; start at address
          goto     main
; Space for interrupt handlers
          org      0x08
main:
; Wiring:
;     LCD data to port D, lines 0 to 7
;     E line -> port E, 1
;     RW line -> port E, 2
;     RS line -> port E, 0
; Set PORTE D and E for output
; First, initialize port B by clearing latches
          clrf     STATUS
          clrf     PORTB
; Select bank 1 to tris port D for output
          Bank1
; Tris port D for output. Port D lines 4 to 7 are wired
; to LCD data lines. Port D lines 0 to 4 are wired to LEDs.
          movlw    B'00000000'
          movwf    TRISD               ; and port D
;
; By default port A lines are analog. To configure them
; as digital code must set bits 1 and 2 of the ADCON1
; register (in bank 1)
          movlw    0x06                ; binary 0000 0110  is code to
                                       ; make all port A lines
digital
          movwf    ADCON1
```

```
; Port B lines are wired to keypad switches, as follows:
;   7 6 5 4 3 2 1 0
;   | | | | |_|_|_|_____ switch rows (output)
;   |_|_|_|_____ switch columns (input)
; rows must be defined as output and columns as input
        movlw    b'11110000'
        movwf    TRISB
; Tris port E for output
        movlw    B'00000000'
        movwf    TRISE              ; Tris port E
; Enable port B pullups for switches in OPTION register
        movlw    b'00001000'
        movwf    OPTION_REG
; Back to bank 0
        Bank0
; Initialize serial port for 2400 baud, 8 bits, no parity
; 1 stop
        call     InitSerial
; Test serial transmission by sending "RDY-"
        movlw    'R'
        call     SerialSend
        movlw    'D'
        call     SerialSend
        movlw    'Y'
        call     SerialSend
        movlw    '-'
        call     SerialSend
        movlw    0x20
        call     SerialSend
; Clear all output lines
        movlw    b'00000000'
        movwf    PORTD
        movwf    PORTE
; Wait and initialize HD44780
        call     delay_5 ; Allow LCD time to initialize itself
        call     initLCD ; Then do forced initialization
        call     delay_5 ; (Wait probably not necessary)
; Clear character counter and line counter variables
        clrf     LCDcount
        clrf     LCDline
; Set display address to start of first LCD line
        call     line1
; Store address of display buffer
        movlw    0x20
        movwf    bufAdd
; Display "Receiving:" message prompt
        call     blank20              ; Clear buffer
        movlw    0x00                 ; Offset in buffer
```

```
            call    storeMS1 ; Store message at offset
            call    display20        ; Display message
; Start address of EEPROM
            clrf    EEMemAdd
; Setup for display in second line
            call    line2
            clrf    LCDline
            incf    LCDline,f ; Set scroll control for line 2
; Initialize I2C EEPROM operation
            call    SetupI2C        ; Local procedure
;==============================================================
;           receive serial data, store, and display
;==============================================================
receive:
; Call serial receive procedure
            call    SerialRcv
; HOB of newData register is set if new data
; received
            btfss   newData,7
            goto    scanExit
; At this point new data was received.
            movwf   EEByte          ; Save received character
; Display character on LCD
            movf    EEByte,w ; Recover character
            call    send8           ; Display in LCD
            call    LCDscroll       ; Scroll at end of line
; Store character in EEPROM at location in EEMemAdd
            call    WriteI2C        ; Local procedure
            incf    EEMemAdd,f      ; Bump to next EEPROM
; Check for <Enter> key (0x0d) and execute display function
            movf    EEByte,w ; Recover last received
            sublw   0x0d
            btfsc   STATUS,Z ; Test if <Enter> key
            goto    isEnter         ; Go if <Enter>
; Not <Enter> key, continue processing
scanExit:
            goto    receive         ; Continue
;=============================
;    display EEPROM data
;=============================
; This routine receives control when the <Enter> key is
; received.
; Action:
;         1. Clear LCD
;         2. Output is set to top LCD line
;         3. Characters stored in EEPROM are displayed
;            until 0x0d code is detected
isEnter:
```

```
            call      clearLCD
; Clear character counter and line counter variables
            clrf      LCDcount
            clrf      LCDline
; Read data from EEPROM memory, starting at address 0
; and display on LCD until 0x0d terminator
            call      line1
            clrf      EEMemAdd ; Start at EEPROM 0
readOne:
            call      ReadI2C             ; Get character
; Store character
            movwf     EEByte              ; Save character
; Test for terminator
            sublw     0x0d
            btfsc     STATUS,Z ; Test if 0x0d
            goto      atEnd               ; Go if 0x0d
; At this point character read is not 0x0d
; Display on LCD
            movf      EEByte,w ; Recover character
; Display character on LCD
            call      send8               ; Display in LCD
            call      LCDscroll           ; Scroll at end of line
            incf      EEMemAdd,f          ; Next EEPROM byte
            goto      readOne
; End of execution
atEnd:
            goto      atEnd


;================================================================
;================================================================
;              L O C A L    P R O C E D U R E S
;================================================================
;================================================================
;==========================
; init LCD for 4-bit mode
;==========================
initLCD:
; Initialization for Densitron LCD module as follows:
;        4-bit interface
;    2 display lines of 16 characters each
;    cursor on
;    left-to-right increment
;    cursor shift right
;    no display shift
;=====================|
;    set command mode  |
;=====================|
            bcf                   PORTE,E_line       ; E line low
```

```
        bcf        PORTE,RS_line     ; RS line low
        bcf        PORTE,RW_line     ; Write mode
        call       delay_125                    ; delay 125
microseconds
        movlw      0x28    ; 0 0 1 0 1 0 0 0 (FUNCTION SET)
        call       send8   ; 4-bit send routine
; Set 4-bit mode command must be repeated
        movlw      0x28
        call       send8
        movlw      0x0e    ; 0 0 0 0 1 1 1 0 (DISPLAY ON/OFF)
        call       send8
        movlw      0x06    ; 0 0 0 0 0 1 1 0 (ENTRY MODE SET)
        call       send8
        movlw      0x14    ; 0 0 0 1 0 1 0 0 (CURSOR/DISPLAY
                           ;                     SHIFT)
        call       send8
        movlw      0x01    ; 0 0 0 0 0 0 0 1 (CLEAR DISPLAY)
                           ;               |___ COMMAND BIT
        call       send8
        call       delay_5 ; Test for busy
        return
.;===========================
;   procedure to clear LCD
;===========================
clearLCD:
        bcf              PORTE,E_line     ; E line low
        bcf              PORTE,RS_line    ; RS line low
        bcf              PORTE,RW_line    ; Write mode
        call       delay_125                    ; delay 125
microseconds
        movlw      0x01    ; 0 0 0 0 0 0 0 1 (CLEAR DISPLAY)
                           ;               |___ COMMAND BIT
        call       send8
        call       delay_5 ; Test for busy
        return


;=======================
;  Procedure to delay
;   42 microseconds
;=======================
delay_125:
        movlw      .105               ; Repeat 105 machine cycles
        movwf      count1             ; Store value in counter
repeat
        decfsz     count1,f           ; Decrement counter
        goto       repeat             ; Continue if not 0
        return                        ; End of delay
```

```
;========================
;   Procedure to delay
;     5 milliseconds
;========================
delay_5:
        movlw    .105            ; Counter = 105 cycles
        movwf    count2          ; Store in variable
delay
        call     delay_125       ; Delay
        decfsz   count2,f        ; 40 times = 5 milliseconds
        goto     delay
        return                   ; End of delay
;========================
;     pulse E line
;========================
pulseE
        bsf      PORTE,E_line    ; Pulse E line
        nop
        bcf      PORTE,E_line
        return


;==============================
;   long delay sub-routine
;==============================
long_delay
                movlw    D'200'   ; w delay count
                movwf    J               ; J = w
jloop:  movwf    K                ; K = w
kloop:  decfsz   K,f              ; K = K-1, skip next if zero
                goto     kloop
                decfsz   J,f      ; J = J-1, skip next if zero
                goto     jloop
                return
;========================
;   send 2 nibbles in
;     4-bit mode
;========================
; Procedure to send two 4-bit values to port B lines
; 7, 6, 5, and 4. High-order nibble is sent first
; ON ENTRY:
;          w register holds 8-bit value to send
send8:
        movwf    store1          ; Save original value
        call     merge4          ; Merge with port B
; Now w has merged byte
        movwf    PORTD           ; w to port D
        call     pulseE          ; Send data to LCD
; High nibble is sent
```

```
            movf      store1,w          ; Recover byte into w
            swapf     store1,w          ; Swap nibbles in w
            call      merge4
            movwf     PORTD
            call      pulseE            ; Send data to LCD
            call      delay_125
            return
;==========================
;       merge bits
;==========================
; Routine to merge the 4 high-order bits of the
; value to send with the contents of port B
; so as to preserve the 4 low-bits in port B
; Logic:
;       AND value with 1111 0000 mask
;       AND port B with 0000 1111 mask
;       Now low nibble in value and high nibble in
;       port B are all 0 bits:
;           value = vvvv 0000
;          port B = 0000 bbbb
;       OR value and port B resulting in:
;                  vvvv bbbb
; ON ENTRY:
;       w contains value bits
; ON EXIT:
;       w contains merged bits
merge4:
            andlw     b'11110000'       ; ANDing with 0 clears the
                                        ; bit. ANDing with 1 preserves
                                        ; the original value
            movwf     store2            ; Save result in variable
            movf      PORTD,w           ; port B to w register
            andlw     b'00001111' ; Clear high nibble in port b
                                        ; and preserve low nibble
            iorwf     store2,w ; OR two operands in w
            return
;==========================
;   Set address register
;       to LCD line 2
;==========================
; ON ENTRY:
;       Address of LCD line 2 in constant LCD_2
line2:
            bcf       PORTE,E_line      ; E line low
            bcf       PORTE,RS_line     ; RS line low, setup for
                                        ; control
            call      delay_5           ; Busy?
; Set to second display line
```

```
        movlw    LCD_2              ; Address with high-bit set
        call     send8
; Set RS line for data
        bsf      PORTE,RS_line      ; RS = 1 for data
        call     delay_5            ; Busy?
        return
;==========================
;   Set address register
;        to LCD line 1
;==========================
; ON ENTRY:
;          Address of LCD line 1 in constant LCD_1
line1:
        bcf      PORTE,E_line       ; E line low
        bcf      PORTE,RS_line      ; RS line low, set up for
                                    ; control
        call     delay_5            ; busy?
; Set to second display line
        movlw    LCD_1              ; Address and command bit
        call     send8              ; 4-bit routine
; Set RS line for data
        bsf      PORTE,RS_line      ; Setup for data
        call     delay_5            ; Busy?
        return


;==========================
;   scroll to LCD line 2
;==========================
; Procedure to count the number of characters displayed on
; each LCD line. If the number reaches the value in the
; constant LCDlimit, then display is scrolled to the second
; LCD line. If at the end of the second line, then LCD is
; reset to the first line.
LCDscroll:
        incf     LCDcount,f         ; Bump counter
; Test for line limit
        movf     LCDcount,w
        sublw    LCDlimit           ; Count minus limit
        btfss    STATUS,Z           ; Is count - limit = 0
        goto     scrollExit         ; Go if not at end of line
; At this point the end of the LCD line was reached
; Test if this is also the end of the second line
        movf     LCDline,w
        sublw    0x01               ; Is it line 1?
        btfsc    STATUS,Z           ; Is LCDline minus 1 = 0?
        goto     line2End           ; Go if end of second line
; At this point it is the end of the top LCD line
        call     line2              ; Scroll to second line
```

```
            clrf      LCDcount          ; Reset counter
            incf      LCDline,f         ; Bump line counter
            goto      scrollExit
; End of second LCD line
line2End:
            call      initLCD           ; Reset
            clrf      LCDcount          ; Clear counters
            clrf      LCDline
            call      line1             ; Display to first line
scrollExit:
            return


;================================
;     LCD display procedure
;================================
; Sends 20 characters from PIC buffer with address stored
; in variable bufAdd to LCD line previously selected
display20:
            call      delay_5           ; Make sure not busy
; Set up for data
            bcf       PORTA,E_line      ; E line low
            bsf       PORTA,RS_line     ; RS line high for data
; Set up counter for 20 characters
            movlw     D'20'
            movwf     count3
; Get display address from local variable bufAdd
            movf      bufAdd,w     ; First display RAM address to w
            movwf     FSR               ; w to FSR
getchar
            movf      INDF,w    ; get character from display RAM
                                ; location pointed to by file select
                                ; register
            call      send8     ; 4-bit interface routine
; Test for 20 characters displayed
            decfsz    count3,f          ; Decrement counter
            goto      nextchar          ; Skipped if done
            return
nextchar:
            incf      FSR,f             ; Bump pointer
            goto      getchar


;================================
;    first text string procedure
;================================
storeMS1:
; Procedure to store in PIC RAM buffer the message
; contained in the code area labeled msg1
; ON ENTRY:
```

```
;               variable bufAdd holds address of text buffer
;               in PIC RAM
;               w register hold offset into storage area
;               msg1 is routine that returns the string characters
;               and a zero terminator
;               index is local variable that hold offset into
;               text table. This variable is also used for
;               temporary storage of offset into buffer
; ON EXIT:
;               Text message stored in buffer
;
; Store offset into text buffer (passed in the w register)
; in temporary variable
        movwf   index           ; Store w in index
; Store base address of text buffer in FSR
        movf    bufAdd,w ; first display RAM address to w
        addwf   index,w         ; Add offset to address
        movwf   FSR             ; w to FSR
; Initialize index for text string access
        movlw   0                       ; Start at 0
        movwf   index   ; Store index in variable
; w still = 0
get_msg_char:
        call    msg1            ; Get character from table
; Test for zero terminator
        andlw   0x0ff
        btfsc   STATUS,Z ; Test zero flag
        goto    endstr1         ; End of string
; ASSERT: valid string character in w
;         store character in text buffer (by FSR)
        movwf   INDF    ; store in buffer by FSR
        incf    FSR,f   ; increment buffer pointer
; Restore table character counter from variable
        movf    index,w ; Get value into w
        addlw   1       ; Bump to next character
        movwf   index   ; Store table index in variable
        goto    get_msg_char    ; Continue
endstr1:
        return

; Routine for returning message stored in program area
; Message has 10 characters
msg1:
        addwf   PCL,f           ; Access table
        retlw   'R'
        retlw   'e'
        retlw   'c'
        retlw   'e'
```

```
        retlw   'i'
        retlw   'v'
        retlw   'i'
        retlw   'n'
        retlw   'g'
        retlw   ':'
        retlw   0


;=========================
;    blank buffer
;=========================
; Procedure to store 20 blank characters in PIC RAM
; buffer starting at address stored in the variable
; bufAdd
blank20:
        movlw   D'20'     ; Set up counter
        movwf   count1
        movf    bufAdd,w ; First PIC RAM address
        movwf   FSR      ; Indexed addressing
        movlw   0x20     ; ASCII space character
storeit
        movwf   INDF     ; Store blank character in PIC RAM
                         ; buffer using FSR register
        decfsz  count1,f         ; Done?
        goto    incfsr           ; no
        return                   ; yes
incfsr
        incf    FSR,f    ; Bump FSR to next buffer space
        goto    storeit


;=============================================================
;                communications procedures
;=============================================================
; Initialize serial port for 2400 baud, 8 bits, no parity,
; 1 stop
InitSerial:
        Bank1              ; Macro to select bank1
; Bits 6 and 7 of Port C are multiplexed as TX/CK and RX/DT
; for USART operation. These bits must be set to input in the
; TRISC register
        movlw   b'11000000'      ; Bits for TX and RX
        iorwf   TRISC,f          ; OR into Trisc register
; The asynchronous baud rate is calculated as follows:
;                    Fosc
;         ABR = -----------
;                    S*(x+1)
; where x is value in the SPBRG register and S is 64 if the high
; baud rate select bit (BRGH) in the TXSTA control register is
```

```
; clear, and 16 if the BRGH bit is set. For setting to 2400 baud
; using a 10 MHz oscillator at a slow baud rate the formula
; is:
; At slow speed (BRGH = 0)
;           10,000,000    10,000,000
;           ----------  = ----------  = 2,403.84 (0.16% error)
;           64*(64+1)        4160
;
        movlw    spbrgVal ; Value in spbrgVal = 64
        movwf    SPBRG    ; Place in baud rate generator
; Setup value: 0010 0000 = 0x20
        movlw    0x20     ; Enable transmission and high baud
                         ; rate
        movwf    TXSTA
        Bank0            ; Bank 0
; Setup value: 1001 0000 = 0x90
        movlw    0x90     ; Enable serial port and continuous
                         ; reception
        movwf    RCSTA
;
        clrf     errorFlags; Clear local error flags register
        return
;==============================
;       transmit data
;==============================
; Test for Transmit Register Empty and transmit data in w
SerialSend:
        Bank0                     ; Select bank 0
        btfss    PIR1,TXIF        ; check if transmitter busy
        goto     $-1     ; wait until transmitter is not busy
        movwf    TXREG   ; and transmit the data
        return


;==============================
;       receive data
;==============================
; Procedure to test line for data received and return value
; in w. Overrun and framing errors are detected and
; remembered in the variable errorFlags, as follows:
;       7  6  5  4  3  2  1  0   <== errorFlags
;       -- not used ----  |  |___ overrun error
;                         |_____ framing error
SerialRcv:
        clrf     newData ; Clear new data received register
        Bank0            ; Select bank 0
; Bit 5 (RCIF) of the PIR1 Register is clear if the USART
; receive buffer is empty. If so, no data has been received
        btfss    PIR1,RCIF        ; Check for received data
```

```
        return                    ; Exit if no data
; At this point data has been received. First eliminate
; possible errors: overrun and framing.
; Bit 1 (OERR) of the RCSTA register detects overrun
; Bit 2 (FERR) of the RCSTA register detects framing error
        btfsc    RCSTA,OERR       ; Test for overrun error
        goto     OverErr          ; Error handler
        btfsc    RCSTA,FERR       ; Test for framing error
        goto     FrameErr ; Error handler
; At this point no error was detected
; Received data is in the USART RCREG register
        movf     RCREG,w          ; get received data
        bsf      newData,7   ; Set bit 7 to indicate new data
; Clear error flags
        clrf     errorFlags
        return
;=========================
;     error handlers
;=========================
OverErr:
        bsf      errorFlags,0     ; Bit 0 is overrun error
; Reset system
        bcf      RCSTA,CREN       ; Clear continuous receive bit
        bsf      RCSTA,CREN       ; Set to re-enable reception
        return
; error because FERR framing error bit is set
; can do special error handling here - this code simply clears
; and continues
FrameErr:
        bsf                errorFlags,1; Bit 1 is framing error
        movf     RCREG,W          ; Read and throw away bad data
        return
;============================================================
;                  I2C EEPROM data procedures
;============================================================
; GPRs used in EEPROM-related code are placed in the common
; RAM area (from 0x70 to 0x7f). This makes the registers
; accessible from any bank.
;============================
;     LIST OF PROCEDURES
;============================
; SetupI2C   ---  Initialize MSSP module for I2C mode
;                 in hardware master mode
;                 Configure I2C lines
;                 Set slew rate for 100kbps
;                 Set baud rate for 10 MHz
; WriteI2C   ---  Write byte to I2C EEPROM device
;                  Data is stored in EEByte variable
```

```
;                        Address is stored in EEMemAdd
; ReadI2C     --- Read byte from I2C EEPROM device
;                        Address stored in EEMemAdd
;                        Read data returned in w register
;=============================
;     I2C setup procedure
;=============================
SetupI2C:
        Bank1
        movlw   b'00011000'
        iorwf   TRISC,f              ; OR into TRISC
; Setup MSSP module for Master Mode operation
        Bank0
        movlw   B'00101000'; Enables MSSP and uses appropriate
;  0  0  1  0  1  0  0  0  Value to install
;  7  6  5  4  3  2  1  0  <== SSPCON bits in this operation
;  |  |  |  |  |__|__|__|___ Serial port select bits
;  |  |  |  |             1000 = I2C master mode
;  |  |  |  |             Clock = Fosc/(4*(SSPAD+1))
;  |  |  |  |_____ UNUSED IN MASTER MODE
;  |  |  |_____ SSP Enable
;  |  |                   1 = SDA and SCL pins as serial
;  |  |_____ Receive overflow indicator
;  |                      0 = no overflow
;  |_____ Write collision detect
;                         0 = no collision detected
        movwf   SSPCON   ; This is loaded into SSPCON
; Input levels and slew rate as standard I2C
        Bank1
        movlw   B'10000000'
;  1  0  0  0  0  0  0  0  Value to install
;  7  6  5  4  3  2  1  0  <== SSPSTAT bits in this operation
;  |  |  |  |  |  |  |  |___ Buffer full status bit READ ONLY
;  |  |  |  |  |  |  |_____ UNUSED in present application
;  |  |  |  |  |  |_____ Read/write information READ ONLY
;  |  |  |  |  |_____ UNUSED IN MASTER MODE
;  |  |  |  |_____ STOP bit READ ONLY
;  |  |  |_____ Data address READ ONLY
;  |  |_____ SMP bus select
;  |                      0 = use normal I2C specs
;  |_____ Slew rate control
;                         0 = disabled
        movwf        SSPSTAT
; Setup Baud Rate
; Baud Rate = Fosc/(4*(SSPADD+1))
;    Fosc = 10 MHz
;    Baud Rate = 24 for 100 kbps
        movlw   .24                  ; Value to use
```

```
            movwf    SSPADD    ; Store in SSPADD
            Bank0
            return


;=============================
;      I2C write procedure
;=============================
; Write one byte to I2C EEPROM 24LC04B
; Steps:
;                 1. Send START
;                 2. Send control. Wait for ACK
;                 3. Send address. Wait for ACK
;                 4. Send data. Wait for ACK
;                 5. Send STOP
; STEP 1:
WriteI2C:
            Bank1
            bsf      SSPCON2,SEN ; Produce START Condition
            call     WaitI2C ; Wait for I2C to complete
; STEP 2:
; Send control byte. Wair for ACK
            movlw    LC04READ           ; Control byte
            call     Send1I2C           ; Send Byte
            call     WaitI2C ; Wait for I2C to complete
            btfsc    SSPCON2,ACKSTAT ; Check ACK bit to see if
                              ; I2C failed, skip if not
            goto     FailI2C
; STEP 3:
; Send address. Wait for ACK
            Bank0
            movf     EEMemAdd,w        ; Load Address Byte
            call     Send1I2C          ; Send Byte
            call     WaitI2C ; Wait for I2C operation to complete
            Bank1
            btfsc    SSPCON2,ACKSTAT ; Check ACK Status bit to see
                              ; if I2C failed, skip if not
            goto     FailI2C
; STEP 4:
; Send data. Wait for ACK
            Bank0
            movf     EEByte,w          ; Load Data Byte
            call     Send1I2C ; Send Byte
            call     WaitI2C ; Wait for I2C operation to complete
            Bank1
            btfsc    SSPCON2,ACKSTAT ; Check ACK Status bit to see
                                  ; if I2C failed, skip if not
            goto            FailI2C
; STEP 5:
```

```
; Send STOP. Wait for ACK
        bsf     SSPCON2,PEN ; Send STOP condition
        call    WaitI2C ; Wait for I2C operation to complete
; WRITE operation has completed successfully.
        Bank0
        return

;==============================
;    I2C read procedure
;==============================
; Procedure to read one byte from 24LC04B EEPROM
; Steps:
;                   1. Send START.
;                   2. Send control. Wait for ACK.
;                   3. Send address. Wait for ACK.
;                   4. Send RESTART + control. Wait for ACK.
;                   5. Switch to receive mode. Get data.
;                   6. Send NACK.
;                   7. Send STOP.
;                   8. Retreive data into w register.
; STEP 1:
ReadI2C
; Send RESTART. Wait for ACK
        Bank1
        bsf     SSPCON2,RSEN ; RESTART Condition
        call    WaitI2C ; Wait for I2C operation
; STEP 2:
; Send control byte. Wait for ACK
        movlw   LC04READ ; Control byte
        call    Send1I2C ; Send Byte
        call    WaitI2C ; Wait for I2C operation
; Now check to see if I2C EEPROM is ready
        Bank1
        btfsc   SSPCON2,ACKSTAT ; Check ACK Status bit
        goto    ReadI2C ; ACK Poll waiting for EEPROM
                        ; write to complete
; STEP 3:
; Send address. Wait for ACK
        Bank0
        movf    EEMemAdd,w          ; Load from address register
        call    Send1I2C            ; Send Byte
        call    WaitI2C ; Wait for I2C operation
        Bank1
        btfsc   SSPCON2,ACKSTAT ; Check ACK Status bit
        goto    FailI2C ; failed, skipped if successful
; STEP 4:
; Send RESTART. Wait for ACK
        bsf     SSPCON2,RSEN ; Generate RESTART Condition
```

```
        call    WaitI2C ; Wait for I2C operation
; Send output control. Wait for ACK
        movlw   LC04WRITE ; Load CONTROL BYTE (output)
        call    Send1I2C        ; Send Byte
        call    WaitI2C ; Wait for I2C operation
        Bank1
        btfsc   SSPCON2,ACKSTAT ; Check ACK Status bit
        goto    FailI2C ; failed, skipped if successful
; STEP 5:
; Switch MSSP to I2C Receive mode
        bsf     SSPCON2,RCEN ; Enable Receive Mode (I2C)
; Get the data. Wait for ACK
        call    WaitI2C ; Wait for I2C operation
; STEP 6:
; Send NACK to acknowledge
        Bank1
        bsf     SSPCON2,ACKDT ; ACK DATA to send is 1 (NACK)
        bsf     SSPCON2,ACKEN ; Send ACK DATA now.
; Once ACK or NACK is sent, ACKEN is automatically cleared
; STEP 7:
; Send STOP. Wait for ACK
        bsf     SSPCON2,PEN ; Send STOP condition
        call    WaitI2C ; Wait for I2C operation
; STEP 8:
; Read operation has finished
        Bank0
        movf    SSPBUF,W ; Get data from SSPBUF into W
; Procedure has finished and completed successfully.
        return


;=============================
;   I2C support procedures
;=============================
; I2C Operation failed code sequence
; Procedure hangs up. User should provide error handling.
FailI2C
        Bank1
        bsf     SSPCON2,PEN ; Send STOP condition
        call    WaitI2C ; Wait for I2C operation
fail:
        goto    fail

; Procedure to transmit one byte
Send1I2C
        Bank0
        movwf   SSPBUF ; Value to send to SSPBUF
        return
```

```
; Procedure to wait for the last I2C operation to complete.
; Code polls the SSPIF flag in PIR1.
WaitI2C
        Bank0
        btfss    PIR1,SSPIF ; Check if I2C operation done
        goto     $-1       ; I2C module is not ready yet
        bcf      PIR1,SSPIF ; I2C ready, clear flag
        return


;===============================================================
        end             ; END OF PROGRAM
;===============================================================
```

```
;  File name: Key2LCD.asm
;  Date: May 11, 2010
;  Authors: Sanchez and Canton
;
;  Description:
;  Decode 4 x 4 keypad and display scan code in LCD.
;  Program uses 4-bit PIC-to-LCD interface.
;  Code assumes that LCD is driven by Hitachi HD44780
;  controller and PIC 16F977. Display supports two lines
;  each one with 20 characters. The wiring and base
;  address of each display line are stored in #define
;  statements. These statements can be edited to
;  accomodate a different set up.
;  Keypad switch wiring (values are scan codes):
;         --- KEYPAD --
;         0   1   2   3    <= port B0 |
;         4   5   6   7    <= port B1 |--- ROWS = OUTPUTS
;         8   9   A   B    <= port B2 |
;         C   D   E   F    <= port B3 |
;         |   |   |   |
;         |   |   |   |_____ port B4 |
;         |   |   |_____ port B5 |--- COLUMNS = INPUTS
;         |   |_____ port B6 |
;         |_____ port B7 |
;
;  Program operations:
;  1. Key press action generates a scan code in the range
;     0x0 to 0xf.
;  2. Program converts scan code to ASCII digit and displays
;     the digit on the LCD.
;  3. When the end of the first LCD line is reached, display
;     continues in the second line. When the end of the
;     second line is reached, LCD is reset to the first line.
;
;  Program uses delay loops for interface timing.
;  WARNING:
;  Code assumes 4 MHz clock. Delay routines must be
;  edited for faster clock.
;
;===========================
;      16F877 switches
;===========================
;  Switches used in __config directive:
;    _CP_ON          Code protection ON/OFF
;  * _CP_OFF
;  * _PWRTE_ON       Power-up timer ON/OFF
;    _PWRTE_OFF
;    _BODEN_ON       Brown-out reset enable ON/OFF
```

```
; *  _BODEN_OFF
; *  _PWRTE_ON        Power-up timer enable ON/OFF
;    _PWRTE_OFF
;    _WDT_ON          Watchdog timer ON/OFF
; *  _WDT_OFF
;    _LPV_ON          Low voltage IC programming enable ON/OFF
; *  _LPV_OFF
;    _CPD_ON          Data EE memory code protection ON/OFF
; *  _CPD_OFF
; OSCILLATOR CONFIGURATIONS:
;    _LP_OSC          Low power crystal oscillator
;    _XT_OSC          External parallel crystal oscillator
; *  _HS_OSC          High speed crystal resonator
;    _RC_OSC          Resistor/capacitor oscillator
;    |                (simplest, 20% error)
;    |
;    |_____  * indicates setup values presently selected

        processor         16f877              ; Define processor
        #include <p16f877.inc>
        __CONFIG _CP_OFF & _WDT_OFF & _BODEN_OFF & _PWRTE_ON &
_HS_OSC & _WDT_OFF & _LVP_OFF & _CPD_OFF

; __CONFIG directive is used to embed configuration data
;  within the source file. The labels following the directive
;  are located in the corresponding .inc file.


;===============================================================
;                   constant definitions
;    for PIC-to-LCD pin wiring and LCD line addresses
;===============================================================
#define E_line 1         ; |
#define RS_line 0        ; | -- from wiring diagram
#define RW_line 2        ; |
; LCD line addresses (from LCD datasheet)
#define LCD_1 0x80        ; First LCD line constant
#define LCD_2 0xc0        ; Second LCD line constant
#define LCDlimit .20; Number of characters per line
; Note: The constants that define the LCD display line
;       addresses have the high-order bit set in
;       order to faciliate the controller command.
;
;===============================================================
;                   PIC register equates
;===============================================================
portd   equ                   0x08
porte   equ         0x09
fsr             equ                   0x04
```

```
status   equ                    0x03
indf     equ        0x00
z                    equ                   2
c                    equ                   0
;================================================================
;                    variables in PIC RAM
;================================================================
; Reserve 20 bytes for string buffer
         cblock  0x20
         strData
         endc
; Leave 16 bytes and Continue with local variables
         cblock  0x34     ; Start of block
         count1           ; Counter # 1
         count2           ; Counter # 2
         count3           ; Counter # 3
         pic_ad           ; Storage for start of text area
                          ; (labeled strData) in PIC RAM
         J                ; counter J
         K                ; counter K
         index            ; Index into text table (also used
                          ; for auxiliary storage)
         store1           ; Local storage
         store2
; For LCDscroll procedure
         LCDcount ; Counter for characters per line
         LCDline          ; Current display line (0 or 1)
; Keypad processing variables
         keyMask          ; For keypad processing
         rowMask          ; For masking-off key rows
         rowCode          ; Row added for calculating scan code
         rowCount ; Counter for key rows (0 to 3)
         scanCode ; Final key code
         newScan          ; 0 if no new scan code detected
         endc
;================================================================
;                         M A C R O S
;================================================================
; Macros to select the register banks
; Data memory bank selection bits:
; RP1:RP0           Bank
;   0:0                0    Ports A,B,C,D, and E
;   0:1                1    Tris A,B,C,D, and E
;   1:0                2
;   1:1                3
Bank0    MACRO                         ; Select RAM bank 0
                   bcf     STATUS,RP0
                   bcf     STATUS,RP1
```

```
               ENDM

Bank1    MACRO                          ; Select RAM bank 1
                    bsf      STATUS,RP0
                    bcf      STATUS,RP1
                    ENDM

Bank2    MACRO                          ; Select RAM bank 2
                    bcf      STATUS,RP0
                    bsf      STATUS,RP1
                    ENDM

Bank3    MACRO                          ; Select RAM bank 3
                    bsf      STATUS,RP0
                    bsf      STATUS,RP1
                    ENDM
;=============================================================
;                   M A I N   P R O G R A M
;=============================================================
                    org            0         ; start at address
                    goto    main
; Space for interrupt handlers
          org               0x08
main:
; Wiring:
;     LCD data to port D, lines 0 to 7
;     E line -> port E, 1
;     RW line -> port E, 2
;     RS line -> port E, 0
; Set ports D and E for output
; First, initialize port B by clearing latches
          clrf      STATUS
          clrf      PORTB
; Select bank 1 to tris port D for output
          Bank1
; Tris port D for output. Port D lines 4 to 7 are wired
; to LCD data lines. Port D lines 0 to 4 are wired to LEDs.
          movlw    B'00000000'
          movwf    TRISD             ; and port D
; By default port A lines are analog. To configure them
; as digital code must set bits 1 and 2 of the ADCON1
; register (in bank 1)
          movlw    0x06              ; binary 0000 0110  is code to
                                              ; make all
port A lines digital
          movwf    ADCON1
; Port B lines are wired to keypad swtiches as follows:
;    7 6 5 4 3 2 1 0
```

```
;     |  |  |  |  |_|_|_|_____ switch rows (output)
;     |_|_|_|_____ switch columns (input)
; rows must be defined as output and columns as input
          movlw   b'11110000'
          movwf   TRISB
; Tris port E for output
          movlw   B'00000000'
          movwf   TRISE              ; Tris port E
; Enable port B pullups for switches in OPTION register
;   7  6  5  4  3  2  1  0 <= OPTION bits
;   |  |  |  |  |  |  |__|__|_____ PS2-PS0 (prescaler bits)
;   |  |  |  |  |  |                Values for Timer0
;   |  |  |  |  |  |                000 = 1:2    001 = 1:4
;   |  |  |  |  |  |                010 = 1:8    011 = 1:16
;   |  |  |  |  |  |                100 = 1:32   101 = 1:64
;   |  |  |  |  |  |                110 = 1:128 *111 = 1:256
;   |  |  |  |  |  |_____ PSA (prescaler assign)
;   |  |  |  |  |                  *1 = to WDT
;   |  |  |  |  |                  0 = to Timer0
;   |  |  |  |  |_____ TOSE (Timer0 edge select)
;   |  |  |  |                     *0 = increment on low-to-high
;   |  |  |  |                     1 = increment on high-to-low
;   |  |  |_____ TOCS (TMR0 clock source)
;   |  |  |                        *0 = internal clock
;   |  |  |                        1 = RA4/TOCKI bit source
;   |  |_____ INTEDG (Edge select)
;   |  |                           *0 = falling edge
;   |_____ RBPU (Pullup enable)
;   |                              *0 = enabled
;   |                              1 = disabled
;
          movlw   b'00001000'
          movwf   OPTION_REG
; Back to bank 0
          Bank0
; Clear all output lines
          movlw   b'00000000'
          movwf   portd
          movwf   porte
; Wait and initialize HD44780
          call    delay_5 ; Allow LCD time to initialize itself
          call    initLCD ; Then do forced initialization
          call    delay_5 ; (Wait probably not necessary)
; Set display address to start of second LCD line
          call    line1
; Clear character counter and line counter variables
          clrf    LCDcount
          clrf    LCDline
```

```
;========================
;       scan keypad
;========================
; Keypad switch wiring:
;           x    x    x    x    <= port B0 |
;           x    x    x    x    <= port B1 |--- ROWS = OUTPUTS
;           x    x    x    x    <= port B2 |
;           x    x    x    x    <= port B3 |
;           |    |    |    |
;           |    |    |    |_____ port B4 |
;           |    |    |_____ port B5 |--- COLUMNS = INPUTS
;           |    |_____ port B6 |
;           |_____ port B7 |
; Switches are connected to port B lines
; Clear scan code register
        clrf    scanCode
;==============================
;  scan keypad and display
;==============================
keyScan:
; Port B lines are wired to pushbutton swtiches as follows:
;   7 6 5 4 3 2 1 0
;   | | | | |_|_|_|_____ switch rows (output)
;   |_|_|_|_____ switch columns (input)
; Keypad processing:
; Switch rows are succesively grounded (row = 0)
; Then column values are tested. If a column returns 0
; in a 0 row, that switch is down.
; Initialize row code addend
        clrf    rowCode            ; First row is code 0
        clrf    newScan            ; No new scan code detected
; Initialize row count
        movlw   D'4'               ; Four rows
        movwf   rowCount           ; Register variable
        movlw   b'11111110'        ; All set but LOB
        movwf   rowMask
keyLoop:
; Initialize row eliminator mask:
; The row mask is ANDed with the key mask to successively
; mask-off each row, for example:
;
;                     |------- row 3
;                     ||------ row 2
;                     |||----- row 1
;                     ||||---- row 0
;           0000 1111 <= key mask
;       AND 1111 1101 <= mask for row 1
;           ---------
```

```
;                  0000 1101 <= row 1 is masked off
;
; The row mask, which is initally 1111 1110, is rotated left
; through the carry in order to mask-off the next row
         movlw    b'00001111'        ; Mask-off all lines
         movwf    keyMask            ; To local register
; Set row mask for current row
         movf     rowMask,w          ; Mask to w
         andwf    keyMask,f          ; Update key mask
         movf     keyMask,w          ; Key mask to w
         movwf    PORTB              ; Mask-off port B lines
; Read port B lines 4 to 7 (columns are input)
         btfss    PORTB,4
         call     col0               ; Key column procedures
         btfss    PORTB,5
         call     col1
         btfss    PORTB,6
         call     col2
         btfss    PORTB,7
         call     col3
; Index to next row by adding 4 to row code
         movf     rowCode,w          ; Code to w
         addlw    D'4'
         movwf    rowCode
;=========================
;     shift row mask
;=========================
; Set the carry flag
         bsf      STATUS,c
         rlf      rowMask,f          ; Rotate mask bits in storage
;=========================
;     end of keypad?
;=========================
; Test for last key row (maximum count is 4)
         decfsz   rowCount,f  ; Decrement counter
         goto     keyLoop
;============================================================
;============================================================
;                  display scan code
;============================================================
;============================================================
; At this point all keys have been tested.
; Variable newScan = 0 if no new scan code detected, else
; variable scanCode holds scan code
         movf     newScan,f ; Copy onto itself (sets z flag)
         btfsc    STATUS,z           ; Is it zero
         goto     scanExit
; At this point a new scan code is detected
```

```
           movf      scanCode,w          ; To w
; If scan code is in the range 0 to 9, that is, a decimal
; digit, then ASCII conversion consists of adding 0x30.
; If the scan code represents one of the hex letters
; (0xa to 0xf), then ASCII conversion requires adding
; 0x37
           sublw     0x09                ; 9 - w
; if w from 0 to 9 then 9 - w = positive (c flag = 1)
; if w = 0xa then 9 - 10 = -1 (c flag = 0)
; if w = 0xc then 9 - 12 = -2 (c flag = 0)
           btfss     STATUS,c ; Test carry flag
           goto      hexLetter ; Carry clear, must be a letter
; At this point scan code is a decimal digit in the
; range 0 to 9. Convert to ASCII by adding 0x30
           movf      scanCode,w          ; Recover scan code
           addlw     0x30                ; Conver to ASCII
           goto      displayDig
hexLetter:
           movf      scanCode,w          ; Recover scan code
           addlw     0x37                ; Conver to ASCII
displayDig:
           call      send8      ; Display routine
           call      LCDscroll ; Auto line scrolling procedure
scanExit:
           call      long_delay          ; Debounce
           goto      keyScan             ; Continue


;===========================
;   calculate scan code
;===========================
; The column position is added to the row code (stored
; in rowCode register). Sum is the scan code
col0:
           movf      rowCode,w ; Row code to w
           addlw     0x00      ; Add 0 (clearly not necessary)
           movwf     scanCode ; Final value
           incf      newScan,f           ; New scan code
           return

col1:
           movf      rowCode,w           ; Row code to w
           addlw     0x01                ; Add 1
           movwf     scanCode
           incf      newScan,f
           return

col2:
           movf      rowCode,w           ; Row code to w
```

```
            addlw      0x02              ; Add 2
            movwf      scanCode
            incf       newScan,f
            return

col3:
            movf       rowCode,w         ; Row code to w
            addlw      0x03              ; Add 3
            movwf      scanCode
            incf       newScan,f
            return


;================================================================
;                   initialize LCD for 4-bit mode
;================================================================
initLCD:
; Initialization for Densitron LCD module as follows:
;         4-bit interface
;     2 display lines of 16 characters each
;     cursor on
;     left-to-right increment
;     cursor shift right
;     no display shift
;======================|
;   set command mode   |
;======================|
            bcf        porte,E_line      ; E line low
            bcf        porte,RS_line     ; RS line low
            bcf        porte,RW_line     ; Write mode
            call       delay_125         ; delay 125 microseconds
;**********************|
;     FUNCTION SET     |
;**********************|
            movlw      0x28    ; 0 0 1 0 1 0 0 0 (FUNCTION SET)
                                ;        | | | |__ font select:
                                ;        | | |    1 = 5x10 in 1/8 or 1/11
                                ;        | | |    0 = 1/16 dc
                                ;        | | |___ Duty cycle select
                                ;        | |      0 = 1/8 or 1/11
                                ;        | |      1 = 1/16
                                ;        | |___ Interface width
                                ;        |      0 = 4 bits
                                ;        |      1 = 8 bits
                                ;        |___ FUNCTION SET COMMAND
            call       send8    ; 4-bit send routine

; Set 4-bit mode command must be repeated
            movlw      0x28
```

```
        call    send8

;***********************|
; DISPLAY AND CURSOR ON |
;***********************|
        movlw   0x0e    ; 0 0 0 0 1 1 1 0 (DISPLAY ON/OFF)
                        ;           | | | |___ Blink character
                        ;           | | |       1 = on, 0 = off
                        ;           | | |___ Curson on/off
                        ;           | |       1 = on, 0 = off
                        ;           | |____ Display on/off
                        ;           |         1 = on, 0 = off
                        ;           |____ COMMAND BIT
        call    send8
;***********************|
;   set entry mode      |
;***********************|
        movlw   0x06    ; 0 0 0 0 0 1 1 0 (ENTRY MODE SET)
                        ;         | | |___ display shift
                        ;         | |       1 = shift
                        ;         | |       0 = no shift
                        ;         | |____ cursor increment
                        ;         |         1 = left-to-right
                        ;         |         0 = right-to-left
                        ;         |___ COMMAND BIT
        call    send8

;***********************|
; cursor/display shift  |
;***********************|
        movlw   0x14    ; 0 0 0 1 0 1 0 0 (CURSOR/DISPLAY
                        ;                   SHIFT)
                        ;       | | | |_|___ don't care
                        ;       | |_|__ cursor/display shift
                        ;       |         00 = cursor shift left
                        ;       |         01 = cursor shift right
                        ;       |         10 = cursor and display
                        ;       |             shifted left
                        ;       |         11 = cursor and display
                        ;       |             shifted right
                        ;       |___ COMMAND BIT
        call    send8
;***********************|
;   clear display       |
;***********************|
        movlw   0x01    ; 0 0 0 0 0 0 0 1 (CLEAR DISPLAY)
                        ;                 |___
COMMAND BIT
```

```
        call    send8
; Per documentation
        call    delay_5  ; Test for busy
        return


;=======================
;   Procedure to delay
;    42 microseconds
;=======================
delay_125:
        movlw   D'42'    ; Repeat 42 machine cycles
        movwf   count1   ; Store value in counter
repeat
        decfsz  count1,f         ; Decrement counter
        goto    repeat           ; Continue if not 0
        return                   ; End of delay


;=======================
;   Procedure to delay
;    5 milliseconds
;=======================
delay_5:
        movlw   D'42'            ; Counter = 41
        movwf   count2           ; Store in variable
delay
        call    delay_125        ; Delay
        decfsz  count2,f         ; 40 times = 5 milliseconds
        goto    delay
        return                   ; End of delay
;=======================
;     pulse E line
;=======================
pulseE
        bsf     porte,E_line     ; Pulse E line
        Nop
        bcf     porte,E_line
        return


;===========================
;   long delay sub-routine
;      (for debugging)
;===========================
long_delay
        movlw   D'200'  ; w delay count
        movwf   J                ; J = w
jloop:  movwf   K                ; K = w
kloop:  decfsz  K,f              ; K = K-1, skip next if zero
        goto    kloop
```

```
            decfsz    J,f                ; J = J-1, skip next if zero
            goto      jloop
            return


;========================
;   send 2 nibbles in
;      4-bit mode
;========================
; Procedure to send two 4-bit values to port B lines
; 7, 6, 5, and 4. High-order nibble is sent first
; ON ENTRY:
;         w register holds 8-bit value to send
send8:
            movwf     store1             ; Save original value
            call      merge4             ; Merge with port B
; Now w has merged byte
            movwf     portd              ; w to port D
            call      pulseE             ; Send data to LCD
; High nibble is sent
            movf      store1,w ; Recover byte into w
            swapf     store1,w ; Swap nibbles in w
            call      merge4
            movwf     portd
            call      pulseE             ; Send data to LCD
            call      delay_125
            return
;==================
;   merge bits
;==================
; Routine to merge the 4 high-order bits of the
; value to send with the contents of port B
; so as to preserve the 4 low-bits in port B
; Logic:
;     AND value with 1111 0000 mask
;     AND port B with 0000 1111 mask
;     Now low nibble in value and high nibble in
;     port B are all 0 bits:
;          value = vvvv 0000
;         port B = 0000 bbbb
;     OR value and port B resulting in:
;                 vvvv bbbb
; ON ENTRY:
;     w contains value bits
; ON EXIT:
;     w contains merged bits
merge4:
            andlw     b'11110000'        ; ANDing with 0 clears the
                                         ; bit. ANDing with 1 preserves
```

```
                                        ; the original value
        movwf    store2             ; Save result in variable
        movf     portd,w            ; port B to w register
        andlw    b'00001111' ; Clear high nibble in port b
                                    ; and preserve low nibble
        iorwf    store2,w ; OR two operands in w
        return


;========================
; Set address register
;    to LCD line 1
;========================
; ON ENTRY:
;        Address of LCD line 2 in constant LCD_2
line1:
        bcf      porte,E_line      ; E line low
        bcf      porte,RS_line     ; RS line low, set up for
                                   ; control
        call     delay_5           ; Busy?
; Set to second display line
        movlw    LCD_1                        ; Address with
high-bit set
        call     send8
; Set RS line for data
        bsf      porte,RS_line     ; RS = 1 for data
        call     delay_5                      ; Busy?
        return


;========================
; Set address register
;    to LCD line 2
;========================
; ON ENTRY:
;        Address of LCD line 2 in constant LCD_2
line2:
        bcf      porte,E_line      ; E line low
        bcf      porte,RS_line     ; RS line low, set up for
                                   ; control
        call     delay_5           ; Busy?
; Set to second display line
        movlw    LCD_2             ; Address with high-bit set
        call     send8
; Set RS line for data
        bsf      porte,RS_line     ; RS = 1 for data
        call     delay_5           ; Busy?
        return


;===========================
```

```
;    scroll to LCD line 2
;===========================
; Procedure to count the number of characters displayed on
; each LCD line. If the number reaches the value in the
; constant LCDlimit, then display is scrolled to the second
; LCD line. If at the end of the second line, then LCD is
; reset to the first line.
LCDscroll:
          incf     LCDcount,f              ; Bump counter
; Test for line limit
          movf     LCDcount,w
          sublw    LCDlimit           ; Count minus limit
          btfss    STATUS,z           ; Is count - limit = 0
          goto     scrollExit         ; Go if not at end of line
; At this point the end of the LCD line was reached
; Test if this is also the end of the second line
          movf     LCDline,w
          sublw    0x01               ; Is it line 1?
          btfsc    STATUS,z           ; Is LCDline minus 1 = 0?
          goto     line2End           ; Go if end of second line
; At this point it is the end of the top LCD line
          call     line2              ; Scroll to second line
          clrf     LCDcount           ; Reset counter
          incf     LCDline,f          ; Bump line counter
          goto     scrollExit
; End of second LCD line
line2End:
          call     initLCD            ; Reset
          clrf     LCDcount           ; Clear counters
          clrf     LCDline
          call     line1              ; Display to first line
scrollExit:
          return
          end
```

Index